SELECTED UNITS AND EQUIVALENTS

Metric	English	Conversions
	Linear Measure	
centimeter (cm)	inch (in.)	1 cm = 0.3937 in.
		1 in. = 2.54 cm
meter (m)	foot (ft)	1 m = 3.28 ft
		1 ft = 0.305 m
kilometer (km)	mile (mi)	1 km = 0.621 mi
		1 mi = 1.61 km
	Area Measure	
square centimeter (cm^2)	square inch (sq in.)	1 cm^2 = 0.155 sq in.
		1 sq in. = 6.4516 cm^2
square meter (m^2)	square foot (sq ft)	1 m^2 = 10.764 sq ft
		1 sq ft = 0.0929 m^2
square kilometer (km^2)	square mile (sq mi)	1 km^2 = 0.3861 sq mi
		1 sq mi = 2.59 km^2
hectare (ha)	acre	1 ha = 10,000 m^2 = 2.471 acres
		1 acre = 43,560 sq ft = 0.4047 ha
	Volume Measure	
cubic centimeter (cm^3)	cubic inch (cu in.)	1 cm^3 = 0.061 cu in.
		1 cu in. = 16.387 cm^3
cubic meter (m^3)	cubic foot (cu ft)	1 m^3 = 35.315 cu ft
		1 cu ft = 0.02832 m^3
liter	gallon	1 liter = 1,000 cm^3 = 0.264 gallons
		1 gallon = 231 cu in. = 3.7853 liters
	Mass	
gram (g)	ounce (oz)	1 g = 0.0353 oz
		1 oz = 28.35 g
kilogram (kg)	pound (lb)	1 kg = 2.205 lbs
		1 lb = 0.454 kg
metric ton (m ton)	short ton	1 m ton = 10^3 kg = 1.102 short tons
		1 short ton = 2000 lbs = 0.907 m tons
	Temperature	
Celsius degree (C°)	Fahrenheit degree (F°)	1 C° = 1.8 F°
		1 F° = 5/9 C°
Kelvin or Absolute degree (K or T)		1 C° = 1 K

TEMPERATURE SCALES

$$K^a = °C + 273$$
$$°C = 5/9 \, (°F - 32)$$
$$°F = 9/5 \, (°C) + 32$$

COMPARISON OF TEMPERATURE SCALES

	Kelvin (K)[a]	Celsius (°C)	Fahrenheit (°F)
Boiling point of water	373	100°	212°
Freezing point of water	273	0°	32°
Absolute zero	0	−273.16°	−459.69°

[a]*The symbol K is not written with a degree mark.*

Fundamentals of Remote Sensing and Airphoto Interpretation

Top—*Mount Saint Helens, Washington, on March 30, 1980, looking north-northeast. Mount Rainier in distance. Distribution of dark ash resulted from wind control of drift plume. Left portion of cone is free of ash; right portion is largely covered. Snowstorms later covered these ash layers, which in turn were covered by new ash deposits. Bottom—Mount Saint Helens on May 18, 1980. Catastrophic eruption began at 8:30 a.m. PDT. The photograph was taken at about noon toward the northeast. The debris avalanche, explosive eruption, and associated mudflows caused 63 deaths and billions of dollars of damage. (Courtesy EROS Data Center, U.S. Geological Survey.)*

Fundamentals of Remote Sensing and Airphoto Interpretation

Formerly entitled **Interpretation of Aerial Photographs**

FIFTH EDITION

Thomas Eugene Avery
Graydon Lennis Berlin

Macmillan Publishing Company
New York

Maxwell Macmillan Canada
Toronto

Maxwell Macmillan International
New York Oxford Singapore Sydney

Cover photograph: Color infrared vertical photograph showing a dense cluster of center-pivot irrigation plots in Morrow County, Oregon. Most of the circular irrigated fields in the photograph are planted with wheat and depicted in various shades of red. Excluding lost corners, each large circle plot occupies 160 acres or 64.75 hectares, and four plots together occupy a section or 1 square mile (2.59 square kilometers). In the center-pivot irrigation scheme, water is applied by a series of sprinklers mounted on a radial pipe that is in turn supported by a slowly advancing row of wheeled towers. The photograph was taken at an altitude of 40,000 feet (12,195 meters) above the ground with a mapping camera; scale is about 1:58,000. (Courtesy EROS Data Center, U.S. Geological Survey).

Editor: **Robert McConnin**
Production supervisor: **Betsy Keefer**
Production manager: **Aliza Greenblatt**
Text designer: **Fred C. Pusterla Design Associates**
Cover designer: **Fred C. Pusterla Design Associates**

This book was set in 10/11 Times Roman by Waldman Graphics, printed by Eusey Press, and bound by Book Press. The cover was printed by Lehigh Press.

Macmillan Publishing Company
866 Third Avenue, New York, New York 10022

Macmillan Publishing Company is
part of the Maxwell Communication
Group of Companies.

Maxwell Macmillan Canada, Inc.
1200 Eglinton Avenue East
Suite 200
Don Mills, Ontario M3C 3N1

Library of Congress Cataloging-in-Publication Data
Avery, Thomas Eugene.
    Fundamentals of remote sensing and airphoto interpretation /
Thomas Eugene Avery, Graydon Lennis Berlin.—5th ed.
        p.   cm.
    Rev. ed. of: Interpretation of aerial photographs. 4th ed. c 1985.
    Includes bibliographical references and index.
    ISBN 0-02-305035-7
    1. Photographic interpretation.   2. Photography, Aerial.
3. Remote sensing.   I. Berlin, Graydon Lennis, 1943–
II. Avery, Thomas Eugene.   Interpretation of aerial photographs.
III. Title.
TR810.A9   1985
778.3′5—dc20   91-18241
                                                       CIP

Printing: 4 5 6 7   Year: 4 5 6 7 8

Preface

Scope and Purpose

The predecessor to this book, the Fourth Edition of *Interpretation of Aerial Photographs,* emphasized photographic remote sensing with an introduction to nonphotographic remote sensing. This new volume, the Fifth Edition, while retaining the philosophy and much of the style of the former book, incorporates many new topics, revisions, and changes that are aimed at providing a more even balance between photographic and nonphotographic remote sensing. Consequently, to better reflect the contents of the Fifth Edition, the book's title has been changed to *Fundamentals of Remote Sensing and Airphoto Interpretation.*

The Fifth Edition, prepared by the co-author, integrates information from many disciplines and covers all essential topics for undergraduate courses in both remote sensing and airphoto interpretation. *Fundamentals of Remote Sensing and Airphoto Interpretation* is intended to provide a thorough and understandable treatment of established and new technologies in the field plus the practical applications of remote sensing as a problem-solving tool in the natural and cultural sciences. To this end, the underlying objective has been to provide an interesting and readable textbook for a student population having varied backgrounds and interests.

The book's 15 chapters are informally divided into five sections. The initial chapter, representing the first section, provides a comprehensive overview of remote sensing systems and applications, along with introductions to electromagnetic radiation, the electromagnetic spectrum and the bands utilized in remote sensing, energy-matter interactions, spectral signatures, and the "multi" concept in remote sensing.

The second section is concerned with photographic remote sensing and includes Chapters 2–5. It starts with a chapter on cameras, films, and filters and is followed by chapters treating the principles and techniques of airphoto interpretation and photogrammetry and the acquisition of aerial photographs.

The third section is concerned with nonphotographic remote sensing and includes Chapters 6 and 7. The remote sensing systems described are video cameras, vidicon cameras, across-track scanners, including multispectral and thermal infrared systems, along-track scanners, imaging radars, imaging sonars, and passive microwave imagers. Also included are discussions on the earth observation satellite programs of the United States, France, India, and Japan.

The fourth section contains seven chapters (8–14) that emphasize the practical applications of aerial photographs and images to specific disciplines and projects. Applications are illustrated in chapters treating land use and land cover, archaeology, agriculture and soils, forestry, geology, engineering, and urban-industrial patterns.

The fifth section, represented by Chapter 15, is designed to introduce students to the concepts of digital image processing. Topics discussed include digital image characteristics, image processing systems, and the four major areas of computer operation in remote sensing—image restoration, image enhancement, image classification, and data-set merging and analysis.

The individual chapters have been written to represent independent units, enabling them to be selected and recombined as necessary to accommodate varied curricula, course lengths, instructor interests and expertise, and student qualifications. For example, all 15 chapters could be used for a two-semester course in remote sensing, while a one-semester course could be based on Chapters 1, 2, 3, 4, 6, 7, and 15. A one-semester airphoto interpretation course could emphasize Chapters 1, 2, 3, 4, 5, 8, 9, 10, 11, 12, 13, and 14.

Fundamentals of Remote Sensing and Airphoto Interpretation is highly illustrated, containing some 127 line drawings, 440 black-and-white photographs and images, including 160 stereopairs, and 50 color photographs and images. At the end of each chapter are review questions, including numerical problems and interpretations, plus selected references and suggested readings. Also, the following reference materials have been placed on the book's front and back endsheets: units of wavelength and frequency, multiples and submultiples of SI (metric) units, SI/English units and conversions, temperature scales and conversions, the Greek alphabet, trigonometric functions in a right triangle, and the geologic time scale. An *Instructor's Manual* has been prepared as an additional teaching aid; contact your local Macmillan sales representative for further information.

Acknowledgments

The co-author wishes to express his sincere appreciation to all of those who have contributed to the successful completion of the Fifth Edition. The manuscript for all 15 chapters was read by four anonymous reviewers. Their suggestions and constructive comments were very helpful, and I am grateful for their efforts. I also wish to thank the following individuals for their valuable technical contributions: Diana F. Elder and Jay O. Cooper, Northern Arizona University; Pat S. Chavez, Jr., Philip A. Davis, and James V. Gardner, U.S. Geological Survey; David M. Lavinge, University of Guelph; Kalman N. Vizy, Eastman Kodak Company; Vern W. Cartwright, Cartwright Research Corporation; George England and Thomas R. Ory, Daedalus Enterprises, Inc.; Michel Hersé, Centre National de la Recherche Scientifique; John P. Ford, Jet Propulsion Laboratory; James H. Everitt and David E. Escobar, Agricultural Resource Service, U.S. Department of Agriculture; Donald J. Bogucki, State University of New York, Plattsburgh; John Robinson and Lonnie Schuepbach, FLIR Systems, Inc.; Thomas M. Holm and Thomas R. Loveland, Technicolor Government Services, Inc.; Kevin C. Horstman and Christopher T. Lee, University of Arizona; and Greg Scharfen and Claire Hanson, National Snow and Ice Center, University of Colorado.

Special illustration materials were provided by a number of colleagues in academia, government, and industry, and I am grateful for their generosity. Individual contributions are acknowledged in the appropriate figure captions.

The staff at Macmillan Publishing Company has been extremely helpful in producing *Fundamentals of Remote Sensing and Airphoto Interpretation.* Special thanks are extended to Robert A. McConnin, Senior Editor; Aliza Greenblatt, Freelance Manager; and Betsy Keefer, Production Supervisor, for their patience, professionalism, encouragement, and editing expertise.

Finally, special recognition is due to the co-author's wife, Judy, and daughter, Jodi, who provided two years of patient understanding and encouragement while this new volume was in preparation.

With any book there is always the potential for errors or mistakes which, despite the best efforts of all involved, somehow manage to appear in the final product. The co-author accepts full responsibility for any such errors and requests that they be brought to his attention. He would also welcome comments and suggestions for improving future editions.

G. L. B.

Contents

Chapter 1

Overview of Remote Sensing

of measurement. A common example of an *in situ* measurement instrument is the soil thermometer.

The detection and recording instruments for this special technology are known collectively as **remote sensors** and include photographic cameras, mechanical scanners, and radar systems. Remote sensors are typically carried on aircraft and earth-orbiting spacecraft, which has led to the familiar phrase "eye in the sky" (Figure 1-1).

Traditionally, the energy measured in remote sensing has been electromagnetic radiation, including visible light, which is reflected or emitted in varying degrees by all natural and synthetic objects (Figure 1-2). The scope of remote sensing has been recently broadened to include acoustical or sound energy, which is propagated under water. With the inclusion of these two different forms of energy, the human eye and ear are examples of remote sensing data collection systems.

To complete the remote sensing process, the data collected by remote sensing systems must be analyzed by interpretive and measurement techniques in order to provide useful information about the subjects of investigation. These techniques are diverse, ranging from traditional methods of visual interpretation to methods using sophisticated computer processing. Accordingly, the two major components of remote sensing are **data collection** and **data analysis**.

The overall focus of this book has two parts. One is directed toward understanding electromagnetic radiation sensors that are in general use today. The other is on how the collected data, commonly in the form of photographs and photolike images, can be used to analyze the physical and cultural character of the earth's surface. Special emphasis is placed on the branch of remote sensing concerned with the acquisition and interpretation of aerial photographs. An excellent review of the history of remote sensing is provided by Fischer et al. (1975).

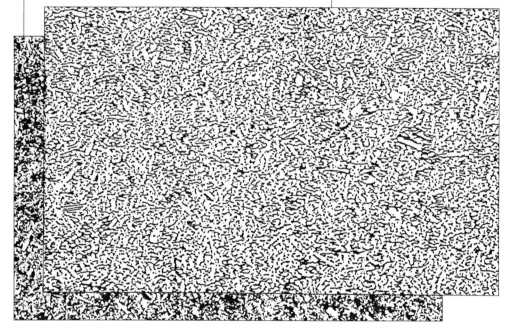

Remote Sensing Defined

Remote sensing is defined as the technique of obtaining information about objects through the analysis of data collected by special instruments that are not in physical contact with the objects of investigation. As such, remote sensing can be regarded as **reconnaissance from a distance**. Remote sensing thus differs from **in situ sensing**, where the instruments are immersed in, or physically touch, the objects

Electromagnetic Energy

Electromagnetic energy is a dynamic form of energy that is caused by the oscillation or acceleration of an electrical charge. This is associated with atomic nuclei during fission and fusion reactions, with electrons as they drop from high- to lower-energy orbits in an atom or molecule, and with the random movement of atoms and molecules. All natural and synthetic substances above **absolute zero** continuously produce and emit a range of electromagnetic energy in proportion to their temperatures. Absolute zero is expressed as 0 K on the Kelvin or absolute scale, −273.16°C on the Celsius scale, and −459.69°F on the Fahrenheit scale.* All random motions of an atom or molecule would cease at a temperature of absolute zero.

There is a wide range of natural and artificial electromagnetic energy in the universe. Visible light is familiar to

*With the Kelvin scale, K is no longer written with a degree symbol.

all of us because it is discernable by our eyes, but other important types include invisible gamma, X ray, ultraviolet, infrared, microwave, and radio energy (Figure 1-2). Practically all the natural electromagnetic energy injected into the **earth system** (i.e., the atmosphere and the earth's surface) is produced by the sun.** It provides the original energy for all forms of terrestrial life and for the natural processes that are operative in the atmosphere, water bodies, and upper layers of the solid earth. Today, a wide variety of electromagnetic energy is artificially produced here on earth, and several familiar sources are listed in Table 1-1. No fundamental physical differences exist between natural and artificial electromagnetic energy.

Methods of Energy Transfer

Energy is defined as the capacity to do work. In addition to electromagnetic, energy takes forms such as mechanical, chemical, electrical, and heat. In the course of work being done, the resulting energy must be transferred from one body to another or from one location to another location. Such energy transfers can be accomplished by one of three methods.

Conduction is the direct transfer of energy when one body, such as an atom or molecule, comes into physical contact with an adjacent body. It can occur in like or unlike substances (e.g., a metal spoon placed in a cup of hot coffee).

**A small amount is produced by internal heat and radioactive decay within the earth.

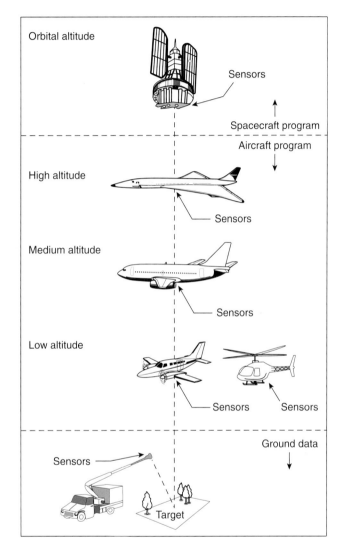

Figure 1-1 Multilevel concept of remote sensors operating at different distances above the earth's surface. Note that in no case does the sensor come into physical contact with the ground target. (Adapted from a NASA drawing.)

Figure 1-2 Major regions of the electromagnetic spectrum with wavelength, frequency, and photon energy scales. Because their ranges are so vast, they are shown graphically on a logarithmic scale. Note the relatively narrow portion occupied by the visible region to which the human eye is sensitive. (Adapted from Feinberg 1968.)

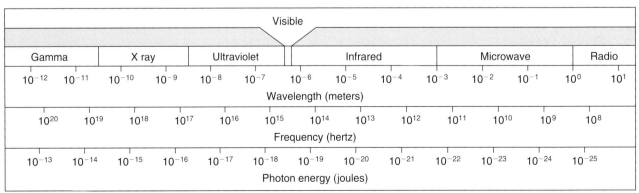

TABLE 1-1 Common Sources of Artificial Electromagnetic Energy	
Source	Primary Energy Type
PET (Positron Emission Tomography) scanner	Gamma
X-ray tube	X ray
CAT (Computer-Assisted Tomography) scanner	X ray
Sunlamp	Ultraviolet
Blacklamp	Ultraviolet
Electronic flash (camera)	Visible
Laser	Visible
Internal combustion engine	Infrared
Heat lamp	Infrared
Microwave oven	Microwave
Weather radar	Microwave
Maser	Microwave
UHF television broadcast	Microwave
VHF television broadcast	Radio
AM and FM radio broadcasts	Radio

The efficiency of energy transmission is directly related to how close the bodies are packed together, which determines the area of surface contacts. Consequently, conduction is most important in solids, of intermediate importance in liquids, and least important in gases, including the earth's atmosphere.

Convection is energy transfer by the physical movement of some energized medium from one location to another. Convection is possible only in liquids (e.g., water) and gases (e.g., the atmosphere) because they alone possess internal mass motion, or the ability to circulate internally.

Radiation is the method by which energy can be transferred from one body to another in the absence of an intervening material medium. If such a medium is present, it must be sufficiently transparent for the energy transfer to occur. Radiation is the only method by which solar energy can cross millions of kilometers of free or empty space (i.e., a vacuum) and reach the earth, and it is the method of energy transfer with which we are primarily concerned in remote sensing. The term radiation is commonly used today to describe both the *process* of energy propagation and the *energy* that flows in this manner (i.e., electromagnetic radiation).

Electromagnetic Radiation

Electromagnetic radiation (EMR) is electromagnetic energy in transit that can be detected only when it interacts with matter. It travels in a straight path at the **speed of light** across empty space and only slightly slower in the atmosphere. The speed of light in a vacuum is a fixed **universal constant**, as postulated by Albert Einstein in 1905. In 1963, the National Bureau of Standards officially set the velocity of electromagnetic radiation in a vacuum at 299,792.8 km/sec, or 186,281.7 mi/sec.* For most applications, EMR velocity is rounded to 300,000, or 3×10^5, km/sec and 186,000, or 1.86×10^5, mi/sec.

Modern physics views EMR as having a dual nature, enabling it to be *independently* described as a wave or a particle. The **wave model** shows EMR is carried by a series of continuous waves that are equally and repetitively spaced in time (**harmonic waves**). As seen in Figure 1-3, the wave pattern is in the form of two fluctuating fields—one **electric** and the other **magnetic**. Each has a sinusoidal shape because their plots resemble sine curves. These paired fields are perpendicular to each other, and both are perpendicular to the direction of wave propagation (**transverse waves**) (Figure 1-3). The refraction or bending of visible light by a glass prism is best explained by assuming the EMR is behaving like waves.

In order to avoid confusion, sec—not s—is used as the abbreviation for second in this textbook.

Figure 1-3 Graphical representation of the electric and magnetic fields of an electromagnetic wave at a given instant in time. The fields always occur together as mutual partners.

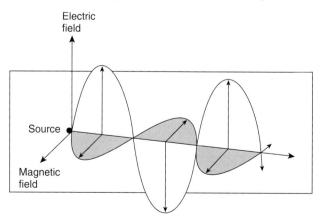

The wave nature of EMR is characterized by **wavelength** (λ), which is the linear distance between two successive wave crests or troughs, and by **frequency** (v), which is the number of wave crests or troughs (cycles) that pass a fixed point per second. They are related to the speed of light (c) in the following manner:

$$c = v\lambda,$$

$$v = \frac{c}{\lambda},$$

$$\lambda = \frac{c}{v}. \qquad (1\text{-}1)$$

From this equation and its variations, it is seen that (1) frequency and wavelength are directly proportional to velocity, which is essentially a constant within a uniform environment, and (2) wavelength and frequency have an inverse, or reciprocal, relationship. When electromagnetic radiation passes from a less dense to a more dense medium (e.g., from outer space to the atmosphere) and vice versa, velocity and wavelength change, but frequency always remains constant.

The **particle model** emphasizes the behavior of EMR as if EMR were composed of a collection of discrete, particle-like objects called **quanta** or **photons**, in which the electromagnetic energy is transferred at the speed of light. Photons carry particlelike properties such as energy and momentum from a source, but unlike other particles, such as protons and electrons, they have no mass at rest.

The operation of an industrial laser that is used for cutting through metals is best explained by assuming the laser's EMR is behaving like a discrete stream of high-energy particles. The particlelike nature of EMR is also illustrated by the activation of a camera's light meter and film when they are exposed to light.

The *intensity* of EMR is directly proportional to the number of photons present, and the *energy content* in a photon or quantum of radiation is related to the frequency and length of a wave by Planck's constant (h):

$$E = hv = \frac{hc}{\lambda}, \qquad (1\text{-}2)$$

where: E = energy of a photon in joules (J),* and
h = 6.626×10^{-34} J · sec.

Equation 1-2, known as the **basic energy equation**, shows (1) the direct relationship between v and E, or the energy of a photon varies directly with frequency, and (2) the inverse relationship between λ and E, or the energy of a photon varies inversely with wavelength (Figure 1-4). These relationships are borne out when one considers the interaction of ultraviolet radiation and visible light with the human skin. Ultraviolet, by nature of its shorter wavelength and higher frequency, possesses more photon energy than visible light, which can cause inflammation of the skin (sunburn) when there is prolonged exposure to sunlight.

Remote sensing is concerned with the measurement of EMR returned by the earth's natural and cultural features that first receive energy from the sun or an artificial source such as a radar transmitter. Because different objects return different types and amounts of EMR, the objective in remote sensing is to detect these differences with the appropriate instruments. This, in turn, makes it possible for us to identify and assess a broad range of surface features and their conditions.

Electromagnetic Spectrum

The entire range of EMR comprises the **electromagnetic spectrum**. This entire spectrum represents a **continuum**, consisting of the ordered arrangement of EMR according to wavelength, frequency, or photon energy (Figure 1-2). For the sun, the electromagnetic spectrum stretches from biologically lethal gamma rays (short wavelength, high frequencies, and high energy content) to passive radio waves (long wavelength, low frequencies, and low energy content). The terminal limits at both the short- and long-wavelength areas of the spectrum are not known precisely.

Largely for convenience of reference, the continuum is subdivided into several named divisions called **spectral bands**, which share similar characteristics (Figure 1-2). A spectral band is composed of some defined group or bundle of continuous **spectral lines**, where a line represents a single wavelength or frequency. The boundaries between most of the bands are arbitrarily defined because each portion overlaps adjacent portions. The boundaries of the visible band are the most precise because they are defined by the wavelength limits of human vision. As seen in Figure 1-2, the

---

Figure 1-4 Four different types of electromagnetic waves, illustrating the inverse relationship between wavelength and frequency.

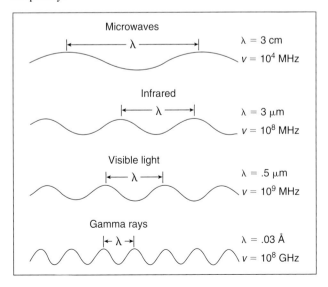

---

*A joule is a unit of work or energy equal to 10 million (10^7) ergs.

visible band falls in the midportion of the spectrum and occupies an extremely small segment of the total continuum.

Wavelength or frequency descriptors are normally used to define the various bands of the spectrum. EMR wavelengths range from kilometers for long radio waves to billionths of a centimeter for gamma rays. For continuity, the entire range of wavelengths can be expressed in terms of meters (Figure 1-2). However, it is often more convenient to specify the shorter wavelengths in **angstroms**, where one angstrom (1 Å) equals 10^{-10} m, in **nanometers** where one nanometer (nm) equals 10^{-9} m, or in **micrometers** where one micrometer (1 μm) equals 10^{-6} m (Figure 1-4).* They are related as follows: 10,000 Å = 1,000 nm = 1 μm. Using these measures, the visible band stretches from 4,000 Å, 400 nm, or 0.4 μm for blue light to 7,000 Å, 700 nm, or 0.7 μm for red light. For most remote sensing applications, the preferred unit of measure in the ultraviolet, visible, and infrared spectral regions is the micrometer. Wavelengths in the microwave band that are applicable to remote sensing are normally expressed in centimeters (Figure 1-4).

The entire frequency range of radiation can be expressed in cycles per second, or **hertz** (Hz). In this unit of measure, EMR frequencies vary from more than 10^{20} Hz for gamma rays to less than 10^{-25} Hz for radio waves (Figure 1-2). It has become common practice in remote sensing to specify the lower frequency ranges in **megahertz** (MHz), where 1 MHz is equal to 10^8 Hz, and the higher frequency ranges in **gigahertz** (GHz), where 1 GHz is equal to 10^9 Hz (Figure 1-4).

*Micrometer is now used in place of the term **micron** (μ).

Figure 1-5 Major spectral regions pertinent to remote sensing, showing atmospheric windows (white) and blinds (gray), the gases responsible for absorption, and the operational range of common remote sensing systems. Note that the wavelength scale is logarithmic. (Adapted from Sherz and Stevens 1970.)

Remote Sensing Spectral Regions

Given in order of increasing wavelengths, the bands of the electromagnetic spectrum that are used in remote sensing are **ultraviolet** (**UV**), **visible**, **infrared** (**IR**), and **microwave** (Figure 1-2). These spectral regions are shown at an expanded scale in Figure 1-5. Remote sensing systems have been designed to detect EMR selectively within one or more of the divisions, but no single instrument is capable of detecting radiation within all the divisions. Although no single remote sensor is an optimum device, as a group they are able to extract information throughout this spectral range, which is several million times wider than the visible band.

The **ultraviolet band** lies between X rays and visible light with wavelength limits of 0.01 and 0.4 μm (Figure 1-2). Because it borders the violet part of visible light on the long-wavelength side, ultraviolet literally means "above the violet." The ultraviolet band may be divided into **far UV** (0.01 to 0.2 μm), **middle UV** (0.2 to 0.3 μm), and **near UV** (0.3 to 0.4 μm). Although humans cannot see UV radiation directly, bees and certain other insects are visually sensitive to near UV. The sun is the natural source of UV radiation, but wavelengths shorter than 0.3 μm are unable to pass through the atmosphere and reach the earth's surface (Figure 1-5). Consequently, only the 0.3- to 0.4-μm wavelength interval, or near UV, is available for terrestrial remote sensing studies.

Traditionally, the most commonly used region of the electromagnetic spectrum in remote sensing has been the **visible band** or the **visible spectrum** (Figure 1-2). Its wavelength span is from 0.4 to 0.7 μm, limits established by the

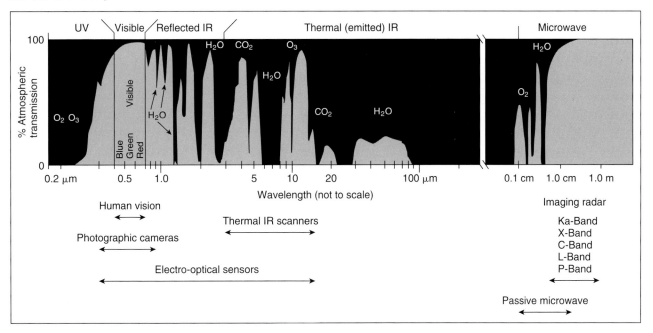

TABLE 1-2 Sun's Radiant Energy Distribution at 6,000 K[a]

Spectral Region	Wavelength (μm)	Percent of Total Energy
Gamma and X rays	<0.01	0.02
Far UV	0.01–0.2	
Middle UV	0.2–0.3	1.95
Near UV	0.3–0.4	5.32
Visible	0.4–0.7	43.50
Near IR	0.7–1.5	36.80
Middle IR	1.5–5.6	12.00
Far IR	5.6–1,000	0.41
Microwave and radio waves	>1,000	
Total		**100.00**

[a]*Adapted from Miller and Thompson (1979).*

sensitivity of the human eye. An interesting question is why, among the multitude of wavelengths comprising the solar spectrum, the retina has become sensitive to radiant energy in this very narrow wavelength band. A probable answer is that the bulk of the sun's radiation is concentrated here (Table 1-2), and it is able to pass relatively unimpeded through the atmosphere to the earth's surface (Figure 1-5).

A large part of the sun's radiant energy reaches the top of the atmosphere as **white light**, which is a term indicating the rather evenly distributed wavelengths of the visible band that have not been separated into their spectral components. When a narrow beam of white light is passed through a **glass prism**, it is separated into different wavelength bands that are made noticeable by their different colors (Figure 1-6). These familiar **rainbow colors** range from violet (shortest wavelengths) through blue, green, yellow, orange, and red (longest wavelengths) (Figures 1-6 and 1-7).

In remote sensing, the visible spectrum is also viewed as being composed of three equal-wavelength segments that represent the **additive primary colors**—blue (0.4 to 0.5 μm), green (0.5 to 0.6 μm), and red (0.6 to 0.7 μm) (Figure 1-7). A primary color is one that cannot be made from any other color. All colors perceived by the human optical system can be produced by combining the proper proportions of light representing the three primaries. This principle forms the basis for the operation of color television.

Most of the colors we see are the result of the preferential reflection and absorption of wavelengths that make up white light. For example, the chlorophyll of healthy grass selectively absorbs more of the blue and red wavelengths of white light and reflects relatively more of the green wavelengths to our eyes. The human optical system perceives fresh snow as white because its surface reflects all wavelengths of the visible spectrum equally well. By contrast, a fresh basaltic lava flow appears black because the wavelengths of white light are absorbed by its surface, leaving essentially no wavelengths to reach the eye from the reflection process. It is commonly said that black is the absence of color and white is the combination of all colors.

The **infrared (IR) band** has wavelengths between red light of the visible band at 0.7 μm and microwaves at 1,000

Figure 1-6 White light separated into its spectral colors by a glass prism. The colors are produced because the short violet wavelengths are slowed more in the glass than the long red wavelengths and thus are bent through a larger angle at each surface of the prism. The prism experiment was devised by the English scientist Sir Isaac Newton in 1666.

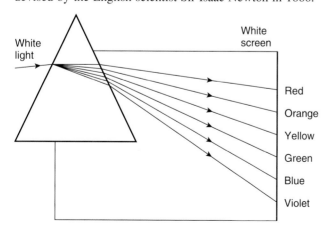

Figure 1-7 Colors of the visible spectrum produced when a beam of white light is passed through a glass prism. White light can also be divided into the three primary colors—blue, green, and red.

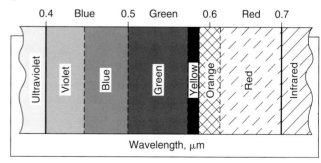

μm, or 0.1 cm (Figure 1-2). Because it is adjacent to red light, infrared means "below the red." In physics, the infrared band is divided into **near IR** (0.7 to 1.5 μm), **middle IR** (1.5 to 5.6 μm), and **far IR** (5.6 to 1,000 μm). In remote sensing, the total IR band is usually divided into two components that are based on basic property differences; the subdivisions are the **reflected IR band** and the **emitted, or thermal, IR band** (Figure 1-5).

The reflected IR band represents reflected solar radiation, which behaves like visible light. Its wavelength span is from 0.7 to about 3 μm (Figure 1-5). The dominant type of radiation in the thermal IR band is heat energy, which is continuously emitted by the atmosphere and all objects on the earth's surface. Its wavelength span is from about 3 μm to 1,000 μm, or 0.1 cm (Figure 1-5). Because of atmospheric attenuation, the thermal IR region beyond about 14 μm is generally not available for remote sensing studies.

The **microwave band** falls between the infrared and radio bands and has a wavelength range extending from approximately 0.1 cm to 1 m (Figures 1-2 and 1-5). It contains the longest wavelengths used in terrestrial remote sensing. At the proper wavelengths, microwave radiation can pass through clouds, precipitation, tree canopies, and dry surficial deposits, such as sand and fine-grained alluvium. There are two types of sensors that operate in the microwave band. **Passive microwave** systems detect natural microwave radiation that is emitted from the earth's surface, whereas **radar** propagates artificial microwave radiation to the surface and detects the reflected component (Figure 1-5).

Solar and Terrestrial Radiation

As described earlier, most remote sensing instruments are designed to detect (1) **solar radiation**, which passes through the atmosphere and is reflected in varying degrees by the earth's surface features and (2) **terrestrial radiation**, which is continuously emitted by these same features. Unlike solar radiation, which is detectable only during daylight, terrestrial radiation can be detected both day and night. The wavelength span and intensity of electromagnetic energy radiated by the sun and earth are primarily a function of their surface temperatures. The spectral distribution curves in Figure 1-8 show how much energy is emitted by the surfaces of the sun and earth per unit wavelength.

The sun's visible surface, or **photosphere**, has a temperature of about 6,000 K (5,727°C, or 10,340°F). At this very high temperature, the radiated energy covers a broad range of wavelengths, extending from very short gamma rays to very long radio waves (Figure 1-2). However, about 99 percent of the sun's radiation falls between wavelengths of 0.2 and 5.6 μm (Table 1-2). Furthermore, some 80 percent is contained in wavelengths between 0.4 and 1.5 μm (visible and reflected IR), to which the atmosphere is quite transparent (Table 1-2). Maximum radiation occurs at a wavelength of 0.48 μm in the visible band. (Figure 1-8).

Slightly less than half of solar radiation passes through the atmosphere and is absorbed in varying degrees by surface features of the surface. Most of this absorbed radiation

is transformed into low-temperature heat (warming the surface), which is continually emitted back into the atmosphere at longer thermal infrared wavelengths.

The earth's land and water surface has an ambient temperature of about 300 K (27°C, or 80°F). At this comparatively cool temperature, the earth radiates about 160,000 times less energy than the sun. Note in Figure 1-8 that essentially all the energy is radiated at invisible thermal IR wavelengths between about 4 and 25 μm, with a maximum emission occurring at a wavelength of 9.7 μm. Because the wavelengths covering most of the earth's energy output are several times longer than those covering most of the solar output, terrestrial radiation is frequently called **longwave radiation**, whereas solar radiation is termed **shortwave radiation**.

Longwave radiation is also emitted by the atmosphere's gases and clouds and from artificially heated objects on the earth's surface, such as from buildings, steam lines, and certain industrial effluents. Very hot earth phenomena (>800 K, or 527°C), such as forest fires and active lava flows, are luminous, which indicates the release of both visible light and invisible infrared radiation.

Radiation-Matter Interactions

Electromagnetic radiation manifests itself only through its interactions with matter, which can be in the form of a **gas**, a **liquid**, or a **solid**. This concept is clearly illus-

---

Figure 1-8 Spectral radiation curves for the sun at 6,000 K and the earth at 300 K, assuming each is a perfect radiation source or blackbody. (Adapted from Dobson 1963.)

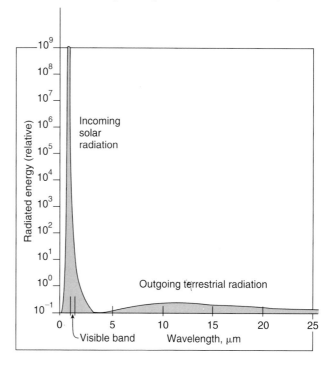

trated by shining a flashlight beam of visible light on a white wall in a darkened room. If we stand at a right angle to the long axis of the beam, the light is visible only at its source and where it strikes the wall and is reflected to our eyes. The beam cannot be seen from the side and can be made visible only when its optical path contains particles large enough to scatter some of the light beam sideways. This can be accomplished by adding chalk dust or smoke to the invisible beam. Their large particles will scatter a portion of the EMR to our eyes, enabling the beam to be seen from the side. This side scattering of visible light along a beam path is known as the **Tyndall effect**.

EMR that impinges upon matter is called **incident radiation**. For the earth, the strongest source of incident radiation is the sun. Such radiation is called **insolation**, a shortening of <u>in</u>coming <u>sol</u>ar r<u>a</u>di<u>a</u>tion. The full moon is the second strongest source, but its radiant energy measures only about one millionth of that from the sun. When EMR strikes matter, EMR may be **transmitted, reflected scattered**, or **absorbed** in proportions that depend upon (1) the compositional and physical properties of the medium, (2) the wavelength or frequency of the incident radiation, and (3) the angle at which the incident radiation strikes a surface. The four fundamental energy interactions with matter are illustrated in Figure 1-9.

Transmission is the process by which incident radiation passes through matter without measurable attenuation; the substance is thus transparent to the radiation. Transmission through material media of different densities (e.g., air to water) causes the radiation to be refracted or deflected from a straight-line path with an accompanying change in its velocity and wavelength; frequency always remains constant. In Figure 1-9, it is observed that the incident beam of EMR (θ_1) is deflected toward the normal in going from a medium of low density to a denser medium (θ_2). Upon emerging from the other side of the denser medium, the beam is refracted away from the normal (θ_3). The angle relationships in Figure 1-9 are $\theta_1 > \theta_2$ and $\theta_1 = \theta_3$.

The change in EMR velocity is explained by the **index of refraction** (n), which is the ratio between the velocity of EMR in a vacuum (c) and its velocity in a material medium (v):

$$n = \frac{c}{v}. \qquad (1\text{-}3)$$

The index of refraction for a vacuum (perfectly transparent medium) is equal to 1, or unity. Because v is never greater than c, n can never be less than 1 for any substance. Indices of refraction vary from 1.0002926 for the earth's atmosphere, to 1.33 for water, to 2.42 for a diamond.

Reflection (also called **specular reflection**) describes the process whereby incident radiation "bounces off" the surface of a substance in a single, predictable direction. The angle of reflection is always equal and opposite to the angle of incidence ($\theta_1 = \theta_2$ in Figure 1-9). Reflection is caused by surfaces that are smooth relative to the wavelengths of incident radiation. These smooth, mirror-like surfaces are called **specular reflectors**. Specular reflection causes no change to either EMR velocity or wavelength.

Scattering (also called **diffuse reflection**) occurs when incident radiation is dispersed or spread out unpredictably in many different directions, including the direction from which it originated (Figure 1-9). In the real world, scattering is much more common than reflection. The scattering process occurs with surfaces that are rough relative to the wavelengths of incident radiation. Such surfaces are called **diffuse reflectors**. EMR velocity and wavelength are not affected by the scattering process.

Absorption is the process by which incident radiation is taken in by the medium (Figure 1-9). For this to occur, the substance must be opaque to the incident radiation. A portion of the absorbed radiation is converted into internal heat energy, which is subsequently **emitted** or **reradiated** at longer thermal infrared wavelengths (Figure 1-9).

The interrelationships between energy interactions, as a function of wavelength (λ), can be expressed in the following manner:

$$E_I(\lambda) = E_T(\lambda) + E_R(\lambda) + E_A(\lambda), \qquad (1\text{-}4)$$

where: $E_I(\lambda)$ = incident radiant energy,
$E_T(\lambda)$ = decimal fraction transmitted,
$E_R(\lambda)$ = decimal fraction reflected (specular and diffuse), and
$E_A(\lambda)$ = decimal fraction absorbed.

Most opaque materials transmit no incident radiant energy; hence, $E_T(\lambda) = 0$ and $E_R(\lambda) + E_A(\lambda) = 1 = E_I(\lambda)$. In regard to visible light, (1) clear glass would have a high transmission value and low reflection and absorption values; (2) fresh snow would have a high reflectance value and low transmission and absorption values; and (3) fresh asphalt would be characterized by a high absorption value and minimal transmission and reflection values. Because only the part of incident radiation that is absorbed by an object is effective in heating it, there would only be a minuscule rise in temperature for glass and snow, whereas the asphalt's temperature would be markedly higher.

Figure 1-9 Four fundamental EMR interactions with matter.

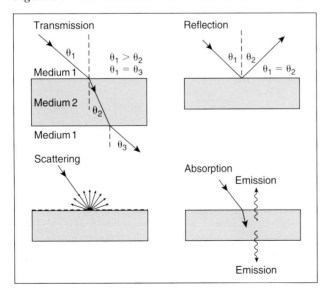

EMR-Atmosphere Interactions

Although electromagnetic radiation travels through empty space without modification, a series of diversions and depletions occurs as solar and terrestrial radiation interact with the earth's atmosphere. This interference is **wavelength selective**, meaning that EMR at certain wavelengths passes freely through the atmosphere, whereas it is restricted at other wavelengths. Areas of the spectrum where specific wavelengths can pass relatively unimpeded through the atmosphere are called **transmission bands**, or **atmospheric windows**, whereas **absorption bands**, or **atmospheric blinds**, define those areas where specific wavelengths are totally or partially blocked (Figure 1-5). When the objective is the study of the earth's surface, different remote sensing instruments have been designed to operate within the windows where the cloudless atmosphere will transmit sufficient radiation for detection.

Electromagnetic radiation interacts with the atmosphere in the following ways: (1) It may be absorbed and reradiated at longer wavelengths, which causes the air temperature to rise; (2) it may be reflected and scattered without change to either its velocity or wavelength; or (3) it may be transmitted in a straight-line path directly through the atmosphere. Figure 1-10 shows how 100 units of shortwave solar radiation interact with the earth's atmosphere and surface. Note that the amount returned to space is represented by 35 units of reflected shortwave radiation and 65 units of longwave radiation emitted from the earth system. Therefore, outgoing shortwave and longwave radiation (35 + 65 = 100) is equal to incoming solar radiation, a phenomenon known as the **radiation balance**.

Atmospheric Absorption and Transmission

Among the numerous gases of the atmosphere, the most significant absorbers of EMR are oxygen (O_2), nitrogen (N_2), ozone (O_3), carbon dioxide (CO_2), and water vapor (H_2O). Figure 1-5 illustrates the absorption-transmission characteristics of the cloud-free atmosphere and the gases responsible for EMR absorption as a function of wavelength. Some 16 percent of shortwave solar radiation is absorbed directly by the gaseous atmosphere as compared to only 2 percent by clouds, which are much better reflectors (Figure 1-10).

The atmosphere's gases are selective absorbers with reference to wavelength. Gamma and X-ray radiation are completely absorbed in the upper atmosphere by oxygen and nitrogen. Ultraviolet radiation, at wavelengths smaller than 0.2 μm, is absorbed by molecules of oxygen; this added energy is sufficient to cause the affected diatomic molecules to split apart, leaving single atoms or monatomic oxygen. Reactions occur where monatomic oxygen (O) and diatomic

Figure 1-10 Radiation budget of the earth and its atmosphere. (Adapted from Critchfield 1974.)

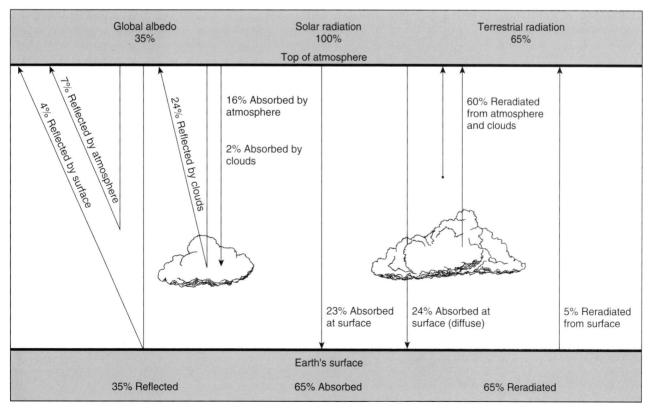

TABLE 1-3   Major Windows of the Gaseous Atmosphere

Window	Radiation Type
0.3–1.1 μm[a]	UV, visible, reflected IR
1.5–1.8 μm	Reflected IR
2.0–2.4 μm	Reflected IR
3.0–5.0 μm	Thermal IR
8.0–14.0 μm (below ozone layer)	Thermal IR
10.5–12.5 μm (above ozone layer)	Thermal IR
>0.6 cm	Microwave

[a]*No absorption bands, but scattering can be very troublesome for UV wavelengths and the shorter wavelengths of visible light.*

oxygen (O_2) are combined to form triatomic oxygen or ozone (O_3). The resulting ozone effectively absorbs ultraviolet radiation with wavelengths between 0.2 and 0.3 μm in the ozone layer of the stratosphere.

Water vapor and carbon dioxide are primarily responsible for several narrow absorption bands in the reflected infrared spectral region between 0.9 and 2.7 μm. In the thermal infrared region, strong water-vapor absorption occurs between 5 and 8 μm and from about 20 μm to the beginning of the microwave region at 1,000 μm (0.1 cm). Carbon dioxide effectively absorbs between 14 and 20 μm and ozone in the 9- to 10-μm wavelength span. This absorbed radiation heats the lower atmosphere, especially in humid areas where there is abundant water vapor.

Three relatively narrow absorption bands occur at the beginning of the microwave region between 0.1 and 0.6 cm; the absorption agents are oxygen and water vapor. Beyond the latter wavelength, the atmosphere's gases generally do not impede the passage of microwave radiation.

In summary, absorption by the atmosphere's gases has its maximum influence on wavelengths shorter than 0.3 μm (ultraviolet, X rays, gamma rays) and a minimum impact on wavelengths greater than 0.6 cm (microwaves) (Figure 1-5). Between these two wavelength extremes, there are a number of important windows that coincide with where most shortwave solar and longwave terrestrial radiation occur. The major atmospheric windows exploited in remote sensing are listed in Table 1-3.

It is emphasized that most of the atmospheric windows become less transparent when the air is moist (i.e., high humidity). In addition, clouds absorb most of the longwave radiation emitted from the earth's surface, essentially closing the thermal infrared windows (Table 1-3). This is why cloudy nights tend to be warmer than clear nights. Only microwave radiation with wavelengths longer than about 0.9 cm is capable of penetrating clouds.

Atmospheric Scattering

Electromagnetic radiation within certain sections of the ultraviolet, visible, and reflected infrared bands is impeded in its direct journey through the atmosphere by the scattering process, which disperses the radiation in all directions. Important scattering agents include gaseous molecules, suspended particulates called aerosols, and clouds. Their influence on scattering solar radiation is illustrated in Figure 1-10. Three types of atmospheric scattering are important to remote sensing.

Rayleigh, or **molecular**, **scattering** is primarily caused by oxygen and nitrogen molecules, whose effective diameters are at least 0.1 times smaller than the affected wavelengths. Rayleigh scattering is most influential at altitudes above 4.5 km, occurring in what is called the **pure atmosphere**. The amount of Rayleigh scattering is highly selective, being inversely proportional to the fourth power of wavelength (λ^{-4}) (Figure 1-11). Consequently, invisible

Figure 1-11 Various magnitudes of atmospheric scattering as a function of wavelength. The shaded portion indicates the range of scattering for typical atmospheric conditions. (Adapted from Slater 1983.)

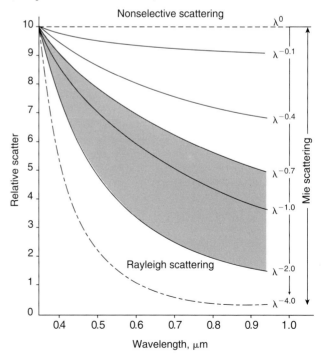

ultraviolet radiation, at a wavelength of 0.3 μm, is scattered 16 times as readily as red wavelengths at 0.6 μm $[(0.6/0.3)^4]$. In the visible spectrum, blue wavelengths at 0.4 μm are scattered about five times as readily as red wavelengths at 0.6 μm $[(0.6/0.4)^4]$. The preferential scattering of blue wavelengths explains why the clear sky (i.e., low humidity and few aerosols) appears blue in daylight. The blue wavelengths reach our eyes from all parts of the sky.

Mie, or **nonmolecular**, **scattering** occurs when there are sufficient particles in the atmosphere that have mean diameters from about 0.1 to 10 times larger than the wavelengths under consideration. Important Mie scattering agents include water vapor and tiny particles of smoke, dust, volcanic ejecta, and salt crystals released from the evaporation of sea spray. The influence of Mie scattering is most pronounced in the lower 4.5 km of the atmosphere, where the larger Mie particles are most abundant (i.e., in the **impure atmosphere**). Mie scattering influences longer radiation wavelengths than Rayleigh scattering. Depending upon the size distribution and concentration of Mie particles, the wavelength dependence varies between λ^{-4} and λ^0.

The clear atmosphere is a medium for both Rayleigh and Mie scattering. For a range of typical atmospheres, their combined influence is from about $\lambda^{-0.7}$ to λ^{-2} (Figure 1-11). Using a λ^{-2} relationship, blue wavelengths at 0.4 μm are scattered about three times as readily as red wavelengths at 0.7 μm $[(0.7/0.4)^2]$. The observation of a red sunrise or sunset is caused by the preferential treatment of Rayleigh and Mie scattering agents on sunlight. At these times, the solar beam, which starts out as white light, passes through its longest atmospheric path. This extended path causes the shorter wavelengths of sunlight to be scattered away (blue and green), leaving only the red wavelengths to reach our eyes.

Nonselective scattering becomes operative when the lower atmosphere contains a sufficient number of suspended aerosols having diameters at least 10 times larger than the wavelengths under consideration. Important agents include the larger equivalents of Mie particles plus the water droplets and ice crystals of which clouds and fog are composed. Nonselective scattering is independent of wavelength (λ^0). Its influence spans the near ultraviolet and visible bands and extends into the reflected infrared band (Figure 1-11).

Within the visible band, colorless water droplets and ice crystals scatter all wavelengths equally well, causing, for example, the sunlit surfaces of clouds to appear brilliant white. Also, large smog particles, if not possessing special absorption properties, will cause the color of the sky to go from blue to grayish white.

Skylight and Haze

The clear sky is a source of illumination because its gases preferentially scatter the shorter wavelengths of sunlight. This diffuse radiation is called **skylight**, or **sky radiation** (Figure 1-12). Consequently, daylight illumination at the earth's surface consists of both direct sunlight and diffuse skylight (Figure 1-10). By comparison, only direct sunlight

Figure 1-12 Oblique photograph showing the luminous atmosphere (bright tones). The atmosphere acts as an illumination source (skylight) because its gases preferentially scatter the shorter wavelengths of sunlight to which the film is sensitive. View is to the south from an 18.2 km altitude and includes the San Francisco volcanic field in north-central Arizona. (Courtesy U.S. Geological Survey.)

Figure 1-13 Visible band photograph showing the moon's surface silhouetted against a black sky. An Apollo astronaut and the Lunar Rover are seen in the foreground. (Courtesy Philip A. Davis, U.S. Geological Survey.)

reaches the lunar surface because the moon has no atmosphere and, hence, no scattering agents to cause skylight. As a consequence, the moon's sky color is black (Figure 1-13).

Within the visible spectrum, skylight is composed primarily of blue wavelengths. It is skylight that prevents absolute darkness in shadows where direct sunlight is absent. To our eyes, sky radiation is manifested as **haze**, which causes a reduction in visibility and also causes distant landscapes to take on a soft, blue-gray appearance.

Atmospheric haze has important ramifications in remote sensing. In the short-wavelength region, radiation reaching an airborne or spaceborne sensor consists of two components: (1) radiation that is scattered by the earth's surface and then reaches the sensor without being affected by the intervening atmosphere (component S_s in Figure 1-14), and (2) radiation that is scattered by the atmosphere, either before or after it reaches the earth's surface (component S_a in Figure 1-14). Thus, the total radiation reaching the sensor (S) is the sum of these two components:

$$S = S_s + S_a. \qquad (1-5)$$

The radiation scattered by the atmosphere (S_a) contains no information about the earth's surface, and it acts as a masking agent when a remote sensing system records these wavelengths. The most strongly affected wavelengths are ultraviolet and blue. The net effect of this extra illumination, or nonimage-forming "haze light," in applicable photographs and images is a loss of detail and a diminution in the scene contrast. Specifically, haze is visualized as a foglike veil in a black-and-white photograph and as an overall bluish tint in a color photograph (Figure 1-15 and Plate 1). The influence of haze increases with increasing flying height for one set of atmospheric conditions.

Figure 1-14 Shortwave radiation reaching an airborne or spaceborne remote sensor consists of a scattered component from the earth's surface (S_s) and a scattered component from the atmosphere (S_a).

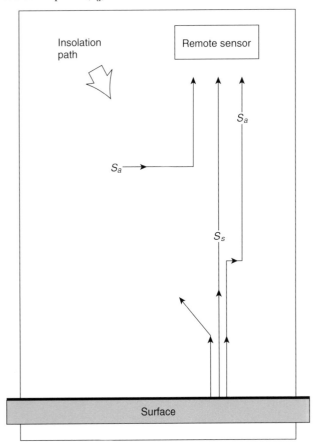

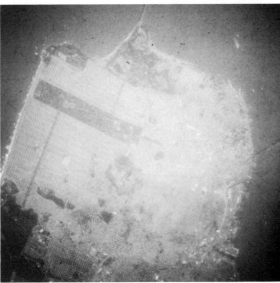

Figure 1-15 Vertical photographs of the San Francisco region in the red (*top*) and blue (*bottom*) portions of the visible spectrum. The fogged appearance of the blue-sensitive photograph was caused by atmospheric haze. Exposures were made simultaneously from a 20-km altitude. (Courtesy U.S. Geological Survey.)

As we see in Chapter 2, the detrimental effects of haze can be eliminated, or at least reduced, in aerial photographs by the judicious use of special lens filters that do not transmit blue and/or ultraviolet radiation to the film. Proper filtration, in essence, removes or reduces the S_a component in Equation 1-5.

EMR-Surface Interactions

The natural and cultural features of the earth's surface differ markedly in their disposal of solar radiation. On the

Figure 1-16 Weather satellite image in the visible band showing cloud patterns over a portion of the Western Hemisphere. Note the pronounced tonal contrast between the clouds (high albedo) and water (low albedo).

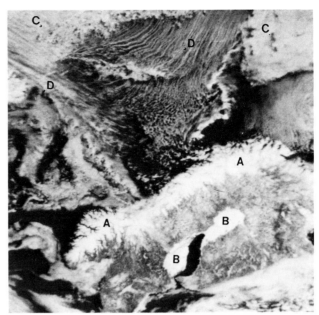

Figure 1-17 Weather satellite image in the visible band showing snow-covered Scandinavia (A), the partially ice-covered Gulf of Bothnia (B), pack ice in the Barents Sea (C), and cloud patterns associated with storm systems in the Barents Sea (D). Note the pronounced tonal contrast between snow, ice, and clouds (high albedo) versus open water (low albedo).

average, 51 percent of shortwave radiation incident on the top of the atmosphere reaches and interacts with the earth's surface features. From this total, 4 percent is reflected directly without transfer of energy, and 47 percent is absorbed (Figure 1-10). It is these two components that drive most terrestrial remote sensing studies, either directly (reflection component) or indirectly (absorption component followed by reradiation).

Essentially all the absorbed shortwave radiation is reradiated or emitted back to the atmosphere as longwave terrestrial radiation (5 percent component in Figure 1-10). Most of this heat energy is emitted at wavelengths falling within the thermal infrared atmospheric windows (Table 1-3 and Figure 1-5). It is this type of radiation that contains information about the different temperature properties of the earth's surface features.

The average amount of incident radiation reflected by an object at some wavelength interval is called its **albedo**, or **spectral reflectance** $R(\lambda)$. Albedo is normally expressed as a percentage and is determined by the equation

$$R(\lambda) = \frac{E_R(\lambda)}{E_I(\lambda)} \times 100, \qquad (1\text{-}6)$$

where: $E_R(\lambda)$ = radiant energy reflected from an object, and
$E_I(\lambda)$ = radiant energy incident upon the same object.

For the incoming solar beam, the albedo of the earth system (including 50 percent cloud cover) is about 35 per-

cent. This means that, on the average, 35 percent of the solar beam is reflected and 65 percent is absorbed (Figure 1-10). When viewed from space, the earth is made visible only by its albedo. From this vantage point, the earth's brightest features are its clouds and its snow and ice surfaces; water bodies are typically among its darkest features (Figures 1-16 and 1-17). Albedos of various materials are listed in Table 1-4.

TABLE 1-4   Albedos of Various Materials at Visible Wavelengths

Material	Percent Reflected
Fresh snow	80–95
Old snow	50–60
Thick cloud	70–80
Thin cloud	20–30
Water (sun near horizon)	50–80
Water (sun near zenith)	3–5
Asphalt	5–10
Light soil	25–45
Dark soil	5–15
Dry soil	20–25
Wet soil	15–25
Deciduous forest	15–20
Coniferous forest	10–15
Crops	10–25
Earth system	35

Figure 1-18 Photographs of white clothing and a toboggan lying on a snow background in the visible (*top*) and ultraviolet (*bottom*) bands. (Courtesy David M. Lavigne, University of Guelph.)

Figure 1-19 Visible photograph (*top*) and thermal infrared image (*bottom*) of wooded terrain. Note how three camouflaged soldiers stand out only in the thermal infrared image. (Courtesy Lonnie Schuepbach and Deborah Hewitt, FLIR Systems, Inc.)

Albedo is not only an important property of an object in regard to its reflection efficiency, but it also helps to explain how warm an object becomes when exposed to sunlight. In general, objects with high albedos are good reflectors but poor absorbers, which dictates slow and small temperature increases. Conversely, objects with low albedos are poor reflectors but good absorbers, dictating rapid and large jumps in temperature when they are exposed to sunlight. This explains the differences in sensible heat we experience when wearing light or dark clothing on a summer day or the difference in walking barefoot on black asphalt versus grass.

Spectral Signatures

Every natural and synthetic object reflects and emits electromagnetic radiation over a range of wavelengths in its own characteristic manner according, in large measure, to its chemical composition and physical state. The distinctive reflectance and emittance properties of these objects and their different conditions are called **spectral signatures**. Within some limited wavelength region, a particular object or condition will usually exhibit a diagnostic spectral response pattern that differs from that of other objects. Remote sensing depends upon operation in wavelength regions of the spectrum where these detectable differences in reflected and emitted radiation occur. It is only then that the features or their different conditions will show enough variation to allow for individual identifications.

The diagnostic response patterns that make it possible to discriminate objects (spectral signatures) often lie beyond the narrow confines of the visible spectrum where no detectable differences occur. In order to make an object's spectral response identifiable outside of the visible band, reliance is placed on special films with extended spectral sensitivities or various types of nonfilm detectors. These latter detectors translate the sensed radiation into electrical energy, which can then be used to drive invisible-to-visible image translation devices. Figures 1-18 and 1-19 illustrate object discriminations in areas outside the short- and long-wavelength boundaries of the visible spectrum.

Figure 1-18 shows visible and ultraviolet photographs of white clothing and a toboggan on a background of white snow. The photograph in the visible band results in a black-and-white representation of what the eye would see. The ultraviolet photograph is a visual record of how well the equipment and snow reflected the ultraviolet component in solar radiation. It is observed that the equipment and snow have bright signatures (i.e., high reflectance) in the visible photograph, which would make their discrimination difficult even from a relatively short distance. However, the ultraviolet photograph demonstrates that the white equipment absorbs much of the ultraviolet radiation (dark tones), whereas the snow is a much better reflector (bright tones). Thus, it is only in the ultraviolet photograph that the camouflaged equipment clearly displays its unique spectral signature against the snow background.

Figure 1-19 shows a visual photograph and thermal infrared image of wooded terrain. In the visible photograph, only one of three camouflaged soldiers can be seen because their clothing and the vegetation have similar colors. All three soldiers can be clearly distinguished in the thermal infrared image because their temperatures were considerably higher than that of the background foliage. The tones in the thermal infrared image correspond to different temperatures, with white representing the warmest temperature.

In the field environment, quantitative measurements of reflected and/or emitted radiation emanating from surface features are often made with portable **radiometers**. The principal types of radiometers are (1) **single-band radiometers**, which measure radiation intensity integrated through one broad wavelength band, (2) **multiband**, or **multispectral**, **radiometers**, which measure radiation intensity in more than one waveband, and (3) **spectroradiometers**, which measure radiation intensity over a continuous range of wavelengths by simultaneously sampling a large number of very narrow wavebands. Figure 1-20 shows a portable spectroradiometer, which acquires a continuous spectrum by measuring radiation intensity in 256 bands between 0.2 and 2.5 μm (UV, visible, reflected IR).

Radiometer measurements provide useful reference data for identifying spectral regions in which various features can be best differentiated for a particular application. This, in turn, allows for the improved interpretation of remote sensing photographs and images and for the selection of spectral bands for future remote sensing imaging devices. Radiometers may be operated in a variety of modes, from close-range, hand-held, and tripod positions to operation from hovering helicopters and truck-mounted "cherry pickers" (Figure 1-1).

Spectroradiometer measurements are used to prepare **spectral signature curves**, which are two-dimensional line plots showing the radiation intensity for various features as a function of wavelength. Typical curves representing the spectral signatures for three common materials, vegetation, soil, and water, are shown in Figure 1-21. Note how distinctive the curves are for each material class. Most importantly, a wide separation between the spectral signature curves relates to large intensity differences. These proportional differences are rendered as distinct tonal or color separations in photographs and images acquired in the same wavelength range. In Figure 1-21, for example, the largest difference in radiation intensity between vegetation, soil, and water occurs near a wavelength of 1 μm. Therefore, we could best discriminate the three materials with a remote sensing system that detects reflected infrared radiation.

Remote Sensors

The term **remote sensor** encompasses all instruments that detect and measure reflected and/or emitted electromagnetic radiation from a distance and reflected underwater sound waves in the case of sonar. Instruments used for collecting remote sensing data fall into two broad categories—**nonimaging**, such as spectroradiometers (Figure 1-20), and **imaging**, from which two-dimensional pictorial representations of the features under investigation can be made.

Imaging devices are designed to provide views of a given surface from **vertical** (downward) or **oblique** (slanted)

Figure 1-20 Spectron Engineering SE 590 portable spectroradiometer showing microprocessor data logger, tape cassette, and detector head. (Courtesy Thomas Hutchcroft. Spectron Engineering.)

perspectives. These points of reference are quite different from our normal ground-level observations. This concept is illustrated in Figure 1-22, where vertical, oblique, and ground-level views of a volcanic cinder cone are shown.

Imaging systems can be divided into five main groups: (1) **photographic camera**, (2) **electro-optical**, (3) **passive microwave**, (4) **radar**, and (5) **sonar**. These systems are further classified as passive or active. **Passive remote sensors**, represented by the first three categories, detect only radiation emanating naturally from an object, such as reflected sunlight or thermal IR and microwave emissions. **Active remote sensors**, represented by radar and sonar, beam their own artificially produced energy to a target and record the reflected component. Thus, a passive system *only* receives, whereas an active system *both* transmits and receives.

Figure 1-21 Typical spectral signature curves for three common surface materials. The abscissa is plotted on a logarithmic scale. (Adapted from a NASA drawing.)

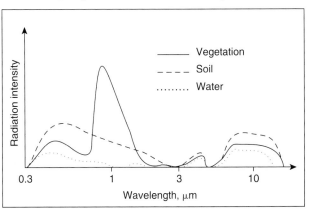

Figure 1-22 Vertical (*top*), oblique (*middle*), and ground-level (*bottom*) views of SP Mountain, a volcanic cinder cone in north-central Arizona.

The critical components of a photographic camera are its lens, which renders a sharply defined image of a given ground scene, and its film, which acts as both the radiation detector and the recording medium. Because of these two components, this type of camera is sometimes referred to as a **photo-optical** system. Because of military restrictions, the photographic camera was the only remote sensor available to the civilian scientific community prior to the early 1960s.

Photographic cameras can respond to wavelengths ranging from about 0.3 to 0.9 μm (Figure 1-5), which includes the near UV band (0.3 to 0.4 μm), all the visible band (0.4 to 0.7 μm), and the shortest wavelengths of the reflected IR band (0.7 to 0.9 μm). The infrared portion is referred to as **photographic infrared**, or **photographic IR**. The camera's wavelength sensitivity is nearly twice as wide as our eye sensitivity and is referred to as the **photographic spectrum** because it contains those radiation wavelengths that can be recorded directly onto film.

Electro-optical sensors use mirrors and/or lenses to collect and focus impinging radiation onto different types of **photon detectors**. These nonfilm detectors include photomultiplier tubes, photodiodes, and metallic crystals that are each sensitive to a limited wavelength range. Photon detectors convert the collected radiation into proportional electrical signals that are generally recorded on magnetic tape. Subsequently, the stored signals are converted to two-dimensional images that can be viewed on a television screen or printed onto photographic film for viewing. With this class of remote sensor, film is used only as the recording medium.

In many cases, particular photon detectors are installed in electro-optical systems to simultaneously collect image data in multiple, narrow wavelength ranges that can be located at various points within the atmospheric windows of the near UV, visible, reflected IR, and thermal IR spectral regions. This total wavelength span extends from about 0.3 to 14 μm and is called the **optical spectrum** because these radiation wavelengths can be reflected and refracted with lenses and mirrors (Figure 1-5). Electro-optical sensors operating in the thermal IR region exhibit a day or night capability and are not restricted by dust and haze.

Two types of antenna-based remote sensors operate in the microwave band, where atmospheric attenuation is negligible. Passive microwave imagers detect the low-level microwave radiation that is emitted from the earth's surface. These devices normally operate at wavelengths between 0.15 and 30 cm (Figure 1-5). Imaging radars, or the active microwave remote sensors, operate within very narrow wavelength bands from about 0.8 to 68 cm. During World War II, these bands were assigned letter designations for security purposes, and this nomenclature continues to be used today (Figure 1-5). The newest types of microwave systems use data recording and image display methods analogous to electro-optical sensors. Both passive and active microwave systems have the important advantage of an all-weather capability at wavelengths exceeding about 3 cm. In addition, their operation is independent of time of day.

Imaging sonar has many operational similarities to active imaging radar, except the aircraft or spacecraft is replaced by a ship, the atmosphere is changed to water, and the electromagnetic energy is replaced by acoustical or sound energy. Sonar images offer tremendous potential for the accurate mapping and assessment of the earth's water-covered floors. Sonar images are analogous to radar images of land areas.

Many electro-optical, passive microwave, radar, and sonar systems are able to place image data directly onto magnetic tape during acquisition. Because of this capability, there is an increasing reliance on computers and auxiliary equipment to process, analyze, and display the voluminous amounts of data that are now being routinely collected by these nonphotographic imaging systems. This branch of remote sensing is known as **digital image processing**.

Photograph Versus Image

In remote sensing, care is exercised in the use of the terms **photograph** and **image**. The term photograph is used exclusively for pictorial representations that are recorded directly onto film by a process known as **photography**. The word photography is derived from the Greek *phos*, meaning "light," and *graphia*, meaning "writing." Image is a more general term referring to all pictorial representations of remote sensing data. Therefore, all photographs are images, but not all images are photographs. The pictorial record of microwave radiation collected by a radar system, for instance, is called a **radar image** and not a radar photograph because photographic film is not used as the original detection medium. The pictorial products of electro-optical, passive microwave, and sonar systems are likewise referred to as images, not photographs.

The Multi Concept in Remote Sensing

The success of many remote sensing studies can be significantly improved by the **multi concept** as applied to data acquisition and analysis. The multi concept revolves around the following components, which are ideally implemented in combinations applicable to a specific application (Colwell 1975; Short 1982). Remote sensing image data can be acquired from different platforms (**multistage**) that are operated from different altitudes above the earth's surface (**multilevel**) (Figure 1-1). Multilevel sampling gives rise to **multiscale images**, which depict a given ground scene in different degrees of detail.

Multistation refers to the acquisition of successive, overlapping photographs taken along a given flight line. When two such photographs are viewed with stereoscopic instruments, surface features are seen in three dimensions. This ability often makes identifications easier than if a photograph from only one of the stations is available for two-dimensional viewing.

For most remote sensing applications, having image data from several spectral regions (**multispectral** or **multiband**) has proved more valuable than having data from only a single region (Figures 1-18 and 1-19). The former type of data collection can be accomplished by different sensors operating simultaneously (**multisensor**) or by a single sensor that operates in several spectral regions simultaneously (**multispectral** or **multiband sensor**).

Multitemporal or **multidate** involves sensing the same surface area at more than one time (e.g., seasonal or annual time scales). The principal advantage of multidate analysis is the increased amount of information that is obtainable compared with the case in which image data are collected on only one date. Multidate images are frequently used to monitor land use changes and seasonal changes associated with agricultural practices.

Because remote sensing is a problem-solving tool, it is best applied in concert with various kinds of supportive information (**multisource**). These multisources of information can include topographical and geological maps, on-site interviews, visual observations, instrumented surface measurements, and material sampling followed by laboratory analyses (e.g., determination of particle-size distributions, moisture levels, or chemical compositions).

TABLE 1-5   Remote Sensing Journals

Journal	Organization
Canadian Journal of Remote Sensing	Canadian Aeronautics and Space Institute, Saxe Building 60–75 Sparks Street Ottawa, Ontario K1P 5A5, Canada
ESA Bulletin *ESA Journal*	European Space Agency Publications ESTEC, P.O. Box 299 2200 AG Noordwijk, The Netherlands
Geocarto International	Geocarto International Centre G.P.O. Box 4122 Hong Kong
IEEE Transactions on Geoscience and Remote Sensing	Institute of Electrical and Electronics Engineers 345 East 47th Street New York, NY 10017
International Journal of Remote Sensing	Taylor & Francis Ltd. 4 John Street London WC1N 2ET, UK
ITC Journal	International Institute for Aerospace Survey and Earth Sciences P.O. Box 6 NL-7500 AA Enschede, The Netherlands
Photogrammetria	Elsevier Science Publishers P.O. Box 211 1000 AE Amsterdam, The Netherlands
Photogrammetric Engineering and Remote Sensing[a]	American Society for Photogrammetry and Remote Sensing 5410 Grosvenor Lane, Suite 210 Bethesda, MD 20814
Remote Sensing of Environment	Elsevier Science Publishers 655 Avenue of the Americas New York, NY 10010
Remote Sensing Reviews	Harwood Academic Publishers One Bedford Street London WC2E 9PP, UK

[a]*Prior to 1975, the name of the journal was* Photogrammetric Engineering.

Sources of Remote Sensing Information

Because of the burgeoning growth of remote sensing since the early 1960s, several journals devoted entirely to topics in remote sensing are now published on a regular schedule. The major English-language journals are listed in Table 1-5. In addition, many articles on remote sensing are found in periodicals devoted to the fields of archaeology, biology, engineering, environmental science, forestry, geography, geology, and meteorology.

The acknowledged reference work in remote sensing is the *Manual of Remote Sensing* (Colwell 1983). This two-volume set is now in its second edition and is published by the American Society for Photogrammetry and Remote Sensing; the society's address is found in Table 1-5.

Questions

1. Use Equation 1-1 to calculate the wavelength in meters of a taxicab broadcast at a frequency of 150 MHz; set c equal to 3×10^8 m/sec, and express 150 MHz in Hz. In what band of the electromagnetic spectrum is the wavelength located?

2. Use Equation 1-1 to calculate the frequency in gigahertz (GHz) of an active radar system operating at a wavelength of 3.2 cm; set c equal to 3×10^8 m/sec and λ equal to 3.2×10^{-2} m.

3. Use Equation 1-2 to calculate the energy content in two photons, one with a frequency of 40 GHz (microwave) and one with a wavelength of 0.3 μm (ultraviolet). Set c equal to 3×10^8 m/sec, and express frequency in hertz and wavelength in meters. The energy content of the ultraviolet photon is how many times greater than that of the microwave photon?

4. Use Equation 1-3 to calculate the velocity of electromagnetic radiation (kilometers per second and miles per second) as it travels through the earth's atmosphere.

5. Complete the following table, which defines the three subdivisions of the ultraviolet (UV) band of the electromagnetic spectrum.

Wavelength	Far UV	Middle UV	Near UV
Å	—	—	3,000–4,000
nm	—	200–300	—
μm	0.01–0.2	—	—

6. Although they are present, explain why we cannot see stars during daylight.

7. Explain why a thick cloud is white as viewed from above and dark gray as seen from beneath.

8. Examine Figure 1-21 and answer the following questions.
   a. Which regions of the spectrum (UV, visible, reflected IR) show the largest reflectances for vegetation, soil, and water?
   b. Which materials have the highest and lowest emittances in the thermal IR region between 8 and 14 μm?
   c. Which materials have the highest and lowest reflectances in the visible band?
   d. In which regions (band name and wavelength span) would separation between the three materials be the most difficult?

Bibliography and Suggested Readings

Colwell, R. N. 1975. Introduction. In *Manual of Remote Sensing*, edited by R. G. Reeves, 1–26. Falls Church, Va.: American Society of Photogrammetry.

———, ed. 1983. *Manual of Remote Sensing*, 2d ed. Falls Church, Va.: American Society for Photogrammetry and Remote Sensing.

Critchfield, H. J. 1974. *General Climatology*, 3d ed. Englewood Cliffs, N.J.: Prentice-Hall.

Dobson, G. M. B. 1963. *Exploring the Universe*. London: Oxford University Press.

Feinberg, G. 1968. Light. *Scientific American* 219: 50–59.

Fischer, W. A., P. Badgley, D. G. Orr, and G. J. Zeissis. 1975. History of Remote Sensing. In *Manual of Remote Sensing*, edited by R. G. Reeves, 27–50. Falls Church, Va.: American Society of Photogrammetry.

Harper, D. 1983. *Eye in the Sky*. Montreal: Multiscience Publications.

Hoffman, R. R., and J. Conway. 1989. Psychological Factors in Remote Sensing: A Review of Some Recent Research. *Geocarto International* 4:3–22.

Hyatt, E. 1988. *Keyguide to Information Sources in Remote Sensing*. London: Mansell Publishing Limited.

Jensen, J. R. 1983. Biophysical Remote Sensing. *Annals of the Association of American Geographers* 73:111–32.

Lo, C. P. 1987. *Applied Remote Sensing*. New York: Longman Inc.

Miller, A., and J. C. Thompson. 1979. *Elements of Meteorology*, 3d ed. Columbus, Ohio: Charles E. Merrill Publishing Company.

Schanda, E. 1986. *Physical Fundamentals of Remote Sensing*. Heidelberg: Springer-Verlag.

Scherz, S. P., and A. R. Stevens. 1970. An Introduction to Remote Sensing for Environmental Monitoring. Report No. 1. Madison: Institute of Environmental Studies, University of Wisconsin.

Short, N. M. 1982. *The Landsat Tutorial Workbook*. Washington, D.C.: U.S. Government Printing Office.

Slater, P. N. 1983. Photographic Systems for Remote Sensing. In *Manual of Remote Sensing*, edited by R. N. Colwell, 2d ed, 231–91. Falls Church, Va.: American Society for Photogrammetry and Remote Sensing.

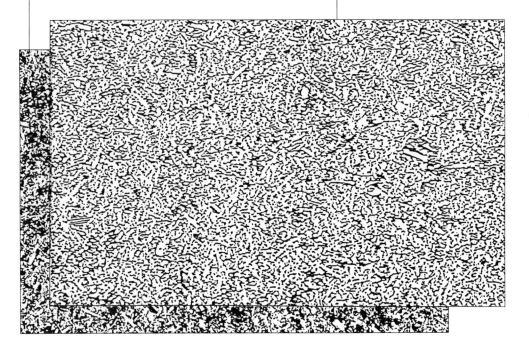

Chapter 2

Cameras, Films, and Filters

tographed Boston from the gondola of a captive balloon (Figure 2-1). Balloon photography proved highly utilitarian during the Civil War when General George McClellan employed this innovation to make photomaps of Confederate positions in Virginia.

The first recorded photographs taken from an airplane were made in 1909 by Wilbur Wright and a Pathé news cameraperson, who took motion pictures of Centotelli, Italy. However, it was World War I that brought together the airplane, improved cameras and films, and a real need for aerial photography. Most advancements in the scientific uses of aerial photography have been made since the end of World War II.

With the advent of the balloon and, more importantly, the airplane, the camera represented a revolutionary new instrument for viewing, measuring, and mapping the earth's surface. Today, modern photographic cameras are routinely operated from many different types of aircraft and earth-orbiting spacecraft (Figure 1-1).

Types of Photographs

In remote sensing, we are concerned with the photographs taken by cameras carried on platforms operating above the earth's surface. These *pictorial products* are called **aerial photographs**, or simply **airphotos**. This nomenclature distinguishes them from **terrestrial**, or **ground**, **photographs**, which are taken exclusively from the surface of the earth. The term **aerial photography** denotes the *process* of taking photographs from above the earth's surface.

Depending upon the orientation of the camera's optical axis with respect to the earth's surface at the time of film exposure, airphotos are classified as vertical or oblique (Figure 2-2). **Vertical airphotos** are those taken with the camera's optical axis oriented in a vertical or nearly vertical angle to the local ground surface (90° ± 3°) (Figure 2-3). **Oblique airphotos** are those taken with the camera's optical axis intentionally tilted away from the vertical by an angular amount that usually exceeds 20°.

Oblique photographs can be further classified as high-oblique or low-oblique (Figure 2-2). **High-oblique airphotos** show the surface, the horizon, and a portion of the sky, regardless of the altitude from which they are taken (Figure 2-4). **Low-oblique airphotos** show only the surface (Figure 2-5).

The Camera as a Remote Sensor

The **photographic camera**, with its associated film, is the oldest and the most frequently used remote sensing instrument. The merging of the camera with an airborne platform, to form a true remote sensing system, dates back to 1859 when Gaspard Felix Tournachon obtained balloon photographs of a small village near Paris. In the United States, it commenced 1 year later when James Wallace Black pho-

Figure 2-1 This 1860 picture of Boston Harbor is thought to be the first aerial photograph taken in the United States. The exposure was made from a balloon at an altitude of about 365 m above the ground. (Courtesy General Aniline and Film Corp.)

Vertical airphotos present relatively undistorted overhead views of the landscape and are essential for accurate mapping and interpretation tasks (Figure 2-3); the vast majority of all airphotos fall into this category. In addition, overlapping vertical airphotos can be viewed stereoscopically to produce three-dimensional views of the landscape

Figure 2-2 Orientation of an aerial camera for vertical, low-oblique, and high-oblique photography. Also shown is the shape and relative size of the ground area associated with each type of photograph.

Figure 2-3 Vertical airphoto of an urban area along the Atlantic coast of southern Florida.

Figure 2-4 High-oblique airphoto (horizon included) of SP Mountain, a volcanic cinder cone in north-central Arizona. Compare with Figure 1-22.

Figure 2-5 Low-oblique airphoto (horizon not shown) of a flooded area near Whakatane, New Zealand. (Courtesy Aero Surveys (New Zealand), Ltd.)

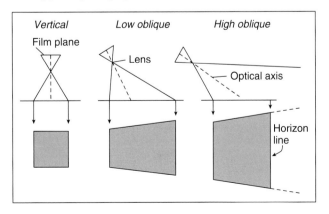

(discussed in subsequent chapters). The vertical photo perspective is entirely different from our normal viewing, and it may take a while to become accustomed to looking at features from a vantage point directly overhead (Figure 2-3).

From the same camera-station altitude, oblique airphotos capture larger ground areas than verticals, and the ground area imaged is trapezoidal in shape (Figure 2-2). Their perspective views are distorted, with distant features appearing smaller than similar-sized features in the foreground. Oblique airphotos incorporate views that are similar to ours when we view the ground through an airplane window (Figures 2-4 and 2-5). As a general rule, oblique airphotos are best employed to show specific features or as supplements to vertical airphotos.

Figure 2-8 Camera lens with between-the-lens shutter and diaphragm. Note shutter speed and *F*/stop settings. (Courtesy Calumet Manufacturing Co.)

Figure 2-6 Basic operating principle of a framing camera, which exposes a total ground area instantaneously. (Adapted from a NASA drawing.)

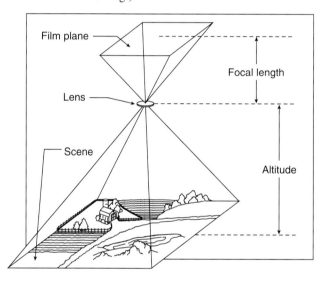

Figure 2-7 Generalized diagram of a single-lens framing camera.

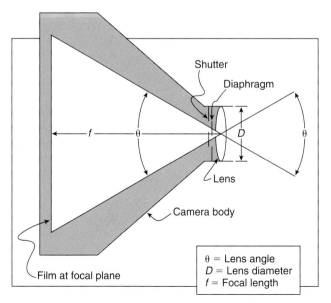

The Framing Camera

The most common type of camera for taking aerial photographs is the **framing**, or **frame**, **camera**. It instantaneously captures an entire ground area with each exposure, and the pictorial record or image of that ground area occupies a "frame" on a roll of film (Figure 2-6). The term **image format** designates the size of the image area occupying a frame. The basic elements of a single-lens framing camera are illustrated in Figure 2-7.

The framing camera consists of a lightproof chamber, called the **camera body**, with light-sensitive **film** at one end, and an opening at the opposite end that houses the **lens**, **shutter**, and **diaphragm**. The function of the lens is to gather the light (i.e., near ultraviolet (UV), visible, and/or near infrared (IR)) that is directed to it from a ground scene and bring it into focus at the **focal plane** (i.e., the **plane of focus**) where the film is positioned. When focused at infinity, the linear distance from the center of the lens to the focal plane is known as the **focal length**; it is usually measured in millimeters or inches. The shutter is an open-close device that controls the length of time the film is exposed to the incident light; the various times of exposure are known as **shutter speeds**. The diaphragm controls the amount of light transmitted to the film while the shutter is open. It admits light through various sizes of circular openings called **apertures**.

In many cameras, the shutter and diaphragm are positioned in the air space between the front and rear elements of the lens (Figures 2-7 and 2-8). In this configuration, the shutter consists of a number of overlapping metal blades or leaves (**leaf shutter**) that rapidly open and close during an exposure. A leaf shutter can accurately expose film for extremely small fractions of a second. The effective diameter of the lens opening, or **relative aperture**, is controlled by adjustment of an adjacent set of overlapping blades that compose the diaphragm.

The camera elements previously described have their counterparts in the human eye. The eyeball is the enclosed chamber. A visual image is formed and focused by a combination of the cornea and the lens. The eyelid is the shutter, and the retina's rods and cones are the light-sensitive material. The iris is the diaphragm that adjusts the size of the pupil to regulate the amount of light entering the eye.

Film Exposure

Film exposure is defined as the quantity of light that is allowed to reach the film. Exposure is regulated by the relative aperture and shutter speed, which are interdependent. The relative aperture, or lens opening, is called the **F/number**, or **F/stop**, and is defined as follows:

$$F/\text{number} = \frac{f}{d},\qquad(2\text{-}1)$$

where: f = focal length, and
d = effective lens diameter.

(Values must be expressed in equivalent units.)

With focal length held constant, Equation 2-1 shows that as the diameter of the lens opening increases, the F/number decreases and vice versa. The international series of F/numbers at full-stop increments are, from larger to smaller openings, $F/1$, $F/1.4$, $F/2$, $F/2.8$, $F/4$, $F/5.6$, $F/8$, $F/11$, $F/16$, $F/22$, $F/32$, and so on (Figure 2-9). These F/numbers have a progression of $\sqrt{2}$ (about $1.4\times$), with occasional rounding (e.g., 11.2 to 11).

Each F/number position changes the area of the aperture opening and the amount of light passing through the lens by a factor of 2. Thus, the aperture area and the light quantity are halved in "stopping down" from one F/number to the next larger number (e.g., from $F/11$ to $F/16$) and doubled in "stopping up" from one F/number to the next smaller number (e.g., from $F/11$ to $F/8$) (Figure 2-9).

Figure 2-9 Aperture openings of a diaphragm at several different F/stops. Note the inverse relationship between the size of the openings and the F/number. (Adapted from Backhouse et al. 1978.)

| F/4 | F/5.6 | F/8 | F/11 | F/16 | F/22 |

To maintain a given quantity of light reaching the film for a correct exposure, the shutter speed must decrease as the F/number gets larger (lens opening decreases) and increase as the F/number gets smaller (lens opening increases). For example, in going from $F/11$ to $F/16$, the amount of light is halved, making it necessary to double the exposure time (e.g., shutter speed from 1/1,000 sec to 1/500 sec). Conversely, the exposure time would be halved in going from $F/8$ to $F/5.6$ (e.g., shutter speed from 1/250 sec to 1/500 sec). To eliminate image blurring, fast shutter speeds (i.e., shorter exposure times) are necessary to minimize the effects of platform vibrations and for making exposures from low-flying aircraft.

Lens Speed

Lens speed, or the light-gathering power of a lens, is the F/number of the maximum effective diameter of the lens when the diaphragm is wide open, a position known as **full aperture**. Sometimes this number falls between two F/numbers in the standard series; examples are $F/3.5$, $F/4.5$, and $F/6.3$ (Figure 2-8). When this occurs, the next marked aperture setting of the lens is that of the nearest full F/number, and the standard series progresses from there (Figure 2-8).

The larger the lens opening at full aperture, the more light the lens will admit in a given time interval. Thus, the smaller the F rating, the faster the lens (Equation 2-1). For example, a lens having a focal length of 40 mm and a 10-mm diameter at full aperture is an $F/4$ lens. If the full-aperture diameter is doubled from 10 to 20 mm, the lens rating is $F/2$. Conversely, if the full-aperture diameter is halved from 10 to 5 mm, the lens rating is $F/8$.

The relative speed of any two lenses varies inversely as the square of their F/numbers. Therefore, an $F/2$ lens is 16 times faster than an $F/8$ lens $[(8/2)^2]$ and four times faster than an $F/4$ lens $[(4/2)^2]$. Fast-speed lenses are particularly effective for low-light photography.

Camera Viewing Angles

What a framing camera lens "views" during an exposure is determined by its **angular field of view (AFOV)**, or **lens angle**. The AFOV is defined as the angle θ subtended by lines drawn through the center of the lens to the extremities of the full image format at the focal plane. The AFOV is shown in two-dimensional space in Figure 2-7 and three-dimensional space in Figure 2-10.

The lens angle is a function of the camera's focal length (f) and the dimensions of the film's image format. Because of the latter parameter, a camera has three AFOVs at a given focal length—one for the length (θ_L), one for the width (θ_W), and one for the diagonal ($\theta_D = \sqrt{L^2 + W^2}$) (Figure 2-9). The three AFOVs may be calculated as follows:

$$\theta_L = 2 \arctan\left(\frac{L}{2f}\right), \tag{2-2}$$

$$\theta_W = 2 \arctan\left(\frac{W}{2f}\right), \text{ and} \tag{2-3}$$

$$\theta_D = 2 \arctan\left(\frac{\sqrt{L^2 + W^2}}{2f}\right). \tag{2-4}$$

(Values must be expressed in equivalent units.)

These equations show that a direct relationship exists between format size and the AFOV, whereas an inverse relationship exists between focal length and the AFOV. In addition, the largest AFOV will always be associated with the diagonal dimension (Equation 2-4). As opposed to a rectangular-image format, either Equation 2-2 or 2-3 can be used for a square-image format, since length equals width. Large AFOVs (short focal lengths) yield larger ground coverage than small AFOVs (long focal lengths).

Framing cameras are often classified according to their AFOVs in relation to the diagonal dimension of the image format (Equation 2-4). A commonly used classification scheme for square-format mapping cameras is presented in Table 2-1.

TABLE 2-1 Classification of Lens Angles[a]

Range of Angles	Name
<60°	Narrow angle
60–75°	Normal angle
75–100°	Wide angle
>100°	Super-wide angle

[a]*Angles are for the diagonal dimension of a square-image format measuring 23 × 23 cm.*

Ground Distance

For vertical camera positions at a given height above the surface (H), the angular field of view can be used to calculate the side dimensions of a ground scene that is recorded on the film during an exposure. For a rectangular-image format, the ground distance for length (D_L) and width (D_W) may be calculated as follows:

$$D_L = 2\left(H \tan \frac{\theta_L}{2}\right) \text{ and} \tag{2-5}$$

$$D_W = 2\left(H \tan \frac{\theta_W}{2}\right). \tag{2-6}$$

From these two equations, it is seen that there is a direct relationship between ground distance and AFOV at a given camera height above the surface (as θ increases, D increases, and as θ decreases, D decreases). In addition, there is a direct relationship between ground distance and camera height at a given AFOV (as H increases, D increases, and as H decreases, D decreases). These relationships are illustrated in Figure 2-11.

Use can be made of either Equation 2-5 or 2-6 for a square-image format. For a rectangular-image format, ground area can be determined by simply multiplying the

Figure 2-10 Lens angles in three-dimensional space.

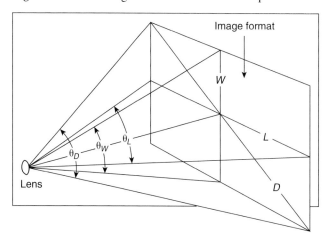

Figure 2-11 Effect of altitude and lens-angle variation on ground distance.

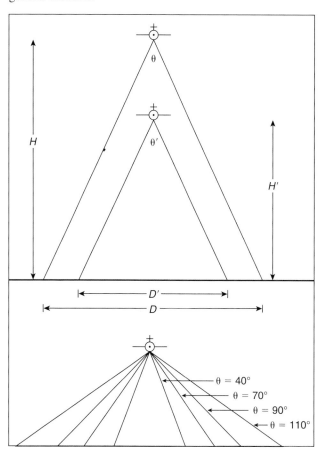

dimensions of length and width; for a square-image format, the length of a side can be squared.

Mapping Cameras

Mapping cameras, also known as **cartographic** or **metric cameras**, are specially designed and calibrated to produce vertical airphotos with high resolution and a minimal amount of geometrical distortion. The photographs from such a camera are used for making precise measurements and accurate maps (e.g., topographic maps) at different scales. Most mapping cameras in use today incorporate an image format measuring 23 × 23-cm (9 × 9-in.). A Zeiss mapping camera employing this format is shown in Figure 2-12. Much less common are mapping cameras with 23 × 46-cm (9 × 18-in.) and 18 × 18-cm (7 × 7-in.) formats.

Figure 2-13 illustrates the principal items of a mapping camera. The **magazine** is a lightproof box that holds the **film**, **supply and take-up spools**, and the **platen**. The film, in the form of a continuous roll, is ordinarily 24 cm (9.5 in.) wide and 120 to 150 m (400 to 500 ft) in length, depending upon its thickness. For the standard 23-cm square format, the longer roll of film will accommodate about 625 individual frames.

The platen is a metal plate, positioned at the focal plane, against which the film is held flat at the instant of exposure. Just prior to an exposure, a vacuum pump draws air through very small holes in the platen. The resulting suction holds the film tightly against the platen to avoid distortions that would be caused by natural wrinkles in the film or air bubbles trapped between the film and platen. Following an exposure, the vacuum is released, and the film is advanced a predetermined distance by the **drive mechanism** for the next exposure.

During the time the shutter is opened for an exposure, the aerial camera is moving relative to the ground. This can cause **image blur**, even with very fast shutter speeds. To negate this problem, many mapping cameras incorporate **image motion compensation**, or **IMC**. One method of achieving IMC during the exposure time is with a moving platen. With this method, the platen and the vacuum-held film move slightly (usually less than 50 μm) in the opposite direction of the flight during the exposure; platen speed is synchronous with the relative ground motion. IMC is especially important for high-speed aircraft.

The **lens cone** houses the **lens assembly** (lens, shutter, and diaphragm) at its narrow end and the **focal-plane frame** and **optical recording instruments** at the opposite end. The cone's outer housing prevents any light, except that transmitted through the lens, from striking the film. The length of the cone is governed by the focal length of the lens. Most mapping cameras have interchangeable cones to accommodate different focal lengths; common focal lengths are 88 mm (3.5 in.), 152 mm (6 in.), 210 mm (8.25 in.), and 305 mm (12 in.). The longer focal lengths are especially useful for high-altitude operations.

The most critical item in a mapping camera is its **compound**, or **multielement**, **lens**. This lens must be capable of

Figure 2-12 Zeiss RMK 8.5/23 mapping camera attached to a suspension mount for placement in the floor opening of an aircraft. Also shown are an intervalometer and a navigation telescope. (Courtesy V. Lippmann, Carl Zeiss, Inc.)

casting an undistorted image onto the film at the instant of exposure. Mapping cameras have **fixed-focus lenses**, which are set to infinity; this is so because of the large distances separating the camera station and the earth's surface. A **filter** is usually placed in front of the lens to control the "character" of light reaching the film.

The focal-plane frame includes rigidly fixed **fiducial marks**. These marks, which are of various designs, are accurately positioned in the middle of the sides, in the corners, or both. Because the frame and film are in contact at the instant of exposure, the silhouette of the frame and fiducial marks are exposed onto the film (Figure 2-14). Lines connecting opposite pairs of fiducial marks intersect to identify the **principal point** (**PP**), or the geometric center, of a vertical airphoto (Figure 2-14). The use of fiducial marks and the principal point for photo-measurement work is discussed in subsequent chapters.

Figure 2-13 Generalized diagram of a mapping camera.

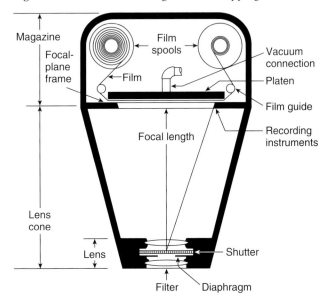

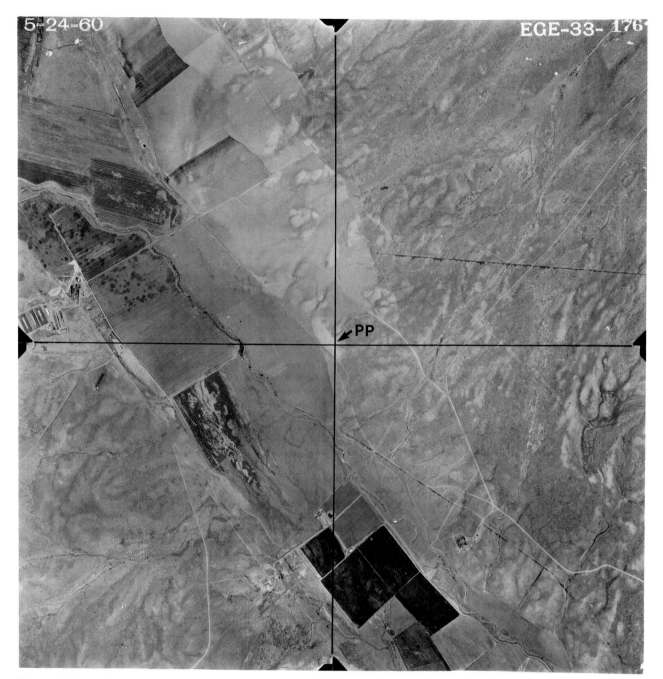

5-24-60   EGE-33-176

PP

Figure 2-14 Vertical airphoto with middle-of-the-side fiducial marks. Lines connecting opposite fiducial marks intersect to identify the principal point (*PP*) of the photograph.

Companion Equipment for Mapping Cameras

Certain mapping cameras are equipped with recording instruments and associative illumination and projection optics for the imaging of their faces onto the film at the time of each exposure. The instruments can include a clock, bubble level, altimeter, and frame counter. An illuminated note panel is usually included; it can be used to indicate such things as the camera's calibrated focal length, a project code, the area being photographed, and the date. The instrument faces and the note panel are projected onto a strip of the film immediately adjacent to the image area, whereas the frame counter is normally imaged within the image format (Figure 2-15). Unfortunately, the data strip is often cut off and discarded when the processed photographs are trimmed.

Figure 2-15 Film data strip showing illuminated instrument faces and the note panel plus the frame counter within the image format.

Figure 2-16 Hasselblad 70-mm film camera that is used for both vertical and oblique photography.

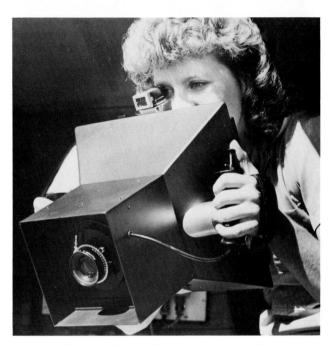

Figure 2-17 Cartwright oblique aerial camera. This camera uses 12 film holders, each containing two sheets of 20 × 35-cm (8 × 10-in.) film. (Courtesy Vern W. Cartwright, Cartwright Research Corp.)

The **camera mount** is a circular frame that supports a mapping camera in the floor of an aircraft for the vertical photography (Figure 2-12). The primary function of all mounts is to isolate the camera from fuselage vibrations. If the vibrations are not isolated at the time of exposure, image blurring can occur, even at fast shutter speeds. Vibration absorption is normally accomplished by rubber buffers or shock absorbers, which provide a cushioned support for the camera.

Many camera mounts incorporate designs that enable the camera to be pivoted in order to correct for the following aircraft motions: (1) **drift**—the horizontal displacement of the aircraft about an azimuthal axis; (2) **pitch**, or **tip**—the rotation of the aircraft about the longitudinal axis (i.e., nose up, nose down); and (3) **roll**, or **tilt**—the rotation of the aircraft about the lateral axis of the wings. Drift is normally caused by crosswinds, whereas a turbulent atmosphere is often the cause for pitch and roll. Corrections for these motions can be made by remote control or by control knobs on the mount itself.

The **intervalometer** is an electronic device that automatically trips the shutter and regulates the time interval between consecutive exposures made along a flight line (Figure 2-12). The exposure interval depends upon flight height, ground speed, the lens angle, and the desired overlap of adjacent photographs in the direction of flight. An overlap of 60 percent or more is usual for stereo photography, whereas

an overlap of 20 to 30 percent may suffice for the preparation of photo mosaics.

A **navigation telescope** is an important accessory for mapping cameras (Figure 2-12). It enables the camera operator to observe the ground area to be photographed at various heights above the surface and, when applicable, to determine the degree of aircraft drift.

Reconnaissance Cameras

Reconnaissance cameras are smaller and less expensive than mapping cameras, but like mapping cameras, they are capable of producing airphotos with a large amount of detailed information. Because their lens systems are of a simpler design and thus not free from distortion-producing defects, their photographs are normally used for interpretation tasks and not mapping tasks.

The most popular reconnaissance cameras are **small-format**, **single-lens cameras** that employ 35- or 70-mm roll film (Figure 2-16). Both 35- and 70-mm film cameras can

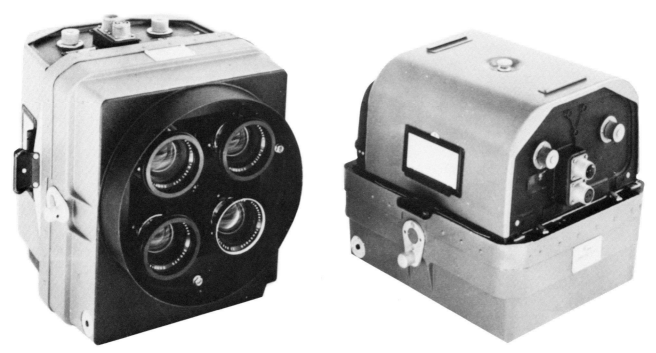

Figure 2-18 Mark I multispectral camera (front and back views) which simultaneously records four spectrally filtered, 9 × 9-cm (3.5 × 3.5-in.) images onto 24-cm (9.5-in.) film during each exposure. (Courtesy International Imaging Systems.)

be operated from hand-held positions for oblique photography or from simple suspension mounts for vertical photography. A special single-lens reconnaissance camera has been recently introduced for the exclusive acquisition of large-format oblique airphotos. Rather than roll film, it uses 20 × 25-cm (8 × 10-in.) sheet film (Figure 2-17).

A second type of reconnaissance camera is the **multiband**, or **multispectral**, **camera**, which is especially designed for the acquisition of **multiband**, or **multispectral**, **photographs**. A multiband camera simultaneously isolates light reflected from a scene into a number of separate wavelength bands from different parts of the photographic spectrum (Figure 1-5). Cameras for taking multiband photographs are available in two basic configurations: (1) the **multiple-lens camera** and (2) the **multiple-camera array**.

The multiple-lens camera uses a number of individual lenses mounted on the same camera body. Although cameras with up to nine lenses have been developed, the most common configuration uses four lenses (Figure 2-18). The lenses have the same focal lengths and are boresighted to a common orientation to assure that the same ground scene is photographed at a common scale. Each lens is fitted with a different filter to isolate specific bands of the photographic spectrum. When the film is exposed with the filters in place, each image provides a visual record of scene reflectance for a certain wavelength interval. Most multiple-lens cameras provide black-and-white photographs that can be interpreted individually, or they produce photographs that can be optically combined in groups of three to produce additive color images to better emphasize subtle differences between features of interest and their backgrounds. The additive color process is described in a later section of this chapter.

The multiple-camera array uses two or more identical

Figure 2-19 Quadricamera assembly employing four Hasselblad 70-mm film cameras. (Courtesy Joseph J. Ulliman, University of Idaho.)

single-lens cameras that are mounted on a common frame and boresighted to a common orientation. Most systems employ a multiple-shutter cock-and-release mechanism plus a synchronized motor drive to advance the film in each camera following an exposure. Multiple-camera arrays have been assembled using both 35- and 70-mm film cameras equipped with different films and filters. The most common configuration uses four matched cameras and is called a **quadricamera** (Figure 2-19).

TABLE 2-2　Kodak Terrestrial and Aerial Films

Aerial Films[a]	Terrestrial Films[b]
Panchromatic	
Plus-X Aerographic 2402	Plus-X Pan
Tri-X Aerographic 2403	Tri-X Pan
Double-X Aerographic 2405	Panatomic-X
Plus-X Aerecon 3411	
Panatomic-X Aerographic 2412	
Panatomic-X Aerecon II 3412	
High Definition Aerial 3414	
Black-and-White Infrared	
Infrared Aerographic 2424	High Speed Infrared
Normal Color	
Aerochrome MS 2448	Kodachrome
Aerial Color SO-242	Ektachrome
Ektachrome EF Aerographic SO-397	
Color Infrared	
Aerochrome Infrared 2443	Ektachrome Infrared 2236
High Definition Aerochrome Infrared SO-131	

[a]*Film widths: 70 mm (2.8 in.), 126 mm (5 in.), and 240 mm (9.5 in.).*
[b]*Film widths: 35 mm (1.4 in.) and 70 mm (2.8 in.).*

Films and Filters

In the broadest sense, the two major groups of photographic film are **black-and-white** and **color**. Black-and-white films can be either **panchromatic** (frequently shortened to **pan**) or **infrared**. Panchromatic literally means sensitive to light of all colors. The two types of color film are **normal**, or **natural**, **color** and **color infrared**. Black-and-white films utilize a single light-sensitive layer, whereas color films utilize three light-sensitive layers (i.e., **tri-emulsion films**). Photographic films have an aggregate sensitivity that includes near UV, visible, and near IR wavelengths. This wavelength span is known as the photographic spectrum. Its width is about twice that of the visible spectrum and human vision (Figure 1-5).

Panchromatic, black-and-white infrared, natural color, and color infrared films are available in two forms—**terrestrial films** and **aerial films**. The so-called terrestrial films are designed for indoor and outdoor photography with 35- and 70-mm film cameras, but they can also be used for taking aerial exposures. By contrast, aerial films are designed exclusively for aerial exposures. They are used in mapping cameras, multiple-lens cameras, and 70-mm film cameras. Table 2-2 lists several types of terrestrial and aerial films marketed by the Eastman Kodak Company.

Lens filters are normally used with black-and-white and color films. Both gelatin and glass filters are designed to absorb selectively certain wavelengths and transmit others. They therefore determine the spectral makeup of radiation reflected from a scene that ultimately reaches and exposes the film. This resultant is best described as "**filtered radiation.**"

Black-and-White Films

Black-and-white films (pan and infrared) are negative materials. Although the detailed structure of a black-and-white film varies with the specific product, each is composed of a relatively thick plastic **base** sandwiched between a thin coating of a light-sensitive material, known as the **emulsion**, and a **backing** layer (Figure 2-20). The total thickness of a

Figure 2-20 Generalized diagram of a black-and-white film.

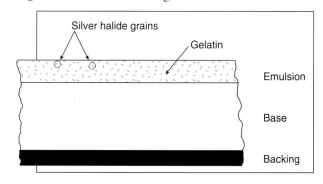

black-and-white film varies from about 0.05 to 0.11 mm; the dimension of the base is 80 to 90 percent of the total film thickness.

The base, which provides support for the emulsion and backing, must satisfy many requirements, among which are optical transparency, a resistance to both moisture and processing chemicals, and a specified degree of flexibility and dimensional stability. The backing is usually a gelatin coating containing a light-absorbing dye. The dye absorbs any light that passes through the emulsion and the base to prevent reflection back to the emulsion layer. The backing also reduces the tendency of the film to curl (Eastman Kodak Co. 1982).

The emulsion consists of **silver halides** deposited in a solidified and inert **gelatin** (Figure 2-20). The silver halides are not deposited as individual molecules, but as tiny crystals or grains; their mean diameters range from about 0.1 to 5 μm. The individual grains have irregular surfaces that favor the absorption of photons striking their surfaces. Pure silver halides are basically sensitive only to ultraviolet and blue radiation (i.e., high-photon-energy wavelengths). However, by adding special **sensitizing dyes** to their surfaces, the sensitivity of the silver halides can be extended to green, red, and near IR wavelengths.

Processing Black-and-White Films

Once processed in the laboratory, black-and-white films yield **negative images**. A negative contains a reversed image of the original scene, both in terms of tone and geometry. During the split second of exposure, photons of radiation reflected from a ground scene strike the emulsion. The number of photons striking a particular area of the film is largely a function of the brightness of the ground objects being imaged on that area of the film. For the affected area, the silver halide grains undergo a **photochemical reaction**, forming an invisible **latent image** in the emulsion.

The latent image is made visible by the process of **film development**. Today, films are processed in high-speed continuous processors. Briefly, the darkroom procedure for producing a negative transparency is as follows.

Developing: Immersion of the exposed film in an alkaline solution, called the developer, causes those silver halide grains that were exposed while the shutter was open to be reduced to metallic silver, which is black. The areas on the film that received the most intense light during the exposure are darkened by heavy silver deposits, whereas areas that received less light are characterized by less-dense silver deposits. Thus, the silver deposits of different density produce shades of gray.

Stop Bath: The film is immersed in diluted acetic acid to stop the developing reaction.

Fixing: Film areas that received no light during the exposure retain undeveloped silver halides, which are dissolved and removed in solution by the fixer; these areas become transparent. In addition, the fixer removes the dye from the backing, which also becomes transparent.

Washing and Drying: The film is washed in running water to remove all processing chemicals and dried to remove residual water deposits.

Once a negative has been processed, it is used to make **positive images** on various materials that have been coated with a light-sensitive emulsion. This is accomplished by passing white light through the negative onto the sensitized material to form a latent image. The amount of light transmitted through the negative is in proportion to the tonal density of its various areas. The exposed material is then subjected to the darkroom procedure previously described. In the processed positive, bright areas occur where no light was transmitted through the negative, whereas dark areas occur where light was transmitted through the negative. The positive image also shows the original scene geometry.

The tones seen in a processed panchromatic or infrared photograph are a qualitative record of how well a particular object reflected radiation in selected portions of the photographic spectrum. Tones can vary in intensity from white to black. Objects appearing in light tones were reflecting relatively large amounts of light in the given wavelength interval at the time of exposure, and objects appearing in dark tones were reflecting relatively small amounts of light in the given wavelength interval at the time of exposure.

Spectral Sensitivity of Black-and-White Films

Most panchromatic films are sensitive to wavelengths ranging from 0.25 to about 0.7 μm. This encompasses near UV, blue, green, and red wavelengths (Figure 2-21). Some pan films have an **extended red sensitivity** for penetrating atmospheric haze. This added sensitivity extends to a wavelength of about 0.72 μm in the near IR region (Figure 2-21). Black-and-white infrared films have a spectral sensitivity that extends from 0.25 to just beyond 0.9 μm, encompassing near UV, blue, green, red, and near IR wavelengths (Figure 2-21).

Figure 2-21 Spectral-sensitivity curves for panchromatic, extended-red panchromatic, and infrared films. (Adapted from Eastman Kodak Co. 1982.)

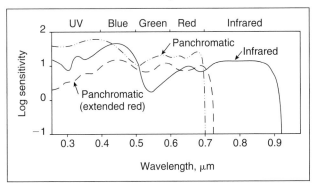

Because of the UV absorption properties of glass, the *effective* spectral sensitivity of both panchromatic and infrared films changes once the films are loaded in a camera. The shortwave cutoff shifts from 0.25 μm to 0.35 μm for conventional lenses and to 0.3 μm for special quartz lenses.

In addition to the spectral modification caused by camera lenses, the sensitivities of black-and-white films are further modified by lens filters. For mapping and general interpretation tasks, panchromatic photographs are normally acquired with a yellow filter (e.g., Kodak Wratten 12) to decrease the negative effect of atmospheric haze. This particular filter absorbs UV and blue light, the main culprits of haze, and transmits only the longer wavelengths to the film (Figure 2-22). The visual effect of atmospheric haze is shown in Figure 1-15. Airphotos obtained with a yellow lens filter are called **minus-blue photographs**.

Greater haze reduction results from the use of a red filter (e.g., Kodak Wratten 25) with panchromatic film. Such a filter absorbs UV, blue, and nearly all the green light (Figure 2-22). This filter is most useful for high-altitude operations.

Several different filters can be used with black-and-white infrared film. If it is desired that only near IR radiation be recorded, an infrared-transmitting filter (visually opaque) is used to absorb UV and essentially all visible wavelengths (e.g., Kodak Wratten 89B and 88A; Figure 2-22). Prints resulting from this type of filtration are referred to as **true infrared photographs**. **Modified infrared photographs** result when the infrared film is exposed through a red filter (e.g., Kodak Wratten 25). They are called *modified* because radiation from the longer-wavelength portion of the visible spectrum as well as the near IR is allowed to react with the film. Both types of infrared airphotos have excellent haze-penetration qualities.

Shadows normally appear black in pan and IR airphotos when the previously mentioned lens filters are used because the filters remove skylight. These terrestrial shadows are equivalent to lunar shadows in the sense that the moon has no atmosphere and hence no skylight (Figure 2-23). By comparison, it is possible to see details within shadowed areas in unfiltered panchromatic airphotos because skylight reaches the film (Figure 2-23).

Color Films

With black-and-white films, contrasting gray tones can provide important clues in object identifications. However, color films offer the additional qualities of **hue** (dominant wavelengths present in the color), **chroma** (strength of the color), and **value** (lightness of the color) as aids to identifying objects. A major advantage of the use of color films is that the human eye can discriminate tens of thousands of color variations but less than 100 tones of gray. There are numerous documented instances where the added dimension of color has improved interpretation reliability while reducing the amount of time required for photo analysis and inference.

Color films are of two types: (1) **reversal films**, which are processed to produce **color-positive transparencies**, and (2) **negative films**, which are processed to yield **color-negative transparencies**. Color negatives incorporate complementary or opposite colors of the original scene and the geometry in reverse; these colors preclude their use for direct interpretation. Rather, the correct colors and geometrical properties of the original scene are achieved only when the negative image is projected onto a photographic print material using white light (**negative-to-positive process**).

Color-reversal films are processed to produce positive images on the film that was originally exposed in the camera. This is the type of film used in 35-mm cameras for color slides. Color-positive transparencies can be viewed directly by transmitted white light or by projection onto a screen. In addition, color prints and duplicate transparencies can be produced directly from the original transparencies (**positive-to-positive process**). Virtually all the color films used in remote sensing are the reversal type, and the mechanics of their operation are described in the following sections.

Normal Color Film

A normal color film is designed to provide a color rendition that approximates the original scene as it would be viewed by the human eye (Plates 1–4). A normal color film is sensitive primarily to wavelengths of the visible spectrum (0.4 to 0.7 μm). These films consist of three separate emulsion layers applied to a backing and a transparent plastic base (Figure 2-24). The thickness of normal color films varies from about 0.09 to 0.12 mm; the dimension of the base is some 70 to 80 percent of the total film thickness.

Each emulsion layer contains silver halide grains that have been treated with sensitizing dyes and color-forming agents called **dye couplers**. The sensitizing dyes make each layer sensitive to one of the **additive primary colors**—blue, green, or red (Figure 1-7). The color-forming agents react

---

Figure 2-22 Small-particle scattering curve and transmission curves for filters commonly used with panchromatic and infrared films. Wavelengths to the left of each filter curve are absorbed.

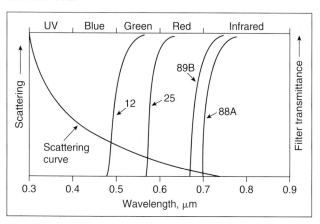

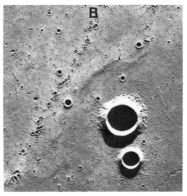

Figure 2-23 Shadow depiction in panchromatic photographs in the absence or presence of skylight: (*A*) minus-blue photo, no skylight; (*B*) lunar photo, no skylight; and (*C*) unfiltered photo, skylight present. (Lunar photograph courtesy of Philip A. Davis, Jr., U.S. Geological Survey.)

with a chemical developer during processing to form specifically colored dyes.

Recall from Chapter 1 that our perception of color depends upon the relative amounts of the three primary colors that are reflected by a particular object to our eyes. In essence, this means all colors can be formed by mixing suitable proportions of the three independent colors of blue, green, and red (**Young-Helmholtz theory**). Thus, a normal color film uses emulsion layers that are independently sensitive to these same colors.

The special arrangement of a normal color film's three emulsion layers is shown in Figure 2-24. The top layer is blue-sensitive. Immediately beneath it is a yellow gelatin layer that acts as a filter. It prevents UV and blue light from reaching the lower two emulsion layers, which are also sensitive to UV and blue light (natural sensitivity of silver halides); the yellow filter is dissolved and removed in solution during processing. The middle layer is green-sensitive, and the bottom layer is red-sensitive.

Figure 2-24 Generalized diagram of a normal color film. A haze filter is placed over the camera lens to stop ultraviolet radiation from reaching the blue-sensitive layer.

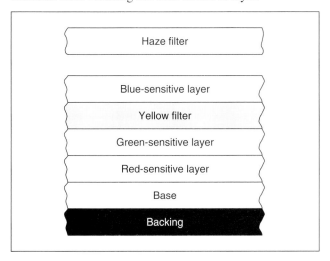

To block UV radiation from reaching the blue-sensitive layer, normal color film is usually exposed with a clear-glass haze filter (e.g., Kodak Wratten HF-3) over the camera lens. This type of filter absorbs wavelengths shorter than about 0.4 µm. When unfiltered, UV radiation can overexpose the blue-sensitive layer, which will impart an excess bluishness to the photographs (Plate 1).

The manner in which a reversal color film operates is shown in Figure 2-25. Part A is a representation of a scene being photographed; it consists of areas that are blue, green, red, white, and black. Upon exposure (Part B), blue light activates the blue-sensitive layer and is prevented from exposing the other two layers by the yellow filter; green light passes through the blue layer and yellow filter and activates the green-sensitive layer; and red light passes through the top two layers plus the yellow filter and activates the red-sensitive layer. The white area provides reflected radiation in blue, green, and red wavelengths and thus activates all three layers of the film. There is no reflectance from the black area and none of the emulsion layers is activated.

Briefly, the darkroom procedure for producing a color-positive transparency is as follows:

First Development: Immersion of the exposed film in a negative black-and-white developer converts the camera-exposed silver halides to metallic silver.

Re-exposure: The film is re-exposed by either white light or a chemical fogging agent, a process known as **flashing**, which exposes all the remaining silver halides.

Second Development: The newly exposed silver halide grains are reduced to metallic silver, and simultaneously the dye couplers form **yellow**, **magenta**, and **cyan** images in the blue-, green-, and red-sensitive layers, respectively. The concentration of each dye is directly proportional to the amount of metallic silver produced during reexposure.

Bleaching and Fixing: The bleaching agent and fixer remove all the silver deposits from the first and second developments plus the yellow gelatin filter and the dye from the backing, leaving only yellow, magenta, and cyan dyes in the three layers, as shown in Part C of Figure 2-25. Also note in Part C that the film is clear where the original exposure occurred.

Washing and Drying: The film is washed in running water to remove all residual chemical deposits and dried to remove residual water deposits.

For a color-reversal film to work properly, yellow, magenta, and cyan dyes must be used. At this juncture it is important to explain their unique properties. As colors, they are formed by an equal combination of two of the three additive primary colors: (1) Red plus green light produces yellow, the complement of blue; (2) red plus blue light produces magenta, the complement of green; and (3) blue plus green light produces cyan, the complement of red. **Complementary colors** are color pairs that produce white light when added together (i.e., yellow plus blue, magenta plus green, and cyan plus red).

As filters, yellow, magenta, and cyan transmit two-thirds of the visible spectrum (blue, green, and red components) and absorb one-third: (1) A yellow filter absorbs blue (minus-blue filter) and transmits green and red; (2) a magenta filter absorbs green (minus-green filter) and transmits blue and red; and (3) a cyan filter absorbs red (minus-red filter) and transmits blue and green. Because of their unique absorption properties, yellow, magenta, and cyan are called the **subtractive primary colors**.

When a positive transparency is viewed with transmitted white light, the dye images from the three layers *subtractively* produce colors that closely match those of the original scene. This is illustrated in Figure 2-25. Blue is visible because the magenta dye absorbs green light and cyan absorbs the red, leaving only blue light to be transmitted through the film. By the same principle, green is produced by the subtraction of blue light by the yellow dye and red light by the cyan dye. Red is produced by the subtraction of blue light by the yellow dye and green light by the magenta

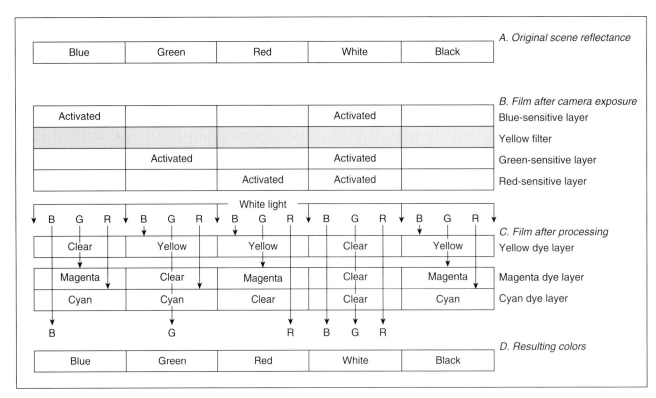

Figure 2-25 Color formation with a color-reversal film. (Adapted from Eastman Kodak Co. 1982.)

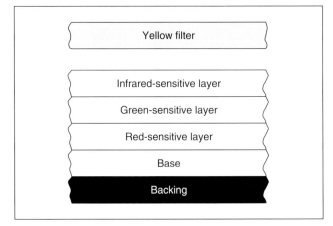

Figure 2-26 Generalized diagram of a color infrared film. A yellow filter must be used to stop ultraviolet and blue radiation from reaching the three emulsion layers.

dye. White is perceived when none of the three dyes is present. By contrast, all three dyes are present for black, and in conjunction they absorb all visible wavelengths. A wide gamut of other colors is produced in accordance with the proportions of blue, green, and red light reflected by the subjects being photographed (Plates 1–4).

Color Infrared Film

Color infrared film is a reversal film originally developed by Eastman Kodak during World War II for use by the military. Its first use was to distinguish highly IR-reflective green vegetation from simulated vegetation, such as green-painted objects and dyed camouflage netting that had low levels of IR reflectance. On the processed film, the painted and dyed objects appeared blue in sharp contrast to red for the true vegetation. Because of this military application, the film was originally called **camouflage-detection film**.

Unlike normal color film, which is sensitive to blue, green, and red radiation, color infrared film has three emulsion layers sensitive to green, red, and near IR radiation (0.7 to 0.9 μm). This film must be used with a yellow filter (e.g., Kodak Wratten 12) to absorb UV and blue light, to which all three emulsion layers are sensitive. Certain films, such as Kodak Aerochrome Infrared Film 2443, require that the yellow filter be placed over the lens. However, with the newest film, Kodak High Definition Aerochrome Infrared Film SO-131, a yellow gelatin filter is coated over the top emulsion layer; it is later removed during processing. With the yellow filter in place, the three emulsion layers are effectively sensitive to green, red, and near IR radiation (Figure 2-26). This wavelength sensitivity extends from 0.5 to 0.9 μm. The filtration ensures that the film possesses excellent haze-penetration capabilities, even from high altitudes.

Color infrared film produces the same dye colors in its three emulsion layers that are developed in a normal color film (i.e., yellow, magenta, and cyan), but each is assigned to different spectral-sensitivity ranges. This means that when the film is correctly exposed and processed, the resulting colors will be *unnatural* or *false* for most objects (Plates 3–6). Because of this, color infrared film is sometimes called **false-color film**.

The manner in which the false colors occur is shown in Figure 2-27. Note that dominant green reflectance reproduces as blue, dominant red reflectance reproduces as green, and dominant near IR reflectance reproduces as red. A blue object will be shown as black because incident blue light is absorbed by the yellow filter. Numerous other colors will be formed, depending upon the proportions of green, red, and near IR reflected by the subjects being photographed (Plates 3–6).

Photographic Products

Black-and-white airphotos are usually printed on photographic paper for viewing with reflected light. Prints may have a shiny or dull finish; the former are called **glossy prints** and the latter are termed **matte prints**. The matte

Figure 2-27 False-color formation with a color infrared-reversal film. (Adapted from Eastman Kodak Co. 1982.)

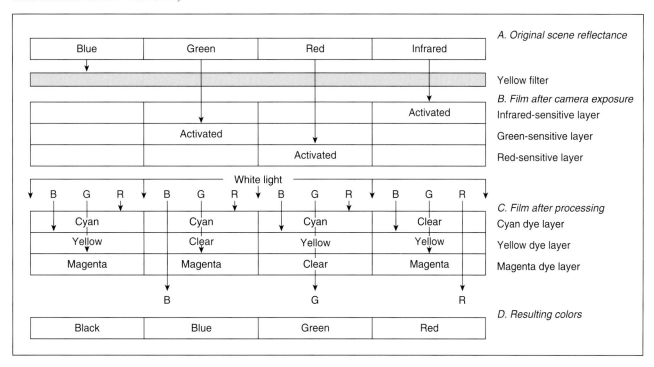

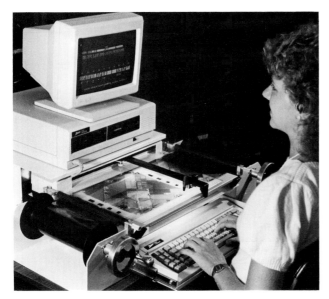

Figure 2-28 Filmaster Aerial Film Titler (Model GS-2000). (Courtesy Vern W. Cartwright, Cartwright Research Corp.)

finish reduces glare from overhead lighting and is more suitable for pencil or ink annotations than the glossy finish. Prints can also be made with an intermediate finish; these are called **semimatte prints**.

Black-and-white prints can also be made on a semiopaque polyester base. Once processed, the prints can be viewed by either reflected or transmitted light. Prints on a polyester base have a matte finish, are waterproof, and do not tear.

Color airphotos can be printed on either paper or nontransparent pigmented acetate or polyester. Paper prints have a matte finish, whereas the plastic prints have a glossy finish. The colors in glossy prints are stable under artificial light or sunlight, and hence the colors do not fade appreciably through time as they do with paper prints.

Both black-and-white and color photographs can be printed the same size as the original negatives or positive transparencies (**contact printing**) or as enlargements or reductions (**projection printing**). Sheets of cut stock are available in a variety of sizes, ranging from 20 × 25 cm (8 × 10 in.) to about 76 × 102 cm (30 × 40 in.).

Black-and-white and color positives may also be prepared on transparent film or glass. These media are called **positive transparencies**, or **diapositives**. Positive transparencies must be viewed with transmitted light, which precludes their use in the field.

Duplicating films are used to produce reproductions from the master transparencies. These copy films cannot be used for making original exposures in an aerial camera. For a color-reversal film, duplicate transparencies are often made for frames of interest when the use does not require the maximum detail offered by the original transparencies. This is done because the master transparencies cannot be replaced if they are damaged or destroyed. It is emphasized that there is *always* a loss in detail with any duplicating or printing operation.

Printed Information on Airphotos

Following the laboratory processing of roll film (70-mm to 24-cm widths), certain types of information are often printed along the margin of each image frame by a machine called a **film titler** (Figure 2-28). The film titler uses fast-drying black ink, which is usually applied to the nonemulsion side of the processed film. This makes it easy to remove the ink with alcohol without damaging the film. The black titling on the negatives appears as white letters and numbers on black-and-white prints (Figure 2-29). On color positive transparencies and associative prints, the titling appears as black letters and numbers.

Typical types of information printed on a film can include the date and time of exposure, approximate acquisition scale, project symbol, roll number, and frame number (Figure 2-29). The latter three descriptors often serve as an identification code for ordering specific frames from the governmental agency or private company holding the original negative or positive transparencies.

Film and Lens Resolution

Resolution, or **resolving power**, is the capability of a film and/or lens to image spatial detail. The most important physical property of a film in determining resolution is the size distribution of its silver halide grains. A film's resolution capability is inversely proportional to grain size (i.e., the larger the grains, the poorer the resolution and vice versa). In general, black-and-white films have better resolution than color films. The resolving power of a lens is largely determined by its optical properties and size. In this context, mapping cameras have a higher resolving power than reconnaissance cameras.

The resolving power of a film or camera lens is determined in a controlled laboratory environment. This is accomplished by photographing a **resolution test target** or **chart** with (1) a standardized camera system and different types of film (**film resolution**) or (2) different lenses and a standardized film (**lens resolution**). A typical **tri-bar resolution target** is shown in Figure 2-30. It consists of a series of progressively smaller three-bar patterns. Each pattern consists of three white bars, five times as long as they are wide, separated by two black spaces of equal width.

Resolution is expressed in **lines per millimeter** or **line pairs per millimeter**. Resolving-power values are determined by the following two-step procedure. Following film development, the smallest pattern in which all the bars are completely distinguishable from the spaces is determined by viewing the target image with a source of white light and under magnification with a stereomicroscope (usually 24× magnification). Then, the reciprocal of the combined width of a bar and adjacent space (referred to as a **line**, or **line**

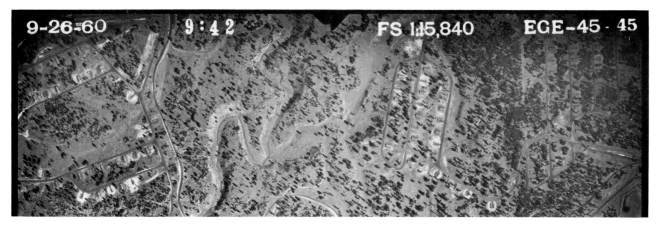

Figure 2-29 Film titler information printed on a vertical airphoto. From left to right, the titling defines the date of exposure, time of exposure, average project scale, project symbol, roll number, and frame number.

pair) from the smallest resolvable grouping yields lines, or line pairs, per millimeter.

If, for example, this combined width was found to be 0.02 mm, the resolving power of the film would be 50 line pairs/mm:

$$\frac{1.0 \text{ line pairs}}{0.02 \text{ mm}} = 50 \text{ line pairs/mm.}$$

Thus, the film would be capable of recording a pattern consisting of 50 bar-and-space pairs, with each pair having a width of 1/50 mm; an observer would be able to ideally distinguish 100 linear elements (50 bars and 50 spaces) per millimeter on the processed film under suitable illumination and magnification (McKinney 1980). *High resolution* (greater line-pair count) refers to the depiction of very fine detail, whereas *low resolution* (lesser line-pair count) implies that the detail is less well defined.

Resolution is highly dependent upon target contrast. Consequently, film manufacturers normally derive resolution data for both high- and low-contrast targets. A **high-contrast target** is one in which there is a large density difference between bright and dark areas (Figure 2-30); a **low-contrast target** has a small density difference between bright and dark areas. For example, a widely used aerial panchromatic film (Kodak Panatomic-X, Type 2412) has a resolution of 400 line pairs/mm for a high-contrast target and 125 line pairs/mm for a low-contrast target. Because most surface materials have relatively low contrasts, the latter derived resolution is a better guide to actual film performance under aerial conditions.

It is emphasized that the resolution of any particular lens-film combination (**system resolution**) can be no better than its lowest-rated component. For example, if a lens is capable of resolving 100 line pairs/mm at a given contrast and the film is capable of resolving 40 line pairs/mm, the effective resolution is established by the film.

The effect of focal length and flying height can be used to determine **ground resolution** from the system resolution just described. The following relationship is used:

$$R_g = \frac{(R_s)(f)}{H}, \tag{2-7}$$

where: R_g = ground resolution in line pairs per meter,

R_s = system resolution in line pairs per millimeter,

f = camera focal length in millimeters, and

H = height of camera above the ground in meters.

Thus, if a 300-mm focal-length camera is operated 6,000 m above the ground and the system resolution is 40 line pairs/mm, ground resolution is 2.0 line pairs/m:

$$R_g = \frac{(40 \text{ line pairs/mm})(300 \text{ mm})}{6,000 \text{ m}}$$

$$= \frac{12,000 \text{ line pairs}}{6,000 \text{ m}}$$

$$= 2.0 \text{ line pairs/m.}$$

This means that the processed film could ideally resolve a resolution pattern on the ground that consists of 2.0 line pairs/m.

The width of an individual line pair in meters is determined by the reciprocal. In our example, this would be 0.5 m:

$$\frac{1.0 \text{ line pairs}}{R_g} = \frac{1.0 \text{ line pairs}}{2.0 \text{ line pairs/m}} = 0.5 \text{ m.}$$

The width of a bar or space in the line pair would be 0.25 m (0.5 m/2).

The width of a bar or space can be used to infer **minimum ground separation**, which is the minimum distance between two objects on the ground that enables them to be resolved as individual entities on the processed film. Thus, in our example, the film could ideally resolve objects that were 0.25 m apart on the ground. This value can also be used to indicate the size of the smallest object that could be detected on the film (i.e., the smallest detectable object would measure about 0.25 m across).

It is emphasized that the calculations for ground resolution represent approximations. For a given lens-film combination, actual ground resolution will be influenced by numerous factors that include: (1) the type of filter used, (2) camera vibration, (3) extent of image motion during exposure, (4) whether the film is the original or a duplicate, (5) the illumination intensity, (6) atmospheric conditions, and (7) object-background contrast.

Film Speed

Film speed is a measure of a film's sensitivity to reflected light. Knowledge of film speed permits the computation of camera settings (F/stop and shutter speed) that will yield properly exposed images. Film speed is directly related to the size distribution of the film's silver halide grains. A "fast" film incorporates large-diameter grains (**coarse-grained film**) and requires only minimal lighting conditions to yield proper exposures. The major disadvantage of fast films is their coarse graininess, which causes a reduction in resolution.

A "slow" film utilizes small-diameter grains (**fine-grained film**) and requires bright sunlight to yield properly exposed images. Generally, slow films require longer exposure times; this can cause image blurring when exposures are made under minimal lighting conditions. However, their fine graininess enables the finer spatial details of a scene to be recorded and makes them especially useful for making photographic enlargements.

Several numerical scales have been introduced to offer a means of comparing the relative speeds of different films. For each scale, the larger the rating number, the more sen-

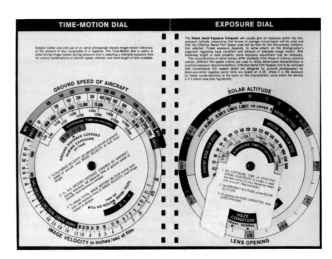

Figure 2-31 Kodak Aerial Exposure Computer, showing its two dial computers.

sitive—or the faster—the film. The rating systems are arithmetic, meaning that a film with a speed of 200 is twice as fast and requires one-half the exposure of a film with a speed of 100. It is important to remember that there is no interrelationship between the different scales because they are determined in different ways.

For terrestrial films, there are two rating scales. In the United States, the American Standards Association rates each film on a relative scale of light intensity. This **ASA exposure index**, as it is called, provides a uniform classification system that can be applied easily under changing light intensity. In European countries, film speeds are rated on the Deutsche Industrie Norm, or **DIN**, scale. Most 35- and 70-mm film cameras provide settings for ASA and DIN scales.

Because of such factors as the wide variation in camera-to-object distance, the small range of contrasts between objects and their surroundings, plus atmospheric haze, a different speed-rating scale is used for aerial films. It is called the **Effective Aerial Film Speed (EAFS)** index.

To assist the photographer in determining the proper exposure for an aerial film, use is made of the **Kodak Aerial Exposure Computer** rather than a light meter. This computer is a six-page folding card with two dial computers, latitude-zone map, and solar altitude tables (Figure 2-31). It is designed for determining the proper combination of F/stop and shutter speed based upon a particular film's EAFS rating, sun angle, haze condition, aircraft altitude, and the amount of tolerable image motion. The latter parameter is based upon aircraft altitude, speed, and the camera's focal length.

Filter Factor

Because lens filters absorb part of the incident radiation, their use generally requires longer exposure times. The number of times by which an exposure must be increased

Figure 2-30 Typical tri-bar resolution chart of high contrast. The actual chart is a reversal of this illustration (i.e., white bars on a black background).

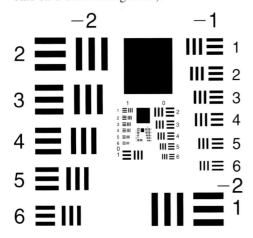

Figure 2-32 Panchromatic (*top*) and infrared (*bottom*) photographs of the Gol-dach region, Switzerland. Note the clear separation of tree types and the differentiation between the small stream and its terrain background in the infrared photograph. (Courtesy Wild Heerbrugg Ltd.)

Figure 2-33 Minus-blue photograph of Richmond, California, and vicinity. Aircraft altitude was 9.1 km, and the photo acquisition scale was 1:60,000.

for a specific filter with a given film is called the **filter factor**. For panchromatic and black-and-white infrared films, the exposure time derived from the Kodak Aerial Exposure Computer does not account for filters. Consequently, to compensate for the loss of light, the exposure time must be *multiplied* by the filter factor or the film speed must be *divided* by the filter factor. Most haze filters for normal color film have a filter factor of 1 (i.e., no exposure increase is required) because they are colorless. For aerial color infrared films, reported film speeds are based on exposures through a Kodak Wratten 12 filter, and no compensation for the filter is required. The Eastman Kodak Company (1982) lists filter factors for specific filters with different aerial films.

Film Selection

Aerial photographs have become a valuable tool for studying the cultural and physical features of the earth's surface. However, the proper choice of film (whether panchromatic, normal color, black-and-white infrared, or color infrared) is mandatory to meet the objectives of a specific investigation. To meet the specified objectives, several factors must be considered. These factors include (1) the resolution required, (2) the reflective characteristics of the features to be studied, (3) the spectral properties of the films and filters being considered, (4) the photographic format (e.g., prints or positive transparencies) needed for the map-

ping or interpretation task, (5) the expected atmospheric conditions, and (6) the time of day or season of the year when the film is to be used. These factors combine to identify the film that will most likely satisfy the end use.

Panchromatic Photography

The most common type of film used for mapping (e.g., topographic maps) and interpretation tasks is panchromatic. Because of the film's long-term popularity, historical panchromatic photographs are available for most of North America and Europe. Pan films are available in a variety of speeds, are conveniently processed, and are the least expensive of the films used in remote sensing.

Panchromatic film provides a tonal rendition that closely approximates the brightness of the scene being photographed. It is capable of distinguishing between objects of truly different colors, but its somewhat lower sensitivity to green light can make separation of vegetation types, such as tree species, difficult (Figure 2-32).

A yellow filter is normally used for exposures on panchromatic film to reduce the fogging effect that atmospheric haze causes. Such filtration is especially important when exposures are to be made from high altitudes (Figure 2-33). Most of the panchromatic airphotos used in this book were obtained with a yellow lens filter.

Figure 2-34 Unfiltered panchromatic photograph of submerged beach rock along the Atlantic coast of southern Florida. Water depth at the end of the pier is about 5 m.

Unfiltered panchromatic film is used for penetration through clear water. For coastal mapping missions, where underwater detail is needed, unfiltered pan photos are valuable for detecting and delineating features such as reefs, shoals, and channel obstructions (Figure 2-34). When unfiltered, panchromatic film is best suited for low-altitude operations under conditions of low humidity.

Standard-speed panchromatic films (e.g., Kodak Plus-X Aerographic 2402) provide good tonal contrast, a wide exposure latitude to compensate for over- and underexposure, excellent resolution, and low graininess. High-speed pan films (e.g., Kodak Tri-X Aerographic 2403) are used when exposures have to be made under minimal levels of illumination. These films are typically about twice the speed of standard panchromatic aerial films. They are exposed through similar haze filters and, except for increased graininess, produce comparable tonal renditions.

Normal Color Photography

When properly exposed and processed, the major advantage of normal color photographs is ease of interpretation because objects appear in their natural colors (Plates 1–4 and 6). Normal color film is especially useful for identifying soil types, rock types and surficial deposits, water-surface patterns, and various forms of polluted water (Plate 2). For clear water, color film has good penetration qualities and is therefore valuable for recording underwater features (Plate 2). Its penetration through clear water can exceed 25 m. Color photography is also useful for detecting forest damage caused by various insects (Plate 3). This type of film is best suited for low- to medium-altitude operations to minimize atmospheric haze effects (Plate 2).

Normal color film has a limited exposure latitude in comparison to panchromatic film, and it is preferably exposed under conditions of bright sunlight. Without the correct exposure and proper filtration, photographs are likely to be of poor quality (Plate 1). Today's color films have faster speeds, better definition, and less granularity than films of previous years.

Because the largest single cost in obtaining new aerial photography is that of aircraft operation, color photography is not much more expensive than panchromatic photography. Although film and processing costs are greater, these factors are not normally significant in terms of the total cost of a photographic survey.

Infrared Photography and Object Spectra

Both black-and-white and color infrared films operate beyond the confines of the visible spectrum, where the eye cannot see directly. It is therefore useful to compare the spectral-reflectance characteristics of various materials in the visible band (0.4 to 0.7 μm) and the photographic IR band (0.7 to 0.9 μm). One such comparison is presented in Figure 2-35.

Several significant points can be deduced by examining the spectral-reflectance curves or spectra for the six materials shown in Figure 2-35. First, it is seen that certain materials have a relatively stable reflectance in both bands. For example, snow has a relatively stable and high reflectance in both bands, whereas asphalt has a relatively stable and low reflectance in both bands. Second, it is observed that certain materials, such as vegetation and water, have considerably different reflectance characteristics in the visible and infrared regions. Third, it is seen that the most appropriate region for differentiating between most of the materials would be the photographic IR region, because it is here where the greatest differences in reflectance generally occur (Figure 2-35).

Maximum reflectance from vegetation occurs in the near IR region of the photographic spectrum (Figure 2-35). The spectral character of light reflected by a normal plant leaf is controlled largely by two groups of cells (Figure

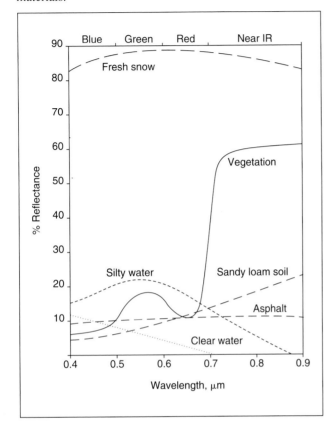

Figure 2-35 Average spectral-response curves for six materials.

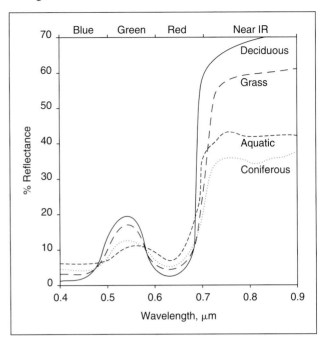

Figure 2-37 Average spectral-response curves for four types of vegetation.

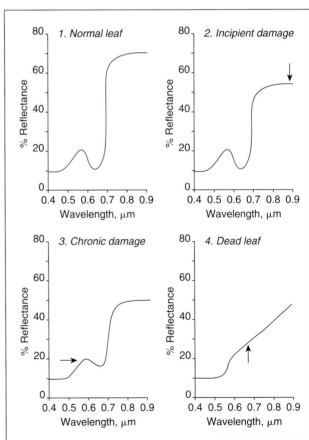

Figure 2-38 Average spectral-response curves for a plant leaf as it progresses from a healthy state through different stages of damage. (Adapted from Murtha 1978.)

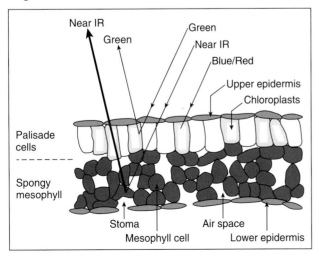

Figure 2-36 Generalized diagram of a leaf's structure and its reflectance characteristics at visible and near IR wavelengths.

2-36). The long, narrow cells below the upper epidermis are the **palisade cells**. They contain many **chloroplasts** with chlorophyll pigments that absorb most blue and red light for photosynthesis and efficiently reflect green light. Up to 20 percent of incident green light is reflected by the chloroplasts in the normal functioning of a plant, causing the green color of leaves (Figure 2-35).

Below the palisade cells is the **spongy mesophyll** layer, where oxygen and carbon dioxide exchange takes place. This layer consists of **mesophyll cells** and numerous **air spaces**. Near IR radiation is unaffected by the chloroplasts, but it is strongly reflected by the spongy mesophyll. Most of the reflection occurs at **cell-wall/air-space interfaces**. It is not uncommon for more than 50 percent of incident near IR radiation to be reflected upward through the leaf, with most of the remainder being transmitted through the bottom of the leaf (Figure 2-35).

Infrared reflectance from various types of healthy vegetation differs because of leaf mesophyll arrangements. For example, a corn leaf has a relatively compact mesophyll, which is responsible for a relatively low infrared reflectance and high transmittance. A leaf with more porous mesophyll (i.e., many wall-air interfaces), such as a maple leaf, will have a higher infrared reflectance and lower infrared transmittance.

Spectra for four vegetation types are shown in Figure 2-37. Although all four spectra have the same general shape, characteristic of green vegetation, there are large reflectance differences in the near IR region. By comparison, the differences in reflectance in the green region are negligible. Consequently, distinct tonal or color separations would be associated with infrared films.

Leaf age can also be a cause of variation in infrared reflectance. For example, an immature citrus leaf is compact, with few air spaces in its mesophyll (lower IR reflectance), whereas the mature leaf is spongy, with many air cavities in its mesophyll (higher IR reflectance) (Gausman 1974). In addition, air spaces do not start to form in a cotton leaf until it reaches one-fourth to one-third its full size. Other conditions that can influence the level of IR reflectance from vegetation include the water content of the leaves, density of growth, time of year, and topographic location.

When a healthy plant becomes *strained* (effect) by some type of *stress* (cause) in its environment, changes occur in its spectra. Types of vegetation stress include insect infestations, plant diseases, freezing temperatures during the growing season, concentration of soil salts, moisture deprivation, prolonged inundation of root zones by floodwater, nutrient deficiency, and pollution. Figure 2-38 shows the changes in the typical reflectance properties of a plant leaf as it progresses from a healthy state through different stages of damage (Murtha 1978).

Stage 1: The leaf is **healthy**; peak visual reflectance is in the green region (about 20 percent), with lower reflectance in the blue and red regions (10 percent each). There is considerably higher reflectance in the near IR region (about 70 percent).

Stage 2: The leaf has **incipient damage** (i.e., damage just beginning to exist) which is accompanied by a drop *only* in near IR reflectance (about 10 percent).

This diminution is caused by the collapse of some of the air cavities in the mesophyll. Incipient damage is commonly called **previsual** damage because the leaf still appears green. It is at this stage that infrared-sensitive films may be able to detect plant decline; this can be days or weeks before the decline is manifested in the visible spectrum.

Stage 3: The leaf has **chronic damage**, and its color changes from green to yellow. This is caused by chloroplast deterioration, which increases red reflection. At near IR wavelengths, the yellow leaf has about 5 percent less reflectance than it had in Stage 2.

Stage 4: The leaf is **dead**, and its color becomes red-brown. The major spectra changes are the loss of the green peak, an increase of red reflectance, and the loss of the high shoulder of near IR reflectance beyond 0.7 μm. At this stage, the leaf's cell structure has collapsed and the leaf becomes dry and brittle.

Black-and-White Infrared Photography

When used with an IR-transmit filter, black-and-white infrared films record tonal gradations that result from the ability of objects to reflect near IR radiation. On infrared photographs, objects of high IR reflectance appear in light tones, whereas objects of low reflectance appear in dark tones. In many instances, the tonal gradations differ significantly from those seen in panchromatic photographs for the same subjects. Table 2-3 lists several tonal signatures associated with panchromatic and infrared photographs.

One of the main values of infrared photographs is their ability to differentiate different types of vegetation. For example, healthy deciduous (broadleaf) vegetation is recorded in light tones because near IR radiation is strongly reflected from the spongy mesophyll of the leaves; coniferous (needle-leaf) vegetation tends to reflect less near IR radiation and consequently registers in darker tones (Figure 2-32). These characteristic tones make the film especially useful for forest inventories.

Because water is a strong absorber (i.e., a poor reflector) of near IR radiation, clear water bodies are rendered very dark in sharp contrast to most land surfaces (Figure 2-35). This tonal contrast is helpful in locating and mapping drainage systems, shallow inundated areas (e.g., marshes and swamps), lake and coastal shorelines, and canals (Figures 2-32, 2-39, and 2-40).

Black-and-white infrared film is also useful for differentiating dry and moist soils (Figure 2-41). Moist soils, because of the presence of pore water, absorb near IR radiation and appear in dark tones, whereas dry soil is more reflective and appears in lighter tones.

Infrared photographs have excellent haze penetration because the filters recommended for use with this film remove atmospheric scattering effects that occur in the UV,

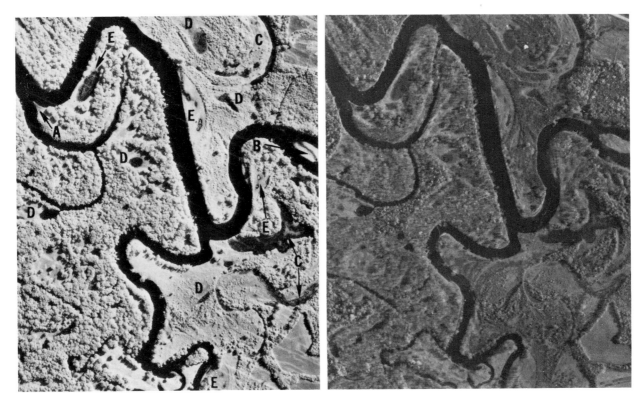

Figure 2-39 Infrared (*left*) and panchromatic (*right*) photographs of the meander floodplain of the Oromucto River, New Brunswick. Annotations are: (*A*) channel bar accretion—darker-toned areas represent the most recent deposits (moist) that have not yet been vegetated, (*B*) minor channel through a channel-bar complex, (*C*) meander cutoff, (*D*) back swamp, and (*E*) point-bar swamp. (Courtesy J.T. Parry, McGill University.)

TABLE 2-3   Object Signatures on Panchromatic and Infrared Photographs

Object	Panchromatic[a]	Infrared[b]
Snow	White	White
Clouds	White	White
Sky (high oblique)	Medium gray	Black
Clear water	Dark gray	Black
Silty water	Light gray	Medium gray
Deciduous foliage	Dark gray	White
Coniferous foliage	Dark gray	Medium gray
Autumn foliage (yellow)	Light gray	Light gray
White sand (dry)	Light gray	Light gray
White sand (moist)	Medium gray	Dark gray
Red sandstone (dry)	Medium gray	Light gray
Red sandstone (moist)	Medium gray	Dark gray
Swamp	Dark gray	Black
Asphalt	Dark gray	Black
Concrete (dry)	Light gray	Medium gray

[a]*Acquired with a Kodak Wratten 12 filter.*
[b]*Acquired with Kodak Wratten 88A or 89B filters.*

Figure 2-40 Infrared photograph of the central Adirondack region, New York, showing lakes created by beaver dams (relatively straight shores). Nonforested areas are beaver-influenced/created wetlands. (Courtesy Donald J. Bogucki, State University of New York, Plattsburgh.)

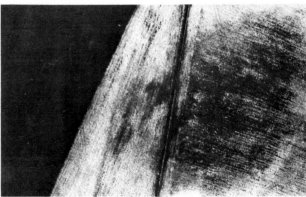

Figure 2-41 Panchromatic (*top*) and infrared (*bottom*) photographs of a plowed field and beet field in France. Dark tones in the infrared photo indicate high soil-moisture levels. (Courtesy Michel Hersé, Centre National de la Recherche Scientifique.)

blue, and green spectral regions. The film provides excellent results for high-altitude photographs, including high obliques where distant details are rendered with remarkable clarity.

Color Infrared Photography

Color infrared photography had its beginnings with the development of a film for detecting camouflaged military targets. However, since the end of World War II, many earth-science applications have been found for this type of photography. Today, its use parallels that of panchromatic photography. Table 2-4 lists several color signatures associated with normal color and color infrared photographs.

It is noted in this table that most of the color assignments are false for the infrared medium (Plates 3–6). However, in certain instances the observed colors are the same as those seen in normal color photographs. This can be explained by recalling the spectral sensitivities of the two films (normal color film = blue, green, and red; color infrared film = green, red, and near IR) and the dye assignments for each film type (Figures 2-25 and 2-27). Clouds and snow, for example, are seen as white because they reflect approximately equal amounts of green, red, and near IR radiation (Figure 2-35). Fresh asphalt is seen as black because it efficiently absorbs green, red, and near IR radiation (Figure 2-35). Sky color is blue because the infrared film, in association with the required yellow filter (minus-blue), records the green wavelength scattering in the atmosphere. As seen in Figure 2-27, dominant green reflectance is seen as blue.

Color infrared film has become extremely valuable for a variety of vegetation studies. For example, infrared spectral reflectances are often unique for given species (Figure 2-36), and this results in color signatures that are separable on infrared photographs (Plates 3–6). Healthy vegetation, while appearing green in conventional color photographs, will appear in red hues in color infrared photographs because of the elevated near IR reflectance from the mesophyllic tissue of the leaves. The large disparity between green and near IR reflectance from plant leaves (Figure 2-36) explains why healthy green foliage appears red instead of blue in color infrared photographs (Figure 2-27).

Because deciduous trees have a higher infrared reflectance than coniferous trees (Figure 2-36), there are distinct differences between the colors of these tree groups as seen on infrared photographs. Healthy deciduous trees normally appear bright red, whereas conifers appear brownish red. There is a close similarity in the green color of deciduous and coniferous trees (Figure 2-37). Deciduous trees whose leaves have turned red or yellow in autumn still retain much of the infrared reflectance. Consequently, on color infrared photographs, red leaves photograph yellow (equal reflectance in red and near IR), and yellow leaves register as white (equal reflectance in green, red, and near IR) (Plate 3).

Fruit and nut trees are recorded in a variety of red gradations, from pink (peach, almond) through red (pear), to reddish brown (walnut). Healthy agricultural crops, at

Object	Normal Color	Color Infrared
Snow	White	White
Clouds	White	White
Sky (high oblique)	Blue	Blue
Clear water	Blue or green	Black or dark blue
Silty water	Red or brown	Light blue or green
Deciduous foliage	Green	Bright red
Coniferous foliage	Green	Brownish red
Aquatic vegetation	Green	Pink
Citrus trees		
Healthy	Green	Red
Previsual stress	Green	Pink
Late stage of stress	Yellow	White
Defoliated trees	Gray	Blue or green
Artificial turf		
Dry	Green	Blue
Wet	Green	Black
Red sandstone	Red	Yellow or green
Asphalt	Black	Black
Damp ground	Slightly darker	Distinctly darker
Shadows	Bluish with details	Black

their peak-growth stage, also yield typical colors (e.g., oats, dark red; safflower, dark pink; alfalfa, bright red) (Eastman Kodak Co. 1978).

Color infrared film has been found useful for detecting losses of plant vigor that result from environmental stress (Plates 3 and 6). Stressed vegetation may first appear in darker or lighter values of red, eventually grading into a number of other false colors (e.g., pink, magenta, yellow, white) (Plates 3 and 6). It is emphasized that the false colors associated with damaged vegetation will vary according to the type of stress, the length of time the vegetation has been stressed, and the particular type of vegetation being stressed.

In the early stages of plant stress, the strain may show up on the color infrared photographs before symptoms are visible from ground observations or in normal color photographs (e.g., loss of green coloration), an effect known as **previsual stress** (stage 2 in Figure 2-38). The early detection of previsual stress may provide an ample time window for a remedial action to be implemented before a cure becomes impossible.

The evidence suggests that color and color infrared films are not good previsual detectors of coniferous tree damage caused by insects. For example, Heller (1968, 1978) found that neither film could detect insect-induced stress in different types of conifers before visible coloration changes occurred in the foliage. Normal color and color infrared photographs of healthy, dying, and dead ponderosa pine trees are shown in Plate 3.

Vegetation that appears blue or green in color infrared photographs indicates low infrared reflectance because the plant is defoliated (Plate 3). These colors result largely from the reflectance of visible light from limbs, branches, and soil or nonliving ground cover exposed through the bare crowns (Knipling 1969).

Color infrared film is also useful for recording wetlands vegetation (e.g., cattail, black rush, marsh grass, wild rice), floating aquatic species (e.g., water chestnut, water hyacinth, water lily), and algae blooms (Plates 5 and 6).

Figure 2-42 Band-pass filters for ultraviolet photography. Both filters are glass; gelatin filters are not suitable for UV photography.

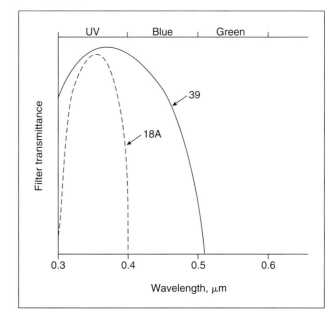

Figure 2-43 Panchromatic (*left*) and ultraviolet (*right*) photographs of the following animal pelts: (1) adult harp seal, (2) polar bear, (3) pup harp seal, (4) gray seal pup, and (5) Arctic hare. (Courtesy David M. Lavigne, University of Guelph.)

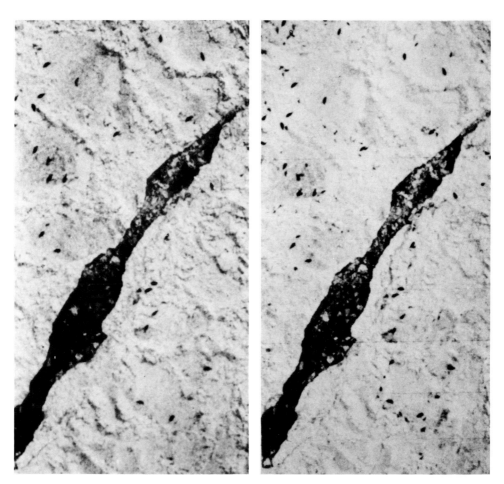

Figure 2-44 Panchromatic (*left*) and ultraviolet (*right*) airphotos of harp seals on the snow and ice surface of the Gulf of St. Lawrence. The dark adults are visible in both photographs, but the pups are visible only in the UV photograph. (Courtesy David M. Lavigne, University of Guelph.)

Bogucki et al. (1980) found that water chestnut had a distinctive pink color in infrared photographs that was rarely confused with other types of aquatic vegetation in the Lake Champlain region of New York (Plate 5). Being able to record algae blooms is important because they are an indirect indicator of phosphorous-rich water. The population explosion of algae leads to water deoxygenation and fish kills. The areal dimensions of algae blooms are often difficult to recognize on normal color photographs (Plate 6).

Because of its high absorption of near IR radiation (Figure 2-35) and the blocking of blue light by a yellow filter, clear water registers dark blue or black in color infrared photographs (Plates 4 and 5). Silty water (brown or brownish red) normally has a light blue or green signature (Plates 4 and 5). Damp ground shows up darker than dry ground because of near IR absorption by soil moisture.

The yellow filtration ensures that the film possesses excellent haze-penetration capabilities, and sharp images are usually attainable, even from high altitudes (see front cover). Shadows are black on color infrared photographs because the yellow filter absorbs UV and blue scattered light, the components of skylight.

Special Types of Photography

Ultraviolet photographs and **additive color photographs** represent two special types of photography that exploit specific spectral sensitivities of panchromatic and black-and-white infrared films, respectively. The wavelength isolations are achieved with special **band-pass filters** that transmit only a limited range of wavelengths to the film.

Ultraviolet photography is accomplished by using a panchromatic film, which is inherently sensitive to near UV radiation (Figure 2-21), and blocking all or most visible light with a UV-transmit lens filter. If the visually opaque Kodak Wratten 18A filter is placed over the lens, only UV wavelengths between 0.3 and 0.4 μm are transmitted to the film (Figure 2-42). The violet Kodak Wratten 39 filter is useful when short exposure times are warranted; this filter has a high transmission in the near UV and blue regions from 0.3 to 0.5 μm (Figure 2-42). Ultraviolet airphotos must be obtained from low altitudes to minimize the negative effects of strong atmospheric scattering of UV and blue radiation.

Special applications of ultraviolet photography center on the detection of white animals on an ice and snow background (Lavigne and Øritsland 1974) and detecting and monitoring oil films on water (Vizy 1974). Figure 2-43 illustrates the reflective characteristics of pelts of several Arctic animals against a snow background in the visible and near UV bands. It is observed in the UV photograph that the pelts absorbed UV radiation (medium to dark tones), whereas the snow was a strong reflector (light tones). The tonal record in the panchromatic photographs shows that the pelts and snow were both strong reflectors of visible light (light tones).

Figure 2-44 illustrates the successful application of UV photography for taking a census of harp seals (Lavigne

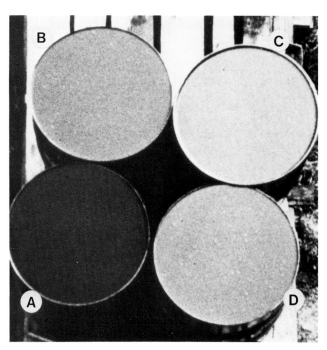

Figure 2-45 Ultraviolet photograph of (*A*) river water, (*B*) diesel fuel, (*C*) gasoline, and (*D*) spent lubricating oil. Oils are floating on water in barrels. (Courtesy Kalman N. Vizy, Eastman Kodak Co.)

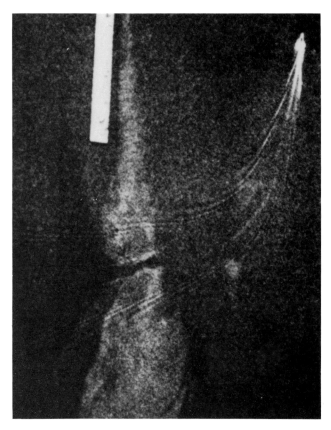

Figure 2-46 Ultraviolet airphoto of a diesel fuel spill in the Genesee River, New York. The opening in the slick was caused by the traversal of a boat. (Courtesy Kalman N. Vizy, Eastman Kodak Co.)

1976). Shown are panchromatic and UV airphotos taken simultaneously of adult and infant (pup) harp seals on the snow and ice surface of the Gulf of St. Lawrence. The dark adults register black against a white background in both photographs. However, the pups are not visible in the panchromatic airphoto because their white fur reflects visible light as efficiently as the snow and ice. In the UV band, the snow and ice are highly reflective, but the pups' white pelts are strong absorbers of UV radiation and photograph black against a white background. Consequently, both the adults and their pups can be detected only in the UV photograph.

The ability of ultraviolet photographs to distinguish between different petroleum products and water is illustrated in Figure 2-45. Note in this photograph that the oils were strong reflectors of near UV radiation, whereas the water was an efficient absorber. The use of UV airphotos to detect oil films is illustrated in Figure 2-46.

With multiband photography, where several images of a common ground scene are recorded separately through different band-pass filters, color reconstruction can be accomplished with the **additive color process**. This technique centers on the projection of three positive transparency images through blue, green, and red filtered light to produce a single, superimposed image in a natural or false color representation.

Multiband photographs used in the additive color process are usually acquired with a four-lens camera that uses black-and-white infrared film (Figures 2-18 and 2-21). Appropriate lens filters are used to individually isolate the blue, green, red, and near IR bands (Figure 2-47). Multiband photographs representing these four spectral regions are shown in Figure 2-48. The infrared negative images are made into positive transparencies for additive color viewing.

The color synthesis is accomplished with a multiple projector system, called an **additive color viewer** (Figure 2-49). This device enables three positive transparency images to be projected individually through blue, green, and

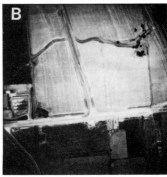

Figure 2-48 Multiband airphotos obtained with a four-lens camera shown in Figure 2-18. The photographs were imaged on black-and-white infrared film through the following Kodak Wratten filters: (*A*) 47B—blue, (*B*) 57A—green, (*C*) 25—red, and (*D*) 88A—near infrared. The original size of each image measured 9 × 9 cm (3.5 × 3.5 in.). (Courtesy of U.S. Geological Survey.)

Figure 2-47 Typical Kodak Wratten filters used for multiband photographs. An infrared blocking filter, such as the Kodak Wratten Infrared Filter 301, must be used in conjunction with the 47B, 57A, and 25 filters because they transmit near infrared radiation. The Wratten 301 is a multilayer interference filter coated on glass that transmits visible light and reflects near infrared radiation. (Adapted from Ross 1973.)

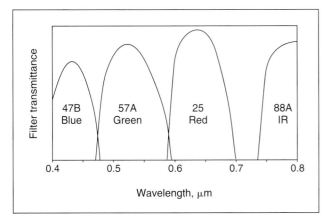

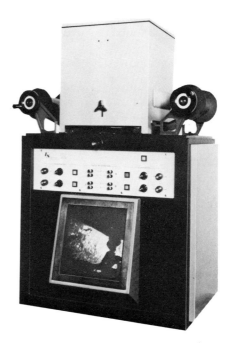

Figure 2-49 Additive color viewer that superimposes individual spectral images onto a rear-projection screen. (Courtesy International Imaging Systems.)

red filtered light and to be coregistered as a single image on a rear-projection screen. For a natural color display, the positive transparency images representing the blue, green, and red spectral regions are projected through blue, green, and red filtered light, respectively. If the green, red, and near IR images are projected, respectively, through blue, green, and red filtered light, a color infrared display is obtained. Other false or "exotic" color displays from the same four-image set can be produced by using different positive transparency or colored-light combinations to enhance objects of special interest.

Figure 2-50 Helicopter photograph of a female polar bear and three cubs. (Courtesy David M. Lavigne, University of Guelph.)

Questions

1. Use Equation 2-1 to calculate the speed of two lenses, with the first having a focal length of 70 mm and a full aperture of 20 mm and the second having a focal length of 305 mm and a diameter of 48 mm at full aperture.
2. In reference to Question 1, which lens is the fastest and by how much?
3. Use Equations 2-3 and 2-4 to calculate the lens angle (side and diagonal dimensions) for a camera system having a focal length of 305 mm and a 230-mm square format. According to Table 2-1, how would this camera system be classified?
4. Use either Equation 2-5 or 2-6 to calculate the *difference* in ground distance for a camera having a lens angle of 62° and operated from 1,800- and 3,100-m altitudes.
5. Upon viewing a tri-bar resolution target on a processed film, an observer is able to distinguish two bars with a 0.015-mm spacing and three bars with a 0.017-mm spacing. Given the preceding, what would be the film's resolving power in line pairs per millimeter?
6. Use Equation 2-7 to calculate ground resolution, given the following: (a) camera height = 6,200 m, (b) focal length = 88 mm, (c) lens resolution = 300 lines/mm, and (d) film resolution = 50 lines/mm.
7. Rank the following film/filter combinations regarding their ability to penetrate atmospheric haze. A rank of 1 would indicate the best penetration.
   a. Panchromatic, Wratten 12
   b. Panchromatic, Wratten 18A
   c. Panchromatic, Wratten 25
   d. Infrared, Wratten 88A
   e. Panchromatic, no filter
8. Upon examining Figure 2-50, state the type of photograph and explain your answer.

Bibliography and Suggested Readings

Backhouse, D., C. Marsh, J. Tait, and G. Wakefield. 1978. *Illustrated Dictionary of Photography*. Watford, England: Fountain Press.

Bogucki, D. J., G. K. Gruendling, and M. Madden. 1980. Remote Sensing to Monitor Water Chestnut Growth in Lake Champlain. *Journal of Soil and Water Conservation* 35:79–81.

Colwell, J. 1974. Vegetation Canopy Reflectance. *Remote Sensing of Environment* 3:175–83.

Eastman Kodak Co. 1978. *Applied Infrared Photography*. Publication M-28. Rochester, New York: Eastman Kodak Co.

———. 1982. *Kodak Data for Aerial Photography*, 5th ed. Publication M-29. Rochester, New York: Eastman Kodak Co.

Gausman, H. W. 1974. Leaf Reflectance of Near-Infrared. *Photogrammetric Engineering* 40:183–91.

Heller, R. C. 1968. Large-Scale Color Photography Samples Forest Insect Damage. In *Manual of Color Aerial Photography*, edited by J. T. Smith, Jr., 394–95. Falls Church, Va.: American Society of Photogrammetry.

———. 1978. Case Applications of Remote Sensing for Vegetation Damage Assessment. *Photogrammetric Engineering and Remote Sensing* 44:1159–66.

Knipling, E. B. 1969. Leaf Reflectance and Image Formation on Color Infrared Film. In *Remote Sensing in Ecology*, edited by P. L. Johnson, 17–29. Athens: University of Georgia Press.

Lavigne, D. M. and N. A. Øritsland. 1974. Ultraviolet Photography: A New Application for Remote Sensing of Mammals. *Canadian Journal of Zoology* 52:939–41.

Lavigne, D. M. 1976. Counting Harp Seals with Ultra-Violet Photography. *Polar Record* 18:269–77.

McKinney, R. G. 1980. Photographic Materials and Processing. In *Manual of Photogrammetry*, edited by C. C. Slama, 4th ed., 305–66. Falls Church, Va.: American Society for Photogrammetry and Remote Sensing.

Murtha, P. A. 1978. Remote Sensing and Vegetation Damage: A Theory for Detection and Assessment. *Photogrammetric Engineering and Remote Sensing* 44:1147–58.

———. 1982. Detection and Analysis of Vegetation Stress. In *Remote Sensing for Resource Management*, edited by C. J. Johannsen and J. L. Sanders, 141–58. Ankeny, Ia.: Soil Conservation Society of America.

Ross, D. S. 1973. Simple Multispectral Photography and Additive Color Viewing. *Photogrammetric Engineering* 39:583–91.

Slater, P. N. 1980. *Remote Sensing: Optics and Optical Systems*. Reading, Mass.: Addison-Wesley.

Strandberg, C. H. 1967. *Aerial Discovery Manual*. New York: John Wiley & Sons.

Ulliman, J. J., R. P. Latham, and M. P. Meyer. 1973. 70-mm Quadricamera System. *Photogrammetric Engineering* 39:583–91.

Vizy, K. N. 1974. Detecting and Monitoring Oil Slicks with Aerial Photos. *Photogrammetric Engineering* 40:697–708.

Wenderoth, S. and E. Yost. 1972. *Multispectral Photography for Earth Resources*. Greenvale, New York: Long Island University.

Chapter 3

Principles of Airphoto Interpretation

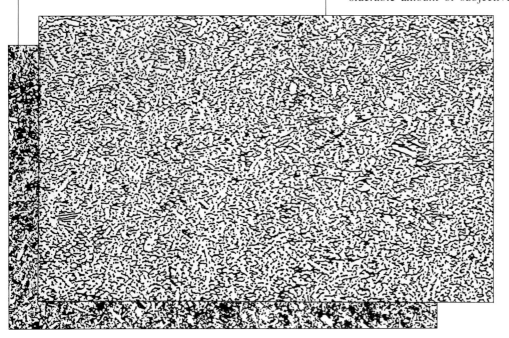

United States before the advent of World War II. During this war, however, countless military decisions were based on intelligence reports derived from aerial reconnaissance. After the war, many air-intelligence specialists converted their newfound knowledge of photographic interpretation into diverse civilian applications.

During the past four decades, the nonmilitary uses of aerial photography have continued to multiply. Today, photo-interpretation techniques are used on such diverse projects as monitoring the changing water levels of lakes and reservoirs, assessing crop diseases, locating new highway routes, identifying and assessing land-use changes, and mapping archaeological sites. In this same time period, significant technical developments have been made in aerial cameras, film emulsions, interpretation equipment, aircraft, and earth-orbiting satellites.

The Interpreter's Task

Because airphoto interpretation often involves a considerable amount of subjective judgment, it is commonly referred to as an art rather than a science. Actually, it is *both*. The interpreter must know how to use scientific tools and methodology to arrive at objective findings; these, in turn, must often be supplemented with deductive reasoning to supply a logical answer to the perennial question, *What is going on here?* Under certain circumstances the mental processes of deduction and association may permit detection of features not actually visible in the photographs—for example, a shallow groundwater reservoir or an archaeological site.

The skilled interpreter must have a large store of information at hand to perform this exacting task adequately. He or she should have a sound general background in the earth and life sciences, particularly geography, geology, forestry, and biology. The value of experience and imagination can never be overemphasized. There are situations in which photographic limitations or a lack of associated information (e.g., information from maps or reference publications) preclude the positive identification of objects. In such cases, the terms *probable* and *possible* are customarily used to qualify the interpreter's findings. The interpreter's job can often be made easier when **comparative**, or **sequential, coverage** is available. This refers to two or more sets of photographs of the same area taken at different times.

Photo Interpretation Defined

Photo interpretation is defined as the process of identifying objects or conditions in aerial photographs and determining their meaning or significance. This process should not be confused with **photo reading**, which is concerned with only identifications. As such, photo interpretation is both *reading the lines* and *reading between the lines*.

Photo interpretation was a little-known skill in the

Figure 3-1 Multirunway airfield used by carrier-based pilots for practicing short takeoffs and landings. Location is Broward County, Florida. Scale is about 1:15,000.

Figure 3-2 Low-altitude photograph of the Pentagon Building, headquarters of the U.S. Department of Defense.

Recognition Elements

Most people have little difficulty in recognizing features or conditions depicted in oblique photographs because such views appear relatively normal to the human eye (Figures 2-4 and 2-5). On the other hand, a vertical or near-vertical view can be quite confusing. This is illustrated in Figure 3-1; try to identify the feature before reading the caption.

An experienced photo interpreter exercises mental acuity as well as visual perception and consciously or unconsciously evaluates a number of characteristics to identify what is being seen in vertical photographs. These characteristics are referred to as **recognition elements** and include shape, size, pattern, shadow, tone or color, texture, association, and site. Although some features or conditions may be identifiable by a single element, their usefulness is enhanced when they are used in combinations. These elements are briefly described here and are covered in detail in Chapters 8 through 14.

Shape: The element of shape describes the external form or configuration of an object. Cultural objects tend to have geometrical shapes and distinct boundaries, whereas natural features tend toward irregular shapes with irregular boundaries. From a vertical perspective, the shape of some objects is so distinctive that they can be conclusively identified by this element alone; the Pentagon Building near Washington, D.C., is a classic example for most U.S. citizens (Figure

3-2). Other cultural features having easily recognized shapes include airport runways, shopping malls, cloverleaf interchanges, center-pivot irrigation systems (Figure 3-3), and certain archaeological sites (Figure 3-4).

Many natural features also have distinctive shapes. Examples include sand dunes, volcanic cinder cones (Figure 1-22), alluvial fans (Figure 3-5), and the various components of a meander floodplain (Figure 3-6).

Size: In two-dimensional space, size is a measure of the surface dimensions of an object. Relative size comparisons can be an important aid in identifying features. For example, there is a relative size difference between a house and an apartment building and between multiple-lane and single-lane streets. For a given photo scale, house size can be used to infer relative value (Figure 3-7). Comparative size is also an important aid for identifying the master and tributary streams of a drainage system (Figure 2-39). Relative size comparisons can be made from unscaled photographs, but absolute size measurements (e.g., length, width, circumference) can be made only when the photographic scale is known.

Pattern: Pattern refers to the overall spatial form of related features. The repetition of certain forms is characteristic of many cultural objects and some natural features. Examples include the orderly arrangement of trees in an orchard or grove (usually a grid pattern) versus the random distribution of trees in a forest (Plates 5 and 6). In geology, recognizing drainage patterns is important because they provide

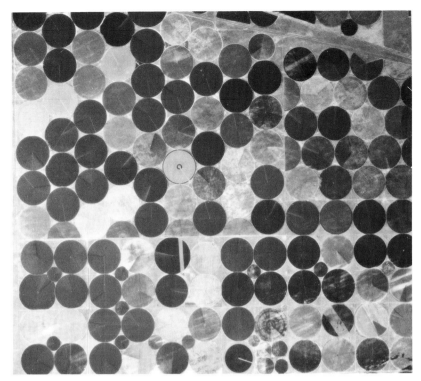

Figure 3-3 Center-pivot irrigation system in Morrow County, Oregon. Most of the fields are planted with wheat. These circular irrigated fields have become common in many arid and semiarid regions of the world.

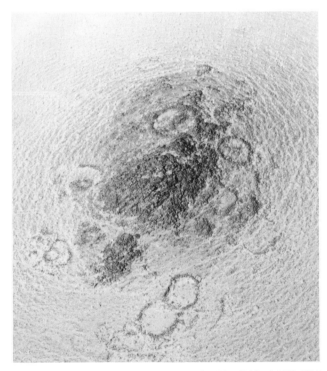

Figure 3-4 Circular wall remains of a Neolithic MAR-TU settlement in north-central Saudi Arabia. The MAR-TU inhabited this region for perhaps four millennia (6,000–2,000 B.C.).

Figure 3-5 Alluvial fans along the east side of Death Valley, California. Alluvial fans can be easily recognized by their fan shape and adjacency to mountain fronts.

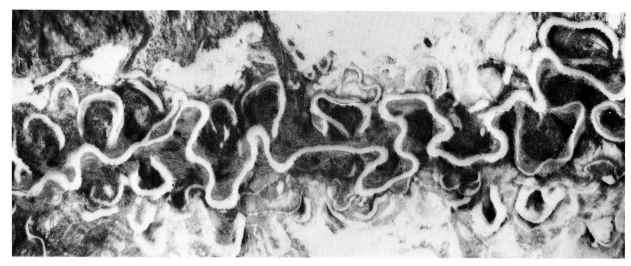

Figure 3-6 Meander floodplain in Alaska. Note the numerous ice-covered oxbow lakes and meander scrolls or loops along the river's serpentine course.

Figure 3-7 Low-altitude photograph of single-family houses in Boca Raton, Florida. Note that the largest houses are adjacent to waterways.

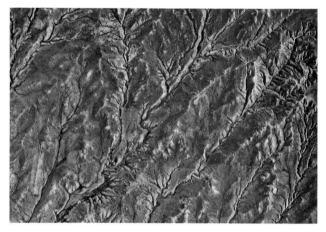

Figure 3-8 Dendritic drainage pattern in flat-lying sedimentary strata. This drainage pattern is characterized by a tree-like branching system in which the tributaries join the gently curving major stream at acute angles.

clues to underlying structure and lithology. For example, the dendritic stream pattern is commonly associated with flat-lying sedimentary rocks (Figure 3-8). The presence of a banding pattern that follows hillside contours is indicative of interbedded sedimentary rocks (Figure 3-9).

Shadow: Shadows cast by oblique illumination are important in photo interpretation because their shapes provide profile views of certain features that can facilitate their identification (Figure 3-10). Features often recognizable by the shadows they cast include water towers, electrical-transmission towers, oil-storage tanks, bridges, and various species of trees (Figure 3-11). Enhancement of subdued topography is one of the most important uses of shadows. To accomplish this, photographs are normally taken with low sun-angles (early morning or late afternoon) to accentuate minor surface irregularities (Figure 3-12).

Shadows are particularly helpful if the objects are small or lack tonal or color contrasts with their surroundings. It is important to remember that extensive shadowing can hinder interpretation by hiding ground detail (e.g., shadows alongside tall buildings or dense tree stands).

Tone or Color: Tone or color relates to the reflective characteristics of objects within the photographic spectrum. The wavelength region of sensitivity is a function of film type and filtration. The ability of an object to reflect radiation at a given wavelength interval is dependent upon its surface composition and physical state plus the intensity and angle of illumination. Tonal contrasts in black-and-white photographs and hue, chroma, and value qualities in color photographs provide important clues for object identification. Numerous examples for a variety of natural and cultural features are presented in Chapter 2.

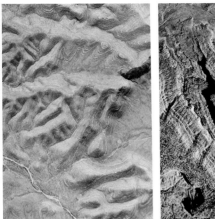

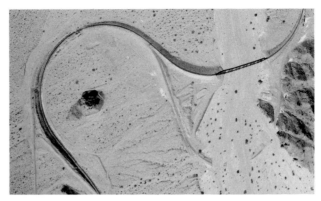

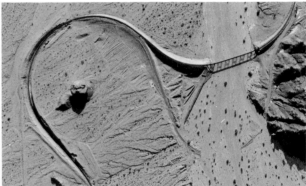

Figure 3-9 Banding associated with interbedded sedimentary rocks. Light tones are often sandstone and limestone layers, whereas the darker bands are often shale layers.

Figure 3-11 High-sun-angle (*top*) and low-sun-angle (*bottom*) photographs of a railroad bridge spanning an intermittent streambed in Riverside County, California. Note that the bridge's superstructure is revealed by its shadow in the low-sun-angle photograph.

Figure 3-10 Low-sun-angle photograph showing topographic and tree shadows cast on a frozen river in Alaska.

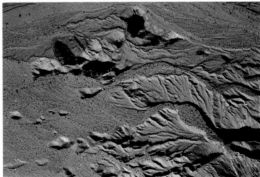

Figure 3-12 High-sun-angle (*top*) and low-sun-angle (*bottom*) photographs of dissected alluvial and bedrock terrain in an arid environment.

Texture: The visual impression of coarseness (roughness) or smoothness caused by the variability or uniformity of image tone or color is known as texture. It is produced by an aggregate of characteristics too small to be detected individually, such as tree leaves and leaf shadows. Smooth textures are associated with cropland (plants at about the same height), bare fields, and calm bodies of water; coarse textures are associated with forestland (mature tree crowns) and young lava flows (Figures 3-13 and 3-14).

It is often possible to distinguish between features with similar reflectance characteristics based upon their textural differences. Texture, just like object size, is directly correlated with photo scale. Thus, a given feature may have a coarse texture in a low-altitude photograph and a smooth texture in a high-altitude photograph.

Figure 3-13 Photographic textures associated with forestland (coarse) and crop-land (smooth). The panchromatic exposure was made during the fall (coniferous trees—dark tones, deciduous trees—light tones).

Figure 3-14 Photographic textures associated with a young lava flow (coarse) and salt evaporation ponds (smooth).

Association: Certain objects are "genetically" linked to other objects, so that identifying one tends to indicate or confirm the other. Association is one of the most helpful clues for identifying cultural features that comprise aggregate components. For example, in Figure 3-15 the four tall cooling towers and two circular-shaped buildings (housing nuclear reactors) are characteristic features of a nuclear-fueled power plant. An example from the natural environment is a strip of phreatophytic vegetation growing along the toe of an alluvial fan in an arid environment; this association indicates a supply of shallow groundwater (Figure 3-16).

Figure 3-15 Nuclear power plant at Three-Mile Island on the Susquehanna River in Pennsylvania. (Courtesy Environmental Protection Agency.)

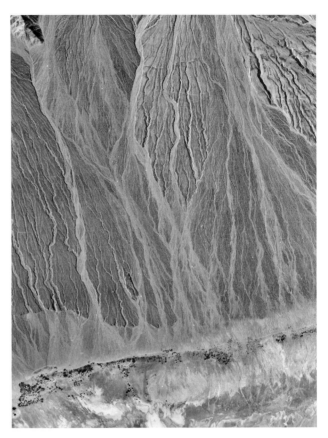

Figure 3-16 Strip of phreatophytes (honey mesquite) growing along the toe of an alluvial fan in Death Valley, California. There is a good supply of shallow groundwater in the alluvium beneath the vegetation because a phreatophyte requires a large and constant supply of water.

Site: The location of an object in relation to its environment is called the site factor and is important for recognizing many cultural and natural features. For example, citrus groves in Florida's central-ridge district are often sited on hillsides to avoid cold-air drainage to low-lying areas. In addition, many types of natural vegetation are characteristically confined to specific locales such as swamps, marshes, and stream banks or to sites differing in elevation and aspect. Thermal and nuclear power plants need an abundant supply of coolant water and are often found near major sources of surface water (Figure 3-15).

Photo-Interpretation Keys

A **photo-interpretation key** is a set of guidelines used to assist interpreters in rapidly identifying photographic features. Keys are valuable as training aids for beginning interpreters and as reference or refresher material for more experienced individuals. Depending on the method of presenting diagnostic features, photo-interpretation keys may be grouped into two general classes—selective keys and elimination keys.

Selective keys are usually made up of typical illustrations and descriptions of objects in a given category. They are organized for comparative use; the interpreter merely selects the key example that most nearly coincides with the feature to be identified. By contrast, **elimination keys** require the user to follow a step-by-step procedure, working from the general to the specific. One of the more common forms of elimination keys is the **dichotomous** type. Here, the interpreter must continually select one of two contrasting alternatives until he or she progressively eliminates all but one item of the category—the one being sought.

When available, elimination keys are sometimes preferred to selective keys. On the other hand, elimination keys are more difficult to construct, and their use may result in erroneous identifications if the interpreter is forced to choose between two unfamiliar image characteristics. Studies have revealed no significant difference between results from the two types of keys as long as the material within each key is well organized.

The determination of the type of key and method of presentation to be used depends on (1) the number of objects or conditions to be recognized and (2) the variability normally encountered within each classification. As a general rule, keys are much more easily constructed and applied in identifications of cultural features than in identifications of natural vegetation and landforms. For reliable interpretation of natural features, training and field experience are often essential to ensure consistent results. Photo-interpretation keys for use in identifying tree species and classes of industry are described in Chapters 11 and 14, respectively.

Three-Dimensional Photography

In many instances, it is entirely feasible to use single photographs for the recognition of specific features. The principal disadvantage of this technique, however, is that only two dimensions (length and width) of most objects can be perceived. This is the equivalent of using only one eye for viewing the surroundings, an effect referred to as **mon-**

Figure 3-17 Old-fashioned parlor stereoscope (the stereopticon). Note paired photographs of the Sphinx. (Courtesy Keystone View Co.)

Figure 3-18 Example of paired photographs that produce a three-dimensional image when viewed through a parlor stereoscope. Scene is the Rock of Gibraltar. (Courtesy Keystone View Co.)

ocular vision. The all-important third dimension of **depth perception** is provided only when objects are viewed with both eyes, a process known as **stereoscopy.** Here, each eye focuses on the same object from a different position and transmits a slightly different image to the brain, where the two images are fused into a three-dimensional counterpart of the original object. This result is known as **binocular**, or **stereoscopic, vision.**

One can quickly compare monocular vision with stereoscopic vision by the ''coin-on-a-table'' trick. If one eye is covered and only the coin's edge is seen from the level of the tabletop, it becomes quite difficult to place the forefinger directly on top of the coin. When one has both eyes open, the difficulty vanishes.

While almost everyone possesses and automatically employs stereoscopic vision, there have been a number of fairly successful business enterprises based on the somewhat startling effects of exaggerated three-dimensional pictures. For example, in the early 1900s the **stereopticon**, or **parlor**

stereoscope, was a standard fixture in many well-to-do American homes. This instrument, shown in Figure 3-17, was used in viewing paired photographs that had been taken from slightly different camera positions. The stereopticon allowed the left eye to see only the left-hand print and the right eye to see only the right-hand print, thus creating the illusion of depth for the viewer. Paired photographs of the Rock of Gibraltar are shown in Figure 3-18. Although the corresponding images are rather widely separated, people experienced in stereo viewing may be able to see this scene three-dimensionally with only their eyes.

When objects farther than 400 to 500 m away are viewed by unaided eyes, the special ability of depth perception is essentially lost. At such distances, lines of sight from each eye converge very little. In fact, they are nearly parallel when the eyes are focused on the horizon. If the human **eye base**, or **interpupillary distance** (i.e., the linear distance between the pupils), were increased from the average 6.4 cm, the third dimension or the perception of depth could be greatly increased. In a manner of speaking, this feat can be accomplished through aerial photography, where the camera exposure positions substitute for the eye positions.

In obtaining aerial photographs, the camera station moves along the line of flight, and photographs are taken at separate locations hundreds to thousands of meters apart. The exposure interval is fixed to produce **overlapping photographs**, so that the same objects will appear on two or more of the photos but from slightly different views (Figure 3-19). When any two of the overlapping photographs are then viewed through a **stereoscope**, each eye in effect occupies one of the widely separated camera exposure stations. The stereoscope aligns the sight of each eye to one of the photographs, thus enabling the area of overlap to be perceived in three dimensions.

This perception of depth extends photo interpretation. The stereoscopic effect usually permits an interpreter to identify objects or conditions that cannot be detected or identified in single photographs. For example, the reason for the unusual loop in Figure 3-20 is not apparent until the photographs are viewed stereoscopically.

Flight Procedures for Stereo Coverage

Aerial flights for stereo coverage are planned so that the photographs are taken frequently enough along a flight line to ensure that all ground objects appear in at least two consecutive photographs (Figure 3-19). The area of common coverage is called **overlap**, **end lap**, or **forward lap**. Two successive overlapping photographs represent a **stereopair**; when viewed with a stereoscope, their common areas yield a three-dimensional image called a **stereo model**.

At least 50 percent overlap is necessary for complete stereoscopic coverage in each flight line (excluding first and last photographs). The standard overlap is 60 percent; the extra 10 percent represents a margin of safety to compensate for slight overlap variations caused by changing aircraft speed, exposure interval, or topographic irregularities. With 60 percent overlap, a series of three photographs ensures that

Figure 3-19 Aerial camera stations are spaced to provide for about a 60 percent forward overlap of aerial photographs along each flight line and a 20 to 30 percent sidelap for adjacent lines.

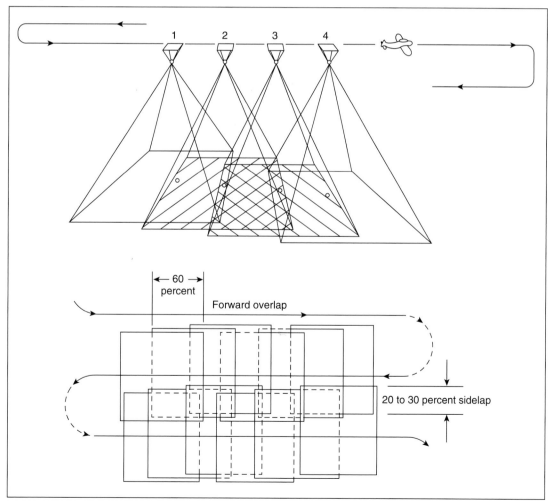

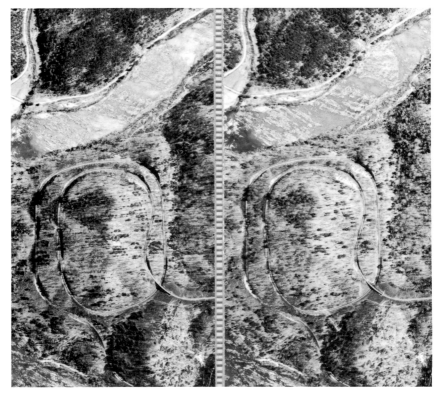

Figure 3-20 Stereopair showing what appears to be relatively flat terrain with an unusual loop. When viewed with a stereoscope, however, hilly terrain is clearly seen; the loop was designed to reduce the grade for tracks of the L & N Railroad in Polk County, Tennessee. Scale is about 1:16,000. (Courtesy Tennessee Valley Authority.)

the two end members will sufficiently overlap the central photograph to provide *complete* stereoscopic coverage for the latter. This series of three overlapping photographs is called a **stereotriplet**.

If more than one flight line is required to provide photographic coverage for a given area, successive flight lines parallel to the initial line are spaced in such a way as to ensure that there are no skipped areas between any two lines. The common marginal strip between two flight lines is called **sidelap** and is usually 20 to 30 percent (Figure 3-19).

Types of Stereoscopes

The function of a stereoscope is to deflect normally converging lines of sight so that each eye views a different photographic image. Parlor stereoscopes accomplished this by the placement of a thin prism before each eye. Ordinarily, no magnification was involved, but the result was a sharply defined, if occasionally distorted, three-dimensional picture.

Instruments used today for three-dimensional study of aerial photographs are of three general types: (1) **lens**, or **pocket, stereoscopes**, (2) **mirror**, or **reflecting, stereoscopes**, and (3) **zoom stereoscopes**. Lens stereoscopes utilize a pair of magnifying lenses to keep the eyes focused at

infinity and their lines of sight approximately parallel. Most lens stereoscopes have $2\times$ to $4\times$ magnification and adjustable eye bases. They are relatively inexpensive and durable and can be quickly folded into a small unit—hence the term "pocket." Two popular lens stereoscopes are pictured in Figure 3-21.

The primary drawback to a lens stereoscope is its narrow field of view. For example, two 23×23-cm photographs with 60 percent overlap will include a common area of about 14×23 cm. Since this type of stereoscope duplicates the eye base, it requires several steps to view the entire stereo model.

Most reflecting stereoscopes can provide a full view of the entire stereo model of two 23×23-cm photographs with 60 percent overlap. This is made possible by a system of prisms and mirrors that, in effect, widens the viewer's eye base from about 6.4 cm to anywhere from 18 to 22 cm. Reflecting stereoscopes offer viewing without magnification, but $3\times$ to $8\times$ binocular attachments are normally available as options. The greater the magnification, however, the smaller the field of view. Reflecting stereoscopes are available in a variety of designs and price ranges. One popular model is illustrated in Figure 3-22.

Zoom stereoscopes are highly versatile instruments that are intended for office or laboratory interpretation work (Figure 3-23). Zoom stereoscopes provide continually variable, in-focus magnification from $2\times$ to $64\times$, depending upon the model. Adjustable arms accommodate conjugate

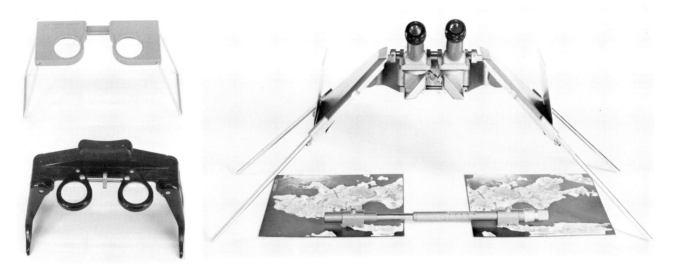

Figure 3-21 *(Above left)* Two types of lens, or pocket, stereoscopes with adjustable eye bases. The model at the top features 2× magnification lenses, an adjustable eye-base range of 5 to 7 cm, and foldaway wire frame legs. The model at the bottom features 2× and 4× magnification by proper positioning of the legs and lenses and an adjustable eye-base range of 5 to 7.5 cm.

Figure 3-22 *(Above right)* Mirror, or reflecting, stereoscope with optional magnifying binoculars. Note the full separation of the stereopair. Positioned on the photographs is a parallax bar for measuring object heights. (Courtesy Wild Heerbrugg Instruments, Inc.)

Figure 3-23 Zoom stereoscope system (includes light table and print illuminator) with variable magnification. The instrument shown here has a magnification range from 2.5× to 20× and is designed for viewing opaque prints or film transparencies of two sizes, 13 × 13 cm or 23 × 23 cm. (Courtesy Bausch and Lomb, Inc.)

image separations on most 70-mm and 24-cm roll films. In addition, each eyepiece assembly can usually be rotated to orient photos properly for stereo viewing. This feature is ideal for studying uncut, roll-film photos where there was drift or crab in the flight line (Figure 3-24).

Stereo Viewing Without Instruments

People with normal or corrected vision in both eyes can often develop a facility for stereoscopic vision without the use of a stereoscope. This is sometimes called **naked-eye stereoscopic viewing**. Seeing stereoscopically with unaided eyes can be practiced with the "sausage-link" exercise sketched in Figure 3-25. The eyes are focused on a distant background, such as a white wall, as the forefingers, at arm's length and horizontal to each other, are brought slowly together into the line of vision. As the spacing closes, a point will be reached where a "sausage link" is seen between the fingertips. The farther apart the fingers and the longer the sausage link when it forms, the more nearly parallel are the lines of sight to the forefingers.

A method of learning to view stereopairs with unaided eyes can be accomplished by holding a card about 15 cm wide and 20 cm high between the left and right photographs of a stereopair such as those shown in Figure 3-20. The stereopair is viewed looking straight down with the forehead and nose touching the top of the card. The card forces the left eye to concentrate on images of the left photo and the right eye on the same images on the right photo. In a few seconds, one of the images will appear to migrate toward its conjugate image on the adjacent photograph. When these images fuse together, the scene should appear in a three-dimensional view. With continued practice, most people can master the technique of keeping the lines of sight parallel and the card is no longer necessary.

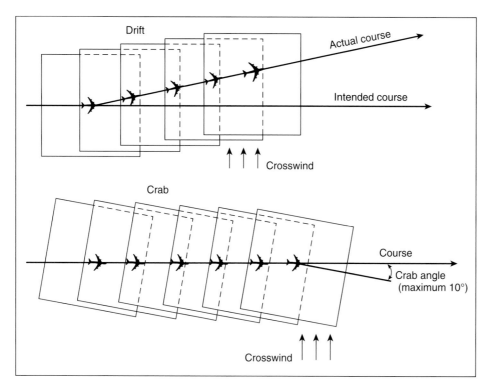

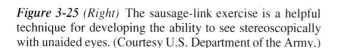

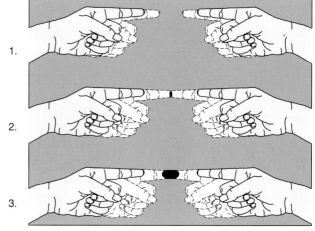

Figure 3-24 (Above) If crosswinds are encountered during a photo flight, the airplane may be blown off course, causing an alignment defect known as **drift**. The pilot can avoid this by heading the airplane slightly into the wind, a technique known as **crabbing**. Excessive drift or crab may reduce overlap to an undesirable level. (Courtesy U.S. Army Engineer School.)

Figure 3-25 (Right) The sausage-link exercise is a helpful technique for developing the ability to see stereoscopically with unaided eyes. (Courtesy U.S. Department of the Army.)

Preparing Photographs for Stereo Viewing

For casual stereoscopic viewing, it is common practice to place two overlapping photographs on a flat surface and simply move them around beneath a stereoscope until the images fuse and the illusion of three dimensions is created. It may help to place the tips of the index fingers on the same object in the left and right photos; the photos are then moved on a trial-and-error basis until the fingers are fused. This superposition indicates the photographs are approximately aligned for stereo viewing, and the fingers are removed from the field of view.

More rigorous photo alignment is essential for precision measurement work or for avoiding eye strain when a stereopair is to be studied for a considerable length of time. To orient the photographs precisely for stereo viewing, it is necessary to establish their positions as they were taken along a flight line. The following procedure is recommended for accomplishing this task with 23 × 23-cm photographic prints.

Both photographs from an overlapping pair of vertical photographs from the same flight line are trimmed, preserving the fiducial marks. Then the **principal point** (**PP**), or optical center, of each photograph is located and marked (e.g., with a small needle hole). It will be remembered from Chapter 2 that lines connecting opposite pairs of fiducial marks intersect to identify the principal point of a vertical photograph (Figure 2-14).

Next the **conjugate principal point** (**CPP**) is located and marked on each photograph. These are the points that correspond to the principal points of adjacent photographs (i.e., **transferred principal points**). The flight-line segment for each print is a line connecting the PP and CPP (Figure 3-26). When the two photos are aligned so that all four points lie on a straight line, the flight line is duplicated and, excluding the spacing or separation factor, the photographs are properly oriented for viewing with a stereoscope.

For a lens stereoscope, slide the two overlapping photographs apart until the separation distance between a reference object on one photo and its conjugate on the second photo is generally equivalent to the eye base of the viewer. The stereoscope lenses are then positioned over the reference object and its conjugate with the long axis of the stereoscope aligned parallel to the reconstructed flight line. In this way, a strip about 6.4 cm wide and 23 cm long can be viewed in three dimensions when the stereoscope is moved up and down over the overlap area at a right angle to the flight line. If good fusion is not immediately attained, it may be necessary to rotate the stereoscope slightly over the prints or change the photo-separation distance slightly. Because the photographs must be positioned in an overlapping position, the edge of one print has to be turned upward to view "hidden" segments of the stereo model.

Unlike with the lens stereoscope, the entire stereo model width can usually be observed at one time with a mirror stereoscope (no magnification). For two 23 × 23-cm photographs with 60 percent overlap, the width of the common area is about 14 cm. Prior to viewing, the two photos are separated parallel to the flight line until a reference feature and its conjugate are centered under each of the mirrors. If stereoscopic vision is not immediately attained, the photo separation can be slowly adjusted along the established flight line or the stereoscope can be slightly rotated over the photographs.

Pseudoscopic Views

Pseudoscopic views are those in which there is an illusional reversal of "highs and lows." Common examples include an elongated hill appearing as a valley, a river appearing to be flowing on top of a sinuous ridge, and a tall building appearing as a depression in the ground. This can occur if shadows in the stereopair fall away from the viewer. It is correctable by rotating the stereopair 180° so that the shadows fall toward the viewer. Most of the stereopairs in

---

Figure 3-26 Method of aligning 23 × 23-cm prints for viewing with a lens stereoscope. Principal points are denoted as PP; conjugate principal points are marked CPP.

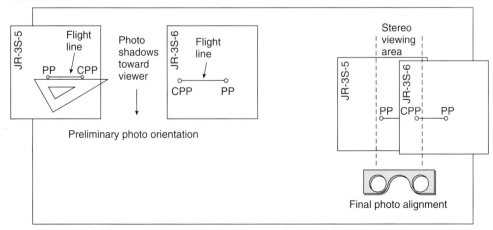

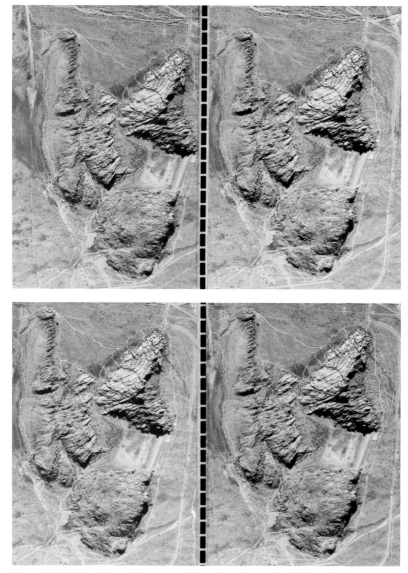

Figure 3-27 Pseudoscopic stereopair (*top*), where the right and left photos have been purposely reversed to create a reversal of relief and a stereopair oriented correctly (*bottom*).

this book have been purposely oriented to make the shadows fall generally toward the observer and thereby aid stereoscopy.

The term pseudoscopic is also used to describe the same illusion seen when viewing single photographs, electro-optical images, and radar images. The correct impression of ground features is obtained when the shadows fall toward the observer.

A second cause of this false illusion is when the two photographs in a stereopair are reversed so that the left eye views the right photo and the right eye views the left photo. In most instances, this reverse view is noted immediately upon stereo examination. A pseudoscopic stereopair is presented in Figure 3-27.

Vertical Exaggeration in Stereo Viewing

Objects seen in stereo normally have exaggerated vertical distances with respect to horizontal distances (Figure 3-28). This distance or scale disparity is called **vertical exaggeration**, or **hyperstereoscopy**, and is especially helpful in the detection and identification of low-lying objects and subtle topographic features (Figures 3-20, 3-28, and 3-29).

Vertical exaggeration is regarded as the product of many variables, including camera and photo characteristics,

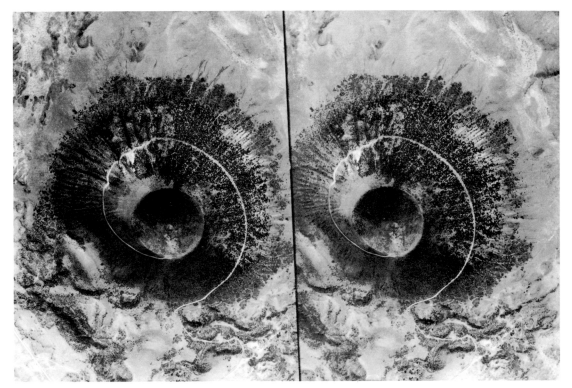

Figure 3-28 (Above) Stereogram of Mount Capulin, New Mexico. Mount Capulin is an almost undissected volcanic scoria cone about 300 m high, which has an unbreached crater about 30 m deep. The cone is surrounded by older basalt flows that still retain much of their initial surface form. The drainage pattern on the cone is radial. Scale is about 1:20,000. (Courtesy U.S. Department of Agriculture.)

Figure 3-29 (Below) Stereogram of a low-relief area in Arizona enhanced by the vertical exaggeration effect when viewed with a lens stereoscope.

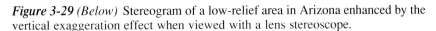

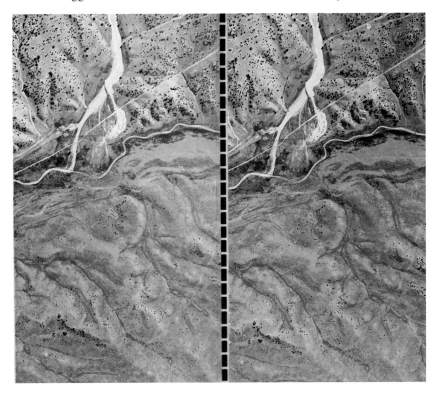

viewing factors, and the physiological and psychological factors involved in stereoscopic vision. Various formulas have been developed for estimating vertical exaggeration, and although none is perfectly accurate, they do permit interpreters to adjust their visual impressions (Roscoe 1960). The following formula is commonly used for *estimating* the vertical exaggeration factor (VE):

$$VE = \left(\frac{AB}{H}\right)\left(\frac{h}{EB}\right), \qquad (3\text{-}1)$$

where: AB = air base, or the horizontal distance between centers of overlapping vertical photographs,
  H = camera height above the ground,
  h = distance from the eyes at which the stereo model is perceived, and
  EB = eye base.

The values for h and EB can be considered constants. For the average adult, EB is 6.4 cm. Although h is difficult to measure, repeated tests with numerous interpreters using a variety of stereoscopes indicate an average value of about 45 cm (Wolf 1983). The value of h is not the distance from the stereoscope's lenses to the photos, but rather the distance to the plane where the stereo model appears to be. If you attempt this experiment, you will find that the perception of the stereo model appears to be somewhere below the table-top on which the photographs are positioned.

With h and EB considered as constants, Equation 3-1 can be simplified to

$$VE = \left(\frac{AB}{H}\right)\left(\frac{6.4 \text{ cm}}{45 \text{ cm}}\right) = \frac{AB}{H} \times 0.14.$$

Thus, the amount of vertical exaggeration is determined by the ratio AB/H, which is called the **base-height ratio**. The larger the base-height ratio, the greater the vertical exaggeration. Figure 3-30 shows the vertical exaggeration factor for various base-height ratios.

Thurrell (1953) considered the relations between several photographic factors (e.g., percent overlap, focal length, and film size) and vertical exaggeration. Figure 3-31 permits

the reading of the vertical exaggeration factor when percent overlap and focal length are known. Examination of this figure shows that vertical exaggeration varies inversely with both overlap and focal length.

Proper Use of the Stereoscope

Beginning photo interpreters should be especially careful to cultivate proper stereoscopic viewing habits to avoid the creation of undesirable distortions and eye fatigue. Some of the important rules to be observed are as follows:

1. For stereoscopes with adjustable eyepieces, the lens separation must match the user's eye base.

2. The long axis of the stereoscope must be kept parallel to the flight line at all times. This includes when the stereoscope is moved up and down over the area of overlap and when separating the photographic pair for stereo viewing.

3. The stereo model must always be viewed looking vertically downward through the eyepieces.

4. The photographs should be oriented so that the shadows extend toward the viewer.

5. Good illumination is essential when using a stereoscope to study prints for extended periods. Some interpreters prefer to use natural lighting, and they work in front of a window whenever possible. Others prefer to use a floating-arm drafting lamp pulled down close to the photographs. The lamp can be moved about to reduce glare or to illuminate particular features of interest.

Figure 3-31 Relationship between percent overlap and vertical exaggeration for several common focal lengths; print size is 23 × 23 cm in all cases. (Adapted from Thurrell 1953 and Miller 1961.)

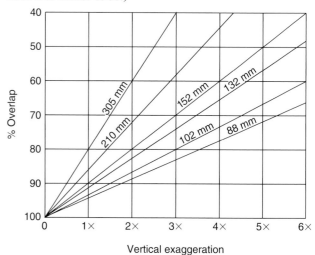

Figure 3-30 Relationship between base-height ratio and vertical exaggeration. (Adapted from Thurrell 1953 and Miller 1961.)

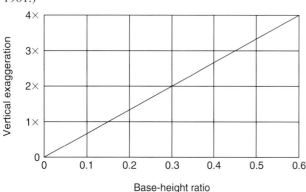

6. Lenses and mirrors must be kept clean. Because the mirrors of a reflecting stereoscope have silvered first, or front, surfaces, they are easily corroded by fingerprints. They should be cleaned only in accordance with the manufacturer's instructions.

Special Problems Affecting Stereovision

Interpreters who have difficulty mastering the use of the stereoscope should be cognizant of the following factors that may affect stereovision:

1. A person's eyes may be of unequal strength. If one normally wears eyeglasses or contact lenses, one should also wear them when using the stereoscope.

2. Poor illumination, misaligned prints, or uncomfortable viewing positions may result in rapid eye fatigue.

3. A pseudoscopic view will be created if there is a reversal of left and right photographs or if the shadows generally fall away from the observer.

4. Objects that change positions between exposures (e.g., moving automobiles, trains, boats) cannot be viewed stereoscopically.

5. In areas of steep topography, scale differences of adjacent photographs may make it difficult to obtain a three-dimensional image.

6. Dark shadows, clouds, or sunspots prohibit stereoscopic study if they obliterate information on one photograph.

Stereograms

A **stereogram** consists of two sections from overlapping airphotos that are properly positioned and mounted on stiff card stock for viewing with a lens stereoscope (Figure 3-20). Stereograms are helpful aids to practice stereoscopic viewing. In addition, one of the best ways for an interpreter to build a file of reference material is to prepare **sample stereograms** that illustrate different features in a given category (e.g., tree species, landforms, industrial types).

The following procedure has been found useful in the preparation of aerial stereograms from 23 × 23-cm contact prints for use with a lens stereoscope (Figure 3-32).

1. Locate principal points and conjugate principal points on the overlapping pair and pinpoint them with small needle holes.

2. Draw in flight lines for the overlapping pair; use a sharpened grease pencil or soft-lead pencil so that the lines can easily be removed later.

3. Delineate and mark the desired view on one of the photos within a space about 5.5 cm wide as measured along the flight line. This width may be expanded, provided it does not exceed the observer's eye base. The lateral lines should be parallel to each other and perpendicular to the flight line.

4. Using a stereoscope, transfer the enclosed view to the overlapping print and mark the boundaries.

5. Mark the strips to be cut out as left (L) and right (R).

6. Cut out the two strips and lay them side by side in the proper order and with the marked flight-line segments properly aligned. Check them with the stereoscope to ensure that they have been properly cut and aligned.

7. Mount the two strips on card stock with rubber cement. After the cement has dried, remove all markings and record locale, feature identifications, and photographic data (e.g., date and scale) on the back of the card.

More than 150 stereograms are employed in this book to illustrate particularly interesting surface features. The airphotos are positioned in accordance with an average eye base of 6.4 cm. Consequently, some people may have to adjust the lens spacing of their stereoscopes to bring the images into "crisp" three-dimensional views.

Figure 3-32 Method of delineating stereogram cutouts on 23 × 23-cm prints. The 5.5-cm viewing width may be varied slightly, provided it does not exceed the observer's eye base.

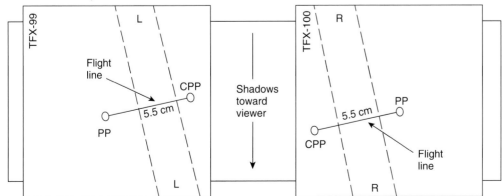

Questions

1. Position a lens stereoscope over the stereoscopic vision test chart shown in Figure 3-33. First, rank the symbols within rings 1, 2, 3 and 6, 7 in height order (highest = 1, second highest = 2, and so on). Second, rank the relative heights of rings 1 through 8. If two or more designs are at the same height, use the same number.

Ring 1

_____ Triangle
_____ Square
_____ Point
_____ Marginal ring

Ring 2

_____ Flanking peaks
_____ Marginal ring
_____ Spotting mark and central peak

Ring 3

_____ Square
_____ Cross
_____ Marginal ring
_____ Circle, lower left
_____ Circle, upper center

Ring 6

_____ Circle, lower right
_____ Circle, upper left
_____ Circle, lower left
_____ Marginal ring
_____ Circle, upper right

Ring 7

_____ Black circle
_____ Black triangle
_____ Black flag with ball
_____ Double cross
_____ Tower with cross and ring
_____ Black rectangle
_____ Marginal ring
_____ Arrow

Relative Heights

_____ Ring 1
_____ Ring 2
_____ Ring 3
_____ Ring 4
_____ Ring 5
_____ Ring 6
_____ Ring 7
_____ Ring 8

2. Position a lens stereoscope over the hidden-word stereograms shown in Figure 3-34. What words appear in the three views?

Top view: _____

Middle view: _____

Bottom view: _____

3. Examine the stereogram in Figure 3-35 with a lens stereoscope and locate each of the following features; designate their location with a letter-number coordinate.

Feature	Coordinate
Highest topographic summit	_____
Hayward fault (inferred)	_____
Expensive housing area	_____
Football stadium	_____
Baseball field (good condition)	_____
Baseball field (poor condition)	_____
Track field	_____
Main entrance to campus	_____
Building on highest ground	_____
Vegetated ravine	_____
Major street	_____
Coniferous vegetation (upland area)	_____
Mixed deciduous vegetation (upland area)	_____
Brushland (upland area)	_____
Road following hillside contour	_____
Terraced slope	_____
Steepest slope	_____

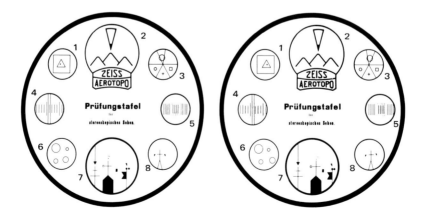

Figure 3-33 Stereoscopic vision test chart. (Courtesy Carl Zeiss, Oberkochen.)

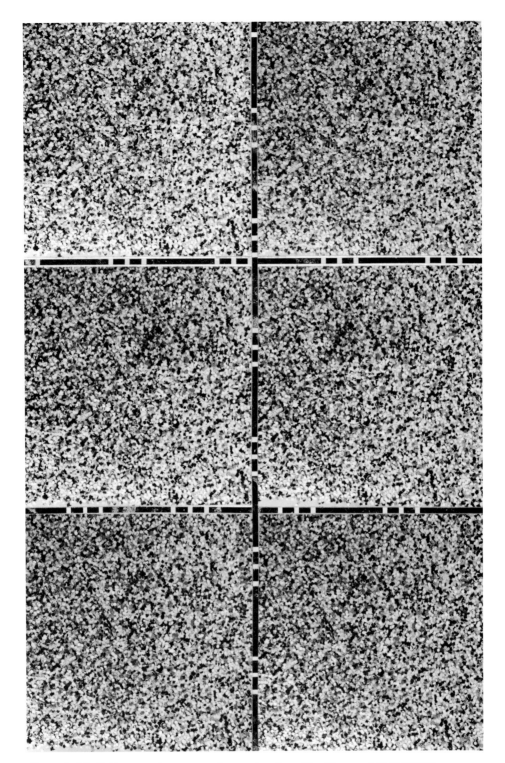

Figure 3-34 Hidden-word stereoscopic test developed by Sims and Hall (1956).

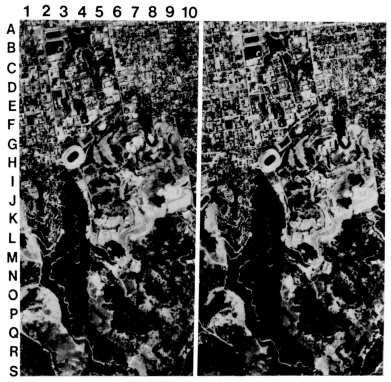

Figure 3-35 Stereogram of the Berkeley Campus of the University of California. (Courtesy William C. Draeger, U.S. Geological Survey.)

---

4. Determine the approximate vertical exaggeration (VE) for the following stereopairs.

	VE
Camera height = 5,000 m, air base = 2,500 m	_____
Percent overlap = 60, focal length = 210 mm, print size = 23 × 23 cm	_____
Camera height = 6,200 m, air base = 2,200 m	_____
Percent overlap = 80, focal length = 152 mm, print size = 23 × 23 cm	_____

5. After examining Figure 3-12, state a use for the high-sun-angle airphoto and a use for the low-sun-angle airphoto.

Bibliography and Suggested Readings

Avery, T. E. 1965. Evaluating the Potential of Photo Interpreters. *Photogrammetric Engineering* 31:1051–59.

————. 1968. Screening Tests for Rating Photo Interpreters. *Photogrammetric Engineering* 34:476–82.

Campbell, J. B. 1983. *Mapping the Land: Aerial Imagery for Land Use Information.* Washington, D.C.: Association of American Geographers.

Colwell, R. N., ed. 1960. *Manual of Photographic Interpretation.* Falls Church, Va.: American Society of Photogrammetry.

Leachtenauer, J. C. 1973. Photo Interpretation Test Development. *Photogrammetric Engineering* 39:1187–95.

Miller, V. C. 1961. *Photogeology.* New York: McGraw-Hill Book Co.

Mollard, J. D., and J. R. Janes. 1984. *Airphoto Interpretation and the Canadian Landscape.* Hull, Quebec: Canadian Government Publishing Centre.

Nash, A. J. 1972. Use of a Mirror Stereoscope Correctly. *Photogrammetric Engineering* 38:1192–94.

Rabenhorst, T. D., and P. D. McDermott. 1989. *Applied Cartography, Introduction to Remote Sensing.* Columbus, Ohio: Merrill Publishing Co.

Roscoe, J. H. 1960. Photo Interpretation in Geography. In *Manual of Photographic Interpretation*, edited by R. N. Colwell, 735–82. Falls Church, Va.: American Society of Photogrammetry.

Sims, W. F., and N. Hall. 1956. *The Testing of Candidates for Training as Airphoto Interpreters.* Canberra, Australia: Forestry and Timber Bureau.

Stanley, R. M. 1981. *World War II Photo Intelligence.* New York: Charles Scribner's Sons.

Thurrell, R. F. 1953. Vertical Exaggeration in Stereoscopic Models. *Photogrammetric Engineering* 19:579–88.

Wolf, P. R. 1983. *Elements of Photogrammetry*, 2d ed. New York: McGraw-Hill Book Co.

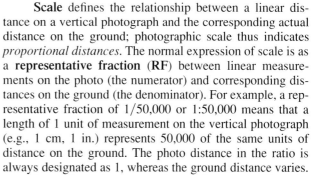

Chapter 4

Principles of Photogrammetry

Scale defines the relationship between a linear distance on a vertical photograph and the corresponding actual distance on the ground; photographic scale thus indicates *proportional distances*. The normal expression of scale is as a **representative fraction** (**RF**) between linear measurements on the photo (the numerator) and corresponding distances on the ground (the denominator). For example, a representative fraction of 1/50,000 or 1:50,000 means that a length of 1 unit of measurement on the vertical photograph (e.g., 1 cm, 1 in.) represents 50,000 of the same units of distance on the ground. The photo distance in the ratio is always designated as 1, whereas the ground distance varies.

Large-denominator numbers refer to *small scale*, whereas small-denominator numbers are indicative of *large scale*. Confusion about this concept can be avoided if one considers the scales of 1:10,000 and 1:100,000. The former RF represents a larger scale because its ratio is large (closer to 1) in comparison to the latter RF. A **small-scale aerial photograph** (e.g., 1:120,000) covers a larger ground area in less detail than a **large-scale aerial photograph** (e.g., 1:20,000), which depicts a small ground area in considerable detail. Comparisons of photographs at two different scales are shown in Figure 4-1.

A representative fraction can be restated by assigning units and applying conversion factors as required. For example, a scale of 1:24,000 can be stated as 1 in. equals 2,000 ft (24,000 divided by 12) or a scale of 1:100,000 can be stated as 1 cm equals 1 km because there are 100,000 cm in 1 km. A scale of 1:60,000 can be stated as 1 in. equals about 0.95 mi because there are 63,360 in. in 1 mi:

$$\frac{1 \text{ mi}}{63,360 \text{ in.}} = \frac{x \text{ mi}}{60,000 \text{ in.}}$$

$$63,360x = 60,000$$

$$x = \approx 0.95 \text{ mi}$$

Similarly, the scales of 1:300,000 and 1:1,000,000 can be stated as 1 in. equals about 4.7 and 15.8 mi, respectively.

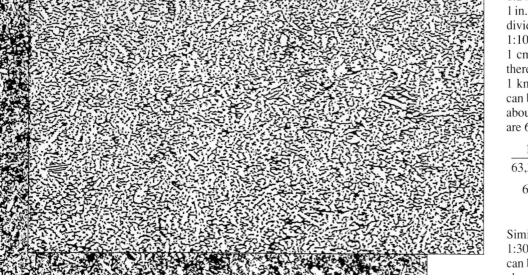

Photogrammetry and the Importance of Scale

Photogrammetry is defined as the technique of obtaining reliable measurements of objects from their photographic images. The word photogrammetry is derived from three Greek roots meaning "light-writing-measurement." To make accurate measurements of distance, area, or height, it is necessary to determine, as accurately as possible, the photographic scale.

Scale Determination from Focal Length and Altitude

As illustrated in Figure 4-2, the scale of a vertical aerial photograph is a function of the focal length of the camera lens (f) and the height above the terrain from which the exposure is made (H). Because of the relationship between similar triangles ($\triangle ABC$ and $\triangle CDE$ in Figure 4-2), the

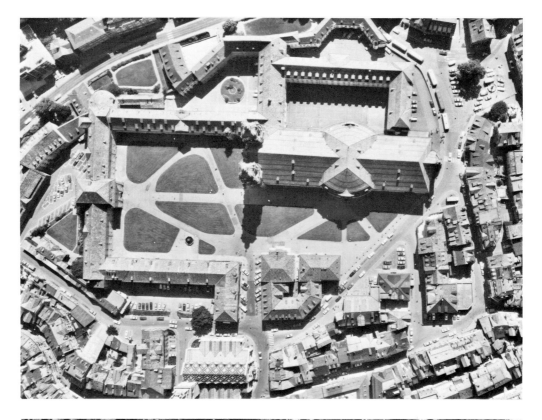

Figure 4-1 Comparison of photographic scales, Saint Gall, Switzerland. Top view is at a scale of about 1:2,000; the scale of the lower view is approximately 1:5,800. (Courtesy Wild Heerbrugg Ltd.)

scale, expressed as a representative fraction, for a vertical airphoto can be determined as follows:

$$RF = \frac{1}{H/f}. \qquad (4\text{-}1)$$

Both H and f must be expressed in the same units, and H indicates flying height above *mean* terrain if one desires an average scale for the entire photograph. For example, with a focal length of 210 mm, a flight altitude of 2,500 m above MSL (mean sea level), and an average ground elevation of 400 m, the scale is computed as follows:

$$RF = \frac{1}{(2,500 \text{ m} - 400 \text{ m})/0.21 \text{ m}}$$

$$= \frac{1}{2,100 \text{ m}/0.21 \text{ m}} = \frac{1}{10,000}, \quad \text{or} \quad 1{:}10,000.$$

It is emphasized that the computed scale of 1:10,000 will be precisely obtained *only* as long as the landscape is uniformly 400 m above sea level. If the elevation decreases, the photographic scale will be smaller. Conversely, if higher features are encountered, photo scale will be increased because the landscape will have "moved closer" to the camera. However, the **average**, or **nominal**, **scale** of the photograph may be regarded as 1:10,000.

It is evident from the relationship described in Equation 4-1 that photo scale varies directly with focal length (as focal length increases, scale increases and vice versa) and

inversely with flying height (as flying height increases, scale decreases and vice versa). Thus, at a given flying height, the use of a long-focal-length lens will yield larger-scale photographs than will a shorter-focal-length lens. It also follows that at a given flying height, more photographs will be needed to cover a given ground area when long-focal-length lenses are used. For example, a doubling of focal length will quadruple the number of photographs required to cover a given ground area at a given flight height.

Scale Determination from Photo-Map Distances

While the preceding method of deriving photo scale is sound, it often happens that either focal length or the exact flight altitude is unknown. Consequently, scale is more often calculated from the relationship between a photo measurement and the same measurement on a map of known scale. The relationship can be expressed as follows:

$$RF = \frac{1}{(MD)(MS)/PD} \qquad (4\text{-}2)$$

where: MD = map distance between two points,
MS = map-scale denominator, and
PD = photo distance between the same two points.

As an example, the distance between two road intersections might be measured as 6 cm on a map at a scale of 1:50,000. If the corresponding photo distance is measured as 3.2 cm, the photo scale is computed as follows:

$$RF = \frac{1}{(6 \text{ cm})(50,000)/3.2 \text{ cm}}$$

$$= \frac{1}{300,000 \text{ cm}/3.2 \text{ cm}} = \frac{1}{93,750}, \quad \text{or} \quad 1{:}93,750.$$

In the application of this technique, the two points selected for measurement on the photo should be diametrically opposed, so that a line connecting them passes through or near the principal point. If the points are also approximately equidistant from the principal point, the effect of photographic tilt upon the scale determination will be minimized. It is preferable that the points also be at the same or similar elevation.

Scale Determination from Photo-Ground Distance

If reliable maps are not available, it is possible to calculate scale from the relationship between the photographic

---

Figure 4-2 Diagram showing the geometry of a vertical airphoto. Note that the terrain distance AB subtends an angle at the camera lens equal to that subtended by the image distance DE. This relationship remains constant regardless of focal length (f) and flight height (H).

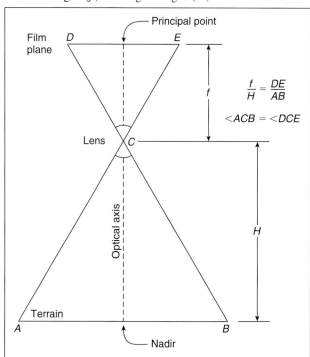

$$\frac{f}{H} = \frac{DE}{AB}$$

$$\angle ACB = \angle DCE$$

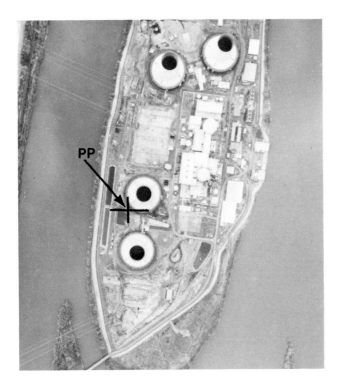

PP

Figure 4-3 Vertical airphoto of four cooling towers of equal height associated with a nuclear power plant at Three-Mile Island on the Susquehanna River in Pennsylvania. Depending upon their distance from the photo center, note how their tops are displaced away from the principal point (PP) along radial lines originating at the PP. (Courtesy Environmental Protection Agency.)

distance between two points (PD) and the ground distance between the same two points (GD):

$$RF = \frac{1}{GD/PD}. \qquad (4\text{-}3)$$

As an example, the distance between two buildings might be measured on a vertical photograph as 5 cm, or 0.05 m. If the corresponding ground distance is measured as 1,584 m, the scale is computed as

$$RF = \frac{1}{1,584 \text{ m}/0.05 \text{ m}} = \frac{1}{31,680}, \quad \text{or} \quad 1:31,680.$$

Scale approximations can also be made by the use of objects or features of *known* ground dimensions, such as athletic fields, aircraft wingspans, automobile lengths, or railroad car lengths. However, because the percentage of error increases as the measured distance decreases, very small objects are apt to produce sizable errors for such small distances or measurements.

Image Displacement on Vertical Airphotos

With the **orthographic projection** of a map, all features are located in their correct horizontal positions and are depicted as though they were each being viewed from directly overhead. This standard cannot be met by the **central projection** of a vertical airphoto because all objects are positioned as though they were being viewed from the same point. This means that the images of most ground objects are shifted or displaced from their correct positions, a phe-

nomenon known as **image displacement**. This type of photo distortion, however, also enables us to view overlapping photographs in three dimensions (Chapter 3) and to determine the heights of objects, as explained later in this chapter.

The most significant source of image displacement is **relief** (i.e., differences in the relative elevations of objects pictured). **Relief displacement** is by no means limited to mountains and deep gorges; all objects that extend above or below a specified ground **datum plane** have their photographic images displaced to a greater or lesser extent. Skyscrapers, houses, and even trees are affected by this characteristic. A vertical airphoto *completely* devoid of relief displacement is uncommon. Perhaps the most common example of such an occurrence is a vertical photograph of a calm water surface (Figure 3-14).

Effects of Relief Displacement

The underlying cause of relief displacement can be traced to the perspective view "seen" by a camera pointed straight down toward the earth's surface. For example, if a tall cooling tower of a nuclear power plant is located near the center of the photograph, it will appear as a doughnut-shaped ring (Figure 4-3). There is little image displacement here, for this is the area where the camera lens affords a truly vertical view. On the ground, this point directly beneath the camera is called the **nadir**, a point that coincides with the principal point of a truly vertical airphoto (Figure 4-2). By contrast, for other cooling towers occurring at greater distances from the photograph's center, the camera tends to look more at their sides. Their recorded images thus appear to lean radially outward from the principal point of the photograph (Figure 4-3).

Whereas the displacements of tall cultural features are relatively easy to see (Figure 4-3), those involving terrain relief are difficult or impossible to discern. Figure 4-4, however, enables us to see relief displacement indirectly by the disruption of a linear utility right-of-way as it traverses terrain with large elevation differences. In the left-hand view, the terrain is almost directly under the camera lens. Therefore, relief displacement is minimized and the right-of-way appears in its true ground configuration as a nearly straight line. In the right-hand view, however, the terrain was imaged near the edge of the photograph. As a result, the same right-of-way is displaced into a nonlinear feature.

Figure 4-5 illustrates radial displacements caused by topographic relief. The datum plane shown is the level of the ground elevation that produces the planned scale of the

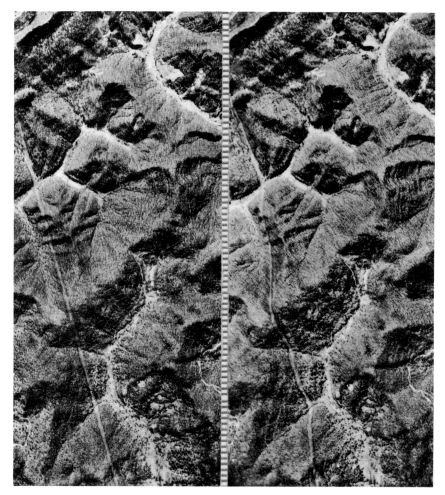

Figure 4-4 Stereogram illustrating the displacement of a utility right-of-way due to topographic relief. Location is in Bell County, Kentucky; scale is about 1:36,000. (Courtesy Tennessee Valley Authority.)

negative and contact print. Four points represent the following ground positions: Point *B* is below the datum plane; point *O* is on the datum plane; point *N* is above the datum plane at nadir; and point *A* is above the datum plane. Their displacement-free locations on the datum plane are shown at *B′*, *O*, *N′*, and *A′* and represent their positions on an orthographic map. If all the ground points were located on the datum plane, their photo positions would be *b′*, *o*, *n*, and *a′*.

However, because of the camera's central projection, the following positions are established on the photographic products (Figure 4-5):

1. Ground point *B* appears not at *b′* but at *b″*; its displacement is **radially inward** toward the photo center because its position was *below* the selected datum plane. The measured distance from *b′* to *b″* is the radial displacement due to relief.

2. Ground point *A* appears not at *a′* but at *a″*; its displacement is **radially outward** from the photo center because its position was *above* the selected datum plane. The measured distance from *a′* to *a″* is the radial displacement due to relief.

3. Ground point *O* appears at *o*; it is *not* displaced because its elevation *coincides* with the selected datum plane.

4. Ground point *N* appears at *n*; it is *not* displaced because it lies *directly below* the camera at nadir. Its location on the vertical airphoto is at the principal point.

The displacements just described are relative with respect to a datum plane at an average height within the terrain. If a datum plane was selected to pass through the lowest point *B* in Figure 4-5, then all points would be displaced outward. Although it is convenient to refer to inward and outward radial displacement with respect to some datum, it should be remembered that any point in the terrain will be displaced outward relative to any lower point (Lattman and Ray 1965).

On a vertical airphoto, the following rules apply to the *amount* of displacement due to relief (Lattman and Ray 1965):

1. For points of the same elevation, radial displacement is greater for a point farther from the principal point.

2. For points of different elevation but equally distant from the principal point, radial displacement is greater for the point of higher elevation.

3. For photographs taken at different flight heights but with the same focal-length lens, radial displacement of any one feature is greater on the photograph taken at the lower flight height.

4. For photographs taken with different focal-length lenses but from the same flight height, radial displacement of any one feature is greater on the photograph taken with the longer focal-length lens.

5. For photographs of the same scale but taken at different flight heights with different focal-length lenses, radial displacement of any one feature is greater on the photograph taken with the shorter focal-length lens at the lower flight height.

Tilt Displacement

A **tilted photograph** represents a slightly oblique view rather than a truly vertical view. Because of tilt, pictured objects are displaced by a small amount (**tilt displacement**) from the positions they would occupy in a precise vertical photograph. The focus of tilt displacement is referred to as the **isocenter**, a point falling between the principal point and nadir (Figure 4-6). The location of the iso-

center can also be viewed as occurring at the "hinge" formed by the tilted airphoto and the plane of a hypothetical vertical airphoto (Figure 4-6).

Features are displaced radially toward the isocenter on the upper side of a tilted airphoto and radially outward or away from the isocenter on the lower side. Along the axis of tilt, there is no displacement relative to an equivalent untilted airphoto.

The exact angle and direction of tilt are rarely known to the interpreter, and precise location of the isocenter is therefore a tedious process. Furthermore, the presence of small amounts of tilt often goes undetected. Because only the central portions of most contact prints are used for interpretation and measurement work, photographic tilt amounting to less than 2° or 3° can usually be ignored without serious consequences. In such cases, it is assumed that the isocenter coincides with the easily located principal point.

Computing Heights Using Object Displacement

The exaggerated displacement of tall objects pictured near the edges of large-scale vertical photographs sometimes permits accurate measurement of object heights on *single* prints. This specialized technique of height evaluation is feasible provided:

Figure 4-5 (Lower left) Effect of topographic relief on image displacement. (Adapted from a U.S. Geological Survey drawing.)

Figure 4-6 (Lower right) Generalized diagram of photographic tilt. (Adapted from Campbell 1987 and Paine 1981.)

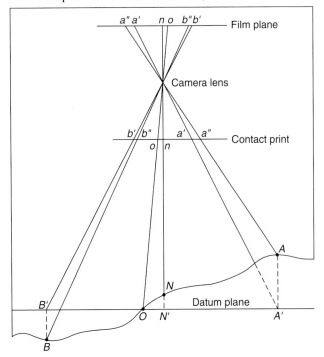

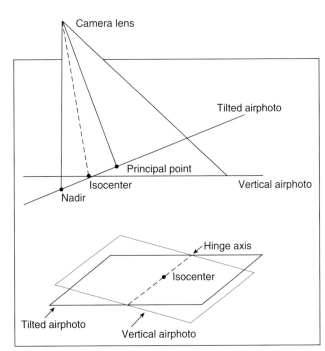

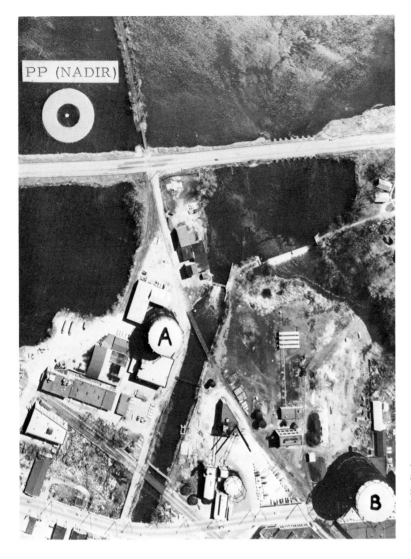

PP (NADIR)

A

B

Figure 4-7 Industrial area, pictured at a scale of about 1:6,000. The tanks marked *A* and *B* are the same height; tank *B* shows more displacement because it is farther from the nadir. (Courtesy Abrams Aerial Survey Corp.)

1. The principal point and the nadir are accepted to be at the same position (i.e., the photograph is vertical).

2. The flight altitude above the base of the object can be precisely determined. If the flying height is unknown, it can be found by multiplying the RF denominator (RF_d) of the photo's scale by the camera's focal length (f):

$$H = (RF_d)(f). \qquad (4\text{-}4)$$

3. Both the top and base of the object are clearly visible.

4. The degree of image displacement is great enough to be accurately measured with available equipment (e.g., an engineer's scale).

When all these conditions can be met, object heights (h) may be determined by the relationship

$$h = \frac{d}{r}(H), \qquad (4\text{-}5)$$

where: d = length of the displaced object from base to top,

r = radial distance from the nadir to the top of the displaced object, and

H = aircraft flying height above the base of the object.

Measurements of d and r must be in the same units; H is expressed in the units desired for the height of the object. For example, if it is assumed that the photograph in Figure 4-7 was taken from an altitude of 914 m, the tank heights are computed as follows:

Tank A: $h = \dfrac{4.5 \text{ mm}}{59.5 \text{ mm}} (914 \text{ m}) = 69 \text{ m},$ and

Tank B: $h = \dfrac{9.5 \text{ mm}}{127 \text{ mm}} (914 \text{ m}) = 68 \text{ m}.$

Thus, both tanks are about the same height. The accuracy of height determinations by this technique is dependent upon vertical (nontilted) photographs, precise values for flying heights, and very careful measurement techniques. If any of these factors is open to question, resulting heights must be regarded only as approximations. In the calculation for tank B, for example, a measurement of 9.0 mm instead of 9.5 mm for the length of the displaced object would change the resulting object height from 68 m to 65 m.

Computing Heights Using Stereoscopic Parallax

Sometimes it is not possible to measure an object's height on a single photograph because the base of the object cannot be seen or its radial displacement is too small to measure accurately. The most commonly used height-measuring technique utilizes **stereoscopic parallax**. This technique requires stereo photographs of the object to be measured, a parallax-measuring device, a stereoscope, and an engineer's scale or similar device for measuring straight-line distances.

If a nearby object is observed alternately with the left eye and right eye, its location will appear to shift from one position to another. This is easily demonstrated by holding a forefinger out in front of the eyes and comparing its change in position when viewed with the left eye and then with the right eye. The apparent displacement, caused by a change in the point of observation, is known as **parallax**. Parallax is a normal characteristic of overlapping aerial photographs, and it forms the basis for three-dimensional viewing. The apparent elevation of an object is due to differences in its image displacement on adjacent photographs.

Two measures of parallax must be obtained when object heights are being determined from stereopairs. **Absolute stereoscopic parallax** (x-parallax) is the sum of the distances between the base of conjugate objects and their respective nadirs. It is always measured *parallel* to the flight line. **Differential parallax** is the difference in the absolute stereoscopic parallax at the top and the base of the object being measured in the stereopair. The basic formula for determining object heights (h) from parallax measurements is

$$h = (H) \frac{dP}{P + dP}, \qquad (4\text{-}6)$$

where: H = aircraft flying height above the base of the object,
P = absolute stereoscopic parallax at the base of the object being measured, and
dP = differential parallax.

The height of the aircraft (H) should be expressed in the units desired for the object height. This is usually in meters or feet. Once a precise photo scale has been ascertained, the flight altitude can be found using Equation 4-4. Absolute stereoscopic parallax (P) and differential parallax (dP) must be in the same units. Ordinarily, these units are millimeters and hundredths of millimeters or inches and thousandths of inches. Because both metric and English instruments may be encountered, the examples that follow illustrate the use of *both* measurement systems in calculations of object heights.

For reasons of convenience and ease of measurement, the **average photo base-length** of a stereopair is commonly substituted as the absolute stereoscopic parallax (P) in the solution of Equation 4-6. The photo base-length is measured as the distance in millimeters or inches from the principal point (PP) to the corresponding conjugate principal point (CPP) on *each* of the photos in the stereopair (Figure 3-26). The mean of the two measurements yields P.

Differential parallax (dP) is measured stereoscopically with a device called a **stereometer**, which incorporates the "floating-mark" principle; the two types of stereometers are the parallax wedge, or ladder, and the parallax bar (described in later sections). The concept of differential parallax can best be illustrated by direct scale measurement of heavily displaced objects, and the stereopair of the Washington Monument supplies an ideal example (Figure 4-8). The nominal photo scale of 1:4,800 is first corrected to an exact scale of 1:4,600 at the base of the monument. Because a 12-in. camera focal length was used, the flying height above the ground (H) is 4,600 ft (Equation 4-4).

Average photo base-length for the stereopair is 4.4 in. and is recorded as the absolute stereoscopic parallax (P). The distance at the base and top of the monument is measured parallel to the line of flight with an engineer's scale to determine the differential parallax of the displaced images; the difference (2.06 − 1.46 in.) is dP (0.6 in.). Because the monument has the shape of an obelisk, measurements were made at the midpoint of the base and vertically above this position at the pyramidal top. Substituting the foregoing values into Equation 4-6, we have

$$h = (4{,}600 \text{ ft}) \frac{0.6 \text{ in.}}{4.4 \text{ in.} + 0.6 \text{ in.}} = 552 \text{ ft.}$$

This is an unusually precise estimate because the actual height of the monument is 555.5 ft (about 169 m). Had the nominal scale of photography (1:4,800) been used instead of the corrected scale, the height would have been computed as 576 ft, an error of 21 ft. Errors of similar magnitude would result from inaccurate parallax measurements. Thus, the necessity for precision can hardly be overemphasized.

A diagrammetric explanation of differential parallax is shown in Figure 4-9. If a flying height of 3,000 m and an average photo base-length of 80.5 mm are assumed for this illustration, the tree height is computed as

$$h = (3{,}000 \text{ m}) \frac{0.52 \text{ mm}}{80.50 \text{ mm} + 0.52 \text{ mm}} = 19.25 \text{ m.}$$

Functions of Stereometers

The interpreter must recognize that the degree of stereoscopic parallax encountered on aerial photographs is often much less than that illustrated by Figures 4-8 and 4-9. Therefore, differential parallax is usually measured under the stereoscope with a stereometer because a precise determination by direct measurement is virtually impossible.

If a small dot is inked at precisely the same location on both prints of a stereopair, the two dots will merge into one when viewed through a stereoscope. Had one pair of dots been placed on level ground and another pair on top of

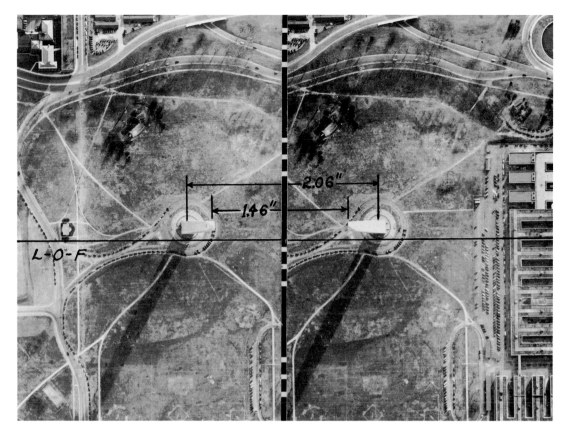

Figure 4-8 Stereogram of the Washington Monument, Washington, D.C. Note displacement of images parallel to line of flight (LOF) and measurements for determination of differential parallax. Scale is 1:4,600 at the base of the monument.

a tree or building, each pair would merge into a single mark; the first fused pair would appear to lie at ground level, whereas the second fused pair would appear to ''float'' in space at the elevation of the object on which the dots were inked. If the horizontal distance between each pair of corresponding dots can be precisely measured, the *algebraic difference* will be a measure of differential parallax. The function of a stereometer is to measure such changes in parallax that are too small to be determined by direct linear scaling.

The Parallax Wedge, or Ladder

The **parallax wedge**, or **ladder**, is the simplest and least expensive device for determining differential parallax (*dP*). Wedges are printed in red or black on transparent plastic or film; most varieties of this device are designed to be used with the lens stereoscope. The wedge may consist of two converging rows of dots, rendering a series of dot pairs with different horizontal separations, or two converging lines with intercepts of specific horizontal separations indicated by tick marks. Graduations on the wedge typically permit readings to hundredths of a millimeter or thousandths of an inch (Figure 4-10).

When properly oriented on a stereopair under the stereoscope, the two converging lines or rows of dots will fuse together for a portion of their length; this fused portion appears to float in space as a sloping line or series of dots. To measure parallax for a feature such as a building, the wedge is moved up or down on the photographs until the sloping line or series of dots appears to intersect or touch the ground at the base of the building. A reading is then taken from the scale alongside the wedge at that exact point. The wedge is again moved up or down until the sloping line or series of

Figure 4-9 Direct measurement of differential parallax for a heavily displaced tree image. Note that base and height measurements are made exactly parallel to the photo base or line of flight. The differential parallax (*dP*) is 0.52 mm.

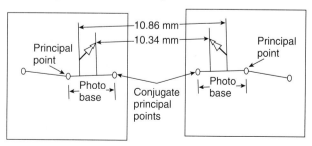

Figure 4-10 Parallax wedge oriented over a stereogram of a large, flat-roofed building. Graduations on the right-hand side indicate the separation of the converging lines to the nearest 0.002 in.

Figure 4-11 Lens and mirror stereoscopes with attached parallax bars for measuring the heights of objects seen in stereopairs. (Courtesy Carl Zeiss, Oberkochen.)

Figure 4-12 Accessory parallax bar for measuring the heights of objects seen in stereopairs with a mirror stereoscope.

dots appears to intersect the top of the building; the scale is then read at this point. The difference between the two readings yields differential parallax.

In Figure 4-10, for example, the difference in parallax (dP) between the ground and the highest roof level of the building was calculated as 0.02 in. Assuming a flight altitude (H) of 5,400 ft and a photo base-length (P) of 1.85 in., the building height is computed as follows:

$$h = (5,400 \text{ ft}) \frac{0.02 \text{ in.}}{1.85 \text{ in.} + 0.02 \text{ in.}} = 58 \text{ ft.}$$

The Parallax Bar

Although more expensive, most photo interpreters prefer the **parallax bar** to the wedge because the floating mark, which can be a cross, dot, or open circle, is movable and thus is easier to place at the base and on the tops of objects. Attached and accessory parallax bars have been designed for use with both lens and mirror stereoscopes (Figures 4-11 and 4-12).

The typical parallax bar consists of two glass plates, each with inscribed marks, that are mounted on a steel bar (Figure 4-12). One of the plates contains the "fixed mark" and is fastened to the bar. The second plate contains the "movable mark," which can be moved laterally along the bar by a micrometer screw. This enables the inscribed marks to be centered over conjugate image points in the stereopair; the amount of separation between the conjugate image points is read from a graduated scale on the support bar to the nearest hundredth of a millimeter or thousandth of an inch.

The parallax bar is operated by placing it under or attaching it to the legs of a stereoscope, which is positioned over a properly oriented stereopair. Both the stereoscope and parallax bar should be positioned so that their long axes are parallel to the line of flight. When viewed through the stereoscope, the two marks on the bar's glass plates will appear as one, which, by adjustment of the micrometer screw, can be made to float up or down within the stereoscopic model. Using this screw, the fused mark is floated downward until it appears to rest at the base of the feature whose height is to be determined; a reading is then made from the graduated scale on the support bar to determine the distance separating the two marks. The same procedure is repeated for the top of the feature. The difference in readings for the top and bottom of the feature is the measure of differential parallax.

Precision of Height Determinations

Precision in measurements of object heights with stereometers depends on a number of factors, not the least of which is the individual's ability to detect and measure small parallax differences. It is also apparent that measurement precision will be improved when objects are clearly depicted on nontilted photographs.

In practice, it is recommended that several parallax readings be taken for both the lower and upper points of the feature being measured and then averaged. This will avoid single measurements of potentially high variability. The photographs should also be securely fastened, since a slip of either photo between parallax measurements can cause highly inaccurate height readings.

Computing Heights Using Shadow Length

It has been previously shown that the heights of objects pictured on aerial photographs can be determined from image displacements on single prints or by measurement of parallax differences on stereopairs. Under special conditions, heights may also be computed from measurements of shadow lengths. First, objects must be vertical (i.e., perpendicular to the local surface); and second, shadows must fall on open, level ground, where they are undistorted and easily measured. Other things being equal, the precision of shadow-height measurements is highly dependent on the scale of the photographs; the precision improves with larger scales because more accurate shadow measurements can be made.

Because of the great distance between the sun and the earth, the rays of the sun are essentially parallel throughout the area shown on a vertical aerial photograph. Thus, at any given moment the lengths of object shadows will be directly proportional to object heights. Two methods are commonly used to compute object heights from shadow lengths.

In the first, an object of *known* height and its shadow are used in a proportional equation to determine the height of other objects in the photo based upon their shadow lengths:

$$h_1 = \frac{(h_2)(s_1)}{s_2}, \qquad (4\text{-}7)$$

where: h_1 = height of object 1,
s_1 = shadow length of object 1 on photo,
h_2 = height of object 2, and
s_2 = shadow length of object 2 on photo.

For example, if a radio antenna is known to be 125 m tall and its shadow length on the photo is 1.68 cm, how tall is a smokestack if its shadow length is 1.25 cm?

$$h_1 = \frac{(125 \text{ m})(1.25 \text{ cm})}{1.68 \text{ cm}}$$

$$= \frac{156.25}{1.68}$$

$$= 93 \text{ m}.$$

A second method is used when there is no object of a known height in the photograph. To calculate object height using this method, it is necessary to know the length of the shadow cast by the object and the sun's **elevation angle** (i.e., angle measured between the horizontal plane at the surface and the sun) at the time the photograph was taken. The elevation angle (a) is shown in Figure 4-13. The tangent of this angle (tan a) multiplied by shadow length (s) provides a measure of object height:

$$h = (\tan a)(s). \qquad (4\text{-}8)$$

The actual ground length of the shadow is established by multiplying the shadow length on the photo by the photoscale denominator. The shadow length should be expressed in the unit desired for the object height.

Determining the sun's elevation angle requires the use of astronomical tables (i.e., a solar ephemeris), such as those published by the U.S. government in the *American Ephemeris and Nautical Almanac*, and knowledge of the month, day, and time (nearest hour) of the exposure plus the latitude and longitude of the photographed area. The elevation angle (a) is determined by the equation

$$\sin a = (\cos x)(\cos y)(\cos z) \pm \\ (\sin x)(\sin y), \qquad (4\text{-}9)$$

where: angle x = sun's declination or latitude on the day of the photography, corrected to Greenwich Mean Time (GMT) and read from the ephemeris;
angle y = latitude of the photography; and
angle z = hour angle, or the difference in degrees of longitude between local noon and the locality of the photograph. For example, if the exposure was made 2 hr after local noon, the angle is 30° (sun "moves" 15° of longitude per hour times 2 hr = 30°).*

*To avoid confusion, the abbreviation hr (rather than h) is used in this textbook.

Figure 4-13 The tangent of angle *a* is the relationship of the height of a vertical object (opposite side) divided by the length of its shadow (adjacent side).

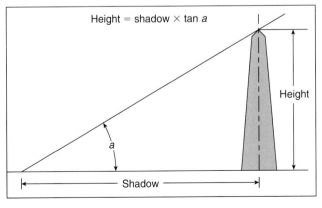

The algebraic sign in Equation 4-9 is *plus* from March 21 through September 23 and *minus* from September 24 through March 20 in the Northern Hemisphere; the signs are reversed for the Southern Hemisphere. For most scientific calculators, once the sine of angle *a* has been determined, the key sequence $\boxed{\text{INV}}$ $\boxed{\text{sin}}$ calculates the angle whose sine is in the display (e.g., $\boxed{\text{INV}}$ $\boxed{\text{sin}}$ 0.5 = 30°, which is equivalent to arcsin 0.5 = 30° or $\sin^{-1}$ = 30°).

Parry and Gold (1972) have prepared a shadow-height nomogram that provides a means of rapidly obtaining the sun's elevation angle by using either the date and time of film exposure or the date and azimuth of the object shadow. With practice, the solution for the elevation angle can be obtained in about 2 min.

Compass Bearings

Vertical photographs present reliable records of angles at or near the principal point and when the terrain is reasonably flat. Therefore, **compass bearings** or **azimuths** can be measured directly on most photographs with a simple protractor (Figure 4-14). Flight lines are usually run north-south or east-west, but the edges of prints are rarely oriented exactly with the cardinal directions. For this reason, a line of true direction must be established before bearings can be accurately measured. Such reference lines can be plotted from existing maps or located directly on the ground using a compass to determine the bearing of any linear feature.

In Figure 4-15, the highway at the top of the photo was ascertained to be a due north-south reference line. To determine the bearing of the buried pipeline (in the direction

Figure 4-14 Relationship of compass bearings and azimuths. (Courtesy U.S. Army.)

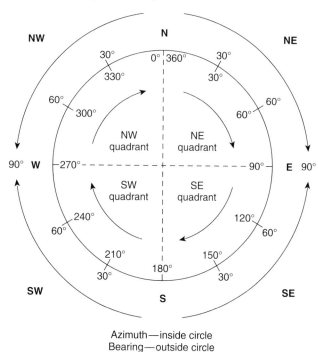

Azimuth—inside circle
Bearing—outside circle

of the arrow), the included angle was measured with a protractor as 29°. Thus, the pipeline bears 29° east of due south. Expressed more conventionally, it has a bearing of S 29° E or an azimuth of 180° − 29° = 151°.

A quick indication of compass direction can also be obtained from large-scale photographs depicting airports. Runways are numbered according to their magnetic-compass direction (azimuth); the compass heading is determined to the nearest 10°, and the zero is dropped. Thus, in Figure 4-16, the runways have magnetic azimuths of 160°, 220°, and 280°. At their opposite ends, these same runways would carry the numbers 34 (340°), 04 (40°), and 10 (100°), respectively. Where the local magnetic declination is known, lines of true compass direction may be approximated from runway numbers.

Area Measurements

Reasonably accurate **area measurements** can be made directly from vertical airphotos when the terrain is level to gently rolling. The reliability of such measurements is dependent upon the precision with which photo scales and area conversion factors are determined. Where topographic changes exceed 5 percent of the flying height (e.g., 100 m if H = 2,000 m or less), large measurement errors will be incurred unless new conversions are computed for each significant variation in land elevation.

For features with regular shapes, such as rectangles, squares, and circles, area determinations can be obtained by the sequence of (1) making linear photo measurements, (2) converting these to ground distances as a function of photo scale, and (3) converting ground distances to ground *areas* (rectangle = lw, square = s^2, circle = $\pi r^2 = \pi d^2/4$). The principal devices used for the area measurement of irregular-shaped features are dot grids and polar planimeters.

Dot Grids

Dot grids are transparent overlays with dots uniformly spaced on a grid pattern (Figure 4-17). The dots *represent* the centers of small, imaginary squares within a given grid cell. Thus, the dots are counted in lieu of the squares themselves. The principal advantage is that squares falling along boundaries are less troublesome because the nondimensional dot determines whether or not the square is to be tallied.

It is extremely important to determine accurately how much area each dot represents. For example, a photo scale of 1:20,000 can be written as 1 cm = 200 m, which is equivalent to 1 cm^2 = 40,000 m^2, or 1 cm^2 = 4 hectares (ha) (1 ha = 10,000 m^2). Thus, on a grid having 4 dots/cm^2, each dot represents 0.5 ha. The simple conversion is represented by the relationship

$$\text{ha/dot} = \frac{\text{number of ha/cm}^2}{\text{number of dots/cm}^2}. \qquad (4\text{-}10)$$

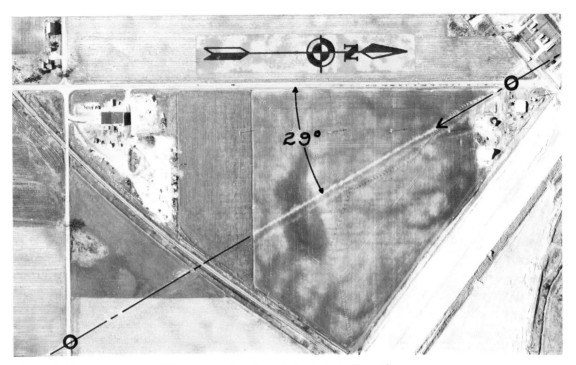

Figure 4-15 Measurement of the compass bearing of a buried pipeline. (Courtesy Abrams Aerial Survey Corp.)

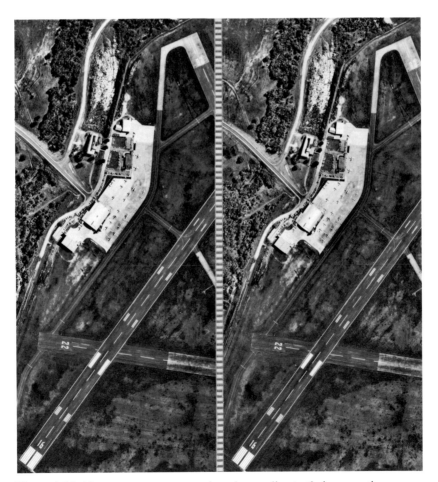

Figure 4-16 Airport runways are numbered according to their magnetic-compass direction. Shown in this stereogram are runways 16 (160°), 22 (220°), and 28 (280°). Photo scale is about 1:10,000. (Courtesy Texas Forest Service.)

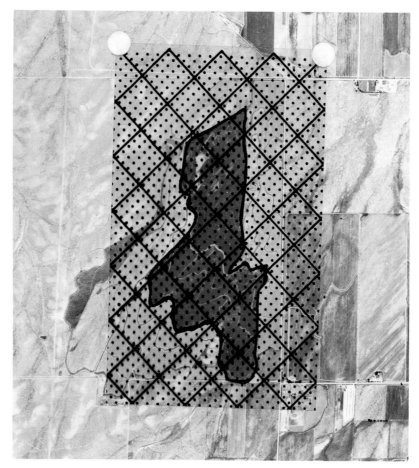

Figure 4-17 Dot grid oriented over a vertical airphoto (scale = 1:20,000). Each square in the grid is 1 cm on a side and contains 25 dots.

The number of dots to be counted depends on the grid intensity, the photo scale, the size of the areas to be measured, and the desired precision. Grids in common use may have from 4 to more than 25 dots per square centimeter. For tracts of 500 to 1,000 ha, it is generally desirable to use a dot-sampling intensity that will result in a conversion of about 0.2 to 1.0 ha/dot.

To measure the area of a tract on a vertical airphoto, the following procedure is recommended:

1. Place the dot grid over the photo by a random drop; *do not* align the grid for a "best fit" with the boundary of the area to be measured. Tape the grid in place so it cannot move.

2. Count all the grid cells that *fall completely within the area being measured.* Convert to total dots and record the number (e.g., 4 cells at 25 dots/cell = 100 dots).

3. For the remaining grid cells, count all the dots that *fall completely within* the area being measured; record this number.

4. Count all the dots that *touch the boundary* of the area being measured. *Divide* by 2 and record the number.

5. *Add* dot counts from steps 2, 3, and 4. This cumulative number of dots represents the total area being mea-

sured. *Multiply* the cumulative count by the area represented by one dot to arrive at the area measurement.

Greater precision can usually be attained by derivation of an *average* dot count based upon several random drops of the grid over a given tract. Also note that thick boundary lines, low-density dot patterns, or large dots will cause less precision.

Polar Planimeters

A standard **polar planimeter** is composed of three basic parts: (1) a weighted polar arm, (2) a tracer arm containing a magnifying tracer lens and tracer point that slides over the photo, and (3) a rolling wheel to which is attached a vernier scale (Figure 4-18). Polar planimeters can read directly in square inches, square centimeters, or in several different scales. The newest planimeters are digital, incorporating built-in microprocessors and LCD digital displays.

In use, the tracer point is run around the boundary of an area in a *clockwise* direction. Usually, the perimeter of a given tract is traced two or three times to obtain an average reading. For planimeters that measure directly in square cen-

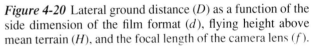

Figure 4-18 *(Left)* Polar planimeter, an instrument used for measuring areas on vertical airphotos.

Figure 4-19 *(Above)* Opisometer, an instrument used for measuring the lengths of curved or irregular features on vertical airphotos.

timeters or square inches, it is an easy task to convert areas to any scale by simply multiplying the measured area by a **scaling factor**. This factor is determined as follows. Suppose the photo scale is 1:10,000 and the result is required in square meters. The scale can be written as 1 cm = 100 m, so 1 cm^2 represents 100 × 100 = 10,000 m^2, or a scaling factor of 10,000. Thus, if the measured area is 14.2 cm^2, the area represented is 14.2 × 10,000 = 142,000 m^2.

It is also a simple procedure to convert the planimeter reading directly into special units of area, such as hectares or acres. For example, if the photo scale is 1 cm = 50 m (1:5,000) and there are 10,000 m^2 in 1 ha, the area of 1 cm^2 on the photo is 50 × 50/10,000 = 0.25 ha (the scaling factor). Thus, if the measured area on the photograph is 6.8 cm^2, the area represented is 6.8 × 0.25 = 1.7 ha.

In the hands of skilled people, dot grids and polar planimeters are generally regarded as devices of comparable precision. However, measurement times are usually less with a polar planimeter than with a dot grid. This has important ramifications if time is limited or if numerous tracts must be measured.

Opisometers

The task of determining distance measurements for linear features depicted on vertical airphotos can be performed easily with an engineer's scale. However, for ascertaining the lengths of curved or irregular features, such as nonlinear transportation routes, shorelines, and drainage systems, **opisometers**, or **map wheels**, are commonly used by an interpreter (Figure 4-19).

An opisometer is composed of a rolling wheel that rests on the photograph and to which is attached a vernier scale. Most models are equipped with a handle for increasing tracing accuracy (Figure 4-19). When a feature has been fully traced, the graduated scale expresses its length in centimeters and/or inches; these are then converted to meters, kilometers, miles, or feet on the basis of photo scale.

Figure 4-20 Lateral ground distance (*D*) as a function of the side dimension of the film format (*d*), flying height above mean terrain (*H*), and the focal length of the camera lens (*f*).

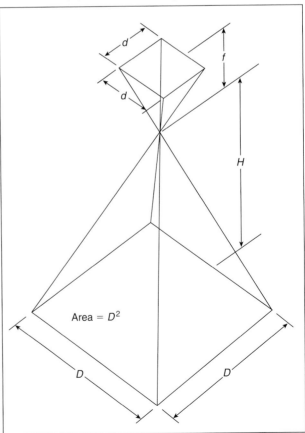

Calculating Ground Distance

The lateral ground distance (D) of a vertical airphoto with a square format (Figure 4-20) can be computed using the following relationship:

$$D = \frac{(d)(H)}{f} \qquad (4\text{-}11)$$

where: d = side dimension of the film format,
H = flying height above mean terrain, and
f = focal length of the camera lens.

Flying height should be expressed in the units desired for the ground distance; d and f must be expressed in the same units. Equation 4-11 shows that lateral ground distance varies directly with format size and flying height and inversely with focal length. Ground area, or coverage, is equal to D^2.

Because of the relationship between H and f in determining photo scale (Equation 4-1), one may also calculate lateral ground distance (D) by using the photo's RF denominator (RF_d) and the side dimension of the film format (d) as follows:

$$D = (RF_d)(d). \qquad (4\text{-}12)$$

Transfer of Photographic Information

For many studies, it is desirable to transfer photographic detail, including interpretative data, to planimetric base maps such as topographic quadrangles. Regardless of the map sheet selected, some enlargement or reduction is ordinarily required before photographic information can be transferred because differences between photo and map scales must be reconciled. This can be accomplished *optically* by the use of special reflection and projection instruments.

The **sketchmaster** is one of the more common reflection instruments used for transferring information from single vertical aerial photographs to planimetric maps. **Vertical** and **horizontal sketchmasters** are shown in Figures 4-21 and 4-22. The instruments are relatively inexpensive, lightweight, and portable.

The vertical sketchmaster employs two glass mirrors that enable the observer to view a photographic image superimposed on a base map (Figure 4-21). The large mirror above the photograph is full-silvered (opaque) and front-surfaced to avoid refraction due to the thickness of the glass. This mirror reflects the photo image into an upright position to a smaller eyepiece mirror mounted at the front of the instrument. Because the eyepiece mirror is half-silvered (semitransparent), it *reflects* the photo image and *transmits* the map image to its outer surface, enabling the observer to

Figure 4-21 Diagram of a vertical sketchmaster. A half-silvered eyepiece mirror enables the operator to view a photograph and map simultaneously.

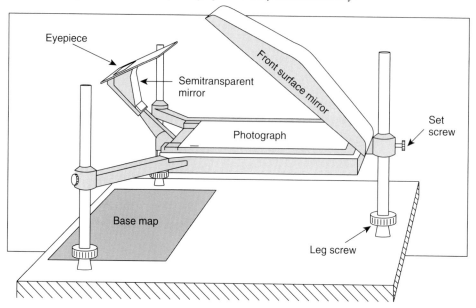

Figure 4-22 (Left) Luz Aero-Sketchmaster with vertical airphoto held in position by magnets. This horizontal sketchmaster operates on the same basic principle as the device pictured in Figure 4-21. (Courtesy Carl Zeiss, Oberkochen.)

Figure 4-23 (Below) Stereo Zoom Transfer Scope™ with Vertical Measurement Module. (Courtesy Ronald J. Martino, Cambridge Instruments, Inc.)

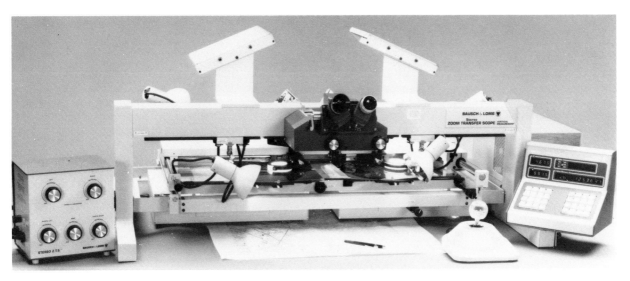

view both images simultaneously and to trace details from the photograph to the map. This operation is known as the **camera lucida principle**.

The vertical sketchmaster may be raised or lowered by adjustment of its three legs until some number of photo image points match those of the map. The instrument can also be tilted by means of leg screws to correct for small amounts of photographic tilt or radial displacement due to relief variation. The photo-map scale ratio can be varied from approximately 1:0.3 to 1:2.7. Interchangeable lenses of various powers are inserted beneath the eyepiece mirror when the instrument is used to transfer detail at a ratio other than about 1:1. These lenses bring the photo plane into focus with the map plane and eliminate parallax, or the apparent movement of the map with respect to the photograph when the operator's eye is moved.

The horizontal sketchmaster employs a double prism with reflecting and transmitting surfaces (Figure 4-22). The operator views a superimposed image of the photograph and map at the 45°-inclined eyepiece. Both the prism holder and photocarrier can be moved to change the viewing scale. The photocarrier is mounted in a ball joint, allowing it also to be tilted and rotated to facilitate fitting the photograph to the map. The photo-map scale ratio can be varied from about 1:0.4 to 1:2.8 for the unit shown in Figure 4-22. As with the vertical sketchmaster, auxiliary lenses are used when the scale ratio is below or above 1:1. A reflection-type lamp is used to illuminate the photo and map; it can be moved horizontally or vertically to provide optimum illumination.

Stereo zoom transfer instruments utilize stereopairs of vertical photographs to superimpose a three-dimensional image on a base map (Figure 4-23). Through zoom magnification, this type of device can accommodate a wide disparity of map and photo scales. The instrument shown in Figure 4-23 is also equipped with a height-measurement module that permits object height and terrain elevation measurements to be made from either transparent or opaque stereopairs.

Figure 4-24 Krones LZK rear-projection device for use with transparent film. This instrument provides variable scale magnification up to 72×. (Courtesy Krones, Inc.)

Reflecting projectors optically transfer images from single photographs onto a map surface. With one type of reflecting projector, a photo is placed in a holder near the top of the instrument so that its surface faces a mirror. The photo is illuminated by artificial light and the light rays carrying the images are reflected from the mirror through a lens and projected onto the map. Another type transfers the photo image to the back side of a frosted glass surface (Figure 4-24). With this method of projection, the map must be at least translucent in order for the photo information to be transferred. Reflecting projectors offer a greater range of scale adjustments than sketchmasters. However, they are not portable and must ordinarily be used in a dark or semidarkened room.

Compilation of Topographic Maps with Stereoplotters

At the turn of the century, the compilation of topographic maps was largely dependent on field surveys. Such maps are now produced by photogrammetric methods, and fieldwork is limited to obtaining a network of horizontal and vertical ground control required for accurate stereoplotting. **Ground control points** (also called **geodetic control points**) are carefully located positions that show longitude and latitude and/or elevation above sea level. The entire United States is covered with a network of more than 1 million of these precisely known points.

Horizontal control is needed for correct scale, position, and orientation of the map to be maintained. For this purpose, the grid coordinates of many points within the area to be mapped must be determined by field surveys. Similarly, vertical control is needed for the correct location of contours. Therefore, elevations of numerous points must also be determined in the field.

Ground control points become the framework on which map detail is assembled. This framework determines the accuracy with which the positions and elevations of map features may be shown and makes it possible to join maps of abutting quadrangles without a break in the continuity of map detail. The ground control points are usually marked by **bronze markers**, or **tablets** (**survey monuments**), set in rock or masonry and are shown on maps by appropriate symbols. Some markers serve for both horizontal and vertical control; markers that show elevation are called **bench marks**.

Stereoplotters are precision instruments that create exact stereo models of the terrain from overlapping photographs. The photos are ordinarily in the form of glass or highly stable film diapositives (usually 23 × 23 cm) because dimensional stability is very important. The original film is normally panchromatic. The use of stereoplotters is mandatory for the preparation of topographic maps that must meet strict standards for horizontal and vertical accuracy. National standards for the accuracy of topographic maps were adopted by the U.S. government mapping agencies in 1941, and maps that meet these standards carry a statement to that effect in the lower margin.

There are three types of stereoplotters that, regardless of type, enable the operator to view the stereo model in three dimensions. In order of increasing complexity, cost, accuracy, and flexibility, they are (1) **optical stereoplotters**, where the stereo model is formed by direct optical projection; (2) **mechanical stereoplotters**, where the stereo model is formed by mechanical projection, which simulates the optical projection of light rays by means of two metal space rods; and (3) **analytical stereoplotters**, where the stereo model is formed by a mathematical solution in real time. A detailed technical discussion of stereoplotters is presented in the *Manual of Photogrammetry* (Slama 1980).

An optical stereoplotter is shown in Figure 4-25. It consists of three basic components: (1) a dual-projection system for creating the stereo model, (2) a viewing system that enables the operator to see the model stereoscopically, and (3) a system for measuring elevations within the model and for tracing features and contour lines onto a map sheet called a **manuscript**.

The glass diapositives are oriented in the projectors so that they occupy the same relative positions in space as the original film frames. To see the stereo model, the operator's eyes must view each image separately. With the **anaglyph system**, for example, each projector has a colored filter—one red, the other blue. In this way, two different images of the overlap area are projected onto the tracing table in a different color (Figure 4-25). The operator then views the model through glasses with one red lens and one blue lens, which give the three-dimensional effect.

The main component of the measuring and tracing system is the **tracing table**, or **platen** (Figure 4-25). In the center of the platen's white round disk is a small hole that transmits a tiny light beam. This beam of light appears to be a floating dot or point. For determining elevations anywhere within the model, the platen is lowered or raised by turning

Figure 4-25 Schematic representation of a three-dimensional terrain model provided by an optical stereoplotter. (Courtesy TRB Associates, Inc.)

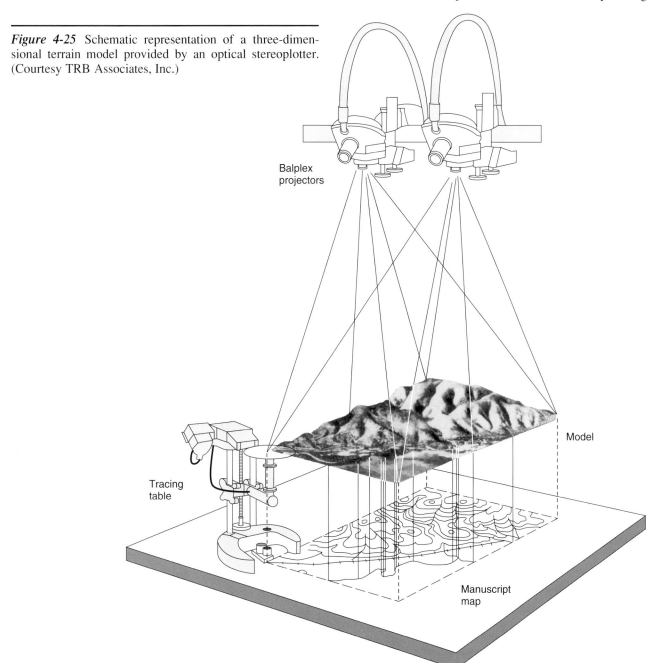

Balplex projectors

Model

Tracing table

Manuscript map

Figure 4-26 The technique of scribing has largely replaced conventional inked drawings in the preparation of finished topographic maps. The polyester film base (scribe-coat) is dimensionally stable and provides lines of high uniformity and sharpness. Maps are reproduced photographically from the scribed manuscript. (Courtesy Abrams Aerial Survey Corp.)

a screw until the dot appears to touch the ground. The screw is attached to a system of gears, which in turn is connected to a dial from which the vertical elevation is read directly.

When drawing contours, the floating dot is located on the ground at the desired elevation. An attached pencil is lowered from the platen, and the operator traces a contour onto the map sheet by moving the entire tracing table over the map while keeping the floating dot on the ground within the stereo model. Natural and cultural features are also added to the map by moving the tracing table while manipulating the floating dot to maintain an apparent contact with the feature being drawn.

When all detail has been transferred from a given stereo model, one diapositive is replaced, and an adjacent model is correctly oriented and tied into the manuscript by a process known as **bridging**. The completed map manuscript is checked for errors and omissions. Detail is then usually traced onto a polyester film base by a technique known as **scribing** (Figure 4-26). Separate scribe sheets and negatives are made for each item to be printed in a different color on the finished map.

In recent years, computer-driven analytical stereoplotters have been introduced to facilitate the bridging process on less powerful plotters (Figure 4-27). With operator input, analytical stereoplotters develop a mathematical stereo model in real time. With no optical or mechanical limitations, they are capable of handling a wide variety of photography, including vertical, tilted, and oblique photographs. In addition, they can correct for errors caused by lens distortion, atmospheric refraction, and earth curvature (Wolf 1983). Output from this type of stereoplotter may be a map generated with an attached *X-Y* plotter or calculated digital elevation data that are recorded on magnetic tape for bridging with other plotters (Figure 4-27).

Orthophotography

An **orthophotograph**, or **orthophoto**, is a reconstructed airphoto that shows natural and cultural features in their true planimetric positions. Geometric distortions and relief displacements are removed from the standard perspective photographs by an **orthophotoscope**, a specialized orthoprojection instrument. One popular type of orthophotoscope scans a stereo model and rectifies the photographic image along individual and contiguous scan lines. Along each line, the operator continuously keeps a moving slit on the ground of the stereo model. The slit acts in the same manner as the floating dot in an optical stereoplotter. The rectified image for each scan line is then transferred optically to film, which becomes the **orthonegative**. Rectification of the original photograph's central projection into an orthographic projection permits the ''corrected'' photo to be used as a planimetric map (Figure 4-28).

The U.S. Geological Survey uses orthophotographs to produce **orthophotoquads**. These **photomaps** are normally produced from single orthophotos that are prepared in a standard quadrangle format with the same positional and scale accuracy as topographic maps. Most orthophotoquads contain selective place names and boundary information, but not contours and elevation data. With their abundance of detail, orthophotoquads are valuable as map substitutes in unmapped areas and as valuable complements to existing maps, especially when the maps are out of date. A portion of the El Mirage, Arizona, 7.5-minute orthophotoquad and its matching topographic map are presented in Figure 4-29.

Figure 4-27 (Above) Computer-driven analytical stereoplotter used by the U.S. Geological Survey at its Flagstaff, Arizona, Field Center. (Courtesy Sherman S. C. Wu.)

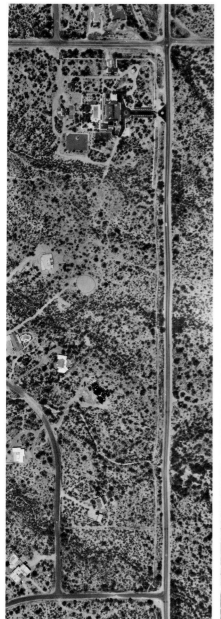
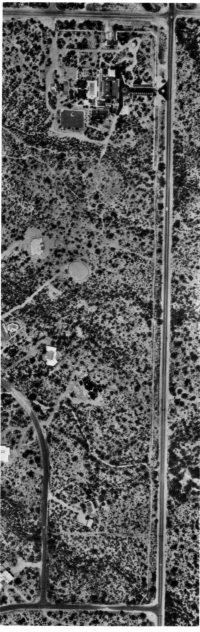

Figure 4-28 (Left) Portion of a perspective airphoto (*left*) and an orthophoto (*right*) showing a road traversing undulating terrain. Note that (1) the crookedness of the road in the perspective photo is eliminated in the orthophoto, and (2) scale is consistent only in the orthophoto. (Courtesy Lyle Slater, The Orthoshop—Tucson.)

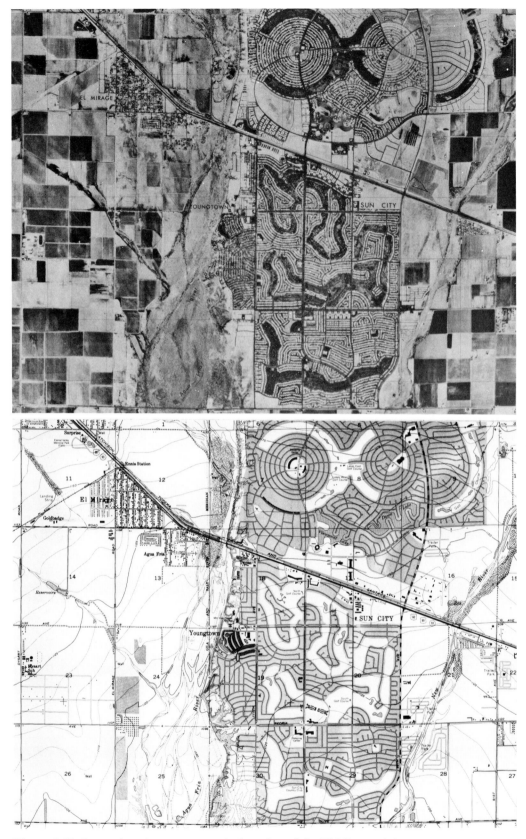

Figure 4-29 Comparison between a matched portion of the El Mirage, Arizona, 7.5-min orthophotoquad (*top*) and topographic map (*bottom*). Both are shown at the original scale of 1:24,000.

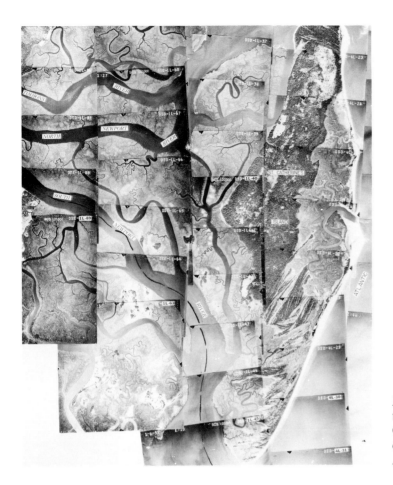

Figure 4-30 Portion of a photo index sheet covering tidal marshes and Saint Catherines Island off the Georgia coast. (Courtesy U.S. Department of Agriculture.)

Airphoto Mosaics

An **airphoto mosaic** is an assemblage of two or more overlapping aerial photographs that form a composite view of the total area covered by the individual photographs. Although several categories of mosaics are recognized, most of them can be grouped into three general categories: (1) **orthophoto mosaics**, (2) **controlled mosaics**, and (3) **uncontrolled mosaics**. A major advantage of all mosaics is that they provide an overview of the landscape where the continuity of features is maintained.

Orthophoto mosaics are the most highly controlled mosaics in terms of scale and positional accuracy. They are made from orthophotographs in which scale variations and displacements due to tilt and relief have been removed. When the orthophotos are properly fitted, cut, and mounted in a continuous representation of the terrain, they represent an **orthophoto map**. This means that precise measurements of distance and area can be made directly on the mosaic because the photos were assembled to match a network of ground control points.

Controlled mosaics rank behind orthophoto mosaics in terms of geometric accuracy. They are assembled by cutting and fitting together the *central* portions of vertical airphotos to minimize radial displacements and to closely match ground control. When the terrain is flat or slightly undulating, the controlled mosaic can be nearly as accurate as a planimetric map at the same scale. However, accuracy decreases in direct proportion to the irregularity of ground relief. Consequently, measurements in mosaic areas depicting rugged terrain should be avoided.

Uncontrolled mosaics are assembled without ground control. Photographs are simply fitted to match pictorially without concern for scale variation or positional relationships. Consequently, uncontrolled mosaics cannot be used for measuring distances or areas. In some instances, only the central area of each photograph is used. Detail is matched with adjacent center areas and the cut assemblage is pasted to a stable base to form the mosaic. In other instances, trimmed photographs are simply overlapped by matching details and secured with transparent tape for temporary use. If a mosaic is to be used for some length of time, it can be attached to a hardboard base (e.g., posterboard, wallboard, plywood) with staples or double-backed tape. Meyer (1962) describes a step-by-step procedure for assembling uncontrolled mosaics.

A special type of uncontrolled mosaic is the **photo index sheet** (also called an **aerial index**, or **index mosaic**). With it, trimmed photographs are placed in overlapping positions in their correct exposure sequence along each flight line and aligned at their sidelap edges; the titling is visible on all or most of the individual frames. Once assembled, the mosaic is photographed and printed at a reduced scale (Figure 4-30). Photo index sheets are normally purchased from private or public organizations that sell aerial photographs. They enable users to quickly identify the photographs that cover their study area, and these are then ordered at their original scale.

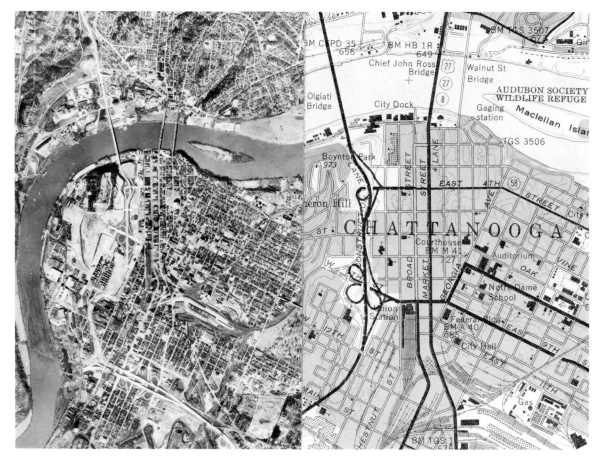

Figure 4-31 Portion of a vertical photograph (*left*) and a topographic map (*right*) of Chattanooga, Tennessee; map scale is 1:24,000. (Courtesy Tennessee Valley Authority.)

Questions

1. Which photo scale is largest, 1:10,000 or 1:30,000? Describe the difference between the scales in terms of areal coverage and the level of displayed detail.
2. In what ways is an orthophotograph superior to a standard airphoto and a planimetric map?
3. Calculate the scale (RF) of a vertical airphoto using the following information: focal length = 305 mm, flying height above sea level = 6,200 m, and average ground elevation = 4,600 m.
4. A road segment shown on an aerial photograph can be located on a 1:24,000-scale topographic map. If the measured distance is 47.6 mm on the map and 94.2 mm on the photograph, what is the scale (RF) of the photograph?
5. Determine the scale (RF) of a vertical airphoto if the photo distance and ground distance between two buildings are 38 mm and 1,250 m, respectively.
6. Determine the area (hectares) of the circular irrigated field shown in Figure 3-3 that looks like a bull's-eye target. Photo scale is 1:80,000.
7. Determine the scale (RF) of the vertical airphoto shown in Figure 4-31; map scale is 1:24,000.

Bibliography and Suggested Readings

Barrett, J. P., and J. S. Philbrook. 1970. Dot Grid Area Estimates: Precision by Repeated Trials. *Journal of Forestry* 68:149–51.

Burnside, C. D. 1985. *Mapping from Aerial Photographs*. London: Collins.

Campbell, J. B. 1987. *Introduction to Remote Sensing*. New York: The Guilford Press.

Landen, D. 1974. Progress in Orthophotography. *Photogrammetric Engineering* 40:265–70.

Lattman, L. H., and R. G. Ray. 1965. *Aerial Photographs in Field Geology*. New York: Holt, Rinehart, and Winston.

Lillesand, T. M., and R. W. Kiefer. 1987. *Remote Sensing and Image Interpretation*, 2d ed. New York: John Wiley & Sons.

Meyer, D. 1962. Mosaics You Can Make. *Photogrammetric Engineering* 28:167–71.

Moffitt, F. H., and E. M. Mikhail. 1980. *Photogrammetry*. New York: Harper and Row.

Paine, D. P. 1981. *Aerial Photography and Image Interpretation for Resource Management*. New York: John Wiley & Sons.

Parry, J. T., and C. M. Gold. 1972. Solar-Altitude Nomogram. *Photogrammetric Engineering* 38:891–99.

Rabenhorst, T. D., and P. D. McDermott. 1989. *Applied Cartography, Introduction to Remote Sensing*. Columbus, Ohio: Merrill Publishing Co.

Slama, C. C., ed. 1980. *Manual of Photogrammetry*, 4th ed. Falls Church, Va.: American Society for Photogrammetry and Remote Sensing.

Ulliman, J. J., and C. R. Hatch. 1978. Test of an Electronic Planimeter. *Journal of Forestry* 76:346–47.

Wolf, P. R. 1983. *Elements of Photogrammetry*, 2d ed. New York: McGraw-Hill Book Co.

Chapter 5

Acquisition of Aerial Photographs

graphic mission, the client still may be responsible for defining project objectives, drawing up preliminary specifications or flight plans, estimating costs, and determining whether the finished product meets interpretation and mapping requirements.

In return for an investment that may amount to many thousands of dollars, the responsible party expects to receive high-quality photographs that are uniquely suited to particular needs. To achieve this goal, it is necessary to define the exact project objectives, become familiar with photographic specifications and contracts, and negotiate only with reputable aerial survey companies. Without close attention to these considerations, special-purpose photography is unlikely to prove cost-effective.

Names and addresses of aerial survey companies can be obtained through city and county engineers, advertisements in telephone directories, or remote sensing specialists at local academic and research institutions. Activities and addresses of many of the larger companies are also described annually in the yearbook issue of *Photogrammetric Engineering and Remote Sensing.* In choosing from among several prospective companies, the purchaser is advised to request photographic samples from each; such samples provide useful guides to the quality of work that may be expected.

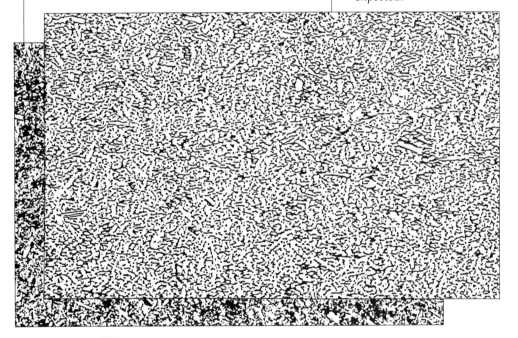

Flight Altitudes and Focal Lengths

For contract aerial photography, the maximum operating ceiling for available aircraft is about 13,700 m (about 45,000 ft) above sea level. This operating altitude can be achieved by the twin-engine "business" jet (Figure 5-1). The lower limit for fixed-wing aircraft can be arbitrarily defined as around 300 m (about 1,000 ft) above mean terrain, depending upon topography, special hazards, and air-safety regulations.

Most aerial survey companies utilize mapping cameras with the standard 23 × 23-cm image format (Figure 2-12). These are usually available with several lens cones to accommodate various focal-length requirements which, along with flying height, determines the scale and ground coverage of vertical aerial photographs. Table 5-1 shows the importance of focal length in determining ground coverage for mapping cameras. Note from this table that there is an inverse relationship between focal length and ground coverage at a given flying height; as focal length increases, ground coverage decreases, and vice versa.

Contract Aerial Photography

Although many people rely largely on existing aerial photographs for interpretation and mapping, such coverage may be unsuitable for certain projects because of age, season, film-filter combination, or scale. As a result, there is considerable interest in the purchase of **special-purpose photography** contracted through commercial aerial survey firms. However, even though a commercial firm may have the technical expertise to handle almost any type of photo-

Figure 5-1 Dual mapping-camera installation in a jet aircraft. (Courtesy Gates Learjet, Inc.)

If we match the shortest focal length with the upper aircraft ceiling and vice versa, the range of photographic scales available (allowing for terrain elevations of up to 3,000 m) might be computed with Equation 4-1 as follows:

$$RF = \frac{1}{13,700 \text{ m} - 3,000 \text{ m}/0.088 \text{ m}}$$

$$= \frac{1}{121,591} \approx 1{:}121{,}600, \text{ and}$$

$$RF = \frac{1}{300 \text{ m}/0.305 \text{ m}} = \frac{1}{983} \approx 1{:}980.$$

In practice, the range of scales actually employed is usually less, namely from 1:1,000 to about 1:50,000. Where simultaneous coverage at two different scales is desired, certain aircraft can be fitted with two mapping cameras employing different focal-length lenses (Figure 5-1).

Focal length is a critical contract specification because it determines the aircraft altitude that must be maintained for exposures of the desired scale to be obtained. Also, its direct effect on the image displacement of objects photographed controls the degree of three-dimensional exaggeration that

the interpreter sees when viewing the exposures stereoscopically. Selecting an optimum scale/focal-length combination is, therefore, an item of paramount importance.

It is generally desirable to specify the smallest photo scale that will meet the requirements of a given study; this approach not only tends to lower costs but also reduces the number of stereo models that must be studied. When a photographic scale is *doubled*, as from 1:20,000 to 1:10,000, *four times* as many photographs are required to cover the same ground area. Consequently, photographic scale and tract size are two of the principal factors affecting the cost of aerial surveys.

Seasonal Considerations

The optimum season for scheduling photographic flights depends upon the nature of the features to be identified or mapped, the film-filter combination to be used, the number of days suitable for aerial photography within a given period, and the minimum sun-angle required. Because

	TABLE 5-1 Ground Coverage Versus Focal-Length Variation[a]	
	Ground Coverage at a Flying Height of 1 km	
Focal Length (mm)	Side Dimension[b] (km)	Area Dimension (km²)
88	2.61	6.83
152	1.51	2.29
210	1.09	1.20
305	0.75	0.57

[a]*For mapping cameras with a 23 × 23-cm image format.*
[b]*Answers derived by use of Equation 4-10.*

Figure 5-2 Summer (leaf on) and winter (leaf off) airphotos for the same ground area in western Pennsylvania.

optimum weather conditions may not prevail during the season when photography is desired, it may be necessary to ascertain the average number of photographic days for a locality during each month of the year. A **photographic day** is defined as one with 10 percent cloud cover or less. This kind of information may be obtained from reports issued by local weather services. Other factors being equal, aerial surveys are likely to be less expensive in areas where sunny, clear days predominate during the desired **photographic season**.

Another consideration in selecting the season for aerial photography is the **project objective** (i.e., the specific type of information to be extracted from the photographs). On the basis of generally divergent objectives, users of aerial photographs may be arbitrarily divided into two main groups—those engaged in topographic mapping, soils mapping, or evaluation of terrain features and those primarily concerned with assessment of vegetation or management of wildland areas. The first group, typified by cartographers, civil engineers, pedologists, and geologists, would be likely to schedule aerial missions during seasons when foliage does not obscure the landscape. By contrast, people in the second category tend to prefer photographs made during the growing season, when the vegetative cover is fully developed. This second group includes foresters, plant ecologists, and range managers.

For topographic mapping, photographs are usually taken either in the spring or the fall, when deciduous foilage is absent (''leaf off'') and the ground is essentially free of snow (Figure 5-2). Only during these periods can terrain features be adequately distinguished and contours precisely delineated. As differences in vegetation are rarely of significance, mapping photographs are commonly made on panchromatic film. Similar coverage would be specified by geologists interested in lithology and structural mapping and by engineers concerned with locating transportation or utility

routes. Although summer photography may suffice in an emergency, dense canopies of foliage greatly inhibit the efficient evaluation of ground detail (Figure 5-2).

Interpreters interested in vegetation analyses will ordinarily specify aerial photography made during the growing season, particularly when deciduous plants constitute an important component of the vegetative cover (''leaf on''). When it is essential that deciduous trees, conifers, and mixtures of the two groups be delineated, either black-and-white infrared or color infrared film is frequently specified (Figure 2-32). Color infrared film should be specified whenever the objective is to detect losses of plant vigor (Plate 6).

Several research projects involving black-and-white photography of forest areas have indicated that the best timing for infrared coverage is from midspring to early summer, a period when all trees have produced some foliage but before maximum leaf pigmentation. Useful panchromatic photos of timberlands can be made throughout the year, but in the northern latitudes the best results have been obtained in late fall, just before deciduous species, such as aspen and oak, shed their leaves. For a brief period of perhaps 2 weeks' duration, foliage color differences will provide good tonal contrasts between most of the important timber species in this region (Figure 3-13).

Time-of-Day Considerations

The time of day is an important contract specification, because the angle at which the sun's illumination strikes the earth's surface affects not only the *quantity* of light being reflected to the aerial camera but also its *spectral quality*.

The sun's elevation angle above the horizon is a function of latitude, season of the year, and the time of day. When extensive shadowing is not wanted, ground scenes should be photographed within about 2 hr of local apparent noon. This is especially important when normal color and color infrared films are to be used, since they both have relatively slow speeds. It will be remembered from Chapter 2 that solar altitude is one of the prime factors incorporated into the Kodak Aerial Exposure Computer (Figure 2-31) for determining proper exposures for an aerial film.

A detrimental consequence of selecting the wrong combination of season and time of day is a phenomenon known as a **hotspot**, or **sunspot**. A loss of photographic detail results when a straight line from the sun passes through the camera lens and intersects the ground inside the area of photo coverage (Figure 5-3). At the center of the hotspot is a saturated area caused by the direct return of reflected light to the camera. The effect on a processed photograph is a washed-out (overexposed) circular area of perhaps several centimeters in diameter where all or most terrain detail is lost (Figure 5-4).

Hotspots are most likely to occur with high sun-angles, at lower latitudes, at higher flight altitudes, and with wide-angle lenses. After the season and latitude of photography have been determined, it is possible to calculate (and thus avoid) those midday hours when hotspots will likely occur. Heath (1973) presents a series of tables and graphs to aid in identifying hotspot-free times.

A major exception to the scheduling of photographic flights within about 2 hr of local apparent noon is when there is a need to enhance certain features (e.g., archaeological structures or topography at the microrelief scale) by the **shadow effect** (Figure 3-12). When shadow enhancement is desired, photo flights should be scheduled for early morning or late afternoon, times when shadow development is maximized by low sun-angles. The former period is usually preferred because the atmosphere is normally less hazy and less turbulent. Best results will be attained with a high-speed panchromatic film. The effect of eight different illumination angles on a topographic model is illustrated in Figure 5-5.

Figure 5-4 Portion of a vertical airphoto containing a hotspot or sunspot.

Technical Specifications for New Photography

When the decision to purchase *new photography* has been made, technical specifications are usually detailed in a formal contract. A specimen contract intended for worldwide use has been developed by Burnside (1985). It is designed to define standards that can be realistically achieved and satisfy the needs of most clients. The following items are usually covered to some degree in all contracts:

Business Arrangements: These include such items as the cost of the aerial survey, a provision for periodic inspection of work completed, criteria for reflights, and schedules for delivery and payment.

Study Area: The tract to be photographed is accurately delineated on maps supplied by the client.

Film-Filter Combination: The specific type of film and filter is specified by the client. Larger firms may be able to operate twin mapping cameras (e.g., for natural color and color infrared photographs (Figure 5-1)).

Mapping Cameras: A report should be furnished to the client that specifies the method of film flattening during exposure, type of shutter, calibrated focal length, and the resolving power of the lens.

Flight Lines: Lines should be parallel, oriented in the correct direction, and within a stated distance from positions drawn on flight maps. When possible, flight lines should run north-south or east-west. For maximum aircraft efficiency, they should parallel the long axis of the study area. As a margin of safety, an extra flight line should be added to each side of the study area to assure complete coverage.

Figure 5-3 Angular relationships between incident solar radiation, camera lens, and hotspot. (Adapted from Heath 1973.)

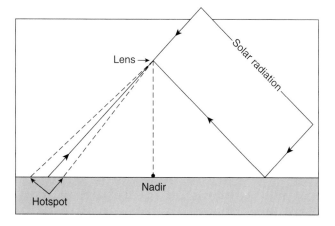

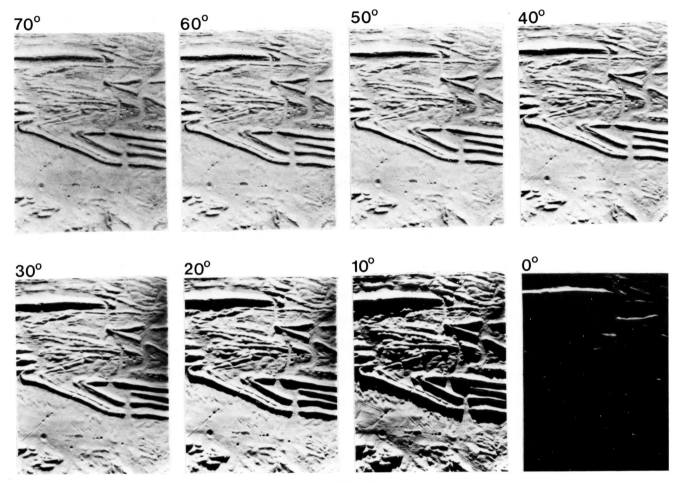

Figure 5-5 Vertical photographs of a plaster topographic model under angles of illumination from 70° to 0°. In this series, the greatest detail of topographic relief is recognizable when the angle of illumination is 10° or 20°. (Adapted from Hackman 1966.)

Overlap and Sidelap: For stereo coverage, overlap is usually 55 to 65 percent (averaging 60 percent) along the line of flight and 15 to 45 percent (averaging 20 to 30 percent) sidelap between adjacent flight lines. At the ends of each flight line, two photos should be added as a margin of safety to assure total coverage.

Photo Alignment: Crab or drift should not affect more than 10 percent of the print width for any three consecutive photographs.

Tilt: Tilt should not exceed 2° to 3° for a single exposure or average more than 1° for the entire project.

Required Timing and Cloud Cover: Both season of year and optimum time of day are usually specified by the client. An allowable cloud cover (percent) should also be specified; cloud-free coverage is desirable for most mapping projects.

Film and Print Specifications: Included are statements defining how the original film (positive or negative) is to be processed plus the type, quantity, and scale of photographic products that are to be delivered to the client.

Film Titling: As a rule, the processed negatives or positives should contain the date, project code, and roll and exposure numbers. Nominal photo scale and local time are ordinarily placed on the first and last exposures in each flight line.

Flight-Plan Map: It is customary for a final version of the flight-plan map to be supplied to the client. It should show the actual locations of the flight lines and the photographic center points (exposure stations) along each line. A copy of the original flight log may also be included.

Inspection of Contract Photography

Following completion of a photographic survey, it is customary for the client, or a representative, to make a technical inspection of all photographic products. Infrequent purchasers of aerial photographs may have difficulty in evaluating the finished products, because acceptance or rejection

Figure 5-6 Photographic quality improved by the dodging process. These photographs, made from the same negative, illustrate the tonal contrasts between an ordinary print (*top*) and a dodged print (*bottom*). (Courtesy Ronald C. Gibson, Log Etronics, Inc.)

often requires checks for such items as tilt, overlap, sidelap, scale, plus film and print quality. As a means of translating technical specifications into guides for the inexperienced inspector, an itemized checklist, as designed by Avery and Meyer (1962), may prove useful.

Photographic quality is usually the most difficult item for an inspector to evaluate because of the lack of standards or criteria for comparison. The inspector must place heavy reliance on subjective judgment in deciding whether a given photographic defect constitutes a reasonable basis for rejection. The underlying objective is that the original film and prints should be free from blurred detail and/or blemishes that detract from their intended use.

If it is determined that the developed film (negative or positive) has poor contrast, it may still be possible to produce acceptable quality prints. This can involve altering exposure and/or development times, using special contrast printing filters, or using print paper with special contrast properties. **Dodging** is a printing technique that is often used to produce high-quality black-and-white prints when the negatives have poor density characteristics (Figure 5-6). The dodging process holds back light from reaching certain areas of the sensitized paper to avoid overexposing those areas. Dodging can be done manually or automatically (e.g., the Log Etronics printer, Figure 5-6).

Determining Photo Coverage

It is to the client's advantage to know the approximate number of airphotos that will be required to cover a study area before the issuance of a photographic contract. The example that follows illustrates the various photo requirements involved in determining the total number of vertical photographs needed to cover a tract measuring 20 *km east-west* by 35 *km north-south*. The basic information required is as follows:

Elevation of Study Area: 500 m above sea level
Desired Photo Scale: 1:25,000
Film Format: 23 × 23 cm, or 0.23 × 0.23 m
Focal Length: 152 mm, or 0.152 m
Overlap: 60 percent
Sidelap: 30 percent

Flight Altitude: The height above the study area at which the aircraft must fly to obtain photographs at a scale of 1:25,000 is determined by Equation 4-4:

$$H = (RF_d)(f)$$
$$= (25,000)(0.152 \text{ m})$$
$$= 3,800 \text{ m above terrain}$$
$$\text{Aircraft altitude} = 3,800 \text{ m} + 500 \text{ m}$$
$$= 4,300 \text{ m above sea level.}$$

Ground Distance: Because the flying height above the terrain has been determined, the lateral ground distance covered by a single photograph can be calculated using Equation 4-12:

$$D = (RF_d)(d)$$
$$= (25,000)(0.23 \text{ m})$$
$$= 5,750 \text{ m, or } 5.75 \text{ km.}$$

Alignment of Flight Lines: The flight lines should be in alternating north-south directions, paralleling the long axis of the study area. This alignment minimizes the number of 180° turns the aircraft would have to make.

Number of Flight Lines: The number of flight lines (NL) required to span the study area is determined by the following relationship:

$$NL = \frac{W}{(D)(S_g)} + 2, \qquad (5\text{-}1)$$

where: W = width of study area,
D = lateral ground distance of single photo,
S_g = sidelap gained by each successive flight line (100 − percent sidelap), expressed as a decimal fraction, and
2 = number of extra flight lines added to the sides of the study area to assure total coverage (i.e., one per side).

W and D must be expressed in the same units. For NL, round any fractional remainders to the next higher whole integer, since a fraction of a flight line cannot be flown:

$$NL = \frac{20 \text{ km}}{(5.75 \text{ km})(0.7)} + 2$$
$$= 4.97 + 2$$
$$= 6.97, \text{ or } 7 \text{ flight lines.}$$

Number of Photos per Flight Line: To determine the number of photographs required to cover a flight line (NP), the following relationship is used:

$$NP = \frac{L}{(D)(O_g)} + 4, \qquad (5\text{-}2)$$

where: L = length of flight line,
D = lateral ground distance of single photo,
O_g = overlap gained by each successive photo (100 − percent overlap), expressed as a decimal fraction, and
4 = number of photos added (two to each end of a flight line) to assure complete coverage.

L and D are expressed in the same units. NP must be in whole numbers because a fraction of a photograph cannot be obtained; when a decimal fraction results, always round up to the next whole integer.

$$NP = \frac{35 \text{ km}}{(5.75 \text{ km})(0.4)} + 4$$
$$= 15.2 + 4$$
$$= 19.2, \text{ or } 20 \text{ photos per flight line.}$$

Figure 5-7 (Left) Meyer sidemount for 35- or 70-mm film cameras. The mount clamps to the left doorframe, and the camera can be slid into the plane to change film and/or filter. (Courtesy Econ, Inc.)

Figure 5-8 (Below) Bogucki wingmount that holds two Hasselblad 70-mm film cameras. Both cameras are triggered simultaneously via an intervalometer in the aircraft cabin. The mount is FAA approved for use under the "restricted" category. (Courtesy Donald J. Bogucki, State University of New York, Plattsburgh.)

Total Number of Photographs: To calculate the total number of photographs required to cover the study area, simply multiply the total number of photos in a flight line (NP) by the total number of flight lines (NL). For the 20 × 35-km study area, the total number of photos would be 20 photos per flight line times 7 flight lines, or *140 photographs.*

Taking Your Own Vertical Photographs

If vertical aerial photographs taken with 35- or 70-mm film cameras (Figure 2-16) are acceptable for coverage of a study area, the *do-it-yourself approach* provides an alternative for aerial surveys. Cameras with 35- or 70-mm formats will commonly have focal-length lenses ranging from about 50 to 100 mm. If we assume the use of such cameras in aircraft without oxygen equipment, the upper altitudinal limit would be about 3,000 m above mean sea level, and the lower limit would be about 100 m above the ground. If we further assume that terrain elevations will not exceed 1,500

m, the range of photographic scales might be computed by the same procedure outlined earlier in this chapter:

$$\text{RF} = \frac{1}{3{,}000\text{ m} - 1{,}500\text{ m}/0.05\text{ m}}$$

$$= \frac{1}{30{,}000} = 1{:}30{,}000\text{, and}$$

$$\text{RF} = \frac{1}{100\text{ m}/0.1\text{ m}} = \frac{1}{1{,}000} = 1{:}1{,}000.$$

This scale range will be adequate for most do-it-yourself aerial surveys.

Exposures made with hand-held cameras are satisfactory for oblique views and for spot coverage with a near-vertical camera orientation. However, if overlapping photographs are needed (e.g., for stereoscopic study), a camera mount should be constructed for best results. It is assumed here that most rented aircraft will not have camera ports in the floor. Consequently, most improvised camera mounts are designed to hold the cameras outside the airframe (Figures 5-7 and 5-8).

Handmade camera mounts have been designed in many configurations, since each must be adapted to a particular aircraft and camera. To combine light weight with maximum stability, most camera mounts are constructed of plywood and/or aluminum; vibration problems are often minimized by the use of foam-rubber shock absorbers.

Availability of Existing Photography

During the past several decades, millions of aerial photographs of the United States and its territories have been produced by various government agencies (federal, state, and local) and private aerial survey firms. Panchromatic film has ordinarily been used, but there is a rapidly growing supply of normal color, color infrared, and black-and-white infrared photographs available for sale. The age of the photography dates from the 1930s to the present, and coverage for many areas is available in several-year cycles. The most common contact format is 23 × 23 cm.

The **National Cartographic Information Center (NCIC)** of the U.S. Geological Survey (USGS) is an important contact source for determining the availability of aerial photography for a given area. Determining photo availability is made through NCIC's **Aerial Photography Summary Record System (APSRS)**, which is a computer bank describing the airphoto holdings of cooperating government agencies and private aerial survey firms. A user of APSRS receives a listing describing the photo coverage for a requested area (e.g., date of photography, amount of cloud cover, scale, film type and format, and camera focal length) plus the name of the entity that holds the original photography. Inquiries regarding APSRS can be made directly to the National Cartographic Information Center, U.S. Geological Survey, 507 National Center, Reston, VA 22092.

U.S. Government Photography

Three agencies serve as the primary sources for purchasing photographs acquired by the federal government. They are (1) the **Earth Resources Observations Systems (EROS) Data Center (EDC)**, USGS; (2) the **Aerial Photographic Field Office**, Agricultural Stabilization and Conservation Service (ASCS); and (3) the **National Archives and Records Library**, General Services Administration.

The EROS Data Center stores and reproduces for sale photographs acquired primarily for USGS programs and NASA's aircraft and manned spacecraft programs. Photography for the USGS's topographic mapping program currently exceeds 2 million frames. The bulk of the stereo photography is obtained with mapping cameras. Contact scales range from about 1:12,000 to 1:80,000, and the film type is ordinarily panchromatic. Acquisitions for the topographic mapping program began in the 1940s and continue on an irregular basis.

Two special programs of the U.S. Geological Survey are the **National High-Altitude Photography (NHAP) Program** and the **National Aerial Photography Program (NAPP)**. Under the NHAP Program (1980–1987), panchromatic and color infrared photographs of the 48 conterminous states were systematically acquired. Consistent acquisition parameters (aircraft altitude of 40,000 ft [about 12,190 m] above mean terrain, sun-angle of at least 30° to minimize shadows, stereoscopic coverage, no cloud cover, and minimal haze) were implemented to assure the quality and comparability of the photographs. NHAP photos were taken with two mapping cameras: (1) a 6-in. (152-mm) focal-length camera loaded with panchromatic film produced photos at a contact scale of 1:80,000 (Figure 5-9), and (2) a 8.25-in. (210-mm) focal-length camera loaded with color infrared film produced photos at a contact scale of 1:58,000 (see front cover).

The primary objective of the first phase of the NHAP Program *(NHAP-1)* was to obtain photographs when deciduous vegetation was largely dormant (i.e., leaf off). Complete NHAP-1 coverage of the conterminous states is now available. For the second phase *(NHAP-2)*, coverage was acquired during the growing season (leaf on) for 11 states.

NAPP, which began in 1987, is designed to provide standardized, high-resolution airphotos of the conterminous states. NAPP photographs are obtained from an aircraft altitude of 20,000 ft (about 6,095 m) above mean terrain. A 6-in. (152-mm) focal-length mapping camera loaded with color infrared film produces photos at a contact scale of 1:40,000.

Since 1964, more than 1.3 million high-altitude photographs have been generated by NASA in support of its earth resources program and are archived at EDC. Many of the photographs were taken from the U-2 aircraft and its enhanced version, the ER-2. The scale of the photography generally falls into the 1:60,000 to 1:120,000 range. The photos cover about 80 percent of the country, and many areas have been photographed several times. Panchromatic, normal color, and color infrared photographs are generally available in a 23 × 23-cm format.

More than 65,000 photographs from the *Mercury, Gemini, Apollo, Skylab,* and *Space Shuttle programs* are also archived at EDC. Many of these photographs were taken by astronauts using hand-held cameras (Figure 5-10). However, "hard-mounted" cameras were used during the Apollo IX, Skylab, and two Space Shuttle missions for vertical photography. Photos from NASA's manned spacecraft program depict many domestic and foreign areas.

Of recent interest to the remote sensing community is the stereoscopic photography produced by NASA's **Large Format Camera (LFC)**. The LFC was designed to operate exclusively from earth-orbital altitudes, and it was first flown on *Space Shuttle Mission 41-G* in October 1984.

The LFC is a high-performance cartographic camera with a 305-mm focal length and a 23 × 46-cm image format (hence the term "large format"), with the long dimension oriented in the direction of flight. From a nominal altitude of 300 km, each frame covers a 225 × 450-km ground area with a ground resolution of 14 to 25 m, depending upon the type of film. Exposures can be timed to provide forward overlaps ranging from 20 to 80 percent. In the 80 percent mode, a stereopair has a base-height ratio of about 1.2, which allows for height perceptions as small as 30 m.

Figure 5-9 NHAP stereogram of Glen Canyon Dam on the Colorado River and the city of Page, Arizona. Scale is 1:80,000.

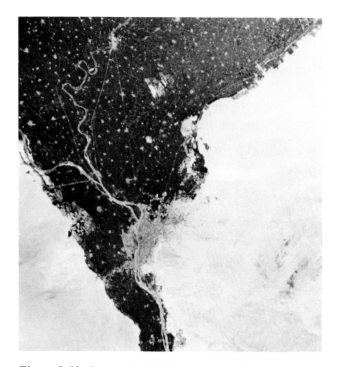

Figure 5-10 *Gemini V* (1965) photograph of Cairo and the Nile Delta, Egypt. (Original in color.)

During *Space Shuttle Mission 41-G,* the LFC acquired a total of 2,160 photographs (minus-blue panchromatic = 1,520 frames, normal color = 320 frames, and color infrared = 320 frames). Photo-acquisition scales range from near 1:750,000 to about 1:1,200,000, but enlargements can exceed 10× with little loss of image quality. Two LFC stereograms are presented in Figures 5-11 and 5-12. It is anticipated that future Space Shuttle missions will routinely carry the LFC.

A final EDC category is a collection of about 1 million frames of aerial photography acquired by nine additional agencies at various scales and film types. The cooperating agencies are Bureau of Indian Affairs, Bureau of Land Management, Environmental Protection Agency, Environmental Research Institute of Michigan, South Dakota State University, U.S. Air Force, U.S. Army Corps of Engineers, U.S. Navy, and Water and Power Resources Service.

At the EROS Data Center, a computerized data base, containing up to 15 descriptors for each archived frame, provides an indexing system similar to APSRS. Inquiries regarding computer searches and ordering procedures can be made directly to the EROS Data Center, USGS, User Services Section, Sioux Falls, SD 57198. LFC photography is also available commercially from Chicago Aerial Survey, Inc., LFC Dept., 2140 Wolf Road, Des Plaines, IL 60018.

The Aerial Photography Field Office is now the de-

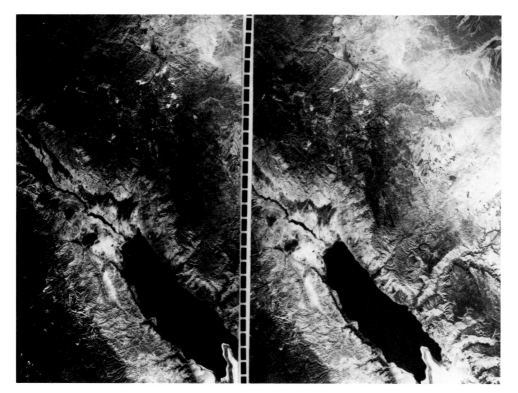

Figure 5-11 LFC stereogram of the Dead Sea region. The Dead Sea is about 396 m below sea level, and its salty water contains a wealth of minerals. The south basin has been transformed into a series of evaporation ponds to extract potash from brine pumped in from the north basin.

pository for photography obtained by the Agricultural Stabilization and Conservation Service, U.S. Forest Service, and Soil Conservation Service. The photography covers about 90 percent of the country. The bulk of the photography is panchromatic, but normal color, color infrared, and black-and-white infrared photographs have been obtained for several national forests. Common scales are 1:12,000 to 1:40,000, but recently the Soil Conservation Service has photographed many areas of the country at scales ranging from 1:31,680 to 1:85,000. Photo index sheets (Figure 4-29) can be viewed at local or regional offices. Written inquiries may be sent to the Aerial Photography Field Office, ASCS—U.S. Department of Agriculture, P.O. Box 30010, Salt Lake City, UT 84125.

When a federal agency's aerial photographs are of no further use for a specific project, they may be stored and reproduced as sale items by the National Archives and Records Library. The leading contributors to the data base have been Agricultural Stabilization and Conservation Service, Department of Defense, Soil Conservation Service, Tennessee Valley Authority, U.S. Army Corps of Engineers, U.S. Forest Service, and USGS. Most of the photographs were taken prior to World War II, and nearly all are panchromatic. Although the majority are verticals, thousands of oblique photographs are also available. For additional information, write to the National Archives and Records Library, Room 2W, 8th and Pennsylvania Ave. N.W., Washington, D.C. 20408.

Airphotos for Other Countries

The **National Air Photo Library** (**NAPL**) is the central storehouse for the Canadian government's aerial photography. More than 4 million airphotos, dating back to the 1920s, are stored in this archive. Panchromatic photos are available for the entire country; natural color and color infrared photos are available for selected regions of the country (Hyatt 1988). Inquiries regarding this photography should be made directly to the National Air Photo Library, Surveys and Mapping Branch, Dept. of Energy, Mines, and Resources, 615 Booth Street, Ottawa, Ontario K1A OE9.

Many other countries have their own collections of aerial photographs and rules on distribution. Initial points of contact for determining availability and restrictions are consulates or embassies, local governmental councils and libraries, and remote sensing or land survey organizations of the country for which the photographs are wanted. Hyatt (1988) discusses the availability and major sources of aerial photographs for South America, Western Europe, Eastern Europe and the Soviet Union, Africa, Asia and the East Pacific, and Australia and New Zealand.

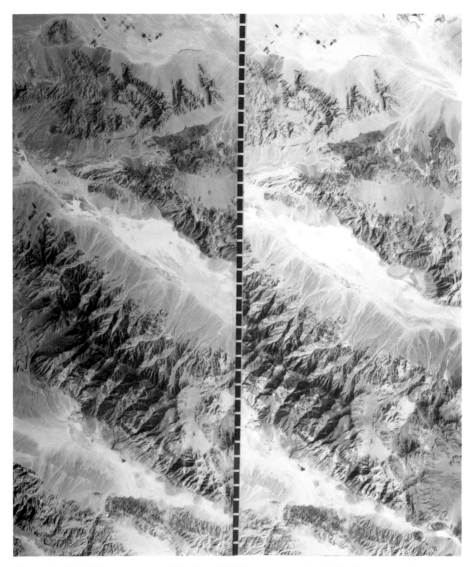

Figure 5-12 LFC stereogram of the Death Valley region. Death Valley, a complex structural basin, drops to an elevation of about 86 m below sea level to rank as the lowest point in North America.

Questions

1. Explain why a structural geologist would be more apt to use low-sun-angle stereo photographs than a cartographer interested in compiling a topographic map.
2. Explain why the photographs used by an individual for crop identifications would be of limited value to an individual interested in mapping soil types.
3. Determine the total number of vertical photographs that would be required to cover a study area measuring 23 km north-south and 51 km east-west, given the following specifications: elevation of study area = 650 m above sea level; desired photo scale = 1:40,000; film format = 23 × 23 cm (or 0.23 × 0.23 m); focal length = 152 mm, or 0.152 m; overlap = 60 percent; and sidelap = 30 percent.

   Flight altitude above terrain: _____

   Flight altitude above sea
   level: _____

   Lateral ground distance per
   photo: _____

   Alignment of flight lines: _____

   Number of flight lines: _____

   Number of photos per flight
   line: _____

   Total number of photos: _____

4. Using Equations *2-2, 2-3,* and *2-4,* calculate the lens angles (angular fields of view) for the LFC, given a focal length of 305 mm, or 30.5 cm, and an image format measuring 23 × 46 cm.

   $$\theta_L = \text{_____}$$
   $$\theta_W = \text{_____}$$
   $$\theta_D = \text{_____}$$

5. Using the answers from Question 4 for θ_L and θ_W plus Equations 2-5 and 2-6, calculate the lateral ground distance for an LFC photograph if the Space Shuttle's altitude was 296 km above mean terrain. How much ground area would the photograph cover?

6. If it takes 650 vertical airphotos at a scale of 1:22,000 to completely cover a square-shaped study area, how many photographs would be required to cover the same area at a scale of 1:65,000?

7. A mapping camera lens has a focal length of 152 mm. To obtain vertical airphotos at a scale of 1:40,000, the camera would have to be flown at an altitude of how many meters?

8. Suppose you wish to fly a photographic mission with a 350-mm focal-length lens over a study area with an average elevation of 1,500 m. If the airplane has a maximum operating ceiling of 13,700 m above sea level, could vertical airphotos be obtained at a scale of 1:40,000?

Bibliography and Suggested Readings

Avery, T. E. 1960. A Checklist for Airphoto Inspections. *Photogrammetric Engineering* 26:81–84.

Avery, T. E., and M. P. Meyer. 1962. Contracting for Forest Aerial Photography in the United States, Lakes States Forest Experiment Station Paper 96. St. Paul, Minn.: U.S. Forest Service.

Burnside, C. D. 1985. *Mapping from Aerial Photographs*, 2d ed. New York: John Wiley & Sons.

Doyle, F. J. 1979. A Large Format Camera for Shuttle. *Photogrammetric Engineering and Remote Sensing* 45:73–78.

Eastman Kodak Company. 1985. *Photography from Light Planes and Helicopters.* Publication M-5. Rochester, New York: Eastman Kodak Company.

Fleming, J., and R. G. Dixon. 1981. *Basic Guide to Small-Format Hand-Held Oblique Aerial Photography.* User's Manual 81-2. Ottawa, Ontario: Canada Centre for Remote Sensing.

Hackman, R. J. 1966. Time, Shadows, Terrain, and Photointerpretation, Technical Letter NASA-22. Washington, D.C.: U.S. Geological Survey.

Heath, G. R. 1973. Hot Spot Determination. *Photogrammetric Engineering* 39:1205–14.

Holz, R. K., ed. 1985. *The Surveillant Science: Remote Sensing of the Environment*, 2d ed. New York: John Wiley & Sons.

Hyatt, E. 1988. *Keyguide to Information Sources in Remote Sensing.* London: Mansell Publishing Limited.

Lillesand, T. M., and R. W. Kiefer. 1987. *Remote Sensing and Image Interpretation*, 2d ed. New York: John Wiley & Sons.

Lund, H. G. 1969. Factors for Computing Photo Coverage. *Photogrammetric Engineering* 35:61–63.

Newhall, B. 1969. *Airborne Camera: The World from the Air and Outer Space.* New York: Hastings House, Publishers.

Paine, D. P. 1981. *Aerial Photography and Image Interpretation for Resource Management.* New York: John Wiley & Sons.

Rabenhorst, T. D., and P. D. McDermott. 1989. *Applied Cartography: Introduction to Remote Sensing.* Columbus, Ohio: Merrill Publishing Co.

Richason, B. F., Jr., ed. 1983. *Introduction to Remote Sensing of the Environment*, 2d ed. Dubuque, Iowa: Kendall/Hunt Publishing Co.

Slama, C. C., ed. 1980. *Manual of Photogrammetry*, 4th ed. Falls Church, Va.: American Society for Photogrammetry and Remote Sensing.

Ulliman, J. 1975. Cost of Aerial Photography. *Photogrammetric Engineering* 41:491–97.

Woodward, L. A. 1970. Survey Project Planning. *Photogrammetric Engineering* 36:578–83.

Zsilenszky, V. G. 1970. Supplementary Aerial Photography with Miniature Cameras. *Photogrammetria* 25:27–38.

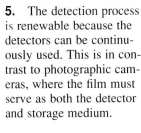

Chapter 6

Electro-Optical Sensors

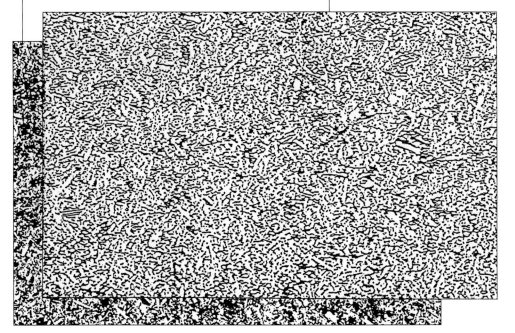

screen. The major types of electro-optical sensors are the video camera, the vidicon camera, the across-track scanner, and the along-track scanner.

Even though they have poorer resolution properties and are more complex, electro-optical imaging systems offer several advantages over photographic cameras:

1. They are capable of operating in numerous bands of the electromagnetic spectrum that lie within and beyond the confines of the **photographic spectrum** (Figure 1-5). Applicable spectral regions are the near ultraviolet (UV), visible, reflected infrared (IR), and thermal IR. This wavelength span is known as the **optical spectrum** (Figure 1-5).

2. For certain systems, the image data can be transmitted over radio links; this telemetry feature is essential for the use of robotic spacecraft as remote sensing platforms.

3. Certain systems employ some type of in-flight display device, which enables a ground scene to be viewed in image form in near real time.

4. Systems operating in the thermal IR region have a day-night capability.

5. The detection process is renewable because the detectors can be continuously used. This is in contrast to photographic cameras, where the film must serve as both the detector and storage medium.

Many of the newer electro-optical sensors are digital systems. These units use an **analog-digital (A/D) converter** that translates the electronic signals from the detectors to discrete **digital numbers (DNs)**, which are placed on magnetic tape. The collection of DNs can then be processed mathematically by a **digital computer** to correct geometric and radiometric errors and to enhance patterns in the original image (discussed in Chapter 15). To create visible images, the numerical data can be transformed into video signals, which enable the images to be seen on a television screen, or they can be converted to visible light, which is then recorded on ordinary photographic film.

Overview

Unlike photographic cameras, which record radiation reflected from a ground scene directly onto film, **electro-optical sensors** use **nonfilm detectors**. These detectors convert the reflected and/or emitted radiation from a ground scene to proportional electrical signals that are ultimately used to construct two-dimensional images for conventional viewing. The images are normally recorded on film, but for some systems it is possible to view the images on a television

Video Cameras

Video cameras used in remote sensing collect reflected radiation typically within the 0.4- to 1.1-μm range of

Figure 6-1 Multispectral video imaging system used by U.S. Department of Agriculture–Agricultural Research Service (USDA-ARS) personnel at Weslaco, Texas. The system comprises (A) four video cameras, (B) an electronic system, (C) recorder for color composite images, (D) recorders for black-and-white images, (E) a color TV monitor to view color composite images, and (F) a TV monitor to view black-and-white images. The equipment rack contains the following: (a) color encoder, (b) time-date generator, (c) color sync generator, (d) pulse distribution amplifier, (e) power supply, and (f) channel-selection panel. The two photographic cameras mounted next to the video cameras are optional. (Courtesy James H. Everitt and David E. Escobar, USDA-ARS.)

Figure 6-2 Color infrared video camera developed by Douglas E. Meisner and manufactured and distributed by E. Coyote Enterprises, Inc. under the name Biovision. (Courtesy Douglas E. Meisner, Interscan, Inc.)

the electromagnetic spectrum (Figure 1-5). A video camera generates NTSC standard television signals (485 lines per image frame), as used in the United States and Japan. This enables the signals to be recorded on videotape by a conventional videocassette recorder (VCR) in VHS or Beta formats. A television monitor is used for in-flight viewing of the data being recorded on videotape and for image playback in the laboratory. Video equipment is lightweight, portable, and easily installed in a single-engine aircraft. The process of obtaining video images is called **videography** to distinguish it from the photographic process of **photography**.

Since the mid-1980s, both black-and-white and color video systems have been commercially available for airborne remote sensing. Many models use **video tubes**, with each tube having a particular spectral sensitivity. The camera lens forms an image onto the faceplate of the video tube that is electronically scanned to generate a video signal. Analogous to photographic cameras, lens filters are used to modify the spectral sensitivity of a particular tube.

For a **multispectral video system**, the simplest approach uses an array of black-and-white video cameras that are equipped with visible-near IR sensitive tubes and properly filtered. The cameras are synchronized electronically and are mounted to provide equivalent fields of view. For a four-camera array, blue, green, red, and near IR spectral bands can be isolated (Figure 6-1). This enables the reflectance data from a ground scene to be recorded and viewed in flight as individual black-and-white images and as normal

color or color IR composite images (Plate 7). Multispectral images can be thought of as video equivalents to multispectral photographs (Figure 2-48), with a major difference being that the color composites can be generated electronically without requiring an additive color viewer (Figure 2-49) (Meisner 1986).

Although normal **color video cameras** have been available for several years, a **color IR video camera** has been recently introduced that produces instant-turnaround color infrared imagery (Figure 6-2). This camera utilizes three tubes with the following spectral sensitivities: (1) 0.5 to 0.6 μm (green), blue on display; (2) 0.6 to 0.7 μm (red), green on display; and (3) 0.7 to 1.1 μm (near IR), red on display. An optical beamsplitter spectrally separates the radiation that passes through the camera's single lens. The camera can be equipped with an automatic zoom lens or one of several fixed focal-length lenses. An image acquired by a color IR video camera is shown in Plate 7.

Solid-state detectors, an alternative to video tubes, are being used increasingly in both black-and-white and color video cameras. These sensors consist of an X-Y array of detectors and readout electronics etched onto a silicon chip using integrated circuit technology. Several acronyms are used for these detectors, depending upon the microcircuit technology employed: **CCD** (charge-coupled device); **CID** (charge-injection device); and **MOS** (metal on silicon). **Solid-state video cameras** offer advantages of compactness, shock resistance, low power consumption, and near-perfect geometry. Presently, the major disadvantage is price; solid-state cameras typically cost up to two times the price of an equivalent tube camera.

A two-band, solid-state video system is shown in Figure 6-3. The cameras have 488 × 754 detector arrays and spectral sensitivities of 0.61 to 0.93 μm (red) and 0.78 to 0.93 μm (near IR). The image processor is capable of producing three types of images during acquisition: (1) false color composite images, (2) normalized ratio images (near IR − red/near IR + red), and (3) standard band im-

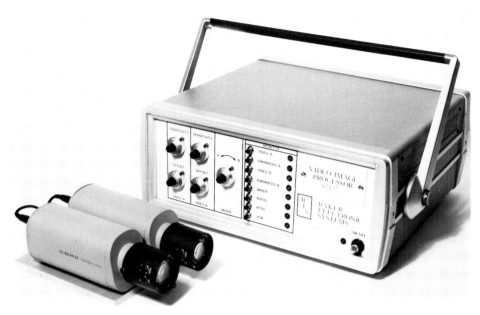

Figure 6-3 Two-channel, solid-state video imaging system. (Courtesy James L. Walsh, Charles F. Hutchinson, Robert A. Schowengerdt, and L. Ralph Baker, University of Arizona.)

ages. A ratio image of an agricultural scene is presented in Figure 6-4.

Additional hardware can be useful when interpreting the video image data in the laboratory. This may include a high-quality VCR that can provide a stable, still-frame capability, allowing for TV screen interpretations or the generation of hardcopy film products by directly photographing the TV screen. The ultimate still-frame is provided by a digital freeze-frame unit, or **frame grabber**; it converts an image frame to digital data for computer processing.

Despite the relatively coarse resolution of video systems (Plate 7), they offer the distinct advantage of providing immediately useful imagery. This capability enables (1) exposures to be adjusted interactively during image acquisition, (2) flight lines to be adjusted in real time, and (3) image acquisition when timeliness is required (e.g., irrigation analysis, water-pollution studies, flood monitoring, crop inventories, detection of vegetation stress).

Vidicon Cameras

Vidicon cameras are very similar to television cameras and cover about the same spectral region as video cameras. They have been normally operated from robotic satellites. Rather than recording an image directly onto film, a latent image is temporarily stored on a photoconductive faceplate and is scanned by an internal electron beam. This process creates a series of electronic signals, which, in a satellite environment, are telemetered to earth receiving stations, where they are used to produce visible images on photographic film or a television monitor. Vidicon cameras are

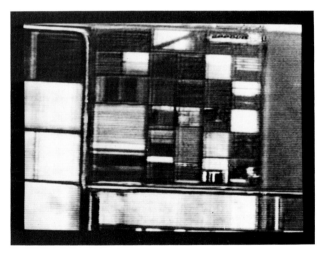

Figure 6-4 Normalized ratio video image (near IR − red/ near IR + red) of a section of the Maricopa Experimental Farm, Arizona. The light-toned fields are heavily vegetated, whereas the dark-toned fields are mostly bare soil. (Courtesy James L. Walsh, Charles F. Hutchinson, Robert A. Schowengerdt, and L. Ralph Baker, University of Arizona.)

equipped with shutters to prevent image blur and to control exposure times (analogous to photographic frame cameras). In addition, they can be designed to operate at very low light levels.

Vidicon cameras have been used successfully for imaging the moon and Mars. **Return Beam Vidicon (RBV) cameras** were carried on the first three *Landsat* satellites. The *Landsat 3* RBV served as a successful high-resolution mapping device; its operation is described later in this chapter.

Across-Track Scanners

Across-track, or whiskbroom, scanners were originally developed for the military, but declassifications commenced in the late 1960s. The two primary types of across-track scanners available today are the multispectral scanner and the thermal IR scanner. A leading manufacturer of across-track scanners is Daedalus Enterprises of Ann Arbor, Michigan. A listing of its current models and their primary applications are presented in Table 6-1.

TABLE 6-1 Across-Track Scanners Manufactured by Daedalus Enterprises, Inc.

System, Primary Applications	Output	Number of Bands	Instantaneous Field of View (mrad)	Operating Wavelength (μm)	Standard Data Recording Mode
AADS1220 Terrain surveillance Forest-fire monitoring Search and rescue Animal census	Analog	2	1.0 or 1.7	3.0–5.5 8.5–12.5	Analog magnetic tape or optional near-real-time hardcopy
AADS1220/MP Monitoring ship traffic Detection of open-sea dumping Detection of offshore oil spills Detection of harbor and coastal pollution Search and rescue	Analog	2	2.5 or 5.5	0.32–0.38 8.50–12.50	Near-real-time hardcopy or optional analog magnetic tape
AADS1230 Environmental monitoring Detection of energy loss Water quality analysis Volcanology Soil moisture studies Animal census	Analog	2	1.7 or 2.5	4.5–5.5 8.5–12.5	Analog magnetic tape
AADS1260 Land management Crop classification and inventory Mineral exploration Water-resource management Pollution studies	Digital	11	2.5 or 1.25	0.38–0.42 0.42–0.45 0.45–0.50 0.50–0.55 0.55–0.60 0.60–0.65 0.65–0.69 0.70–0.79 0.80–0.89 0.92–1.10 8.50–12.50	High-density digital magnetic tape
AADS1268 Oil and mineral exploration Land management Forest and crop classification and inventory Water-resource management Pollution studies	Digital	11	2.5 or 1.25	0.42–0.45 0.45–0.52 0.52–0.60 0.60–0.62 0.63–0.69 0.69–0.75	High-density digital magnetic tape

System, Primary Applications	Output	Number of Bands	Instantaneous Field of View (mrad)	Operating Wavelength (μm)	Standard Data Recording Mode
				0.76–0.90	
				0.91–1.05	
				1.55–1.75	
				2.08–2.35	
				8.50–13.00	
AADS1280 Agricultural surveys Forest and crop classification and inventory	Digital	5	0.54	0.54–0.59 0.62–0.67 0.74–0.80 0.84–0.92 0.95–1.06	High-density digital magnetic tape
AADS1285 Geologic mapping for discrimination of: Silicate rocks Carbonate rocks Certain altered rocks	Digital	6	2.5	8.2–8.6 8.6–9.0 9.0–9.4 9.4–10.2 10.2–11.2 11.2–12.2	High-density magnetic tape

Courtesy George England, Daedalus Enterprises, Inc.

Rather than viewing an entire ground scene instantaneously, as do photographic and vidicon cameras, across-track scanners, through a rotating or oscillating mirror, scan a contiguous series of narrow ground strips at right angles to the flight path. The forward motion of the platform causes new ground strips to be covered by successive scan lines. In this way, a two-dimensional record of reflectance and/or emittance information is built up along the line of flight (Figure 6-5). This technique is referred to as **whiskbroom scanning**.

For a given altitude above mean terrain (H), the collection mirror's **angular field of view (AFOV)**, or **scan angle** (θ), determines the **ground swath**, which is the length of the **ground strip** recorded as a scan line (Figure 6-5).

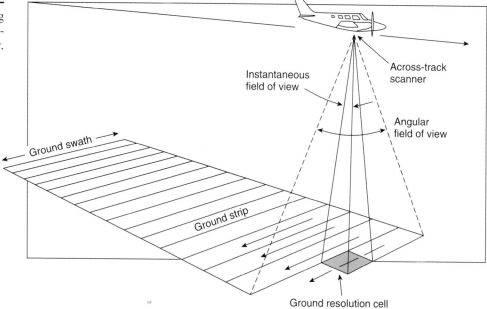

Figure 6-5 Basic operating configuration of an across-track, or wiskbroom, scanner.

Instantaneous field of view

Across-track scanner

Angular field of view

Ground swath

Ground strip

Ground resolution cell

Two common AFOVs for an aircraft operation are 90° and 120°. Ground swath (S) is calculated as follows:

$$S = 2\left(\tan \frac{\theta}{2}\right)(H).$$ (6-1)

H should be expressed in the units desired for S. The equation shows that a direct relationship exists between θ and S and between H and S.

The **instantaneous field of view (IFOV)** determines how much ground area the scanner "sees" at any given instant in time; this ground area is called the **ground resolution cell** (Figure 6-5). The IFOV for aircraft scanners typically varies from 0.5 to about 5.5 milliradians (mrad). One side of the ground resolution cell at nadir (D) is a function of flight altitude above mean terrain (H) and the scanner's IFOV (β):

$$D = H\beta.$$ (6-2)

H should be expressed in the units desired for D; β must be expressed in radians (1 mrad = 10^{-3} rad). Equation 6-2 shows a direct relationship between H and D and between β and D. The *area* of the ground resolution cell at nadir is equal to D^2.

Because the distance measured from the scanner to the ground increases from nadir to the margins of the ground swath (Figure 6-5), ground resolution cells are larger toward the margins than at the center of the ground swath. This results in a scale distortion that can be corrected during image generation; correction techniques are discussed later in this chapter.

A **small IFOV** is mandatory when there is a need for spatial detail because the sensor integrates measurements over the full IFOV. Thus, if two objects with different reflectance or emittance properties occupy the same ground resolution cell, they will not be resolved as separate entities. Instead, one integrated response from the cell will be recorded. An object can be independently resolved only when its size is equal to or larger than the ground resolution cell.

A small IFOV, although providing spatial detail, restricts the amount of radiation received by the scanner. A **large IFOV**, in effect, results in a longer time for the scanner's mirror to sweep across a ground resolution cell. This longer **dwell time**, or **residence time**, in each IFOV allows more radiation to impinge upon a detector, thus creating a stronger signal (with an adequate **signal-to-noise ratio (S/N)**). Consequently, large IFOVs lend themselves to detecting small reflectance or emittance variations, but at the expense of spatial resolution.

Multispectral Scanners

Multispectral scanners are capable of operating simultaneously in the near UV, visible, reflected IR, and thermal IR regions of the electromagnetic spectrum (Figure 6-6 and Table 6-1). This means that both reflected and emitted radiation can be collected by a multispectral scanner. The number of spectral channels can range from fewer than 5 to more than 10. Typical radiation detectors used in multispectral scanners and their wavelength sensitivities are (1) **photomultiplier tubes**, 0.3- to 0.9-μm range; (2) **silicon photodiodes**, 0.9- to 2.5-μm range; and (3) **mixtures of certain metallic elements**, 3- to 14-μm range. The ability to operate within this broad spectral region presents the possibility of identifying objects whose identifiable spectral signatures lie beyond the wavelength limits of the relatively narrow confines of the visible spectrum. Note in Figure 6-7, for example, that the region around 1.5 μm would be particularly useful for distinguishing snow from clouds and that the region around 2 μm would be useful for discriminating between red basalt, gray basalt, and granite. In addition, black-and-white positive transparencies of three spectral bands can be used to produce color composites to take advantage of the eye's sensitivity to subtle color variations. Multispectral scanners are used routinely from both aircraft and robotic spacecraft.

Figure 6-8 shows the essential components of an aircraft-based multispectral scanner. Operational features are as follows:

1. A rotating or oscillating mirror scans across the terrain at right angles to the flight path and collects emitted and/or reflected radiation line by line from the ground scene. (Satellite-mounted scanners usually collect multiple lines of ground data during each sweep of the mirror.)

2. Through a series of secondary mirrors, the radiation beam from each scan line is directed to spectrum-separation devices (e.g., prisms and dichroic gratings), where it is divided into a number of discrete bands or channels. The system shown in Figure 6-8 incorporates eight channels.

3. The scene radiation for each band is directed to and focused on very small detector elements that change the fluctuating radiation into an electrical signal that varies in intensity according to the strength of the radiation. Each detector is designed to have an optimum response over a specific wavelength interval. For longer wavelengths, especially in the thermal IR region, the detectors must be cooled to very low temperatures to suppress molecular motions. In aircraft operations, this can be accomplished by a liquid coolant such as liquid nitrogen. Satellite-based scanners may make use of closed-cycle coolers or passive methods (e.g., heat dissipation to space).

4. The electrical impulses from each detector are amplified and recorded on magnetic tape in analog or digital form for input to ground-based computer-processing and auxiliary equipment. This type of storage provides for maximum versatility in data output because the data can be processed to remove scanner distortions and enhance information of special interest (see Chapter 15). In addition to a ground-based image processing unit, the **Matra Thematic Scanner** system incorporates an on-board, quick-look visualization unit (digital to video) that enables an analyst to view color composite images on a television monitor made from

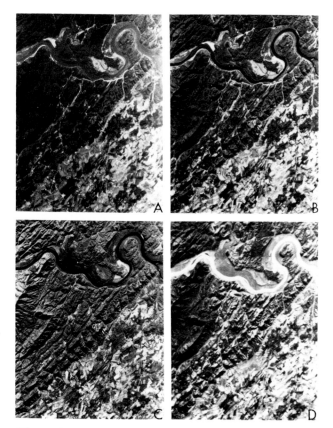

a selection of any three of its six channels (three visible, one reflected IR, and two thermal IR). This enables one to have in-flight control over the recording process and to exploit the images in near real time (Plate 8). For a robotic satellite system, the sensor signals are transmitted to ground receiver stations for processing.

Thermal IR Scanners

Thermal IR scanners function in essentially the same manner as multispectral scanners, but their operation is confined to the thermal IR atmospheric windows at 3 to 5 μm and 8 to 14 μm (Figure 6-9). Thermal IR scanners are commercially available that operate within one or both of the atmospheric windows (e.g., AADS 1220 and AADS 1230 in Table 6-1) or within multiple bands of the 8- to 14-μm window (e.g., AADS 1285 in Table 6-1).

Broadband measurements over wavelength intervals of 2 μm or more have a strong dependence on surface temperature in both the 3- to 5-μm and 8- to 14-μm regions.

Figure 6-6 (*Above*) Four of eleven images of a rural area near Knoxville, Tennessee, acquired by the Daedalus AADS 1260 multispectral scanner (Table 6-1). Spectral sensitivities are as follows: (*A*) 0.45 to 0.50 μm (blue); (*B*) 0.65 to 0.69 μm (red); (*C*) 0.80 to 0.89 μm (reflected IR); and (*D*) 8.5 to 12.5 μm (thermal IR). The images were generated in December 1977 from an altitude of about 6,100 m above ground level. (Courtesy George England, Daedalus Enterprises, Inc.)

Figure 6-7 (*Right*) Average spectral-response curves for five materials. (Adapted from Bowker et al. 1985.)

Figure 6-8 (*Below*) Basic components of a multispectral scanner. (Adapted from Landgrebe 1974.)

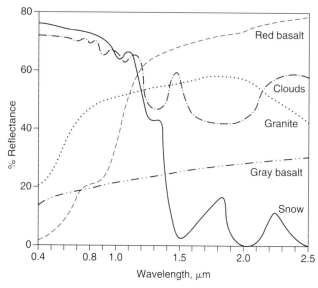

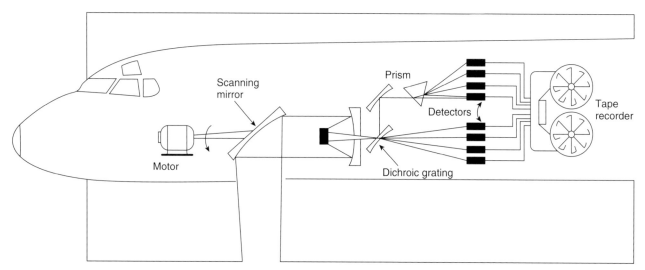

Measurements over wavelength intervals of 1 μm or less in the 8- to 14-μm region have important geologic applications. This is because the narrow thermal IR bands contain important compositional information about silicate rocks, carbonate rocks, and certain altered rocks. For this reason, NASA is using an AADS 1285 six-channel system to acquire multispectral IR images for a number of different geologic environments (Table 6-1). The acronym **TIMS** (Thermal Infrared Multispectral Scanner) is used to designate this system.

With the use of special bandpass filters, thermal IR detectors are designed to sense radiation at wavelengths ranging from 3 to 5 μm and from 8 to 14 μm; radiation at these wavelengths is readily transmitted through the lower

Figure 6-9 Spectral bands used in the thermal IR region. Gases responsible for absorption are indicated.

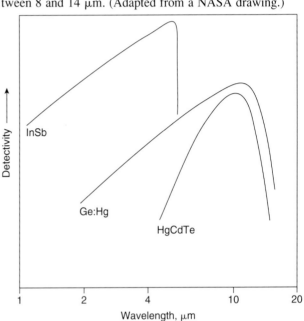

Figure 6-10 Spectral detectivity of three thermal IR detectors. With special band-pass filters, the sensitivity of InSb (indium antimonide) is narrowed to 3 to 5 μm, whereas that of Ge:Hg (mercury doped germanium) and HgCdTe (mercury cadmium telluride) are narrowed to sensitivities between 8 and 14 μm. (Adapted from a NASA drawing.)

Figure 6-11 Basic operating configuration of a calibrated thermal IR scanner, incorporating both film and magnetic tape recording components.

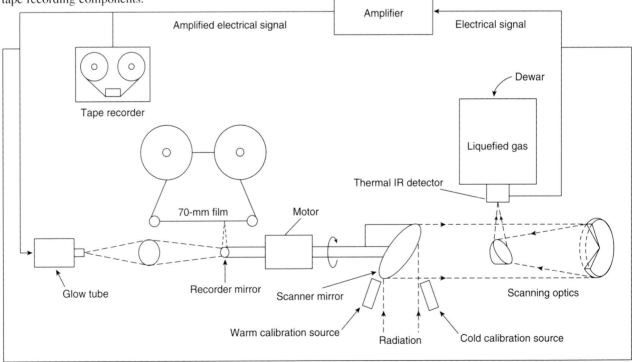

layers of the gaseous atmosphere (Figure 6-9). Because of ozone absorption in the upper atmosphere, thermal IR detectors in satellite systems are typically filtered to operate in the 10.5- to 12.5-μm band (Figure 6-9). The detectors are combinations of certain metallic elements that are sensitive to radiation of these wavelengths. For example, **indium antimonide (InSb)** is used for detections in the 3- to 5-μm wavelength region, whereas **mercury doped germanium (Ge:Hg)** and **mercury cadmium telluride (HgCdTe)** are used for detections in the 8- to 14-μm wavelength region (Figure 6-10). These detectors convert incident thermal IR radiation to electrical signals in direct proportion to the intensity of the radiation.

For maximum sensitivity, the detectors must be cooled to very low temperatures to minimize their own thermal IR emissions. Coolants for aircraft systems can be **liquefied gases**, which are enclosed in a special vacuum flask called a **dewar** (Figure 6-11). **Liquid nitrogen** (77 K, $-196°C$) is used to cool InSb and HgCdTe detectors, whereas **liquid helium** (30 K, $-243°C$) is used to cool Ge:Hg detectors.

Thermal IR scanners introduced in the late 1960s for use in aircraft were usually configured with a direct film recording system, but newer systems usually incorporate both film and magnetic tape recording capabilities (Figure 6-11). The latter configuration is preferable because the photographic record provides imagery for immediate use and the tape record can be later manipulated in a variety of ways in the laboratory to provide higher-quality images.

For direct film recording, the amplified detector signal modulates the brightness of a small light source, such as a glow tube (similar to a light bulb) or a single-line cathode-ray tube (CRT) with a moving electron beam. The intensity-modulated spot of light is focused onto the surface of the photographic film (e.g., 70-mm roll film) and swept across the film line by line in synchronism with the speed of the scanner mirror. The film moves across the exposure station at a rate proportional to the velocity and altitude of the aircraft, thereby producing a continuous photographic record (Figure 6-11). For an aircraft operation, it is not uncommon for 50 lines of data to be placed on the film per second. Certain systems have an inflight film processor and printer that enable an analyst to see the strip image almost immediately.

Broadband thermal IR scanners can be either **uncalibrated** or **calibrated**. Gray tones in uncalibrated images represent relative **radiant**, or **noncontact**, **temperatures**. The darkest tones represent the coolest radiant temperatures, whereas the lightest tones denote the warmest radiant temperatures (Figure 6-12). Such images are well suited to qualitative analyses where the interest is in discriminating among different surface features.

Calibrated scanners are equipped with two operator-controlled temperature-calibration sources (electrical elements)—a **cold reference** and a **warm reference**. These elements are mounted on either side of the scanner mirror's AFOV (Figure 6-11). For each scan line, the mirror views the first calibration element, scans the terrain, and then views the second calibration source. The resulting analog signals are amplified by solid-state circuitry and sequentially recorded on magnetic tape. The two reference standards provide a scale for determining the radiant temperatures of ground objects resolved during the aerial survey. The tape

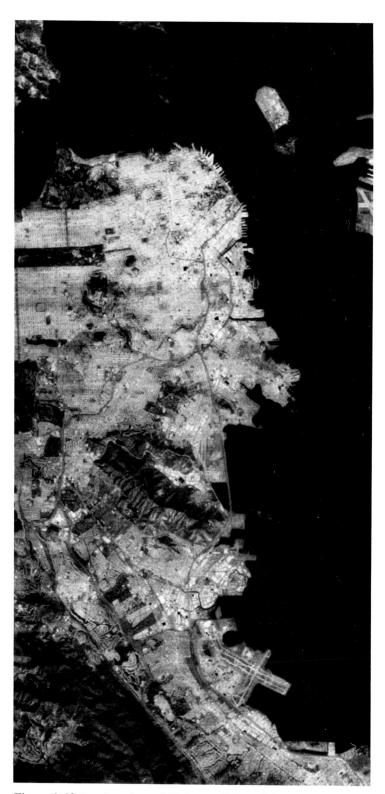

Figure 6-12 Daytime thermal IR image (8.5 to 13.0 μm) of San Francisco and vicinity, April 1980. The aircraft altitude was 20,000 m above ground level. (Courtesy George England, Daedalus Enterprises, Inc. and NASA/Ames Research Center.)

Figure 6-13 Panoramic thermal IR image incorporating scale compression and one-dimensional relief displacement. The image was acquired in darkness; the warm linear signatures depict buried heating lines.

is later played back in the laboratory to produce images, with each tone representing the same radiant temperature throughout the image.

With the Daedalus **DIGICOLOR**™ process, the analog tape record can also be ''sliced'' into radiant temperature ranges and displayed in a color-coded image (Plate 8). The basic slice divides the analog signal into six linear temperature increments and assigns an output color to each. The sequential thermal relationship from warm to cool is red, yellow, green, cyan, blue, and magenta. Black and white indicate that the radiant temperatures for terrain features were lower or higher than the temperatures of the calibration elements, respectively. The temperature interval between each color is 1°C in Plate 8.

Under clear, dry weather conditions, radiant temperature accuracy for the Daedalus calibrated system is on the order of 0.3°C. However, internal calibration cannot account for adverse atmospheric effects, such as high humidity. These effects can bias the scanner-derived temperatures by as much as 2°C (Lillesand and Kiefer 1987).

Image Distortions

Unless corrected, across-track scanner images have an inherent geometric distortion caused by the scanning function. Because the distance between the mirror and the terrain increases with increasing distance from the center of the ground swath, the ground resolution cell, covered per unit of time, is larger at the ends of a scan line than at the nadir (Figure 6-5). For direct film recording with a glow tube, this dictates that each ground resolution cell, regardless of size, is recorded as an equal area in the image. This results in an effect called **scale compression**, and it gives ground features the appearance of being wrapped around a cylinder (Figure 6-13).

With an unrectified, or **panoramic**, **image**, there is severe scale compression in the scan or lateral direction (i.e., at right angles to the flight line or the nadir line). The effect increases nonlinearly toward the margins of the image. Thus, the **lateral scale** varies, whereas the **longitudinal scale** remains constant. Precise measurements are, therefore, not possible, and panoramic images are best suited for interpretation purposes.

With a rectified, or **rectilinear**, **image**, scale is constant in both the lateral and longitudinal directions (Figure 6-12). This makes the full width of the image useful for both interpretation and measurement tasks. For a unit employing a CRT device, scale compression is removed by increasing the sweep velocity of the electron beam in a nonlinear fashion away from nadir. For compensation with digital systems,

Figure 6-14 Basic operating configuration of an along-track, or pushbroom, scanner.

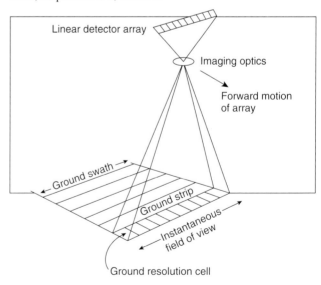

picture elements (pixels) are duplicated in a nonlinear fashion away from nadir, effectively "stretching out" the length of each scan line.

A second type of image distortion, which is not correctable, is **one-dimensional relief displacement**. For a vertical airphoto, relief displacement for vertical features above a specified datum plane is radially outward from the principal point (Figure 4-3). For an across-scanner image, however, relief displacement for vertical features is outward at right angles from the flight line or the nadir line (Figure 6-13). Thus, displacement occurs in only a single direction (i.e., along the scan line). Its effect increases with increasing object height, with increasing distance from the nadir line, and with decreasing flying height. In Figure 6-13, note how the tops of the buildings are displaced orthogonally away from the nadir line.

Along-Track Scanners

Along-track, or **pushbroom**, **scanners** represent a new generation of electro-optical sensors that form images without a scanning mirror. This is accomplished with the use of **line**, or **linear**, **arrays** of very small (e.g., 13 × 13 µm) **charge-coupled devices (CCDs)**. In this configuration there is a dedicated detector element for each across-track ground resolution cell (Figure 6-14). There is one linear array of detectors for each spectral band, and a single array may contain as many as 10,000 individual detectors. Each array is located in the focal plane of the instrument's imaging optics so that an *entire* ground strip in the across-track direction is focused onto the detector elements at the same time without any mechanical scanning (Figure 6-14). In effect, this means that, unlike the case for an across-track scanner, the instantaneous field of view is synonymous with the angular field of view (compare Figures 6-5 and 6-14).

Pushbroom scanning describes the technique of using the forward motion of the platform to sweep the linear array of detectors across the ground scene. Electronic sampling of the detectors in the across-track dimension provides the orthogonal "scan" component to form an image, and the platform motion along its line of flight produces successive lines of the image—the along-track dimension. The detector array is electronically sampled at the appropriate rate to ensure that contiguous image lines are produced (Figure 6-14). The *length* of the array projected through the optics defines the swath width, and the *size* of the individual detectors determines the ground resolution cell (Figure 6-14).

Some of the advantages that an along-track scanner offers over an across-track scanner are as follows:

1. Improved spectral resolution or sensitivity due to the longer dwell time on each ground resolution cell; this enables a stronger signal to be recorded.

2. Greater reliability and longer operating life due to the elimination of moving components.

3. High geometric accuracy in the across-track direction because of the fixed geometry of the detector arrays; this simplifies image reconstruction and processing tasks.

4. Lighter weight and lower power requirements; this makes it especially well suited for small satellites and airplanes.

A current disadvantage of pushbroom technology is that commercially available CCD detectors cannot operate at wavelengths longer than about 1.1 µm.

A popular, aircraft-based pushbroom scanner is the **Multispectral Electro-Optical Imaging Scanner (MEIS)**, which was developed by MacDonald Dettwiler and Associates for the Canada Centre for Remote Sensing. MEIS incorporates five independent channels that cover the 0.4- to 1.0-µm spectral region. Each channel uses a CCD linear array with 1,728 detector elements. The sensor has a real-time processing capability to provide geometric and radiometric corrections and interchannel registration. A color composite image produced by the MEIS system is shown in Plate 8.

Pushbroom sensor technology was first space-tested during *Space Shuttle Mission STS-7* in 1983. The experimental instrument, called the **Modular Optoelectronic Multispectral Scanner (MOMS)**, was developed by the German Ministry for Research and Technology. The sensor successfully produced two-channel images at 20-m resolution for selected ground areas. The remote sensing satellite programs of France, Japan, and India employ along-track scanners. These programs are described later in the chapter.

Thermal IR Characteristics

So far in this book, our discussions have been principally concerned with **reflected, shortwave radiation** (i.e., UV, visible, and reflected IR). With thermal IR, however, we are concerned with **emitted longwave radiation**. Therefore, an overview of its unique properties is given in the following several sections of this chapter. Such a background is necessary for properly interpreting thermal IR images.

It will be recalled from Chapter 1 that thermal IR radiation (heat energy) results from random atomic and molecular motions and is emitted by all substances having a temperature above absolute zero (0 K, $-273.16°C$, $-459.69°F$).* As the temperature of an object increases, a greater amount of energy is radiated, and the spectra of the radiation shifts to shorter wavelengths. Thermal IR radiation is sensed within two atmospheric windows at 3 to 5 µm and 8 to 14 µm where there is minimal absorption and, consequently, very little emission (Figure 6-9). Because all substances continuously emit thermal IR radiation, it can be detected both day and night.

Most thermal IR sensors record the *concentration* of the **radiant emittance**, or **radiant flux**, of surface features. This concentration defines an object's **radiant temperature**

With the Kelvin scale, the symbol K is not written with a degree mark.

Figure 6-15 Battery-powered thermal IR radiometer. Radiant temperatures (°C or °F) are read from an LCD display on the control unit.

where: E_λ = spectral emission in W/m²·m at a wavelength λ,
λ = wavelength in m,
C_1 = 3.74 × 10⁻⁶ W·m² (first radiation constant),
C_2 = 1.44 × 10⁻² m·K (second radiation constant),
e = 2.718 (base of natural logarithms), and
T = absolute temperature (K).

To obtain E_λ in W/m²·μm, as is more customary in remote sensing, W/m²·m should be multiplied by 1 × 10⁻⁶.

Equation 6-3 indicates that at any given wavelength, the total energy of emitted blackbody radiation increases as temperatures increase. It also indicates that the intensity distribution of the radiation varies with wavelength at a given temperature. Values for E_λ are commonly used to construct **energy distribution curves** for objects at various temperatures (Figure 6-16). Such curves enable one quickly to assess the proportions of radiation being emitted between selective wavelength intervals.

The magnitude of radiation emitted from a blackbody

(T_{rad}).* It may be measured remotely (i.e., without physical contact) by a **thermal IR radiometer** (Figure 6-15). This instrument collects the radiation within its field of view and transforms it into an electrical current, which is then converted to a temperature reading for display.

Thermal IR sensors cannot measure an object's **kinetic temperature** (T_{kin}).** Rather, kinetic temperature is measured with a **thermometer** placed in direct physical contact with the object. The radiant temperatures of objects are always *less* than their kinetic temperatures because of a thermal property called emissivity, which is described in the next section. The radiant temperature resolution of state-of-the-art thermal IR scanners is on the order of 0.1°C.

Radiation Laws

Radiation is emitted by all objects in a magnitude and spectral range that is governed by two properties—**temperature** and **emissivity**, the latter being a measure of an object's efficiency as an absorber and emitter. **Plank's radiation law** relates the spectral characteristics and magnitude of the emission to the temperature of the emitting body; the expression for a theoretically perfect emitter or **blackbody** at any given wavelength is

$$E_\lambda = \frac{C_1}{\lambda^5[e^{(C_2/\lambda T)} - 1]},\qquad (6\text{-}3)$$

*Radiant temperature is also called **external**, **apparent**, and **noncontact**.
Kinetic temperature is also called **internal, **real**, **contact**, and **thermodynamic**.

Figure 6-16 Spectral distribution curves of blackbody radiation from objects at different temperatures in accordance with Plank's radiation law. The total amount of energy emitted is defined by the area under each curve. Dashed lines represent wavelengths of peak emission defined by Wien's displacement law. (Adapted from Lillesand and Kiefer 1987.)

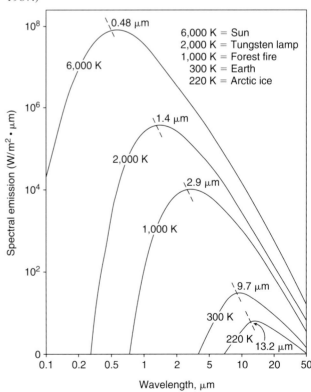

over the entire spectrum (area under each curve in Figure 6-16) is explained by the **Stefan-Boltzmann law**:

$$E_{bb} = \sigma T^4, \qquad (6\text{-}4)$$

where: E_{bb} = radiant emittance from a blackbody in W/m^2,
σ = 5.67×10^{-8} $W/m^2 \cdot K^{-4}$ (Stefan-Boltzmann constant), and
T = absolute temperature (K).

Equation 6-4 shows that the total energy emitted from a blackbody, over all wavelengths, is directly proportional to the fourth power of its absolute temperature. For example, if the temperature of a blackbody is raised from 300 K (earth's ambient temperature) to 600 K (melting point of lead), its temperature is doubled, but the radiant emittance increases 2^4, or 16, times. If the same blackbody is heated from 300 K to 6,000 K (temperature of the sun's photosphere), temperature increases twentyfold, but the radiant emittance increases 20^4, or 160,000, times.

Wien's displacement law identifies the wavelength at which the maximum amount of energy is radiated (λ_{max}) from a blackbody:

$$\lambda_{max} = \frac{W}{T}, \qquad (6\text{-}5)$$

where: W = 2,897 $\mu m \cdot K$ (Wien's constant), and
T = absolute temperature (K).

Wien's displacement law shows that the wavelength of maximum energy emission is inversely proportional to the absolute temperature of the blackbody. Thus, as temperature increases, λ_{max} shifts to progressively shorter wavelengths. For example, substituting the sun's surface temperature of 6,000 K and the earth's ambient temperature of 300 K into Equation 6-5 results in λ_{max} values of 0.48 μm and 9.7 μm, respectively (indicated as dashed lines in Figure 6-16). This means that the radiant power peak of the sun is experienced as *light* and the earth's radiant power peak is experienced as *heat*.

In a practical sense, Wien's displacement law identifies the atmospheric window to use for sensing thermal IR emissions. For example, the radiant power peak for very hot targets, such as forest fires and molten lava flows, is within or very close to the 3- to 5-μm window (Figure 6-9). The 8- to 14-μm window contains the radiant power peaks for most of the earth's passive features, since their temperatures are in the neighborhood of 300 K. For this reason, most thermal IR surveys are performed within this window.

In addition to temperature, the quantity of radiant emission is controlled by an object's emissivity (ϵ). Emissivity is a dimensionless number that describes the actual absorption and emission properties of real objects, or **graybodies**. It is a ratio expression of the radiant emittance from a graybody (E_{gb}) at a given temperature to that from a blackbody (E_{bb}) at the same temperature:

$$\epsilon = \frac{E_{gb}}{E_{bb}}. \qquad (6\text{-}6)$$

A blackbody is a theoretical object that is both a perfect absorber of electromagnetic radiation at all wavelengths ($\alpha = 1$) and a perfect emitter of radiation at all temperatures ($\epsilon = 1$). Therefore, $\alpha = 1 = \epsilon$ and $\alpha = \epsilon$. The latter relationship defines **Kirchhoff's radiation law**, which, when paraphrased, states *good absorbers are good emitters and good reflectors are poor emitters.* We can write $\epsilon = 1$ as the emissivity of a blackbody.

Graybodies are nonperfect absorbers and emitters, and their emissivities are always less than 1 but greater than 0 (Table 6-2). A perfect reflector would be a theoretical **whitebody** ($\epsilon = 0$). The closer ϵ is to 1, the more efficient the real object is as an absorber and emitter. Conversely, the closer ϵ is to 0, the more efficient the real object is as a reflector. For example, polished brass ($\epsilon = 0.1$) is an extremely poor emitter because the radiant energy within the body is reflected back away from the surface; its surface also looks shiny from the inside. Black coal spoil ($\epsilon = 0.99$), by comparison, is a poor reflector but an excellent absorber and emitter of radiation (i.e., "easy in, easy out").

TABLE 6-2	Emissivities in the 8- to 14-μm Wavelength Region
Material	Emissivity (ϵ)
Coal spoil	0.99
Dolomite, rough	0.96
Basalt, rough	0.95
Dolomite, smooth	0.93
Basalt, smooth	0.92
Granite	0.90
Dunite	0.89
Obsidian	0.86
Concrete	0.94
Asphalt	0.90–0.98
Water, distilled	0.99
Water, natural	0.92–0.96
Oil film	0.97
Ice	0.96–0.98
Snow	0.83–0.85
Soil	0.92–0.96
Sand	0.90
Grass	0.97
Coniferous vegetation	0.97
Deciduous vegetation	0.95
Black paint	0.98
White paint	0.90
Human skin	0.98
Glass	0.90–0.95
Polished brass	0.10
Aluminum foil	0.05
Mirror	0.02

The values of ϵ vary according to an object's composition, including its color, and the roughness of its surface. Dark-colored objects are usually better absorbers and emitters than light-colored objects (e.g., black and white paints in Table 6-2). For the same material, a surface that is smooth relative to the wavelengths of emitted radiation has a lower emissivity than a surface that is rough. A rough surface favors increased absorption because of multiple reflections along its surface. The same roughness enables more radiation to be emitted because of a larger surface area (e.g., smooth and rough dolomite or basalt in Table 6-2).

Emissivity is also wavelength-dependent; the values of ϵ, for example, are smaller for a given object in the 3- to 5-μm wavelength region. Emissivity is considered to be constant for a given object in the 8- to 14-μm wavelength region.

Emissivity establishes the radiant temperature of an object. Two materials may have the *same* kinetic temperature, but their radiant temperatures will be *unequal* if their emissivities are different. This is extremely important in remote sensing because the output from a thermal IR sensor is a measurement of an object's radiant temperature. The radiant temperature of an object is related to its kinetic temperature by the following relationship:

$$T_{rad} = \epsilon^{1/4} T_{kin}. \qquad (6\text{-}7)$$

Given a constant T_{kin}, Equation 6-7 shows that T_{rad} varies directly with ϵ (as ϵ increases, T_{rad} increases and vice versa). This relationship is illustrated in Table 6-3.

In practical usage, if we can measure both T_{kin} (thermometer) and T_{rad} (thermal IR radiometer) of a given object, we can estimate its emissivity by rewriting Equation 6-7 to the following form and solving for ϵ:

$$\epsilon^{1/4} = \frac{T_{rad}}{T_{kin}} \quad \text{and}$$

$$\epsilon = \left(\frac{T_{rad}}{T_{kin}}\right)^4.$$

We may also use the emissivity factor to determine the radiant emittance from a graybody (E_{gb}) in the following manner:

$$E_{gb} = \epsilon\sigma T_{kin}^4. \qquad (6\text{-}8)$$

Equation 6-8 shows that a graybody can never radiate as much energy as a blackbody at the same temperature because emissivity is always less than 1 and that, at a given temperature, E_{gb} varies directly with ϵ (as ϵ increases, E_{gb} increases and vice versa).

Internal Thermal Properties

The **thermal properties** of objects play an important role in controlling the internal distribution of heat energy as a function of time and depth. This distribution pattern, in turn, controls the surface temperature of a given object. Typical values of the thermal properties for several common materials are presented in Table 6-4.

Thermal conductivity (K) is a measure of the rate at which heat will flow through a material; it is measured as the calories delivered in 1 sec across a 1-cm^2 area through a thickness of 1 cm at a temperature gradient of 1°C (cal/cm·sec·°C). Rocks are generally poor conductors of heat when compared to metals (Table 6-4). However, rocks are much better conductors of heat than dry, loosely consolidated materials (Table 6-4). This is attributable to the presence of air-filled pore spaces in unconsolidated materials; air has a thermal conductivity of 0.00005.

Thermal capacity (c) is a measure describing the ability of a material to store heat; it is measured as the number of calories required to raise the temperature of 1 gram of a substance by 1°C (cal/g·°C). The thermal capacity of most geologic materials is quite similar; water has a very high thermal capacity compared to other materials (Table 6-4).

With the inclusion of **density (ρ)**, or the ratio of mass to volume (g/cm^3), additional thermal measures can be derived from the variables previously defined. **Thermal storage (C)** is the product of density times thermal capacity,

		Kinetic Temperature		Radiant Temperature[a]	
Object	Emissivity	K	°C	K	°C
Blackbody	1.00	300	27	300.0	27.0
Water, distilled	0.99	300	27	299.2	26.2
Basalt, rough	0.95	300	27	296.2	23.2
Basalt, smooth	0.92	300	27	293.8	20.8
Obsidian	0.86	300	27	288.9	15.9
Mirror	0.02	300	27	112.8	− 160.2

TABLE 6-3   Kinetic and Radiant Temperatures for Six Different Objects

[a]*Answers derived by use of Equation 6-7.*

which defines the number of calories that are stored in a 1-cm^3 volume of a material per 1°C (cal/cm^3·°C):

$$C = \rho c. \qquad (6\text{-}9)$$

It is seen in Table 6-4 that marble and quartzite have a uniform density of 2.7 g/cm^3. If it is assumed that the rocks are at the same temperature, a unit volume of quartzite would contain less stored heat than the same unit volume of marble because of differences in thermal capacity (Table 6-4). Even though basalt and basaltic ash have a uniform thermal capacity of 0.21, a unit volume of the basalt would contain more stored heat than the same unit volume of basaltic ash because of differences in density (Table 6-4). Water has a very high thermal storage compared to other materials (Table 6-4).

Thermal diffusivity (κ) is a measure of the rate of internal heat transfer within a substance. It has units of cm^2/sec and is determined as follows:

$$\kappa = \frac{K}{\rho c} = \frac{K}{C}. \qquad (6\text{-}10)$$

In a remote sensing context, thermal diffusivity controls a material's ability to transfer heat from the surface to the subsurface during the day heating period and from the subsurface to the surface during the night cooling period. For example, granite has a higher thermal diffusivity than basalt, owing in large measure to its significantly higher thermal conductivity (Table 6-4). This means granite can transfer a larger proportion of its surface heat downward into deeper regions during daylight than basalt; hence, the basalt's surface will be warmer. However, basalt will cool relatively quickly during the night, whereas the granite will remain warmer because the heat stored at depth will flow upward to the surface. Unconsolidated materials normally have lower diffusivities than rocks because of their large pore volumes (Table 6-4).

Thermal inertia (P) measures the tendency of a material to resist changes in temperature. It is measured in cal/cm^2·sec$^{1/2}$·°C and is determined as follows:

$$P = \sqrt{K\rho c}. \qquad (6\text{-}11)$$

Materials with low thermal inertia have small conductivity, density, and capacity values and hence possess a weak internal resistance, or impedance to temperature fluctuations. Dry, porous materials usually have low thermal inertias (e.g., basaltic ash and pumice in Table 6-4). Such materials reach high maximum temperatures in daylight and low minimum temperatures at night; this is associated with a large temperature difference, or ΔT ($T_{max} - T_{min}$), during a 24-hr heating-cooling cycle (i.e., diurnal cycle).

Materials with high thermal inertia exhibit a strong internal resistance to temperature changes (e.g., peridotite

TABLE 6-4   Thermal Properties of Various Materials[a]

Material	Density (ρ) g/cm^3	Thermal Conductivity (K) cal/cm·sec·°C	Thermal Capacity (c) cal/g·°C	Thermal Storage (C) cal/cm^3·°C	Thermal Diffusivity (κ) cm^2/sec	Thermal Inertia (P) cal/cm^2·sec$^{1/2}$·°C
Geologic						
1. Basalt	2.6	0.0045	0.21	0.546	0.008	0.049
2. Basaltic ash	1.1	0.0007	0.21	0.231	0.003	0.013
3. Granite	2.6	0.0075	0.16	0.416	0.018	0.056
4. Limestone	2.5	0.0048	0.17	0.425	0.011	0.045
5. Marble	2.7	0.0055	0.21	0.567	0.010	0.056
6. Obsidian	2.4	0.0030	0.17	0.408	0.007	0.035
7. Peridotite	3.2	0.0110	0.20	0.640	0.017	0.084
8. Pumice, loose	1.0	0.0006	0.16	0.160	0.003	0.009
9. Quartzite	2.7	0.0120	0.17	0.459	0.026	0.074
10. Rhyolite	2.5	0.0055	0.16	0.400	0.014	0.047
11. Sandstone, quartz	2.5	0.0120	0.19	0.475	0.025	0.075
12. Shale	2.3	0.0042	0.17	0.391	0.011	0.040
13. Tuff, welded	1.8	0.0028	0.20	0.360	0.008	0.032
Miscellaneous						
14. Copper	8.9	0.9410	0.092	0.819	1.149	0.878
15. Glass	2.4	0.0020	0.12	0.288	0.007	0.024
16. Ice (0°C)	0.92	0.0053	0.49	0.451	0.012	0.049
17. Water	1.0	0.0013	1.00	1.000	0.001	0.036
18. Wood, oak	0.82	0.0005	0.33	0.271	0.018	0.012

[a]*Values are mostly for room temperatures, about 18 to 20°C.*
Adapted from Janza et al. (1975) and Forsythe (1964).

and sandstone in Table 6-4). Such materials are relatively cool in daylight and relatively warm at night. They thus exhibit more uniform surface temperatures in a diurnal cycle (small ΔT) than materials of low thermal inertia.

From the previous discussion it is seen that there is an inverse relationship between thermal inertia and the diurnal change in surface temperature. As thermal inertia increases, ΔT decreases, and as thermal inertia decreases, ΔT increases.

Thermal Behavior of Natural Water Bodies

Although water in the laboratory has thermal properties that are similar to solids (Table 6-4), the thermal behavior of water bodies in natural conditions is influenced by an energy-transfer process that is not operative on land surfaces. Energy transfer in solids is primarily by **conduction** because such substances do not possess internal mass motion. Therefore, heat energy tends to be concentrated near their surfaces (e.g., 50 to 100 cm), causing relatively high daytime temperatures and relatively low nighttime temperatures (large ΔT) (Figure 6-17).

With natural water bodies, however, energy transfer is primarily by **convection**. This mechanism efficiently transfers heat energy to depths of 100 m or more. The mixing action is responsible for a relatively uniform temperature along the surface of a water body both day and night (small ΔT) (Figure 6-17). A water body therefore acts as if it has a very high thermal inertia. Consequently, water, under most environmental conditions, will be cooler than bounding land surfaces during the day heating period and warmer than the bounding land surfaces during the night cooling period (Figure 6-17).

Environmental Considerations

Natural materials can show considerable variability in their radiant temperatures because of **external environmental conditions**. The following conditions can make an important contribution to the radiant temperatures detected by a thermal IR sensor and hence to the quality of the resulting images:

1. *Surface winds* cause an increase in the convective rather than the radiative heat loss from surface materials. The amount of convective heat transfer to the air increases with increasing wind speed. Consequently, high-velocity winds lower the thermal contrast within the affected area. In addition to lowering contrast, wind can also cause ''thermal shapes.'' For example, **wind shadows** will form on the downwind, or leeward, side of protruding features; wind shadows are typically warmer than adjacent areas because of a reduction in wind velocity (Figure 6-18). **Wind streaks** are lines of alternating lighter and darker signatures

that parallel the wind direction (Figure 6-18). Streaking is thought to be the result of convective heat transfer at the boundary layer.

2. *Rain* tends to force all objects it strikes to cooler and more uniform temperatures. The eventual evaporation of the water provides additional cooling. Only hot targets such as forest fires or artificially heated objects may stand out for several hours following a rainstorm. *Dew* and *frost* will produce similar conditions.

3. *Clouds* and *fog* will usually completely mask thermal IR emissions from surface features (Figure 6-18). An image will depict cloud tops and fog as cold materials (dark tones). *Dust* and *smoke,* however, do not constitute a serious problem as they are normally penetrated by the relatively long wavelengths of thermal IR radiation (Figure 6-19). These aerosols pose a problem when they become **active condensation nuclei**, because they then absorb thermal IR radiation.

4. *Total cloud cover* will cause an overall reduction in the thermal contrast between surface features. Therefore, when an airborne mission is conducted beneath a cloud deck, the resulting images may lack sufficient tonal variation for making accurate interpretations.

5. During daylight, ground areas within *cloud shadows* are cool compared to their immediate surroundings (Figure 6-18). Because less solar radiation is available for absorption, surface materials that normally heat

Figure 6-17 Variations in radiant temperature of four materials for a diurnal cycle. (Adapted from Fagerlund et al. 1970.)

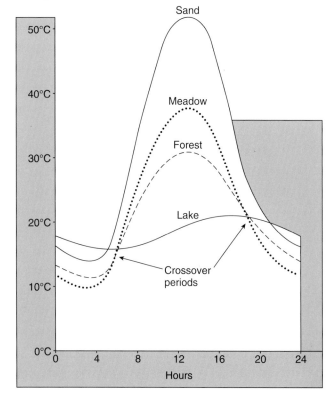

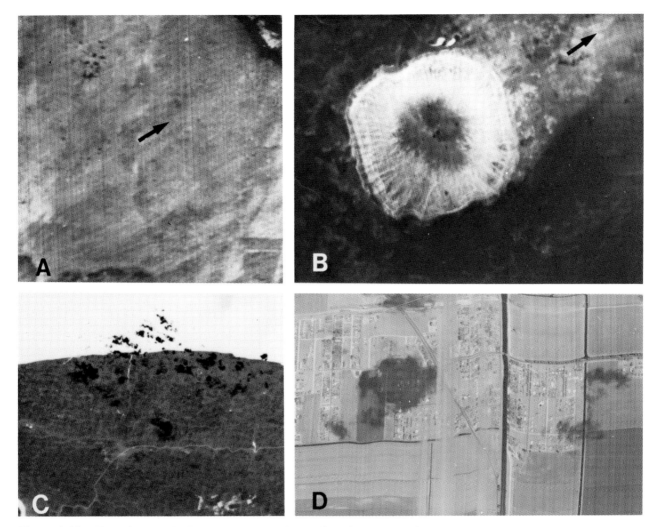

Figure 6-18 Effect of atmospheric phenomena on the quality of thermal IR images: (*A*) wind streaks, day image; (*B*) wind shadow, day image; (*C*) cloud tops, night image; and (*D*) cloud shadows, day image. Arrows show wind direction. Image B shows Meteor Crater, Arizona; its rim projects about 60 m above the surrounding plain.

and cool at different rates may have had time to approach or reach temperature equilibrium.

6. *High humidity* makes surface objects appear to have cooler and more uniform temperatures. The water molecules effectively absorb and reradiate thermal IR radiation (at longer wavelengths) that was emitted by surface objects.

7. In hill and valley terrain, the sinking and collection of cold air in low-lying areas minimizes ground temperature differences (*temperature inversions*), and the hill summits can become "warm islands" (Figure 6-20). Temperature inversions can be an especially serious problem for predawn and early morning missions.

All else remaining equal, the best thermal IR images are acquired under clear skies with no surface wind and low humidity.

Time-of-Day Considerations

For many applications, it is preferable to acquire thermal IR images at night to minimize differential heating patterns caused by **thermal shading** (i.e., **thermal shadows** from topography, trees, or buildings) (Figures 6-20 and 6-21). *Predawn images* are often preferred because they show the residual heat energy remaining in surface materials at the end of the night cooling period. This is the one time in the diurnal cycle when differences in thermal inertia (or thermal diffusivity) play a dominant role in controlling surface temperature. However, the largest thermal contrast during the night will normally occur in the early evening. Cloud shadows and surface winds are less likely to be a problem

Figure 6-19 Thermal IR image (*top*) and panchromatic airphoto (*bottom*) of the Nuns Canyon Fire, Sonoma District, California. Note that the thermal IR sensor "sees" through the thick smoke and permits a pinpoint location of the fire. The image also shows the extent of the burned area while smoke still lingers over the area. Note that it is impossible to locate the fire on the airphoto because of the extensive cover of smoke. (Courtesy U.S. Forest Service.)

for nighttime imaging, but the occurrence of fog, dew, and frost must be considered for predawn missions in certain regions.

Daytime images are usually obtained in *midafternoon,* when radiant temperatures and the thermal contrast between many terrain components are at their maximum levels (Figure 6-17). Topography is enhanced on daytime images because of the differential heating patterns on slopes; such enhancement may be an important interpretation aid for structural geology studies. However, thermal shadows associated with trees and cultural objects can complicate interpretations by creating distracting patterns of coolness and warmth.

Because of differential heating and cooling effects, there are two **temperature crossover periods** in the diurnal cycle when land-water temperatures approach each other, become equal, and then diverge. The two crossovers occupy a relatively small time window and usually occur around *dawn* and *sunset* (Figure 6-17). Because of minimal thermal contrasts, overflights are scheduled in order to avoid the crossover periods.

Interpretation of Thermal IR Images

Thermal IR images depict radiant temperature contrasts of a given ground area as tonal variations. Lighter tones represent warm features, and darker tones represent cooler features. The following descriptions provide a set of general guidelines for interpreting thermal IR images.

Water Versus Soil and Rock: Water bodies are generally cooler (darker tones) than soil and rock during the day, but surface temperatures are reversed at night with water being the warmest (lighter tones) (Figures 6-20 and 6-21). This is primarily because convection does not operate to transfer heat energy

Figure 6-20 Midafternoon (*top*) and predawn (*bottom*) thermal IR images (8 to 14 μm) of the Arbuckle Mountains and Washita River in south-central Oklahoma; north is to the bottom of the images. Note the tonal reversal associated with the Washita River and that thermal shadows are present only in the day image. The valleys are dark (cool) in the predawn image because of cold-air drainage. The predawn image records the smallest streams even though they are beneath a vegetation canopy. This occurred because the heat emitted by the water rose to warm the overhanging vegetation. (Courtesy P. Jan Cannon, Geology/Remote Sensing Consultant.)

in soil and rock. If the time of data collection is not known, the tonal signatures of water bodies are a reliable index: The image was generated at night if water bodies have lighter image tones than adjacent terrain; the image was made during the day if water bodies are depicted in darker tones than the adjacent terrain.

Vegetation: Tree foliage normally appears cool (dark tones) during daylight and warm (light tones) at night (Figure 6-21). Transpiration is at its maximum during the day, and this process lowers leaf temperature. This type of vegetation appears warm in nighttime images because of the high water content of the leaves and because of convective air warming associated with nighttime inversions (i.e., air temperature at crown height is warmer than at ground level). Because trees are essentially decoupled from the

Figure 6-21 Day and night thermal IR images (8 to 14 μm) of Capitol Mall, Sacramento, California. Image-acquisition times were 11:25 A.M. (*top*) and 12:20 A.M. (*bottom*). These correlative images illustrate the tonal reversals that occur for several types of targets. For example, a small circular water feature (right center) imaged black (cold) during the day but white (warm) at night. The same tonal reversal also occurred for various types of trees and ornamental shrubs, but the degree of tonal reversal was somewhat less intense (dark gray to light gray). An opposite tonal reversal (white to black) occurred for grass areas and most of the rooftops. Note that building and tree shadows are present in only the day image. (Courtesy George England, Daedalus Enterprises, Inc.)

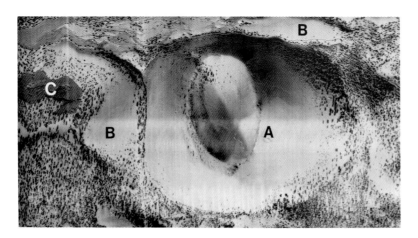

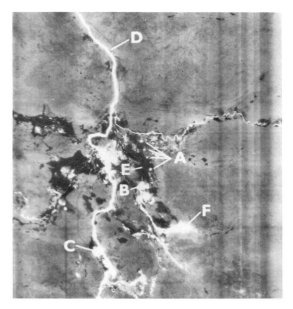

Figure 6-22 (*Above*) Midafternoon thermal IR image (8 to 14 μm) of a volcanic cinder cone named Sunset Crater (*A*), ashfall deposits (*B*), and a lava flow (*C*) in north-central Arizona. Although the basaltic ash and flow rocks have similar albedos and emissivities, they have very different internal thermal properties (Table 6-4), which account for the tonal disparity. Ponderosa pine trees are depicted in dark tones. (Courtesy U.S. Geological Survey.)

Figure 6-23 (*Right*) Predawn thermal IR image (8 to 14 μm) of Yellowstone National Park. Annotated features are: (*A*) test targets (600°C); (*B*) Kaleidoscope Geyser; (*C*) Excelsion Geyser; (*D*) Fire Hole River (10°C); (*E*) Snowfield (−14°C); and (*F*) Hot Lake (52°C). (Courtesy U.S. Forest Service.)

ground, they are isolated from the influence of the ground's thermal properties.

Grass and other low-lying vegetation are warmest during the day but rapidly approach local air temperature after sunset. On a calm night, the air next to the ground is apt to be cooler than a few meters aloft; consequently, low-lying vegetation will image in darker tones than deciduous or coniferous trees (Figure 6-21). Low-lying vegetation is well coupled to the ground, which tends to make it mimic the ground's thermal properties.

Damp Ground: All else remaining equal, damp ground is cooler (darker tones) than dry ground during both day and night because of evaporative cooling of contained moisture.

Consolidated Versus Unconsolidated Materials: Exposed rock surfaces and loosely consolidated materials generally appear in contrasting tones because of significant differences in thermal inertia or thermal diffusivity (Table 6-4). In daytime images, rock surfaces appear in darker tones than unconsolidated materials, whereas a tonal reversal occurs in nighttime images (Figure 6-22).

Pavement Materials: Materials such as concrete, asphalt, and packed dirt appear relatively warm (light tones) both day and night. They are generally good absorbers of solar radiation during the day, and because of their relatively high thermal capacities are able to radiate strongly for many hours after sunset. Packed earth appears in the darkest tones of the three materials on nighttime images because it loses its internal heat at the highest rate.

Metal Surfaces: Bare metal surfaces appear in dark tones on both day and night images because of their cold radiant temperatures. Their shiny surfaces have much lower emissivities than other substances found in aerial reconnaissance; consequently, they emit much less thermal IR radiation. Painted metal surfaces, however, assume the emissivity properties of a particular paint (e.g., white and black paints, Table 6-3).

High-Temperature Sources: Thermal IR emissions from targets such as forest fires, geothermal sources, or active volcanoes are relatively unaffected by time of day. The emitted radiation from these targets remains fairly constant, appearing hot (light tones) at all times.

Figure 6-24 Thermal IR images (8 to 14 μm) of the Connecticut River near Haddam, Connecticut. The images show the flow patterns of the thermal discharge from a nuclear power plant. The top image shows the effluent flowing downstream during low tide (night), whereas the bottom image shows the effluent being carried upstream during high tide (day). (Courtesy Ronald W. Stingelin, Resource Technologies Corp.)

Covered Features: In certain instances, a covered feature may be detected by a thermal IR sensor even though it has no penetration capability. For example, buried heating lines can often be detected on nighttime winter images by virtue of conductive heat transfer to the overlying materials along the alignment of the subsurface lines (Figure 6-13). Nighttime images may also record small streams beneath a vegetation canopy. This occurs when heat energy from the water rises and warms the overhanging vegetation, effectively superimposing the drainage pattern on the vegetation (Figure 6-20).

Ghosts: The ghost impressions of certain objects may appear in thermal IR images when the object that produced a temperature differential with the ground has been moved. For example, airplanes or automobiles that have been parked on asphalt or concrete during the day shield the surface from solar radiation. When removed, they leave ghost impressions that may be detectable for several hours.

Some Uses of Thermal IR Images

Various types of information can be derived through the analysis of thermal IR images that are applicable to many different disciplines. Several applications of thermal IR sensing are described in this section.

In geology, thermal IR images have been used for (1) structural mapping, (2) rock type discriminations and identifications (rocks must be exposed), (3) monitoring volcanic activity, and (4) identifying surface and subsurface hydrothermal features (Figure 6-23). In addition, multispectral sensing in the 8- to 14-μm wavelength region offers potential for detecting compositional differences in certain rock types.

Thermal IR images can be used to extract different types of information from surface water bodies. For example, hot effluents that result in water pollution when discharged into lakes, rivers, and estuaries are easily detected on night and day images because of the large temperature gradients between the effluents and the receiving water (Figure 6-24 and Plate 8). Thermal IR images have also been used to detect natural circulation patterns in water bodies and to locate areas where there is groundwater discharge into an ocean or estuary; freshwater sources have been discovered along the coast of Hawaii by analysis of thermal IR images. Oil slicks can be detected by virtue of temperature or emissivity differences between the oil and the surrounding water (Table 6-2 and Figure 6-25). Overflights at night are useful for surveying coastal areas and estuaries to detect illegal dumping from tanker ships.

Thermal IR imaging is also employed for the detection of water on the land surface. For instance, springs, seeps, and spring-fed streams are usually detectable on images acquired during subfreezing nights. Under this condition, spring water is much warmer than the ground and is shown in an image in light tones against a dark background. Figure 6-26 illustrates the nighttime detection of springs and spring-fed streams for an area in central Pennsylvania.

Figure 6-25 (*Left*) Nighttime thermal IR image (8 to 12 μm) of an oil slick in a marina. (Courtesy Lonnie Schuepbach and Deborah Hewitt, FLIR Systems, Inc.)

Figure 6-26 (*Below*) Predawn (5:35 A.M.) thermal IR image (8 to 14 μm) of springs and spring-fed streams in central Pennsylvania during the winter (leaf off) season. (Courtesy Ronald W. Stingelin, Resource Technologies Corp.)

Due to the evaporative cooling of contained water in a porous medium, thermal IR images are particularly useful for identifying subtle differences in soil moisture. For this reason, thermal IR images have been used to (1) detect leaks in earthen dams and irrigation canals, (2) monitor the delivery effectiveness of irrigation systems, (3) identify cropland areas that are excessively damp and need field drainage, and (4) delineate moisture anomalies in arid and semi-arid environments that can be related to potential shallow groundwater deposits. Figure 6-27 illustrates the ability of a thermal IR image to show different levels of soil moisture.

Thermal IR images, especially from the 3- to 5-μm wavelength region, have long been used to detect and monitor forest fires. Fire information is discernible in daylight or darkness and through dense smoke (Figures 6-19 and 6-28). The U.S. Forest Service routinely uses thermal IR scanners that are capable of producing in-flight paper images, which are dropped to the ground in tubes within minutes of data acquisition, or telemetering the data signals via a microwave downlink to a portable ground station, where the images are printed in near real time.

Two Daedalus scanners are ideally suited for fire detection and control operations. The AADS 1220 unit (Table 6-1) produces simultaneous images in the 3.0- to 5.5-μm region (**fire image**) and the 8.5- to 12.5-μm region (**terrain image**). The AADS 1230 system (Table 6-1) generates a thermal IR image with **event markers** (Figure 6-29). Two thermal IR detectors are employed that have maximum re-

sponses across the 3- to 5-μm and 8- to 14-μm regions. The HgCdTe detector (Figure 6-10) is used to produce the terrain image (ambient temperatures around 20°C) for locating the fire. The boundaries of the fire are clearly delineated in this image, as are spot fires occurring ahead of the active fire front and hotspots within the perimeter (Figure 6-29). The InSb detector (Figure 6-10) has a peak response that covers

Figure 6-27 (*Top left, page 131*) Visible (0.65 to 0.69 μm) (*top*) and thermal IR (8 to 14 μm) (*bottom*) images of three fields in different stages of drying. The three sections are as follows: (1) dry surface, (2) moist surface, and (3) wet surface. Note that sections 1 and 2 both appear light in the visible image, whereas section 3 is dark. However, in the thermal IR image, there is a definite tonal difference between all three sections, with 1 being the lightest, 2 being intermediate, and 3 being the darkest. (Courtesy Ray D. Jackson, U.S. Dept. of Agricuture.)

Figure 6-28 (*Top right, page 131*) Nighttime thermal IR image (3 to 5 μm) of wildfires in Montana. (Courtesy Northern Forest Fire Laboratory, U.S. Forest Service.)

Figure 6-29 (*Bottom, page 131*) Nighttime (4:50 A.M.) thermal IR image (8.5 to 12.5 μm) with event markers of wildfires in Canada. The image was produced by the Daedalus AADS 1230 system (Table 6-1). (Courtesy George England and Thomas Ory, Daedalus Enterprises, Inc.)

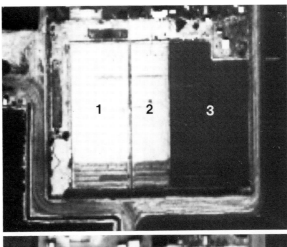

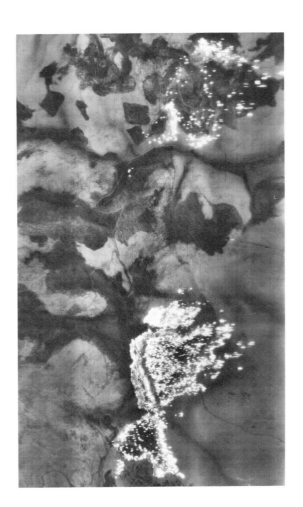

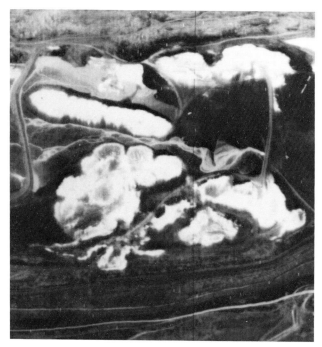

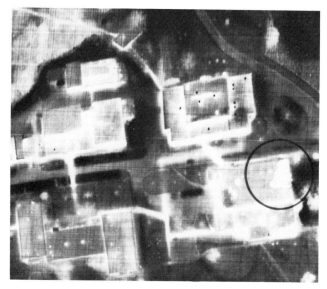

Figure 6-30 (*Above*) Nighttime thermal IR image (3 to 5 μm) of subsurface fires buring in coal refuse piles near Wilkes-Barre, Pennsylvania. (Courtesy Ronald W. Stingelin, Resource Technologies Corp.)

Figure 6-31 (*Right*) Thermal IR image (8.5 to 12.5 μm) showing energy losses from a roof and buried heating lines; an airphoto is included for comparison. Both were acquired on November 18, 1976, at the following times: airphoto, 3:00 P.M.; thermal IR image, 7:40 P.M. (Courtesy George England and Thomas Ory, Daedalus Enterprises, Inc.)

the normal temperature of a grass or timber fire (between 200° and 600°C). It is used to trigger an event threshold, which produces a mark on the edge of the image corresponding to the location of the high-temperature source (Figure 6-29). Any response above the threshold is a high-temperature source and is assumed to be a fire.

Airborne thermal IR surveys have proven useful for detecting and locating subsurface fires in landfills and coal refuse piles or culm banks. Visual observation is unreliable as a means of locating these subsurface fires because there are frequently no visible signs, such as smoke, at the surface. Figure 6-30 shows several subsurface fires burning in coal refuse piles in eastern Pennsylvania.

One of the fastest-growing applications of aerial thermal IR sensing is to detect heat losses from building rooftops and buried line distribution systems (Figure 6-31). The images identify problem areas (e.g., inadequate or missing insulation or pipe breaks) where retrofits or repairs can be made to minimize heat loss and thus reduce energy costs. The optimum conditions for data collection are calm, clear winter nights when all surfaces are dry and free of snow and ice. The value of a thermal IR heat-loss survey is illustrated in Figure 6-31.

With aerial surveying, it is not possible to monitor heat losses from the sides of buildings. A variety of stationary or mobile ground-based imaging equipment is available to accomplish this task. An example of a "drive-by" ground-based system is **VANSCAN**®, developed and patented by Daedalus Enterprises. This system consists of a special calibrated thermal IR scanner mounted in a van with complete film processing and tape recording equipment (Figure 6-32). The van traverses streets at about 16 km/hr (10 mi/hr) during data acquisition. **VANSCAN**® can produce hardcopy, black-and-white paper images a few seconds after data acquisition (Figure 6-33). In the laboratory, the analog tape record can be "sliced" into temperature ranges and displayed in color-coded images by the **DIGICOLOR**® process (Plate 8).

FLIR Systems

FLIR (for Forward Looking InfraRed) systems are real-time scanning instruments that were originally devel-

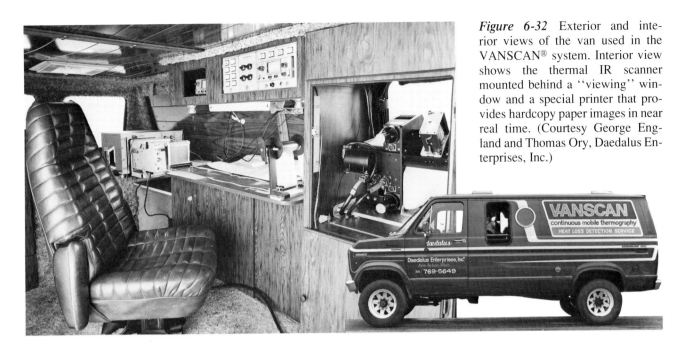

Figure 6-32 Exterior and interior views of the van used in the VANSCAN® system. Interior view shows the thermal IR scanner mounted behind a "viewing" window and a special printer that provides hardcopy paper images in near real time. (Courtesy George England and Thomas Ory, Daedalus Enterprises, Inc.)

Figure 6-33 (*Above*) Nighttime thermal IR image (8.5 to 12.5 μm) showing the fronts of four houses in Grand Haven, Michigan. VANSCAN® equipment was used to collect and process the data in near real time (see Figure 6-32). (VANSCAN® Thermogram by Daedalus Enterprises, Inc.; courtesy George England and Thomas Ory.)

Figure 6-34 (*Right*) FLIR system showing sphere-mounted imager (contains the thermal IR detectors, dewar, and scanning optics); the hand-held controller, which enables the operator to choose between two fields of view and to move the imager while tracking an object of interest; and the video monitor for viewing real-time images. Not shown is a videocassette recorder. (Courtesy John Robinson, FLIR Systems, Inc.)

oped for the military as a night surveillance tool. FLIRs are now being used for a wide range of applications, from search and rescue to police work to environmental monitoring. FLIRs are extremely portable and can be operated from a variety of light airplanes and helicopters, as well as from ground-based mobile platforms.

The FLIR systems manufactured by FLIR Systems, Inc. (Figure 6-34) utilize a detector system consisting of two HgCdTe elements mounted in an evacuated glass dewar (coolant is compressed argon gas). The incident thermal IR radiation is focused by the optical system, which includes scan mirrors (vertical and horizontal) necessary to move the image scene over the detectors. The detectors produce small variable-voltage signals, which are amplified and processed

Figure 6-35 Nighttime FLIR images (8 to 12 μm) that were helicopter-collected (*top pair*) and ground-collected (*bottom pair*). (Courtesy Lonnie Schuepbach and Deborah Hewitt, FLIR Systems, Inc.)

into standard video signals. These are then sent directly to a TV monitor, which displays high-resolution thermal IR images to the system operator in real time (Figure 6-34). The signals can also be recorded on videotape by a conventional videocassette recorder (VCR) for later playback.

Several FLIR images are shown in Figures 6-25 and 6-35. The thermal IR image of oil pollution (Figure 6-25) represents the wide field of view (15° × 28°). The superimposed reticle pattern indicates the portion of the image that can be instantly displayed on the TV monitor in the narrow field of view (3.25° × 7°).

Electro-Optical Sensors and Earth Observation Satellites

Although astronaut spacecraft programs have provided photography of the earth's surface, the bulk of remote sensing data has been obtained from robotic **earth observation satellites**. The sensor systems carried by these automated satellites are primarily of the electro-optical type. The two major types of earth observation satellites are **earth resources satellites** and **meteorology satellites**. The remainder of this chapter is devoted to a discussion of the international and national satellite programs that employ electro-optical sensors. Satellite programs employing radar sensors are described in Chapter 7.

Landsat Program

Landsat (for **land satellite**) represents the first international satellite program designed specifically for collecting synoptic and repetitive multispectral image data of the earth's surface that could be used to analyze and monitor the world's resources and environments. It was initiated by NASA with assistance from the Departments of Interior and Agriculture. All Landsat images were to be collected under an "open skies" policy, meaning that the images would be acquired on a worldwide basis and users from any country would have access to any of the collected data. Until January 1975, the Landsat program was known as the *Earth Resources Technology Satellite (ERTS) program.*

In 1982, when it was determined that the Landsat program was no longer experimental, NASA transferred the operation and management of the Landsat program to the Department of Commerce's *National Oceanic and Atmospheric Administration (NOAA).* In September 1985, in accordance with the *Land Remote Sensing Act,* the Landsat program was transferred to the *Earth Observation Satellite Corporation (EOSAT),* a private firm formed as a partnership by the RCA Corporation and Hughes Aircraft Company. EOSAT currently operates the existing Landsat satellites, receives and disseminates image data, and designs new platforms and sensors for future Landsat missions.

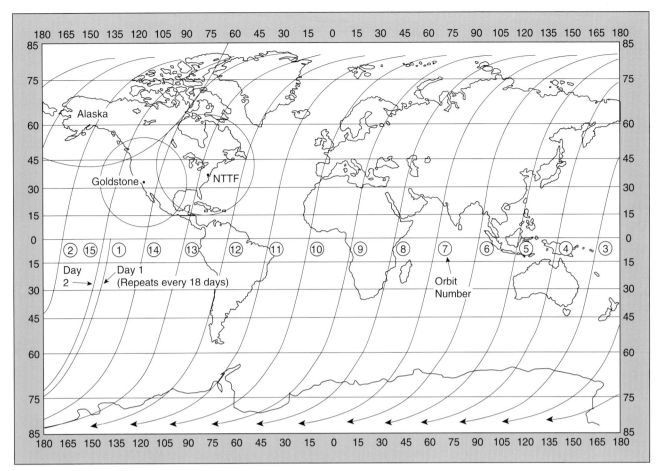

Figure 6-36 Daylight ground tracks for *Landsats 1, 2,* and *3* for a single day (orbits 1 through 14). The earth rotates 2,875 km to the east at the equator between each pass. Ground receiving stations for the United States are also shown. (Courtesy National Aeronautics and Space Administration.)

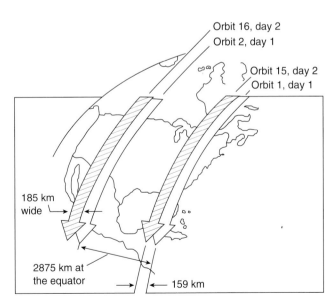

Figure 6-37 Coverage tracks of *Landsats 1–3* for successive orbits on the same day and for the following day. (Adapted from Short 1982.)

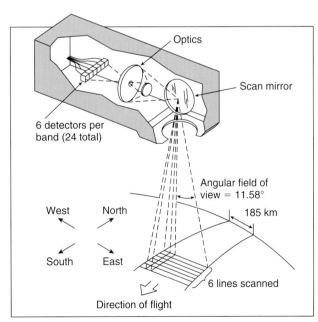

Figure 6-38 Schematic representation of the Landsat MSS. (Adapted from Short 1982.)

To date, five Landsats have operated successfully from earth orbit (Table 6-5). Landsat launches were from Vandenberg Air Force Base in California. Their electro-optical sensors have acquired several million images since the first Landsat was launched in 1972. Three different types of sensors have been used in various combinations on the five Landsats. They are the **Return Beam Vidicon (RBV)** camera, the **Multispectral Scanner (MSS)**, and the **Thematic Mapper (TM)**. Scientists from more than 100 countries have used images from these sensors for a broad range of applications.

Landsats 1, 2, and *3* far exceeded their design life of 1 year (Table 6-5). Although it was planned that *Landsat 4* would have a 3-year mission life, power problems affected its early performance, and it was placed in a "reduced-mission mode" in 1983. Its backup, *Landsat 5,* was placed in orbit in March 1984, more than 1 year ahead of schedule. *Landsat 5* is currently used for routine data collection, but *Landsat 4* is used only for MSS acquisitions. At the first signs of further failure, *Landsat 4* may be placed in a lower "parking" orbit, where it could be retrieved for repairs during a future Space Shuttle mission. *Landsat 6* is scheduled for launch sometime in 1992.

Landsat Orbital Characteristics

Landsats 1, 2, and *3* were placed in earth orbit at a nominal altitude of 915 km. The orbits are **near-polar** in that they pass within 8° of the North and South Poles. Each satellite orbits the earth 14 times per day (Figure 6-36). Daylight (imaging) passes are southbound (**descending**), whereas northbound passes cover the dark side of the earth (**ascending**). In 252 orbits, or every 18 days, each satellite covers the entire earth, excepting polar areas above 82° (Figure 6-36). This means that a satellite passes over the same point on the earth's surface once every 18 days.

Figure 6-37 shows the spatial characteristics of the ground tracks or swath widths (185 km wide) for *Landsats 1, 2,* and *3*. From one orbit to the next, the satellite moves 2,875 km to the west at the equator (Earth is rotating west to east beneath the satellite). The following day, on orbit 15, the satellite is approximately back to its original position but displaced westward from orbit 1 by 159 km at the equator. This procedure continues for 18 days, wherein orbit 252 falls

directly over orbit 1. The sidelap is 26 km (185 − 159 km), or 14 percent, at the equator from adjacent swaths on consecutive days. The amount of sidelap increases with increasing latitude, becoming 34 percent at 40° and 85 percent at 80°. Sidelap coverage enables portions of adjacent Landsat images to be viewed stereoscopically.

The orbits of *Landsats 1, 2,* and *3* are **sun-synchronous**, meaning that each satellite passes over all places on the earth having the same latitude at approximately the same local time. The daylight crossing on each pass at the equator occurs at about 9:30 A.M. local time. Other points are passed slightly after this time in the northern hemisphere and slightly before this time in the southern hemisphere. A sun-synchronous orbit ensures that solar-illumination conditions are repeatable during specific seasons.

The orbits of *Landsats 4* and *5* are also near-polar and sun-synchronous, but the nominal altitude of each satellite is 705 km. The lower orbit results in an earth-coverage pattern that differs from that of the earlier Landsats. For example, the coverage cycle is 16 days (233 orbits), descending equatorial crossings occur at about 9:45 A.M. local time, and sidelap between adjacent swaths is reduced to 7.6 percent at the equator. The lower orbit is necessary to achieve a ground resolution element of 30 m for the Thematic Mapper carried on *Landsats 4* and *5*. The following discussion summarizes the characteristics of the Landsat electro-optical sensors.

Multispectral Scanner

The **Multispectral Scanner (MSS)** is an across-track scanning device carried by all five Landsats. The MSS collects earth-reflected sunlight in four contiguous spectral bands between 0.5 and 1.1 μm (Table 6-6).* From Table 6-6 one will note that two numbering systems are used to designate the four bands. What are known as MSS bands 4, 5, 6, and 7 for *Landsats 1, 2,* and *3* are called, respectively, MSS bands 1, 2, 3, and 4 for *Landsats 4* and *5*. The former numbering system remains in place for identifying MSS images from the earlier satellites. The MSS was the primary, or lead, sensor on the first three Landsats.

*The MSS on Landsat 3 *also contained a thermal IR band (10.4 to 12.6 μm), but it never became operational because of a gas contamination problem.*

TABLE 6-5	Landsat Periods of Operation	
Landsat	Launch	Deactivation
1	July 23, 1972	January 6, 1978
2	January 22, 1975	July 27, 1983
3	March 5, 1978	September 7, 1983
4	July 16, 1982	—
5	March 1, 1984	—

A schematic representation of the MSS sensor is shown in Figure 6-38. The MSS incorporates six detectors per band at the focal plane (24 total detectors). For each sweep of the collection mirror, six lines of the ground are scanned simultaneously in each of the four bands; image data are recorded only during the eastward sweep of the oscillating mirror. The across-track swath is 185 km wide and perpendicular to the orbital path. The forward motion of the satellite provides the along-track buildup of scan lines.

For the first three Landsats, the angular field of view (AFOV) for the MSS was 11.58°. However, the AFOV was increased to 14.9° for *Landsats 4* and *5* in order to maintain the nominal 185-km swath width (altitude decreased from 915 to 705 km). The instantaneous field of view (IFOV) for the MSS on the earlier satellites was 0.086 mrad; this yielded a ground resolution cell measuring 79 × 79 m. The IFOV of the MSS on *Landsats 4* and *5* was changed to 0.116 mrad, which produces an 82 × 82-m ground resolution cell.

Strips of reflectance data are collected by the MSS along a ground track at a fixed width, or swath. Once telemetered to ground stations, the MSS data strips are "framed" into individual **scenes** that cover a nominal ground area measuring 185 × 185 km. A standard MSS image deviates from a rectangle by the distance the earth rotated beneath the spacecraft during the 25 sec the data were being collected from the top to the bottom of a MSS scene. Consequently, the image incorporates **rotational skew** and is represented as a **slanted parallelogram**. MSS images of a portion of New Zealand's South Island are presented in Figure 6-39.

The following discussion describes the responses of each MSS band to various surface materials; band designations are for *Landsats 1–3*. Many of these responses are depicted in image form in Figure 6-39 and Plate 9. It is also useful to compare each spectral band with the reflectance curves shown in Figures 2-35 and 2-37.

Band 4 (0.5 to 0.6 μm, green) is most useful for studying water features. It has the greatest depth penetration in clear water of the four MSS bands, and it is sensitive to turbidity (suspended sediments) and pollution patterns. A major drawback is its sensitivity to atmospheric haze, which causes standard band 4 images to generally lack tonal contrast.

Band 5 (0.6 to 0.7 μm, red) spans one of the chloro-phyll absorption regions and thus shows good contrast between vegetated and most nonvegetated surfaces. Haze penetration is better than for band 4, but water penetration is significantly less. Band 5 provides a good spectral region for differentiating silt-laden and clear water and many natural and cultural features.

Spectral responses in the wavelength regions represented in band 6 (0.7 to 0.8 μm, near IR) and band 7 (0.8 to 1.1 μm, near IR) tend to be similar for most surface features. They are responsive to the amount of vegetation, or biomass, present in a scene; this can facilitate the identification of different crop or timber types, separating bare soil from cropland, or discriminating planted and natural vegetation. Because water is a strong absorber of near IR radiation, band 6 and 7 images accentuate the contrast between land and water surfaces. In addition, moist ground registers in darker tones than dry land. Both bands have an excellent haze-penetration capability. For visual interpretations, band 7 images are usually preferred over band 6 images because they show greater tonal contrast.

Certain surface materials have similar spectral responses over the wavelength interval of the four MSS bands, causing them to appear in similar tones in each image. For example, clouds and dry snow are rendered in light tones, whereas young volcanic features and asphalt register in dark tones.

In practice, any three of the four black-and-white positive transparencies from the MSS bands may be combined into a **false color composite** by the **additive color process** (see Chapter 2). This is commonly done with bands 4, 5, and 7 (bands 1, 2, and 4 for *Landsats 4* and *5*) to form a **color infrared composite**. With one popular photographic method, film positives of bands 4, 5, and 7 are punch registered with a sheet of color positive film in darkness. The transparencies are then separately contact printed onto the color film with blue, green, and red light. Band-color combinations are as follows: band 4 is projected onto the color film through blue light, band 5 through green light, and band 7 through red light. After processing, the result is a false color image that incorporates the same color assignments as those employed in color infrared photography: Dominant green reflectance reproduces as blue; dominant red reflectance reproduces as green; and dominant near IR reflectance reproduces as red (Figure 2-27).

TABLE 6-6   Characteristics of the Landsat Multispectral Scanner

Landsats[a] 1–3	Landsats 4–5	Wavelength Range (μm)	Spectral Location	Resolution (m)	
				Landsats 1–3	Landsats 4–5
Band 4	Band 1	0.5–0.6	Green	79 m	82 m
Band 5	Band 2	0.6–0.7	Red	79 m	82 m
Band 6	Band 3	0.7–0.8	Near IR	79 m	82 m
Band 7	Band 4	0.8–1.1	Near IR	79 m	82 m

[a]*The MSS on* Landsat 3 *contained a thermal IR band (band 8), but it never became operational.*

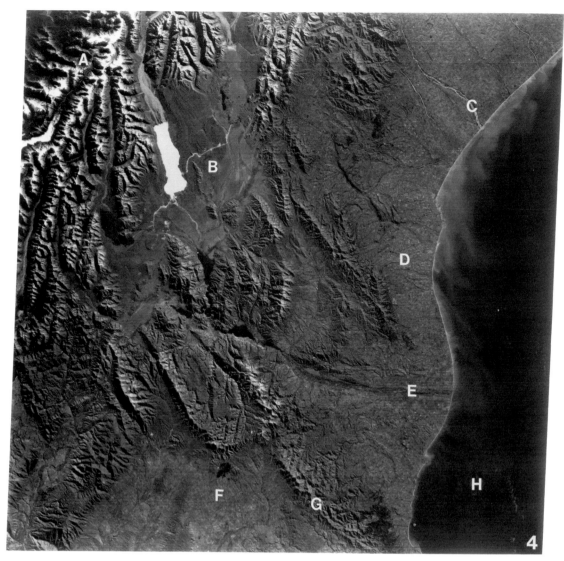

Figure 6-39 (*Pages 138–141*) *Landsat 2* MSS images of a portion of New Zealand's South Island, acquired April 6, 1977. Image numbers correspond to MSS bands 4–7. Annotations are as follows: (*A*) Southern Alps, (*B*) Mackenzie Basin, (*C*) Rangitata River, (*D*) Canterbury Plain, (*E*) Waitaki River, (*F*) Waikouaiti Downs, (*G*) Kakanui Mts., and (*H*) South Pacific Ocean. Each image measures 185 × 185 km. Compare with Plate 9.

Such composite images resemble conventional color infrared photographs but at a diminished resolution (Plate 9). Typical colors are as follows:

Clear water	Black
Silt-laden water	Light to dark blue
Deciduous forest	Dark red
Coniferous forest	Brownish red
Grassland	Pinkish red
Red soils and rocks	Yellow or green
Sand	White or yellow
Cities	Bluish gray
Snow and clouds	White
Lava and asphalt	Black

Return Beam Vidicon Camera

Landsats 1 and *2* carried a multispectral Return Beam Vidicon (RBV) camera system, consisting of three boresighted vidicon tube cameras that collected reflectance data in three spectral bands (0.47 to 0.57 μm, 0.58 to 0.68 μm, and 0.69 to 0.83 μm). However, electrical problems, experienced within weeks after the launch of *Landsat 1*, curtailed operation of the RBV system. The RBV unit was not activated on *Landsat 2*, largely because of the success of the MSS sensor.

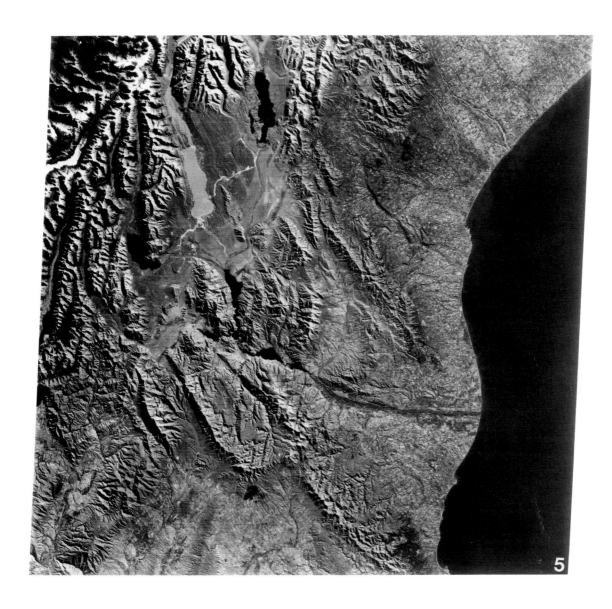

On *Landsat 3*, a panchromatic RBV system served as a successful high-resolution mapping device. The unit consisted of two vidicon tube cameras that viewed 98 × 98-km ground scenes with 16-km overlap and 13-km sidelap. The side-by-side mounted cameras operated alternately, enabling four subscene images to closely approximate the same ground coverage of a single MSS image. The two RBVs had a single sensitivity of 0.50 to 0.75 μm, and ground resolution was about 30 m.

An RBV subscene image of the Grand Canyon is shown in Figure 6-40. Note that, unlike an MSS image, the field of view is not skewed. This is because an image was "frozen" on the vidicon faceplate during shuttering.

Thematic Mapper

The lead sensor carried by *Landsats 4* and *5* is the **Thematic Mapper** (TM), which is a second-generation, earth-observation across-track scanner. Although relying heavily on the technology of the MSS, the TM incorporates improved resolution, additional spectral bands, higher radiometric sensitivity, and improved geometric fidelity over its MSS predecessor. TM is designed to operate simultaneously with the MSS.

The TM simultaneously collects radiance data in seven narrow spectral bands between 0.45 and 12.5 μm. Band designations and spectral ranges are presented in Table 6-7. Of the seven bands, three are in the spectral range of the MSS (green through near IR), whereas the remaining four provide new spectral coverage. The following discussion describes responses of each TM band to various surface materials. Many of these responses are depicted in image form for the San Francisco Bay region shown in Figure 6-41 and Plate 10. It is also useful to compare each spectral band with the reflectance curves in Figures 2-35, 2-37, and 6-7.

Band 1 (0.45 to 0.52 μm, blue-green) is a new band that was included for water penetration studies, making its images especially useful for bathymetric and coastal studies. This band is also capable of differentiating soil and rock surfaces from vegetation and for detecting cultural features. Smoke plumes are the most apparent in band 1 images. It

6

is the most sensitive of the TM bands to atmospheric haze, and consequently, standard band 1 images may lack tonal contrast.

The three bands that approximate the spectral region of the four MSS bands are band 2 (0.52 to 0.60 μm, green), band 3 (0.63 to 0.69 μm, red), and band 4 (0.76 to 0.90 μm, near IR). The green and red bands are narrower than their MSS counterparts to improve the sensitivity to spectral changes associated with chlorophyll reflection and absorption in this wavelength interval. Band 4 is narrower than the combined widths of MSS bands 6 and 7, having its center in a spectral region of maximum sensitivity to vegetation vigor (peak mesophyll reflectance).

Band 2 is sensitive to water turbidity differences plus sediment and pollution plumes. Because it covers the green reflectance peak from leaf surfaces, it can be useful for discriminating broad classes of vegetation. It is also useful for identifying cultural features.

Band 3 senses in a strong chlorophyll absorption region and a strong reflectance region for most soils. It is therefore an excellent band for discriminating vegetation and soil. It is also in a good spectral region for delineating snow cover and discriminating urban and rural areas.

Band 4 operates in the best spectral region to distin-guish vegetation varieties and conditions. Because water is a strong absorber of near IR radiation, it is also useful for locating and delineating water bodies, distinguishing between dry and moist soils, and providing information about coastal wetlands, swamps, and flooded areas.

Band 5 (1.55 to 1.75 μm, mid IR) is responsive to changes in leaf-tissue water content (reflectance decreases as water content increases), which can be related to plant vigor or different species (e.g., succulent versus woody species). Band 5 is also useful for discriminating moisture content in soils. This band is especially sensitive to the presence or absence of ferric iron or hematite in rocks (reflectance increases as ferric iron content increases). Band 5 images can be used to discriminate between snow (light tones) and clouds (dark tones); this capability provides a very useful tool for monitoring snow-pack areas and forecasting the snowmelt runoff.

Band 6 (10.4 to 12.5 μm, thermal IR) is useful in a wide range of heat-mapping applications (e.g., estimates of soil moisture, identifying different types of rocks; detecting thermal pollution in water bodies). The detectors for this band are designed to measure radiant surface temperatures from about $-100°C$ to $+150°C$ and are capable of sensing a radiant temperature difference of about 0.6°C. Normally,

7

Band No.	Wavelength Range (μm)	Spectral Location	Resolution (m)
1	0.45–0.52	Blue-green	30
2	0.52–0.60	Green	30
3	0.63–0.69	Red	30
4	0.76–0.90	Near IR	30
5[a]	1.55–1.75	Mid IR	30
6[a,b]	10.40–12.50	Thermal IR	120
7[a,b]	2.08–2.35	Mid IR	30

TABLE 6-7 Characteristics of the Landsat Thematic Mapper

[a]*The detectors for bands 5, 6, and 7 are cooled to very low temperatures to suppress electronic noise caused by molecular motions in the detectors.*
[b]*Bands 6 and 7 are out of wavelength sequence because band 7 was added to the TM instrument late in the design process.*

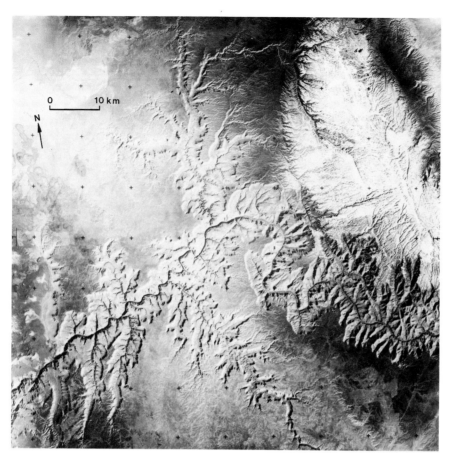

Figure 6-40 *Landsat 3* RBV subscene image (0.50 to 0.75 μm) of the Grand Canyon, Arizona, acquired March 22, 1981. The image crosses are reseau marks (etched on the camera's vidicon faceplate) used to correct image geometry.

band 6 acquires data during daylight passes, but it can also be activated on the nighttime side of each orbit because it is not dependent upon reflected sunlight (Figure 6-42).

Band 7 (2.08 to 2.35 μm, mid IR) coincides with an important absorption band caused by hydrous minerals (e.g., clay, mica, and some oxides and sulfates). This makes band 7 images valuable in lithologic mapping and for detecting clay alteration zones (dark image tones) associated with mineral deposits, such as copper. Analogous to band 5, this band is sensitive to moisture content variation in both vegetation and soils.

Bands 4–7 are not affected by atmospheric haze. Band 6 and band 7 images normally do not show smoke because the wavelengths in question are larger than most smoke particles, thus facilitating penetration. In addition, thin clouds are penetrated by band 5 and band 7 wavelengths, and they do not, consequently, register in the images.

For the six nonthermal TM bands, 20 different three-band color composite images can be produced.* A color infrared composite is produced by projecting band 2, 3, and 4 positive images through blue, green, and red light, respectively. A natural color composite can be made by projecting band 1, 2, and 3 positive images through blue, green, and red light, respectively. Other three-band combinations can be color-composited to enhance features of special interest. A common one for rock discriminations is a false color composite of bands 1 (printed blue), 4 (printed green), and 5

The procedure for determining the number of color composites that can be produced by combining single-band black-and-white images is described in a later section of this chapter.

(printed red). These three color composites for the San Francisco Bay region are shown in Plate 10.

The AFOV for the TM instrument is 14.9°, which produces a 185-km swath width from a Landsat altitude of 705 km. The standard full-scene TM image measures 185 km across-track by 170 km along track and is depicted as a slanted parallelogram.

The IFOV for the visible and reflected IR bands (bands 1–5 and 7) is 0.0425 mrad; this produces a ground resolution cell of 30 × 30 m. The IFOV for the thermal IR band (band 6) is 0.17 mrad; this produces a ground resolution cell measuring 120 × 120 m. The larger cell size for band 6 translates to a longer dwell time, ensuring that a signal of adequate strength can be collected to maintain its temperature resolution of 0.6°C.

The TM instrument scans in both directions normal to the ground track (i.e., **bidirectional scanning**). A 16-element detector array is used for each of the 30-m resolution bands, and 4 detectors are used for the thermal IR band. Therefore, 16 scan lines are generated for bands 1–5 and 7 and 4 are generated for band 6 during each sweep of the mirror.

Landsat 6

At the time of this writing, *Landsat 6* was scheduled for launch sometime in 1992. It will carry a redesigned Thematic Mapper called the **Enhanced Thematic Mapper**

Figure 6-41 (*Pages 143–146*) *Landsat 4* TM subscene images of the San Francisco Bay region, acquired December 31, 1982. Image numbers correspond to TM bands 1–7. Compare with Plate 10. (Courtesy U.S. Geological Survey.)

(ETM). This across-track scanner will contain a new visible–near IR band operating in the 0.5- to 0.9-μm range with a spatial resolution of 15 m. To ensure continuity with the TMs on *Landsats 4* and *5*, the ETM will acquire data in the same seven spectral bands. *Landsat 6* will not carry the standard MSS sensor. Instead, it will carry a **Multispectral Scanner Emulator** (EMSS) to process on-board TM data as the functional equivalent of MSS data.

Obtaining Landsat Image Data

The data collected by *Landsats 4* and *5* are transmitted to one of more than 14 ground receiving stations located around the world. The foreign stations process final Landsat film and tape products locally, whereas U.S. stations deliver the data directly to EOSAT. Once processed, the data are archived at the EROS Data Center (EDC) in Sioux Falls, South Dakota (see Chapter 5). Data from *Landsats 1–3* are also archived at EDC. To obtain data not in the archives at EDC or a foreign station, a user can request EOSAT to collect specific data for a specific area during a future pass of the satellite.

Foreign stations are located in the following countries: Argentina, Australia, Brazil, Canada, Ecuador, India, Indonesia, Italy, Japan, People's Republic of China, Saudi Arabia, South Africa, Spain, Sweden, and Thailand. A receiving station is planned for Pakistan. These stations collect MSS and TM data within their zones of ''visibility''; this is about a 2,100-km radius circle centered on each location.

''Raw'' Landsat data undergo a process of annotation, correction (geometric and radiometric), and conversion to film and digital tape products. Landsat film images incorporate a gray-scale wedge and an annotation block that contains a variety of information (e.g., date, geographic coordinates, sun elevation angle, sun azimuth, type and band of sensor, and scene identification number). The digital products, called **computer-compatible tapes**, or **CCTs**, enable the image data to be subjected to special computer processing routines (discussed in Chapter 15).

Color and black-and-white prints are also available from EOSAT at scales of 1:1,000,000; 1:500,000; and 1:250,000. Black-and-white negatives and color positives are normally produced at a scale of 1:1,000,000.

Because of the large amount of data generated by the Landsat program, a computerized data base is used at EDC to record essential data concerning each scene. The data base includes information concerning date, center point coordinates, type of sensor, band quality, cloud cover, and film and

Figure 6-41
(*Continued*).

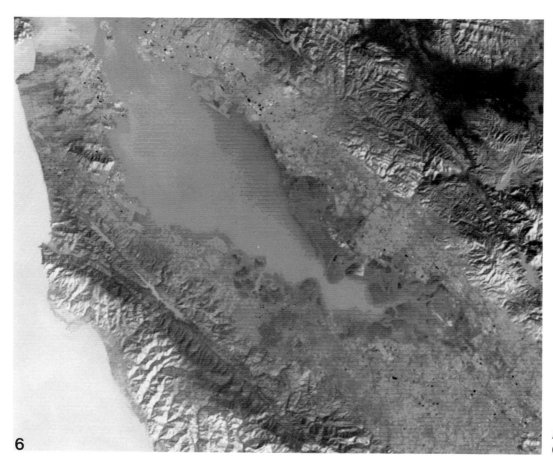

6

Figure 6-41
(*Continued*).

7

0   10 km

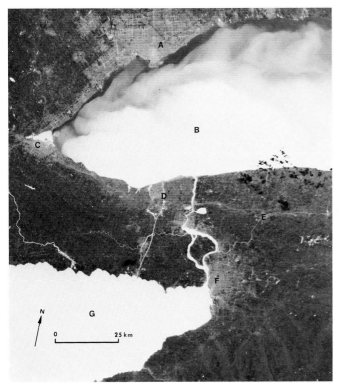

Figure 6-42 TM band 6 thermal IR image (10.4 to 12.5 μm) of the Lake Erie–Lake Ontario region, acquired August 22, 1982 by the *Landsat 4* TM at about 2:00 A.M. local time. Annotations are as follows: (*A*) Toronto, (*B*) Lake Ontario, (*C*) Hamilton, (*D*) Welland Canal, (*E*) Erie Canal, (*F*) Buffalo, and (*G*) Lake Erie.

CCT availability. For users interested in a particular area, EOSAT will compile a computer listing of available scenes; this inquiry service is free and covers all types of Landsat data acquired since 1972. Inquiries regarding computer searches, product prices, and ordering procedures should be directed to Customer Service Department, Earth Observation Satellite Co., 4300 Forbes Boulevard, Lanham, MD 20706.

Heat Capacity Mapping Mission Satellite

The **Heat Capacity Mapping Mission (HCMM)** was an experimental space project sponsored by NASA. The *HCMM* program was to investigate the usefulness of day and night thermal IR images for discriminating different surface materials and ground states from earth orbit. The satellite carried a two-channel across-track scanner called the **Heat Capacity Mapping Radiometer (HCMR)**. The visible–near IR band had a spectral sensitivity of 0.5 to 1.1 μm (equals the combined wavelength span of Landsat MSS bands 4–7), whereas the thermal IR band's sensitivity was 10.5 to 12.5 μm. Ground resolution was 500 m for the visible–near IR band and 600 m for the thermal IR band. The scanner's swath width was 700 km.

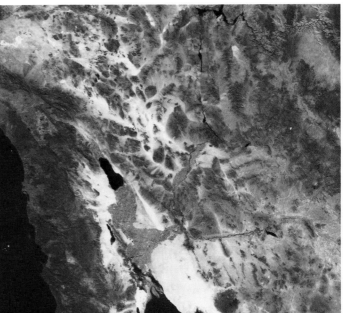

Figure 6-43 *HCMM* subscene images of the southwestern United States and northern Mexico, acquired October 24, 1979: thermal IR image (10.5 to 12.5 μm) (*top*); visible–near IR image (0.5 to 1.1 μm) (*bottom*). The east-trending dark streaks in the thermal IR image are interpreted as wind-oriented condensation.

The HCMM satellite was launched on April 26, 1978, and placed in earth orbit at an altitude of 620 km. HCMM flight operations were ended on September 30, 1980, because of battery deterioration. Although HCMM had a design life of 1 year, it worked for nearly 2.5 years.

Because of HCMM's orbital characteristics, both day and night thermal coverage was possible for the midlatitudes (16-day revisit cycle for both day and night orbits). For example, at 40° N latitude, coverage occurred at about 1:30 P.M. and 2:30 A.M. local time. The standard image area is 700 × 700 km, which is equivalent to about 16 Landsat MSS images. Subscene images covering the southwestern United States and northern Mexico are shown in Figure 6-43.

Because *HCMM* was an experimental program, it was not designed to obtain global coverage. However, image coverage is available for extensive areas of North America, western Europe, northern Africa, and eastern Australia (about 38,000 different images). Image data in film and CTT formats and user guides are available from the National Space Science Data Center, World Data Center-A, Code 601, NASA Goddard Space Flight Center, Greenbelt, MD 20771.

SPOT Program

In the late 1970s, the French government initiated the **Système Pour l'Observation de la Terre (SPOT)**, or "Earth Observation System" program. SPOT was designed to be a fully operational and commercial remote sensing program, as opposed to a government-sponsored research and development program. Analogous to the Landsat program, the data were to be collected on a global basis, but unlike the Landsat program, the satellites were to be equipped with high-resolution along-track scanners with across-track pointable optics that would provide both nadir and off-nadir viewing.

The first SPOT satellite *(SPOT 1)* was launched from the Kourou Launch Range in French Guiana on February 21, 1986, onboard an Ariane launch vehicle. It was designed and is operated by the Centre National d'Études Spatiales (CNES), the French space agency. *SPOT 1* is in a sun-synchronous, 832-km orbit, with descending equatorial crossings occurring at 10:30 A.M. local time. Its orbital path consists of 369 tracks, by which the satellite covers the earth every 26 days. Mission control is at Toulouse, France.

The sensor payload for *SPOT 1* consists of two identical **High-Resolution Visible (HRV)** along-track scanners. The HRV instruments use closely packed one-dimensional linear arrays of charge coupled devices (CCDs) that are aligned perpendicular to the orbital track (Figure 6-14). There is one detector array for each spectral band.

Each HRV is designed to operate in either a black-and-white panchromatic mode or a multispectral mode. In the **panchromatic mode**, a 6,000-element array produces a 10-m ground resolution element for nadir views. This is currently the best resolution available for any civilian remote sensing satellite. The panchromatic band is sensitive to the spectral region extending from 0.51 to 0.70 μm (green-red). An HRV panchromatic subscene image of Long Island, New York, is shown in Figure 6-44.

In the **multispectral mode**, the HRV operates in three narrow spectral regions: band 1 = 0.50 to 0.59 μm (green), band 2 = 0.61 to 0.68 μm (red), and band 3 = 0.79 to 0.89 μm (near IR). These band sensitivities were selected to optimize the capability of discriminating different plant species. For each band, a 3,000-element array is used; this sam-

pling mesh corresponds to a ground resolution element that measures 20 × 20 m. Black-and-white images are normally used to make color infrared composites. Band color assignments are as follows: band 1 printed blue, band 2 printed green, and band 3 printed red. Plate 11 shows a color infrared composite image of the same area on Long Island as the panchromatic image (Figure 6-44).

Each instrument's angular field of view is 4.13°, which produces a 60-km swath width for nadir viewing conditions. Single SPOT images for vertical views are formatted to cover a 60 × 60-km ground area. The two HRVs can also be oriented separately to cover adjacent ground segments. In this "twin-vertical" configuration, the combined swath width is 117 km with a 3-km sidelap.

HRV viewing angles can be adjusted by ground command up to 27° to either side of the satellite's vertical track. Such off-nadir viewing gives imaging access to any area of interest within a 950-km-wide corridor centered on each ground track. Thus, any location can be "revisited" several times within the normal 26-day vertical coverage cycle. The number of revisits is latitude-dependent. For example, at the equator, a single location can be viewed seven times in 26 days, whereas at 45° N or S latitude the number of revisits can total 11. This capability is important for studying time-varying phenomena (e.g., floods, industrial accidents, volcanic eruptions) or obtaining cloud-free coverage.

A second unique capability of its off-nadir viewing is to produce stereoscopic image pairs of a given ground area. This is done by recording the same ground area from two different viewing angles. A stereopair of a coastal region near Marseille, France, is shown in Figure 6-45.

The second satellite, *SPOT 2*, was successfully launched into its 832-km, sun-synchronous orbit on January 21, 1990. Although *SPOT 2* was originally intended to replace *SPOT 1*, the latter is still operating flawlessly, and both satellites are being operated concurrently. *SPOT 2* follows an identical orbit to *SPOT 1*, though 180°, or half a revolution, out of phase; this means that one of the SPOT satellites is over the same locale every 13 days. Both satellites carry identical HRV imaging instruments.

SPOT 3, currently under construction, is identical to the first two satellites and will be launched when needed as a replacement for *SPOT 1* or *2*. *SPOT 4*, for which final funding was approved in 1989, is tentatively scheduled for launch in the mid-1990s. *SPOT 4* will provide an increased spectral capability by the addition of a middle IR band to the two HRVs; it will also carry a third imaging instrument for small-scale ocean and vegetation studies. The design life for *SPOT 4* will be 4 years.

The SPOT program has established a global network of ground receiving stations and processing centers. As of late 1991, there were 13 SPOT receiving stations operating in 12 countries. Worldwide marketing and distribution operations are anchored by private companies (e.g., SPOT Image Corp. in the United States, SPOT Image in France, Satimage in Sweden, and SPOT Imaging Services in Australia) and distributors in about 50 countries.

SPOT markets computer-compatible tapes (CCTs) and a variety of photographic products (prints and transparencies). There are currently more than 800,000 scenes in the worldwide SPOT archive. SPOT will, free of charge, provide a user with a computer listing of available scenes for a particular area. In the United States, inquiries regarding

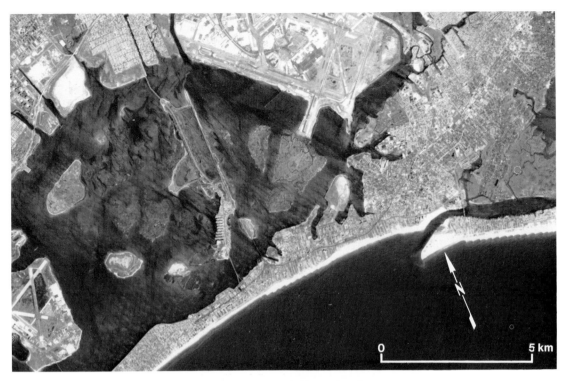

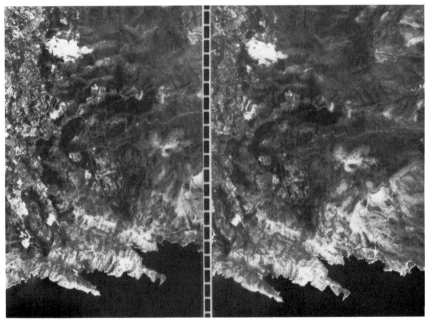

Figure 6-44 (*Above*) *SPOT 1* HRV panchromatic subscene image (0.51 to 0.70 μm) of a portion of Long Island, New York, showing John F. Kennedy International Airport and Jamaica Bay. Compare with Plate 11. (Copyright © 1988 CNES, provided courtesy of SPOT Image Corp., Reston, Virginia.)

Figure 6-45 (*Left*) *SPOT 1* HRV panchromatic stereogram of a coastal region near Marseille, France. The off-nadir images were acquired 7 days apart and incorporate incidence angles of 22.4° W and 22.0° E. (Copyright © 1988 CNES, provided courtesy of SPOT Image Corp., Reston, Virginia.)

computer searches, product prices, and ordering procedures should be directed to SPOT Image Corp., 1897 Preston White Drive, Reston, VA 22091.

National Earth Resources Satellite Programs

Several nations have launched or plan to launch their own earth resources satellites. The People's Republic of

China has launched several satellites in the **Chinasat** series with earth observation payloads since 1975. Although specific payloads are not disclosed, it is presumed that both photographic cameras with recoverable photocapsules and electro-optical sensors have been used. Data from their remote sensing satellite program are not disseminated.

The Soviet Union began using earth-observing electro-optical sensors with its **Meteor** series of satellites in 1980. In addition to operational meteorological sensors, three experimental earth observation instruments have been used. The **MSU-E** is an along-track scanner that senses three spectral bands lying between 0.5 and 1.0 μm; it has a ground resolution element of 30 m and a 30-km ground swath. The **MSU-SK** is an across-track scanner that is sensitive to four spectral bands within the wavelength span of 0.5 to 1.0 μm;

Figure 6-46 LISS-IIA band 3 image (0.62 to 0.68 µm) of a region along India's east coast. The scene size is 74 × 87 km. (Courtesy S. Adiga, National Remote Sensing Agency of India.)

its resolution is 170 m with a 600-km swath width. The third sensor, known as **Fragment**, is an across-track scanner with eight spectral bands lying between 0.4 and 2.4 µm; resolution is 80 m and the ground swath is 85 km. Only a few images from these sensors have been released to the international scientific community.

The first Japanese earth observation satellite **MOS 1 (Marine Observation Satellite 1)** was successfully launched on February 23, 1987. The MOS program is directed and operated by the National Space Development Agency (NASDA). *MOS 1* is in a 909-km, near-polar, sun-synchronous, 17-day recurrent orbit. The satellite's primary mission is to measure oceanographic phenomena such as sea surface color and temperatures for the Japan region and concurrently to serve as an experimental platform for earth resources observation technology development. *MOS 1* carries three sensors: (1) the **Multispectral Electronic Self-Scanning Radiometer (MESSR)**, (2) the **Visible and Thermal Infrared Radiometer (VTIR)**, and (3) the **Microwave Scanning Radiometer (MSR)**.

The MESSR instrument uses CCD detectors to perform electronic scanning in approximately the same four bands as the Landsat MSS sensor: band 1 = 0.51 to 0.59 µm (green), band 2 = 0.61 to 0.69 µm (red), band 3 = 0.72 to 0.80 µm (near IR), and band 4 = 0.8 to 1.0 µm (near IR). The ground swath is 100 km, and resolution is 50 m.

The VTIR is an across-track scanner that measures a 1,500-km swath width in four spectral bands: band 1 = 0.5 to 0.7 µm (green-red), band 2 = 6.0 to 7.0 µm (thermal IR), band 3 = 10.5 to 11.5 µm (thermal IR), and band 4 = 11.5 to 12.5 µm (thermal IR). Note that band 2 operates in an atmospheric blind (Figure 6-9); this band is designed for assessment of the atmosphere's water-vapor content and identifying cirrus clouds. Resolution is 0.9 km for the visible band and 2.7 km for the three thermal IR bands.

The MSR system measures microwave radiation emitted from the earth's surface and atmosphere at two frequencies: 24 GHz (1.25 cm) and 31 GHz (0.97 cm). It is designed to measure the liquid water content of the atmosphere and sea-ice and snow distributions. The swath width is 320 km, and resolution is 31 km for the 24-GHz band and 21 km for the 31-GHz band.

Image data from *MOS 1* are offered in both CCT and photographic forms. Products may be ordered from the Remote Sensing Technology Center (RESTEC), Yuni Roppongi Building, 7-15-17, Roppongi, Minato-Ku, Tokyo 106, Japan.

The **Indian Remote Sensing (IRS)** program was initiated in the early 1980s to form the satellite remote sensing segment of the National Natural Resources Management System (NNRMS). The first spacecraft, *IRS-1A*, was placed in earth orbit by a VOSTAR launcher on March 17, 1988, from Baikanur, USSR. *IRS-1A* is in a 904-km, near-polar, sun-synchronous, 22-day recurrent orbit.

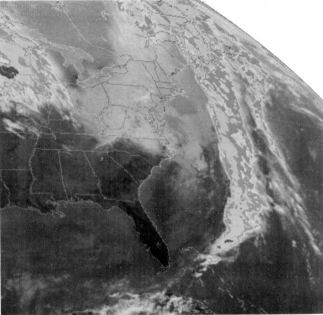

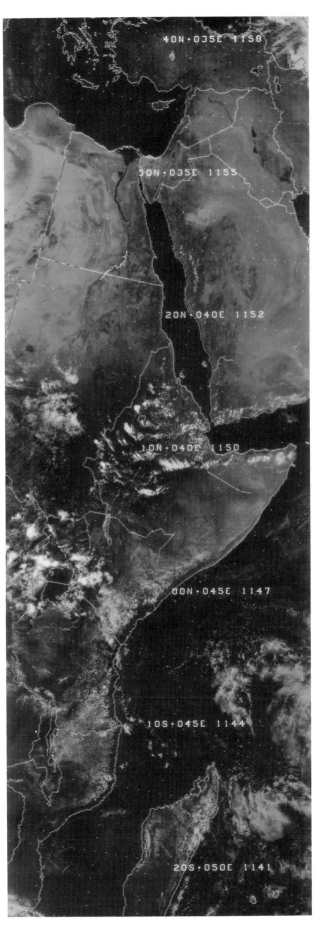

Figure 6-47 *GOES* 7 VISSR visible (0.55 to 0.70 μm) (*top*) and thermal IR (10.5 to 12.5 μm) (*bottom*) images of eastern North America and a large frontal system over the western Atlantic, acquired April 1, 1989. (Courtesy NOAA Satellite Data Services Division.)

Figure 6-48 *NOAA* 7 AVHRR band 1 image (0.58 to 0.68 μm) of the Middle East and eastern Africa, acquired September 21, 1981. Acquisition times (GMT) are shown next to the latitude and longitude coordinates. (Courtesy NOAA Satellite Data Services Division.)

IRS-1A carries two types of along-track scanners, one with a ground resolution of 72.5 m and designated as the **Linear Imaging Self-Scanning Sensor-I (LISS-I)** and the other with two separate imaging sensors with a ground resolution of 36.25 m and designated **LISS-IIA** and **LISS-IIB**. LISS-I images a ground swath of 148 km, whereas LISS-IIA and LISS-IIB image adjacent swaths each 74 km wide with a sidelap of 1.5 km.

Each of the three LISS instruments provides data in four spectral bands: band 1 = 0.45 to 0.52 μm (blue-green), band 2 = 0.52 to 0.59 μm (green), band 3 = 0.62 to 0.68 μm (red), and band 4 = 0.77 to 0.86 μm (near IR). These spectral sensitivities are essentially identical to Landsat TM bands 1–4. A LISS-IIA band 3 image is shown in Figure 6-46.

The IRS program is designed to meet the primary mission objective of India coverage. *IRS-1A* image data are offered in a variety of film products at several scales and as computer-compatible tapes. Products may be ordered from the National Remote Sensing Agency, Department of Space, Balanagar, Hyderabad-500 037, India.

Both Japan and India plan to launch additional remote sensing satellites in the future. Other earth-observation satellite programs are being planned by Brazil, Canada, and the Netherlands in cooperation with Indonesia. Specific plans for programs in the People's Republic of China and the Soviet Union have not been disclosed.

U.S. Meteorology Satellites

Since 1960, the United States has placed more than 35 civilian **meteorology satellites**, or **metsats**, into earth orbit. The earlier satellites carried small, vidicon-tube TV cameras that produced only visible band images with coarse resolution. From these relatively crude sensors have evolved increasingly sophisticated scanning instruments with improvements in resolution and expansion into spectral bands beyond the visible spectrum. Today, metsat image data are being used for hydrology, oceanography, and vegetation studies in addition to meteorological applications.* Civil meteorology satellite programs are directed by NOAA and NASA. The U.S. Air Force operates its own program called the **Defense Meteorological Satellite Program (DMSP)**.

NOAA Program

NOAA operates the nation's civil *operational* weather satellite program and is responsible for processing, archiving, and distributing image data. The prime user is NOAA's **National Weather Service**, which uses the data to create forecasts and weather advisory services. Images, catalogs, and price lists of its satellite data are available from NOAA Satellite Data Services Division, Room 100, Princeton Executive Center, Washington, D.C. 20233.

Meteorology satellites carry a wide range of measurement sensors in addition to imaging instruments.

NOAA currently operates two types of weather satellites. Those of the **GOES series (Geostationary Operational Environmental Satellite)** are essentially "parked" in an orbit about 38,500 km above the equator. They travel at the same speed and in the same direction as the earth's rotation. At that altitude and speed, they appear to remain stationary over a fixed point on the equator. This type of orbit is known as **geosynchronous**. The orbit is high enough to allow the satellites to have a full-disk view of the earth, which represents about one-fourth of the earth's surface area.

Satellites of what are called the **NOAA series** are, in contrast to the GOES series, **polar orbiters** in that they pass over both poles. For this reason, they are sometimes referred to as **Polar Orbiting Environmental Satellites (POES)**. They circle in a sun-synchronous orbit at the nominal altitude of 833 km, and each observes the complete earth twice a day. Even-numbered satellites (e.g., *NOAA 10*) cross the equator at 7:30 P.M. and 7:30 A.M. local time, whereas the odd-numbered satellites (e.g., *NOAA 11*) cross the equator at 2:30 P.M. and 2:30 A.M. local time. Two NOAA-series satellites (one even, one odd) are normally operational.

The imaging sensor aboard GOES is the **Visible-Infrared Spin-Scan Radiometer (VISSR)**. This instrument acquires one band of visible data (0.55 to 0.70 μm) and one band of thermal IR data (10.5 to 12.5 μm). VISSR is designed primarily to provide images of cloud patterns, but VISSR images are also used for other environmental applications, such as monitoring volcanic eruptions, snow and sea-ice mapping, and surface-temperature mapping. Resolution for both bands is approximately 8 km. Full-disk images are generated routinely every 30 min during daylight for the visible band and every 30 min day and night for the thermal IR band. Images of smaller areas can be collected at rates of up to once every 3 min when there is a need to monitor severe weather events.

GOES visible and thermal IR subscene images are presented in Figure 6-47. It is important to remember that the *tonal scheme is reversed for meteorological thermal IR images.* Thus, such thermal IR images show high, cold clouds as white, low-level clouds as gray, and warm land or water as black.

Until the mid-1980s, the *GOES* system consisted of *GOES-East (GOES 5)* and *GOES-West (GOES 6)*. *GOES-East* was positioned over the equator at 75° W longitude and monitored North and South America and most of the Atlantic Ocean. *GOES-West* was positioned over the equator at 135° W longitude and monitored North America and the Pacific Ocean. Neither of these satellites is operating at present. Currently, the GOES system is comprised of a single satellite, *GOES-Central (GOES 7)*. Its longitudinal position is changed throughout the course of a year. During the spring and summer tornado season it is at about 98° W longitude; it is moved eastward during the fall hurricane season and westward during the winter months to monitor Pacific storms.

The current NOAA series of polar-orbiting satellites carry the **Advanced Very High Resolution Radiometer (AVHRR)**. AVHRR systems consist of four or five spectral bands, depending upon the individual instrument. Sensitivities fall in the visible, near IR, and thermal IR regions. Table 6-8 lists the spectral bands of AVHRR and their principal applications. Band 1 (visible red) and band 4 (thermal IR) AVHRR images are shown in Figures 6-48 and 6-49.

The AVHRR scans continuously, acquiring image data with a nominal ground swath of 2,800 km and a ground resolution cell of 1.1 × 1.1 km at nadir. AVHRR images are available at full, 1.1-km resolution (**Local-Area Coverage**, or **LAC**) and at a subsampled resolution of 4 km (**Global-Area Coverage**, or **GAC**).

AVHRR data from band 1 (0.58 to 0.68 μm) and band 2 (0.725 to 1.10 μm) are playing an increasingly important role in monitoring vegetative growth over large areas. Various mathematical combinations of band 1 and 2 digital data have been found to be sensitive indicators of the presence and condition of planted and natural vegetation and are referred to as **vegetation indices**. This effect is due to the differential reflectance properties of vegetation in these two bands (see Figures 2-35 and 2-37).

A popular index incorporating the two AVHRR bands is the **Normalized Difference Vegetation Index (NDVI)**, also referred to as the **Normalized Vegetation Index (NVI)**; it is defined by the following equation:

$$\text{NDVI} = \frac{B_2 - B_1}{B_2 + B_1} = \frac{\text{near IR} - \text{red}}{\text{near IR} + \text{red}}. \qquad (6\text{-}12)$$

The summed denominator largely compensates for changing illumination conditions, surface slope, and viewing aspect. NDVI values range from about 0.1 to 0.6 for vegetation; the higher values are associated with greater coverage of healthy vegetation. Clouds, snow, and water have higher reflectances in band 1 than in band 2, which yield negative NDVI values. Rocks and bare soil have similar reflectances in both bands, which result in NDVI values near zero.

Plate 11 shows NDVI color-coded image mosaics of the northern Great Plains for early June 1987 (normal year) and early June 1988 (drought year). Red, brown, and yellow represent low NDVI vegetation values, whereas light green, dark green, and blue represent high NDVI vegetation values; gray depicts excluded areas. Comparison of the images illustrates the effects of the 1988 drought on the condition of the vegetation. The vegetation condition in 1987 was typical of a normal spring growing period. NDVI values were typically in the 0.35 to 0.55 range (greens and blue). However, NDVI values in 1988 dropped to the 0.15 to 0.34 range for many of the same areas (red, brown, and yellow). The impact of the 1988 drought on livestock grazing and forage production was significant.

Since 1984, NOAA has been preparing and sending NDVI maps to its National Weather Service field offices to help farmers monitor the growing season. Scientists from many developing countries have been trained in the NDVI technique under a program conducted by NOAA for the U.S. Agency for International Development. NDVI data have been used successfully to determine crop vigor and locate areas of drought, desertification, and deforestation (Johnson et al. 1987; Philipson and Teng 1988).

Besides monitoring weather and surface conditions around the world, NOAA weather satellites perform several other functions. These include the following:

1. They receive data from surface instruments that measure tide conditions, river levels, and precipitation, which are relayed to ground receiving stations.

TABLE 6-8 Characteristics of the NOAA Advanced Very High Resolution Radiometer

Band Number	Wavelength Range (μm) NOAA 6, 8, 10[a]	Wavelength Range (μm) NOAA 7, 9, 11[a]	Primary Uses
1[b]	0.58–0.68	0.58–0.68	Daytime cloud and surface mapping, snow and ice extent
2[b]	0.725–1.10	0.725–1.10	Surface water delineation, snow and ice extent
3	3.55–3.93	3.55–3.93	Detecting hot targets (e.g., forest fires), nighttime cloud mapping
4	10.50–11.50	10.30–11.30	Determining cloud and surface temperatures, day or night cloud mapping
5[c]	None	11.50–12.50	Determining cloud and surface temperatures, day or night cloud mapping, water vapor correction

[a]NOAA 10 and NOAA 11 are currently operational.
[b]An indication of ice or snow melt inception is possible when the data from bands 1 and 2 are compared. In addition, the data from bands 1 and 2 can be used in combination to assess vegetation vigor; this technique is described in the text.
[c]This band provides the capability for removing radiant contributions from water vapor when determining temperatures.

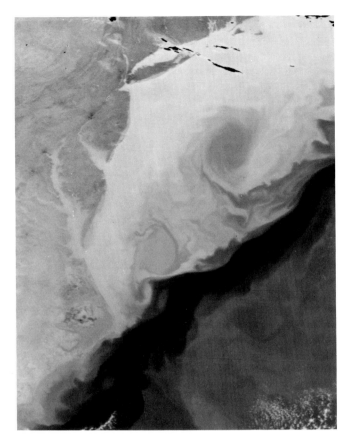

Figure 6-49 *NOAA 6* AVHRR band 4 thermal IR image (10.5 to 11.5 μm) showing the warm meandering Gulf Stream and two cold-core eddies to the north of it. Dark tones represent warm radiant temperatures and light tones represent cool radiant temperatures. This image was acquired April 24, 1982, at about 7:30 A.M. local time. (Courtesy NOAA Satellite Data Services Division.)

2. They receive weather data from ocean buoys, weather balloons, and aircraft in flight, relaying these data to ground receiving stations.

3. The satellites carry search and rescue instrumentation that either relays or locates emergency signals from ships and aircraft in distress.

4. Using a facsimile technique called **Weather Facsimile**, or **WEFAX**, they broadcast weather images and

charts directly to amateur and professional users, who include foreign and commercial American weather services plus research and educational institutions throughout the world. There are over 1,000 WEFAX users who avail themselves of this free service.

NASA Program

NASA operates the nation's civil *experimental* weather satellite program. **Nimbus**, the Latin word for cloud, identifies the current NASA research and development spacecraft. Since 1964, seven Nimbus satellites have been successfully placed into earth orbit. A host of experimental sensors have operated on the different missions. When a sensor is deemed to be operational, it can be carried by a NOAA weather satellite.

The *Nimbus 7* satellite, launched on October 23, 1978, is equipped with nine sensing instruments. In keeping with its primarily meteorological role, eight of the sensors were designed to make measurements within the atmosphere. The ninth instrument, the **Coastal Zone Color Scanner (CZCS)** represents the first satellite sensor specifically designed to measure subtle color variations in the near-shore waters of the oceans. *Nimbus 7* is in a sun-synchronous, near-polar, 955-km orbit; its repeat cycle is 6 days.

Subtle variations in ocean color can be caused by the presence of chlorophyll *a,* suspended inorganic and organic sediments, and various pollutants. Phytoplankton, the microscopic plant organisms that constitute the bottom link in the ocean food chain, contain chlorophyll *a,* which absorbs strongly in the blue and red regions of the visible spectrum. Hence, increasing concentrations of phytoplankton (chlorophyll *a*) change the color of ocean water to green hues from the bluish color of its pure state (Gordon et al. 1980).

The six-channel CZCS has (1) four high-sensitivity bands, each 0.02 μm wide, in the visible spectrum, (2) a lower sensitivity and broader near IR band, and (3) a thermal IR band (Table 6-9). Image resolution is 825 m, and the image swath is 1,800 km. To avoid sun glint, the scanner can be tilted up to 20° from nadir to look either forward or behind the spacecraft. CZCS images of the Bay of Bengal are shown in Figure 6-50.

The four visible bands are centered to enhance the discrimination of very subtle water reflectance differences, which means that they usually saturate over land (Figure 6-50). The visible band images have been used to estimate

Band Number	Wavelength Range (μm)	Measurement
	TABLE 6-9 **Characteristics of the *NIMBUS 7* Coastal Zone Color Scanner**	
1	0.43–0.45	Chlorophyll concentrations
2	0.51–0.53	Chlorophyll absorption
3	0.54–0.56	Dissolved organic material and phytoplankton
4	0.66–0.68	Chlorophyll concentrations
5	0.70–0.80	Surface vegetation
6	10.50–12.50	Sea-surface temperatures

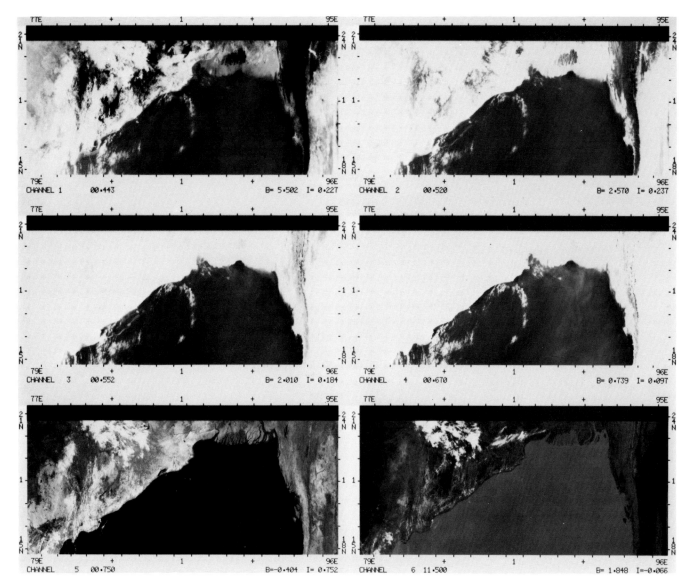

Figure 6-50 *Nimbus 7* CZCS subscene images of the Bay of Bengal. Each image measures 700 × 1,640 km. Refer to Table 6-8 for the wavelength boundaries of each channel. (Courtesy C. Scott Southworth and C. J. Robinove, U.S. Geological Survey.)

the concentrations of phytoplankton, dissolved organic materials, and suspended matter such as silt. The near IR data have been used to map surface vegetation and to aid in separating water from land areas prior to processing the data from the visible bands. The thermal IR data have been used to measure and monitor sea-surface temperatures.

The CZCS instrument ceased operations in December 1986. Inquiries regarding CZCS image and CCT data can be made to NOAA's Satellite Data Services Division at the address given earlier in this chapter.

U.S. Air Force Program

The U.S. Air Force's Defense Meteorological Satellite Program (DMSP) consists of two operational satellites in near-polar, sun-synchronous orbits at a nominal altitude of 825 km. Each satellite utilizes an **Operational Linescan System (OLS)** which provides images in a visible–near IR band (0.4 to 1.1 μm) and a thermal IR band (10.5 to 12.5 μm). The Air Force's Global Weather Control uses the images to forecast weather for the Department of Defense. The image data are then released and archived for the public, usually 45 to 90 days later.

OLS data have the highest resolution for both local and global coverage for any meteorology satellite (0.6 to 2.7 km, respectively) and the only visible–near IR band sensor carried on any satellite that can obtain both daytime and nighttime images. Through an amplification process, the OLS's visible–near IR channel is able to obtain images at night by moonlight; if there is no moonlight, only city lights and other phenomena such as forest fires, oil and gas field fires, and auroral displays will be depicted. Examples of

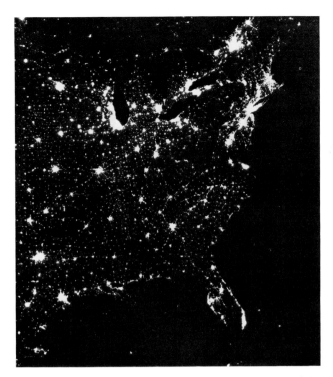

Figure 6-51 DMSP nighttime image (0.4 to 1.1 μm) of the eastern United States, acquired December 16, 1985. (Courtesy Greg Scharfen, National Snow and Ice Data Center, University of Colorado.)

Figure 6-52 DMSP image (0.4 to 1.1 μm) of western Africa on a full-moon night, acquired January 20, 1986. Apparent in the image are city lights along the Mediterranean coast, gas flares in Algeria, hundreds of small agricultural slash and burn fires in the "middle belt," and a sandstorm moving out over the Atlantic Ocean from the Sahara. (Courtesy Greg Scharfen, National Snow and Ice Data Center, University of Colorado.)

nighttime images with and without moonlight are shown in Figures 6-51 and 6-52.

DMSP images are available to the public through the National Snow and Ice Data Center (NSIDC), CIRES, Campus Box 449, University of Colorado, Boulder, CO 80309. NSIDC maintains the entire DMSP image archive from 1973 to the present. The archive consists of more than 1.3 million film transparencies in three formats: (1) local-area coverage direct readout images with 0.6-km resolution, (2) global-area coverage single-orbit swaths with 2.7-km resolution, and (3) global coverage gridded mosaics with 5.4-km resolution.

Imaging Spectrometers

Most recently, the Jet Propulsion Laboratory has developed and successfully tested several **imaging spectrometers** for NASA. The imaging spectrometer is a new generation of electro-optical sensor that employs cooled linear array detectors to simultaneously collect image data in a hundred or more very narrow and contiguous spectral bands throughout the visible, near IR, and mid IR portions of the electromagnetic spectrum. Their purpose is to detect the typically narrow diagnostic absorption and reflection characteristics of the earth's surface features that are often masked within the much broader bandwidths of the various channels of conventional across-track and along-track multispectral scanners.

The instrument currently being used for NASA research projects is the **Airborne Visible/Infrared Imaging Spectrometer** (**AVIRIS**) which commenced operations aboard the NASA ER-2 research aircraft in 1987. The

AVIRIS instrument employs an across-track scan mirror that is connected by optical fibers to four separate **spectrometers**. Blocking filters help to partition the reflected radiation to the appropriate spectrometer where it is dispersed spectrally and focused onto line-array detectors mounted in dewars. Utilizing both silicon and indium antimonide line-array detectors, AVIRIS acquires 224 contiguous digital images, each with a spectral bandwidth of 9.6 nm or 0.096 μm in the spectral region extending from 0.4 to 2.45 μm. When flown at the standard altitude of 20 km, AVIRIS covers a ground swath 10 km wide with 20-m ground resolution cells; the scan rate is 12 scans/sec. Computer processing of the data can produce an image for a given ground scene in any of the 224 spectral bands or the spectrum corresponding to any of the single ground resolution cells within the scene. Images in groups of three may also be used to produce color composites. Figure 6-53 illustrates three images acquired by AVIRIS of desert terrain in southeastern California.

AVIRIS is expected to be the major source of high spatial and spectral resolution images until the **High-Resolution Imaging Spectrometer** (**HIRIS**) is launched on the polar-orbiting NASA Earth Observation System (EOS) in the late 1990s. HIRIS has been designed to acquire simultaneous images in 192 spectral bands between 0.4 and

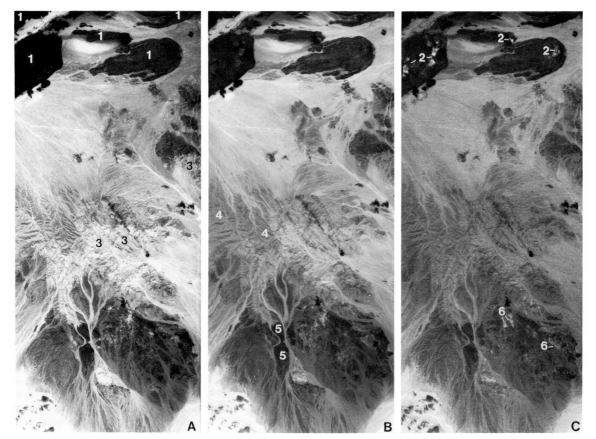

Figure 6-53 AVIRIS images of the Cima Volcanic Field (*top*) and Kelso Mountains (*bottom*), southeastern California. Center wavelength bands are as follows: (*A*) 0.505 μm, (*B*) 1.003 μm, and (*C*) 2.036 μm. Annotations are as follows: (1) basalt lava flows, (2) altered cinder cone deposits (hematite rich), (3) granite, (4) eroded gravel deposits, (5) varnished quartzite, and (6) limonite alteration zones. Note that the lava flows are elongated down the topographic slope. (Courtesy National Aeronautics and Space Administration.)

2.5 μm at a spectral sampling interval of 10 nm. From an orbital altitude of 824 km, the swath width will be 30 km with 30-m ground resolution cells. HIRIS is intended to be a *targeting instrument*, meaning that it will collect data only at specified times and places.

Commercial imaging spectrometers have just recently become available for operation from small aircraft. One such instrument is ITRES Research Limited's **Compact Airborne Spectrographic Imager** (**CASI**) (Figure 6-54). This along-track instrument can collect up to 288 spectral bands of reflectance data in the 0.4- to 1.0-μm range at a spectral resolution of 1.4 to 1.8 nm. CASI's band locations and widths are programmable ''on the fly''; this makes it possible to collect data with different configurations on a single flight.

Determining Number of Color Composites

The number of different color composites that can be produced by combining single-band black-and-white images in groups of three (blue, green, and red components) can be easily determined for any multiband imaging system by use of the following equation:

$$N = \frac{n!}{3!(n-3)!},\qquad (6\text{-}12)$$

where: N = number of different color composites,

$\quad n$ = number of single-band images available,

$\quad 3$ = number of color assignments (blue, green, and red components), and

$\quad !$ = factorial.*

*The exclamation point after a symbol or number means **factorial**. A factorial is the product of all positive integers from 1 to the given number. Thus, 4! = (1)(2)(3)(4) = 24; similarly, 5! = (1)(2)(3)(4)(5) = 120.*

Earlier in this chapter, it was stated that 20 different three-band color composites could be produced from the six nonthermal TM images. This is easily confirmed by using Equation 6-12:

$$N = \frac{6!}{3!(6 - 3)!}$$

$$= \frac{6!}{3!(3)!}$$

$$= \frac{720}{(6)(6)}$$

$$= \frac{720}{36}$$

$$= 20.$$

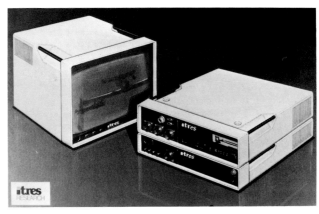

Figure 6-54 Compact Airborne Spectrographic Imager (CASI) manufactured and distributed by ITRES Research Limited. (Courtesy Richard J. Adamson, ITRES Research Limited.)

Questions

1. Calculate the lateral ground coverage and the ground resolution cell of an across-track scanner, given an AFOV of 100°, an IFOV of 2.5 mrad, and a platform height of 4,200 m.
2. Calculate the wavelength where maximum energy release would occur for the following phenomena:

   a. Forest fire   600°C λ_{max} = _____

   b. Asphalt runway   35°C λ_{max} = _____

   c. Sea ice   −55°C λ_{max} = _____

   d. Molten lava   1,100°C λ_{max} = _____
3. Rough lava (aa) has an emissivity of 0.95 in the 8- to 14-μm region, and smooth lava (pahoehoe) has an emissivity of 0.92. If both materials had a kinetic temperature of 25°C and a thermal IR scanner had a temperature resolution of 0.5°C, could the two types of lava be differentiated?

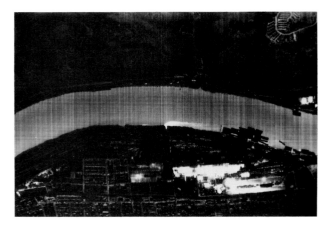

Figure 6-55 Thermal IR image (8 to 14 μm) of the Monongahela Valley in southwestern Pennsylvania. (Courtesy Ronald W. Stingelin, Resource Technologies Corp.)

4. Examine the thermal IR image shown in Figure 6-55, and answer the following questions:

   a. Was the image generated during the daytime or nighttime?

   b. State three reasons that justify your selection.

   c. What type of industrial facility is located next to the river? A review of Chapter 14 will provide important clues.
5. Examine the Landsat MSS images in Figure 6-39 and select which band allows you to perform the interpretation functions listed here. Indicate your first through fourth choices by placing 1, 2, 3, or 4 under the appropriate band.

Interpretation Function	MSS Bands			
	4	5	6	7
a. Evaluating coastal circulation patterns	____	____	____	____
b. Mapping drainage patterns	____	____	____	____
c. Mapping snow cover	____	____	____	____
d. Mapping lowland vegetation	____	____	____	____
e. Evaluating turbidity levels in lakes and rivers	____	____	____	____
f. Mapping landforms	____	____	____	____
g. Mapping exposed bedrock	____	____	____	____
h. Mapping linear features expressed by landforms	____	____	____	____

6. Examine the Landsat TM images in Figure 6-41, and answer the following questions:

a. Is the large white area snow cover or clouds? Explain.

b. Why are the "breaker" waves visible in bands 1 through 3 but not in bands 4, 5, and 7?

c. Why is there a tonal variation between the ocean water and bay water (examine all seven bands)?

d. In general, what are the coldest and warmest features in the TM band 6 image?

e. Why is the Golden Gate Bridge detectable in the TM band 6 image given a resolution of 120 m?

7. Examine the Landsat TM color composite images in Plate 10 and give two uses for each composite.

a. TM natural color composite (bands 1, 2, 3)

b. TM color IR composite (bands 2, 3, 4)

c. TM false color composite (bands 1, 4, 5)

8. Calculate the number of different color composites that could ideally be produced from the seven TM bands (Table 6-7).

9. Examine the Landsat TM images in Figure 6-56 and answer the following questions:

a. Explain why the smoke is visible in some images but not in others.

b. If the fires have a temperature of about 710°C, why can they be pinpointed (light tone) on the band 7 image? *Hint:* Apply Wien's displacement law.

c. Which three-band color composite image would be the most useful for showing the smoke plume?

d. Which three-band color composite image would be the most useful for showing the terrain beneath the smoke plume?

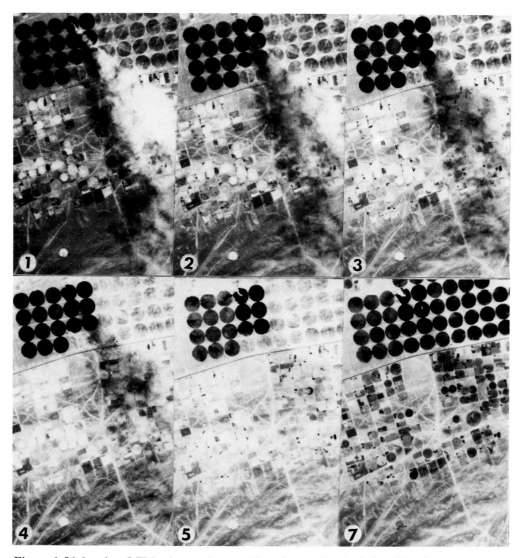

Figure 6-56 *Landsat 5* TM subscene images (six reflective bands, 1–5 and 7) of an agricultural area in north-central Saudi Arabia. Fires have been set to clear the fields of wheat stubble.

Bibliography and Suggested Readings

Allison, L. J., and A. Schnapf. 1983. Meteorological Satellites. In *Manual of Remote Sensing,* edited by R. N. Colwell, 651–79, 2d ed. Falls Church, Va.: American Society for Photogrammetry and Remote Sensing.

Bowker, D. E., R. E. Davis, D. L. Myrick, K. Stacy, and W. T. Jones. 1985. *Spectral Reflectances of Natural Targets for Use in Remote Sensing Studies,* Reference Pub. 1139. Washington, D.C.: National Aeronautics and Space Administration.

Brandli, H. W. 1978. The Night Eye in the Sky. *Photogrammetric Engineering and Remote Sensing.* 44:503–505.

Centre Spatial de Toulouse. 1988. *SPOT User's Handbook.* Reston, Va.: SPOT Image Corp.

Deutsch, M., D. R. Wiesnet, and A. Rango, eds. 1981. *Satellite Hydrology.* Minneapolis, Minn.: American Water Resources Association.

Dokken, D., ed. 1991. *EOS Reference Handbook.* Greenbelt, Md.: NASA Goddard Space Flight Center.

Fagerlund, E. B., B. Kleman, L. Sellin, and H. Svensson. 1970. Physical Studies of Nature by Thermal Mapping. *Earth-Science Reviews* 6:169–80.

Forsythe, W. E. 1964. *Smithsonian Physical Tables,* 9th ed. Washington, D.C.: The Smithsonian Institution.

Francis, P., and P. Jones. 1984. *Images of Earth.* Englewood Cliffs, N.J.: Prentice-Hall.

Freden, S. C., and F. Gordon. 1983. Landsat Satellites. In *Manual of Remote Sensing,* edited by R. N. Colwell, 517–70, 2d ed. Falls Church, Va.: American Society for Photogrammetry and Remote Sensing.

Garrett, W. E., ed. 1985. *Atlas of North America, A Space Age Portrait of a Continent.* Washington, D.C.: National Geographic Society.

Goetz, A. F. H., G. Vane, J. E. Soloman, and B. N. Rock. 1986. Imaging Spectrometry for Earth Remote Sensing. *Science.* 228:1147–53.

Gordon, H. R., D. K. Clark, J. L. Mueller, and W. A. Hovis. 1980. *Nimbus-7* Coastal Zone Color Scanner: System Description and Initial Imagery. *Science* 210:60–66.

Hutchinson, C. F., R. A. Schowengerdt, and L. R. Baker. 1990. A Two-Channel Multiplex Video Remote Sensing System. *Photogrammetric Engineering and Remote Sensing* 56:1125–28.

Janza, F. J. et al. 1975. Interaction Mechanisms. In *Manual of Remote Sensing,* edited by R. G. Reeves, 75–179. Falls Church, Va.: America Society of Photogrammetry.

Johnson, G. E., A. van Dijk, and C. M. Sakamoto. 1987. The Use of AVHRR Data in Operational Agricultural Assessment in Africa. *Geocarto International* 2:41–60.

Landgrebe, D. A. 1974. Machine Processing of Remotely Sensed Data. *ERTS Image Interpretation Workshop Syllabus,* Open-File Report 75-196. Sioux Falls, S.D.: U.S. Geological Survey.

Lillesand, T. M., and R. W. Kiefer. 1987. *Remote Sensing and Image Interpretation,* 2d ed. New York: John Wiley & Sons.

Markham, B. L., and J. L. Barker. 1983. Spectral Characterization of the Landsat-4 MSS Sensor. *Photogrammetric Engineering and Remote Sensing.* 49:811–33.

Markham, B. L., and J. L. Barker. 1985. Spectral Characterization of the Landsat Thematic Mapper Sensors. *International Journal of Remote Sensing.* 6:697–716.

Meisner, D. E. 1986. Fundamentals of Airborne Video Remote Sensing. *Remote Sensing of Environment* 19:63–79.

Norwood, V. T., and J. C. Lansing. 1983. Electro-Optical Imaging Systems. In *Manual of Remote Sensing,* edited by R. N. Colwell, 335–67, 2d ed. Falls Church, Va.: American Society for Photogrammetry and Remote Sensing.

Peng, C. S. 1986. *Atlas of Geo-Science Analyses of Landsat Imagery in China.* Beijing: Science Press.

Philipson, W. R., and W. L. Teng. 1988. Operational Interpretation of AVHRR Vegetation Indices for World Crop Information. *Photogrammetric Engineering and Remote Sensing* 54:55–59.

Robinson, I. S. 1985. *Satellite Oceanography.* New York: John Wiley & Sons.

Sabins, F. F., Jr. 1987. *Remote Sensing Principles and Interpretation,* 2d ed. New York: W. H. Freeman & Co.

Sheffield, C. 1981. *Earthwatch—A Survey of the World from Space.* New York: Macmillan.

Short, N. M., P. D. Lowman, Jr., S. C. Freden, and W. A. Finch, Jr. 1976. *Mission to Earth: Landsat Views the World.* Washington, D.C.: U.S. Government Printing Office.

Short, N. M. 1982. *The Landsat Tutorial Workbook.* NASA Reference Pub. 1078. Washington, D.C.: U.S. Government Printing Office.

Short, N. M., and L. M. Stuart. 1983. *The Heat Capacity Mapping Mission (HCMM) Anthology,* NASA SP-465. Washington, D.C.: U.S. Government Printing Office.

Short, N. M., and R. W. Blair, Jr. 1986. *Geomorphology from Space,* NASA-SP 486. Washington, D.C.: U.S. Government Printing Office.

Szekielda, K. H. 1986. *Satellite Remote Sensing for Resource Development.* London: Graham & Trotman.

USGS/NOAA. 1984. *Landsat 4 Data Users Handbook.* Reston, Va.: U.S. Geological Survey.

Vane, G. 1987. *Airborne Visible/Infrared Imaging Spectrometer (AVIRIS),* JPL Publication 87-38. Pasadena, Calif.: Jet Propulsion Laboratory.

Weisnet, D. R., et al. 1983. Remote Sensing of Weather and Climate. In *Manual of Remote Sensing,* edited by R. N. Colwell, 1305–69, 2d ed. Falls Church, Va.: American Society for Photogrammetry and Remote Sensing.

Williams, R. S., and W. D. Carter, eds. 1976. *ERTS-1, A New Window on Our Planet,* Geological Survey Professional Paper 929. Washington, D.C.: U.S. Geological Survey.

Wolfe, W. L., and G. J. Zissis, eds. 1982. *The Infrared Handbook.* Washington, D.C.: U.S. Government Printing Office.

Chapter 7

Microwave and Acoustical Sensors

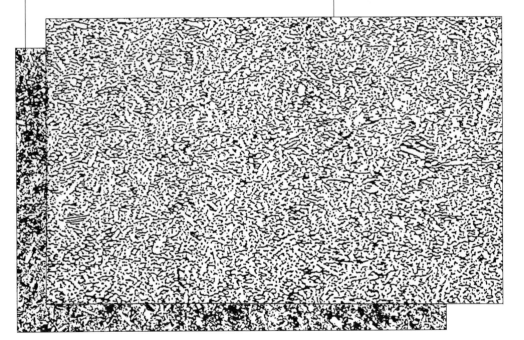

microwave radiation to the surface and detects the reflected component. **Sonar** has many operational similarities to radar except that it uses acoustical, or sound, energy, which is propagated in water. All three systems produce images for interpretation and measurement tasks.

Radar Principles

The term radar is an acronym of the phrase <u>ra</u>dio <u>de</u>tection <u>a</u>nd <u>r</u>anging. The term radio is used because the first radar systems utilized very long wavelengths of radiation (1 to 10 m) that fell in the radio band of the electromagnetic spectrum (Figure 1-2). Although radar systems now utilize microwave energy (**microwave radar**), the acronym was not changed to <u>m</u>icrowave <u>d</u>etection <u>a</u>nd <u>r</u>anging (**midar**).

A radar unit transmits short pulses or bursts of microwave radiation and then receives reflections of the signal from a target. The reflected component is called the **echo**, or **backscatter**. By providing its own radiation, radar operates entirely independent of sunlight, and equally effective missions can be conducted during the day or during the dark of night. In addition, the angle and direction of microwave illumination can be controlled to enhance features of special interest.

Analogous to lasers, radars are **monochromatic** in that they use radiation of single wavelengths. Table 7-1 lists the wavelength subdivisions, or bands of the microwave spectrum, that are used in remote sensing and the wavelengths commonly used in imaging radars. The random letter designations for the various radar bands are a carryover from World War II, when the bands were assigned letter codes for security purposes. This nomenclature continues to be used today (Table 7-1 and Figure 1-5). Long-wavelength radars (e.g., L- and P-bands) are unhindered by clouds and precipitation and are capable of penetrating surface materials such as sand, snow, and vegetation canopies.

The *tonal record* of a radar image is a measure of *microwave echo strength* and is a product of many ground and radar system properties (Figure 7-1). Objects that are good reflectors are depicted in light tones, whereas objects that are poor reflectors are denoted in dark tones; objects that are responsible for no measurable echo are depicted in black. Objects that are moderate reflectors are depicted in medium tones.

Introduction

Two types of remote sensing devices, one passive and one active, operate in the microwave band of the electromagnetic spectrum, which has a wavelength range extending from about 0.1 cm to 1 m (Figures 1-2 and 1-5). A **passive microwave radiometer** detects the low level of natural microwave radiation that is emitted by all objects in the earth system; in this sense it is a **passive sensor**. **Radar**, by comparison, is an **active sensor**, which propagates artificial

Figure 7-1 Westinghouse Corp. K_a-band (0.86-cm) radar image of the San Francisco Peninsula, California. The San Andreas fault is clearly depicted along the length of the image at midrange. San Francisco International Airport is on the left, and the Stanford Linear Accelerator building is on the right just to the east of the San Andreas fault. (Courtesy U.S. Geological Survey.)

Development of Side-Looking Imaging Radar

Most people are familiar with a **plan position indicator** (**PPI**), or the type of radar that is used today for weather forecasting, navigation, and air-traffic control. This type of radar uses a rotating antenna and a circular display. PPI radar was developed during World War II and was used extensively for detecting and tracking airplanes and ships. When mounted in a bomber, PPI was able to image the ground and detect enemy targets in darkness or overcast weather.

Based in large measure on PPI's ability to detect relatively large surface features, **side-looking airborne radar** (**SLAR**), or simply **side-looking radar** (**SLR**), was devel-oped after the war specifically as a high-resolution, wide-swath imaging system. The side-look concept was adopted by the military so that extensive areas behind enemy lines could be safely imaged from an aircraft flying over friendly territory, a special capability that was known as **long-range standoff**. The military began using SLARs in the 1950s, and certain systems and associated images were declassified for civilian uses beginning in the mid-1960s.

There are two types of SLAR: **real-aperture radar** (**RAR**), also known as **brute-force radar**, and **synthetic-aperture radar** (**SAR**), where aperture is used to mean antenna. The primary difference between the two systems is in their resolving power. SAR, the newest system, was specifically developed to achieve fine resolution from great distances, including orbital altitudes.

In the United States, Westinghouse, Motorola, and Goodyear Aerospace have each built SLAR systems. Westinghouse RAR operations ceased in 1973. Although the U.S. Army still uses Motorola RARs, their civilian use has been discontinued. Goodyear Aerospace SARs are currently used

TABLE 7-1 Band Designations and Radar Wavelengths		
Band Designation	Wavelength Range (cm)	Common Wavelengths for Imaging Radars (cm)
K_a	0.8–1.1	0.86
K	1.1–1.7	
K_u	1.7–2.4	
X	2.4–3.8	3.0, 3.2
C	3.8–7.5	6.0
S	7.5–15.0	
L	15.0–30.0	23.5, 24.0, 25.0
P	30.0–100.0	68.0

by the U.S. Air Force, and a civilian model is operated by the Aero Service Division of the Western Geophysical Corporation of America. In Canada, SAR systems are built and marketed commercially by MacDonald Dettwiler and Associates and Intera Technologies. Research SAR systems are operated by the Jet Propulsion Laboratory (JPL) and the Environmental Research Institute of Michigan (ERIM).

To date, SARs have operated successfully from two NASA spacecraft—the robotic *Seasat* and the *Space Shuttle*. *Seasat* operated in 1978 and was designed primarily to investigate ocean phenomena (hence "sea satellite"). The first **Shuttle Imaging Radar mission (SIR-A)** occurred in November 1981, whereas the second (**SIR-B**) occurred in October 1984.

Definitions of Terms for SLAR

Before discussing the operation of an imaging radar system, it is appropriate to discuss the terminology and char-acteristics of SLAR operations. As shown in Figure 7-2, an airplane or spacecraft moves at some velocity and at some altitude in an **azimuth**, or **along-track direction**. Through a **fixed antenna**, **pulses** of microwave radiation are propagated outward in a perpendicular plane at the speed of light in the **range**, **look**, or **across-track direction**. **Slant range** is the line-of-sight distance measured from the antenna to the terrain target, whereas **ground range** is the horizontal distance measured along the surface from the **ground track**, or **nadir line**, to the target. The area closest to the ground track at which a radar pulse intercepts the terrain is the **near range**, and the area of pulse termination farthest from ground track is the **far range**.

The angle measured from a horizontal plane *downward* to a specific part of the radar beam defines the **depression angle** (β). The depression angle varies across the **image swath** from relatively steep (large angle) at near range to relatively shallow (small angle) at far range. The angle measured from a vertical plane *upward* to a specific part of the radar beam defines the **look angle** (θ). The look angle varies across the image swath from a relatively small angle at near range to a relatively large angle at far range. When measured to the same part of the beam, the depression angle and the look angle are **complementary angles** ($\beta + \theta = 90°$).

Figure 7-2 Geometric characteristics of side-looking airborne radar (SLAR).

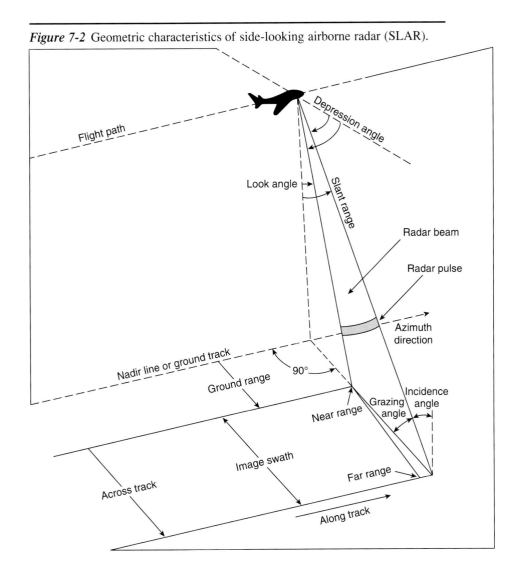

The **incidence angle (ϕ)** is the angle measured between the axis of the radar beam and a line perpendicular to the local ground surface that the beam strikes; the complement of the incidence angle is called the **grazing angle (γ)**. Consequently, the incidence angle and the grazing angle are a function of both the illumination angle (β or θ) and the slope of the terrain. When the terrain is horizontal, the depression and grazing angles are equal ($\beta = \gamma$) and the look and incidence angles are equal ($\theta = \phi$).

A large-area view in a small format is a prime capability of a SLAR system. An aircraft-operated, **single-look** SLAR can produce in excess of 20,000 km^2 of terrain coverage in an hour by illuminating continuous or overlapping swaths. Individual image swaths can exceed 75 km. A few systems are **dual-look**, wherein two antennas simultaneously image to the left and right of ground track. SLAR images are usually acquired at regional scales (e.g., 1:250,000 to 1:500,000), but the negatives can often be enlarged to scales as large as 1:50,000 or 1:25,000 without loss of detail.

Real-Aperture Radar

In the simplest terms, a real-aperture radar (RAR) system consists of (1) a **transmitter**, (2) a **receiver**, (3) a **TR (transmit-receive) switch**, or **duplexer**, (4) an **antenna** that serves in both transmission and reception, (5) a **cathode-ray tube (CRT) display**, and (6) a **photographic recording device** (Figure 7-3). During operation, the following processes take place:

1. The transmitter generates a short burst or pulse of polarized microwave energy at a discrete wavelength, which is propagated in a vertical fan-shaped beam by the directional antenna; propagation is perpendicular to the ground track. (The polarization parameter is discussed in a later section.)

2. As the pulse strikes a narrow strip of the terrain, a portion of it is reflected back to the aircraft, where it is intercepted by the same antenna and sent to a sensitive radio receiver. The receiver converts the detected pulse into an amplified video (electrical) signal.

3. Because the same antenna is used for both transmitting and receiving, the TR switch disconnects the transmitter following pulse propagation and connects the receiver to the antenna for reception of the return echo. This alternating process is continuously repeated as the aircraft advances along the flight path. With this forward motion, the radar beams are moved to new positions, enabling succeeding pulses to intercept adjacent strips of terrain (''scanning'' function).

4. The receiver produces a fluctuating video signal for each returned pulse. Its amplitude is directly proportional to the intensity of the backscatter received at any instant.

5. The fluctuating signal proportionally modulates the intensity of a moving spot of light (a small electron beam) on a CRT display. The fluctuating light is focused on the surface of a photographic film and swept across the film; a single line is traced for each returned pulse. Ultimately, the individual pulse echoes are recorded side by side as the film is advanced,

Figure 7-3 Representation of a real-aperature radar (RAR) system. (Courtesy Goodyear Aerospace Corp.)

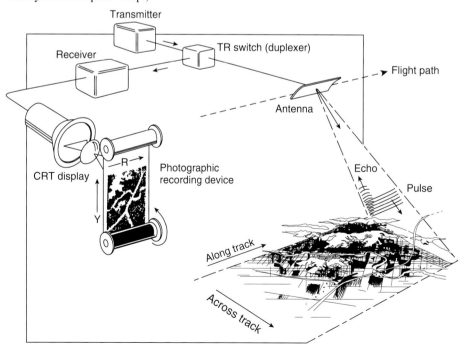

Figure 7-4 Slant-range image of an agricultural area that incorporates scale compression in the near range (top).

forming a two-dimensional radar image. The film advances at a rate proportional to the aircraft ground speed. When processed, the film exhibits tonal densities that are a measure of the backscatter intensity returned from the target scene; in positive form, strong returns equate to light tones while weak returns are portrayed in dark tones (Figure 7-4). Radar return energy varies in response to a complex combination of ground and radar system properties (discussed in a later section).

6. Based upon the sweep velocity of the CRT's electron beam, the image display can be either slant range or ground range for the across-track ordinate. If the sweep rate is linear, the horizontal separation of targets in near range will be compressed, resulting in a **slant-range image** (Figure 7-4). However, by applying a hyperbolic (nonlinear) waveform to the CRT's circuitry, the sweep rate can be increased in near range relative to far range. As a result, a **ground-range image** portrays targets from level terrain in their correct ground positions.

Radar Resolution Versus Radar Detection

A key parameter often used to judge the quality of a radar image is resolution. **Radar resolution** is defined as the minimum separation between two objects of equal reflectivity that will enable them to appear individually in a processed radar image. The most important criterion for es-

tablishing resolution is the size of the **pulse rectangle** projected onto the ground at a given instant of time. The pulse rectangle is similar to the ground resolution cell associated with across-track scanners (Figure 6-5).

When two or more objects fall within the same pulse rectangle, they cannot be resolved as separate entities. Rather, they are presented as one echo to the radar system. If objects are separated by a distance exceeding the corresponding dimension of the pulse rectangle, they will be imaged separately. The size of the pulse rectangle is controlled by two independent resolutions: (1) Range resolution determines resolution cell size perpendicular to the ground track and (2) azimuth resolution establishes the cell size parallel to the ground track. These resolutions are discussed in the following section.

Radar detection is a measure of the smallest object that can be discerned on an image as a result of its ability to reflect microwave radiation. Detection is often associated with highly reflective metal objects such as vehicles, railroad tracks, fences, and power lines and poles, which are physically much smaller than the pulse rectangle. For example, a vehicle is normally a much better reflector of microwave than its surroundings, making it the dominant reflector in the pulse rectangle. When this composite reflectance value differs from those in surrounding cells, the radar system, in essence, detects the vehicle. However, to the radar, the vehicle is as large as the cell size. Users are often amazed by the detail seen in radar images because many small features are detected when resolution is not required to distinguish them from surrounding objects (Matthews 1975).

Range Resolution

Range, or **across-track**, **resolution** in **slant range** (R_{sr}) is determined by the physical length of the radar pulse that is emitted from the antenna; this is called the **pulse length** (τ). If not given, pulse length can be determined by multiplying the pulse duration (δ), or the length of time in microseconds (1 μsec = 1 $\times$ 10^{-6} sec) that the pulse was emitted from the antenna by the speed of light ($c = 3 \times 10^8$ m/sec):

$$\tau = \delta c. \qquad (7-1)$$

For a radar system to discern two targets in the across-track dimension, all parts of their reflected signals must be received at the antenna at different times or they will appear as one large entity in an image. In Figure 7-5 it is seen that objects separated by a slant-range distance equal to or less than $\tau/2$ will produce reflections that arrive at the antenna as one continuous pulse, dictating that they be imaged as one large object (targets A, B, and C). If the slant-range separation is greater that $\tau/2$, the pulses from targets C and D will not overlap, and their signals will be recorded separately. Thus, slant-range resolution measured in the across-track dimension is equal to one-half the transmitted pulse length:

$$R_{sr} = \frac{\tau}{2}. \qquad (7-2)$$

To convert R_{sr} to **ground-range resolution** (R_{gr}), the formula is

$$R_{gr} = \frac{\tau}{2 \cos \beta}, \quad (7\text{-}3)$$

where: τ = pulse length as distance, and
β = antenna depression angle.

It can be noted from Equation 7-3 that (1) ground-range resolution improves as the distance from the ground track increases (i.e., across-track resolution is better in far range than in near range because β is smaller) and (2) resolution can be improved by shortening the pulse length. However, a point will be reached when a drastically shortened

Figure 7-5 Characteristics of range, or across-track, resolution. (Adapted from Barr 1968.)

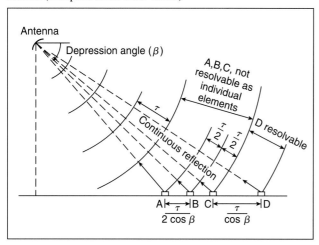

Figure 7-6 Characteristics of azimuth, or along-track, resolution. (Adapted from Barr 1968.)

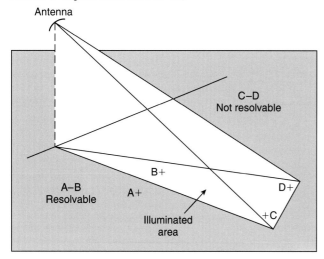

pulse will not contain sufficient energy for its echoes to be detected by the receiver. A minimum pulse length is about 21 m (pulse duration = 0.07 μsec).

Azimuth Resolution

Azimuth, or **along-track resolution** (R_a) is determined by the width of the terrain strip illuminated by a radar pulse, which is a function of the **beamwidth** of a real-aperture radar (RAR). In Figure 7-6 it is shown that the beamwidth increases with range. Thus, two objects at C and D (at the same range) are in the beam simultaneously, and their echoes will be received at the same time. Consequently, they will apear as one extended object in an image. Two objects at A and B are separated by a distance greater than the beamwidth; their returns will be received and recorded separately. Thus, to separate two objects in the along-track direction, it is necessary that their separation on the ground be greater than the width of the radar beam.

The equation for determining azimuth resolution (R_a) is

$$R_a = \frac{0.7 \lambda R_s}{D_a}, \quad (7\text{-}4)$$

where: λ = operating wavelength,
R_s = slant range to the target, and
D_a = length of antenna.

The relationships expressed in Equation 7-4 show that (1) azimuth resolution decreases in proportion to increasing range (i.e., resolution is best in near range, where the width of the beam is narrowest) and (2) a long antenna or a short operating wavelength will improve azimuth resolution. The latter two parameters both enable the radar beam to be focused into a narrower angle; beam spreading is inversely proportional to antenna length and directly proportional to wavelength.

There are several ways to obtain improved azimuth resolution with conventional RARs: (1) a long antenna, (2) a short operating wavelength, or (3) a close-in range interval. However, the practical limit of antenna length for aircraft stability is about 5 m, and the all-weather capability of radar is effectively reduced when the wavelength is decreased below about 3 cm. Because of these limitations, RARs are best suited for low-level, short-range operations.

Synthetic-Aperture Radar

The principal disadvantage of real-aperture radar is that its along-track or azimuth resolution is limited by antenna length. Synthetic-aperture radar (SAR) was developed to overcome this disadvantage. However, to gain improvement in azimuth resolution, a SAR system is much more complex and expensive to build and operate than a RAR system. The discussion that follows describes the fundamental operational principles of SAR; further details are given by Ulaby et al. (1982) and Moore et al. (1983).

SAR produces a very long antenna synthetically or artificially by using the forward motion of the platform to carry a relatively short real antenna to successive positions along the flight line (Figure 7-7). These successive portions are treated electronically as though each were an individual element of the same antenna. The synthetic antenna's length is directly proportional to range—as across-track distance increases, antenna length increases. This produces a synthetic beam with a constant width, regardless of range (Figure 7-7). Consequently, azimuth resolution remains constant throughout the range interval.

Synthetic antennas can reach staggering lengths. For example, a 1- or 2-m-long aircraft-mounted antenna can be synthesized to produce an antenna 600 m long. An 11-m real antenna can be synthesized to an effective length of about 15 km from an orbital platform.

In operation, the relatively short SAR antenna transmits wide beams of microwave radiation in the across-track direction at regular intervals along the line of flight. Due to the wide beamwidth, features will enter the beam, move through the beam, and finally leave it; the time period that an object is illuminated increases with increasing range (Figure 7-7). The SAR receiver measures and records the time delay between transmission and reception of each pulse that establishes range resolution. However, to discriminate different objects at the same distance from the ground track, the azimuths to all features must also be measured and recorded to establish azimuth resolution.

The azimuth details are determined by establishing the position-dependent frequency changes or shifts in the echoes that are caused by the relative motion between terrain objects and the platform. To do this, a SAR system must unravel the complex echo history for a ground feature from each of a multitude of antenna positions. For example, if we isolate a single ground feature, the following frequency modulations occur as a consequence of the forward motion of the platform: (1) The feature enters the beam ahead of the platform and its echoes are shifted to higher frequencies (**positive Doppler**); (2) when the platform is perpendicular to the feature's position, there is no shift in frequency (**zero Doppler**); and (3) as the platform moves away from the feature, the echoes have lower frequencies (**negative Doppler**) than the transmitted signal.

The Doppler shift information is then obtained by electronically comparing the reflected signals from a given feature with a reference signal that incorporates the same frequency of the transmitted pulse. The output is known as a **phase history**, and it contains a record of the Doppler frequency changes plus the amplitude of the returns from each ground feature as it passed through the beam of the moving antenna. Phase histories are recorded either on film, which is called the **data**, or **signal, film**, for latter optical processing or on **high-density digital tape** (HDDT) for later digital processing.

SAR Optical Correlation

To produce images that can be interpreted, the phase histories stored on the data film must be processed by an

---

Figure 7-7 Operating principle of a synthetic-aperature radar (SAR) system. (Courtesy Goodyear Aerospace Corp.)

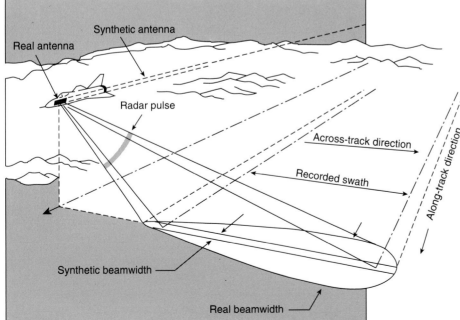

optical correlator, or processor (Figure 7-8). The data film contains phase histories from literally millions of single ground scatterers, and each is represented on the data film as a horizontal line of dark- and light-toned dashes of different lengths (Figure 7-8). With optical correlation, the data film is illuminated with the coherent light of a laser, and each phase history acts as a one-dimensional interference pattern (**radar hologram**) of a point source. Behind the data film there is a single point where the transmitted light waves constructively interfere from each phase history. At that one point, light from the entire length of the interference pattern is focused to form a miniature image of the original object in its true ground location. This plane of focus is where the **image film** is positioned (Figure 7-8).

To reduce the loss of dynamic range during optical correlation with black-and-white film, the Goodyear Aerospace Corporation uses, as an option, color-positive film to produce **color SAR images**. The implementation of this method requires the use of a single color of laser light (blue, green, or red) to illuminate the data film and expose the color film (Figure 7-8). In practice, only two of the color film's three emulsion layers are activated in the correlation process. Because of the slight overlap in the color film's emulsion layers, saturation in one of the layers will be followed on increasing exposure by activation of the second layer. This results in a **two-color image**.

The most commonly used procedure is to expose the color film with laser light having a wavelength of 0.59 μm. A moderate exposure produces shades of red because of activation of the red-sensitive, or cyan, layer, whereas added exposure causes the green-sensitive, or magenta, layer to be exposed (Figure 2-25). The addition of green to red results in a color shift to yellow. This combination produces a **red-yellow image**. A red-yellow image for an area in the Mojave Desert, California, is shown in Plate 13. The colors are interpreted as follows: yellow = strong backscatter, red = moderate backscatter, and black = weak or no backscatter.

SAR Digital Correlation

For **digital correlation**, or **processing**, the transformation from raw to interpretable data is performed mathematically by using special high-speed computers. Unlike optical correlation, digital correlation produces no intermediate data film. Digitally correlated SAR data normally incorporate radiometric and geometric corrections and are stored on computer-compatible tapes (CCTs). The CCT-stored data can be used to create radar images on a TV monitor or produce hardcopy images with special film-writing equipment (see Chapter 15). Although digital correlation is time-consuming and expensive (e.g., more than 500 million complex arithmetic operations may be required to process 1 sec of raw signal data), the output is amenable to digital enhancement techniques, and unwanted artifacts contributed by the hardware and film processing of optical correlation are eliminated.

The MacDonald Dettwiler and Associates's **Integrated Radar Imaging System (IRIS)** incorporates a digital correlator, which produces hardcopy strip images on board the aircraft in real time. The real-time images are especially useful for environmental surveillance (e.g., oil-spill monitoring, flood monitoring, and sea-ice reconnaissance) and for correcting or adjusting system variables during the flight (e.g., aircraft drift and antenna depression angle). In addition IRIS can downlink images to terminals at ground facilities or on ships for immediate interpretation and simultaneously store the raw signal data on HDDTs for ground-based precision processing following a flight (Figure 7-9).

A downlink system for the near-real-time display of digitally correlated SAR images will be an important component of **Radarsat**, a satellite planned for launch by Canada in late 1994. It will carry a C-band (6 cm) SAR, and scaled and annotated images will be available within 4 hr of overflight for prime users.

Figure 7-8 Optical correlation system for SAR data processing. As the data film is advanced through the beam of coherent light, the reconstructed image is recorded on another strip of moving film. Azimuth resolution in the along-track direction is established by optical correlation. The range imaging optics, in the form of cylindrical lens, focuses the signals in the across-track dimension to establish range resolution. The clear and opaque dashes in the circle represent a phase history for a single ground feature. (Courtesy Goodyear Aerospace Corp.)

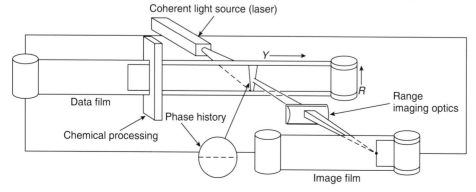

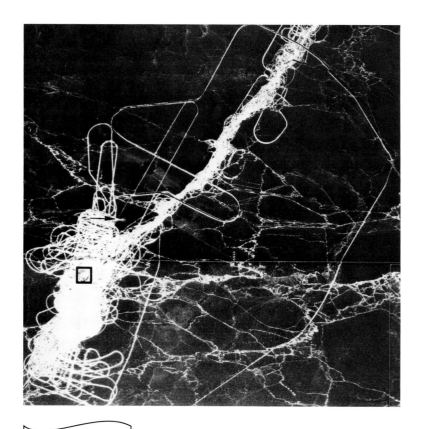

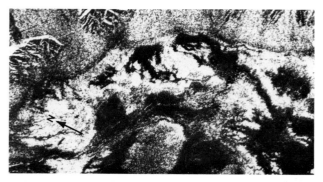

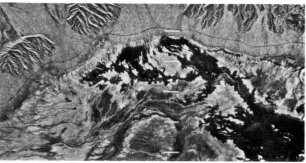

Figure 7-9 X-band (3-cm) SAR image of new ice in the Beaufort Sea. The curvilinear features are tracks of an icebreaker (marked by an arrow) engaged in breaking a path in the ice for a drill ship (within square). New ice is rather smooth and is imaged as a weak return. The tracks comprise blocks of jagged rough ice that produce strong return signals. The image was generated by a real-time digital processor onboard the aircraft and downlinked by radio directly to the drill ship and the icebreaker so that the operation could be easily coordinated. (Courtesy R. Keith Raney, Canada Centre for Remote Sensing.)

SAR Azimuth Resolution

The theoretical limit of azimuth resolution for SAR is equal to one-half the length of the physical antenna ($L/2$). Consequently, the smaller the physical antenna, the wider the physical beamwidth. This, in turn, relates to longer portions of the flight path, where echoes from a given object can be processed to form an extremely large synthetic antenna. However, because of such things as antenna vibrations and imperfections in both the optical and digital correlation processes, azimuth resolution is always somewhat greater than $L/2$. Even with this constraint, SAR systems are capable of producing high-resolution images, which makes them ideally suited for high-altitude, long-range applications. Civilian SARs are capable of producing images of 3-m resolution from aircraft altitudes (Figure 7-10) and images of 25-m resolution from earth-orbiting satellites (Figure 7-11).

Polarization of Microwave Energy

Modern radar imaging systems transmit microwave energy *(electrical field)* that is vibrating in either a horizontally (H) or vertically (V) polarized plane in relation to the long axis of the antenna. On striking the terrain, the polarized pulse interacts with the surface and is **depolarized** (rotated) to varying degrees. Depolarization of the return signal is

Figure 7-10 X-band (3-cm) SAR images showing a portion of Death Valley, California. *Top* image has 15-m resolution (range and azimuth); *bottom* image has 3-m resolution. Bars on *bottom* image are electronically produced range marks to indicate increments of distance. (Courtesy U.S. Geological Survey.)

Figure 7-11 Seasat L-band (23.5-cm) SAR image of Lake Ontario and Niagara Falls (*arrow*). Image textures include (A) smooth = calm water, (B) grainy = rough water, and (C) speckled = urban areas. The image was digitally correlated with 25-m range and azimuth resolutions. (Courtesy David Okerson, MacDonald Dettwiler and Associates.)

caused by several terrain parameters, including surface roughness and object geometry. When depolarized, the pulse vibrates in several planes, but in most circumstances it returns to the antenna in the same polarized plane that it was transmitted in and is so recorded. This results in a **parallel-polarized**, or **like-polarized, image**: an **HH image** (horizontal transmit, horizontal receive) or a **VV image** (vertical transmit, vertical receive). The most common SLAR design is HH, and all images used in this chapter are HH presentations unless otherwise specified.

Some radar systems are equipped with a second antenna element that simultaneously receives the cross-polarized component of the depolarized return signal; the cross-polarized component vibrates at right angles to the polarization of the transmitted pulse. This results in a **cross-polarized image**: an **HV image** (horizontal transmit, vertical receive) or a **VH image** (vertical transmit, horizontal receive).

A radar system containing two antenna elements allows for the simultaneous recording of return echoes from the identical ground area in like and cross-polarizations. This results in two image-pair scenarios: HH-HV and VV-VH. The former configuration has been employed most frequently (Figure 7-12). This type of SLAR is called a **dual-polarization**, or **multipolarity, radar**.

Multichannel Radar

A state-of-the-art extension of a single-wavelength, dual-polarization radar is a **multispectral, multipolarity SAR**. This is commonly called a **multichannel SAR**. The first operational multichannel SAR was developed by the Environmental Research Institute of Michigan (ERIM). The ERIM multichannel SAR can produce four simultaneous images at two wavelengths (X-band = 3 cm, L-band = 25 cm) and two polarizations. Figure 7-13 shows a set of four images generated by the ERIM system for a mixed forest-agriculture area near Saginaw, Michigan. The multichannels are X-band HH, X-band HV, L-band HH, and L-band HV.

The Jet Propulsion Laboratory (JPL) is currently testing a prototype multichannel SAR that incorporates three wavelengths (C-band = 6 cm, L-band = 24 cm, and P-band = 68 cm) and two polarizations. Multichannel SARs provide advantages over single-wavelength, single-polarity SARs similar to the advantages of multispectral over single-channel sensing in the visible and reflected infrared spectral regions.

Figure 7-12 Dual-polarized RAR images of SP Mountain and lava flow, Arizona, generated by the Westinghouse Corp. K_a-band (0.86-cm) system: (A) parallel-polarized image, HH; (B) cross-polarized image, HV. The lava flow is significantly brighter on the HH image because of minimal rotation of the horizontally polarized incident beam. Look direction (east) is indicated by the arrow. (Courtesy Gerald G. Schaber, U.S. Geological Survey.)

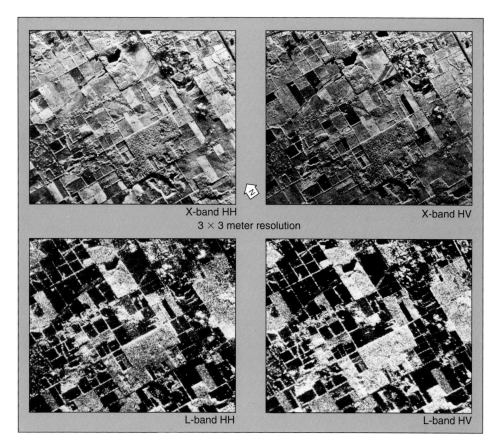

X-band HH
3 × 3 meter resolution

X-band HV

L-band HH

L-band HV

Figure 7-13 ERIM multichannel SAR images collected simultaneously over a mixed forest–agricultural area near Saginaw, Michigan. (Courtesy Eric S. Kasischke, Environmental Research Institute of Michigan.)

Radar Shadows

Because radar illuminates the terrain along the line of sight (i.e., **unidirectional illumination**), protruding topographic features can prevent the beam from striking their **backslopes**, or the slopes facing away from the radar beam. These obscured backslope areas represent **radar shadows**. They will appear black on a positive image because they are not illuminated by the radar beam, and consequently no backscatter is returned to the antenna from their surfaces.

Figure 7-14 illustrates a terrain feature illuminated by a radar pulse. Part of the pulse is reflected from the **foreslope** (slope facing the radar beam) or surface AB; the remainder of the pulse continues along a line of sight and intercepts the terrain at D, which produces the next return signal. Consequently, the surface BCD (backslope to beyond the base) is not illuminated, and there will be no return signal. This time lapse record will appear as a no-return, black void on the resulting image. It is important to remember that an area of nonillumination can extend well beyond the base of a backslope, masking all down-range features within the radar shadow (Figure 7-14).

The creation of a radar shadow is determined by the backslope angle of the feature (α^-) and the antenna depression angle (β). Slopes are obscured and thus in shadow when the backslope angle is larger (steeper) than the depression angle ($\alpha^- > \beta$). A backslope is just illuminated or grazed by the radar beam when $\alpha^- = \beta$; this condition is known as **grazing illumination**, or **grazing incidence**. A backslope

---

Figure 7-14 Radar shadow of surface BCD projected to the image plane as B_1D_1; shown are look angle (θ), depression angle (β), foreslope angle (α^+), and backslope angle (α^-). (Adapted from Ford et al. 1980.)

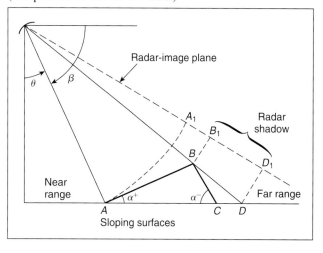

Figure 7-15 Westinghouse Corp. K_a-band (0.86-cm) RAR image of the eastern Grand Canyon, Arizona. Note the excellent enhancement of topography by the radar shadows. The Bright Angel fault is clearly discernible (A-A'). The effect of topographic inversion (pseudoscopic effect) can be demonstrated by rotating the book 180° (i.e., so the shadows fall away from the viewer). Arrow shows look direction. (Courtesy Gerald G. Schaber, U.S. Geological Survey.)

is fully illuminated when $\alpha^- < \beta$. Generally, shadowing occurs most often at shallow depression angles because the radar illumination becomes more oblique in the far range.

The factors that determine the length of the radar shadow are as follows: (1) the height of a feature above the datum plane (h); (2) the slant-range distance from the antenna to the feature (D); (3) the radar platform height above the datum plane (H); and (4) the antenna depression angle (β) (Simonett and Davis 1983). The influence of the preceding factors on establishing shadow lengths (L) can be expressed as follows:

$$L = \frac{hD}{H} \quad \text{and} \quad (7\text{-}5)$$

$$L = \frac{h}{\sin \beta}. \quad (7\text{-}6)$$

Examination of these equations reveals that shadow length is directly related to feature height and slant-range distance and inversely related to platform height and depression angle. Equation 7-6 also shows that small or shallow depression angles would be best for discriminating subtle topography by shadow enhancement but unacceptable in mountainous terrain because the shadows would be too large.

When analyzing radar images, it is important to remember the following shadow rules:

1. A radar shadow can "fall" only in the range or across-track direction, away from the antenna. When shadows are present, this azimuth-shadow phenomenon always makes it possible to identify near- and far-range portions of a radar image (Figure 7-14).

2. At a common range, the highest feature will produce the longest shadow.

3. A feature that casts an extensive shadow at far range can have its backslope completely illuminated at near range.

4. The shape of radar shadows mimics topographic profiles, which can be a useful interpretation tool.

Because SLAR's relatively low angle of illumination produces *a highlighting and shadowing effect*, topographic features with a strike or orientation oblique or parallel to the beam front are often enhanced (Figure 7-15). The juxtaposition of a strong return (light tone) from a foreslope and no return from the backslope results in an image that has a **pseudo-stereoscopic**, or three-dimensional, appearance. This clear expression of relief is of particular advantage in geological interpretations.

Radar Foreshortening and Layover

The presence of topographic relief in a scene can introduce distortions known as **foreshortening** and **layover**. The length of time it takes the radar beam to illuminate a foreslope from its base to its summit determines its length on a radar image. In Figure 7-16, it is seen that the length of foreslope *AB* is shortened or compressed relative to the length of the backslope *BC* when projected to the image plane. This foreshortening effect occurs because it takes less time to illuminate the foreslope, and the echo would be correspondingly shorter relative to that from the backslope.

With radar imaging, all foreslopes are shortened relative to their true lengths. The degree of shortening is a function of the illumination geometry and the foreslope angle: (1) Slope length decreases with increasing β or decreasing θ (most severe compression occurs in the near range) and (2) slope length decreases with increasing α^+. The greater the foreshortening, the more energy per unit area is displayed on the image, until so much is available that it saturates the receiver (Simonett and Davis 1983).

Foreshortening is the most severe when the angle of incidence is equal to 0°, meaning that the radar beam is perpendicular to the foreslope. Under this condition, known as **normal incidence**, the entire slope will be illuminated at the same time. This means that the slope has been shortened to zero length, and it will be displayed as a very bright line on the radar image.

Layover is an extreme case of foreshortening that occurs whenever the look angle is smaller than the foreslope angle ($\theta < \alpha^+$) or, conversely, whenever $\beta > 90° - \alpha^+$. In this situation, the echo from the foreslope summit will be received first because the slant range distance is shorter to the top of the feature than it is to the base. This concept is illustrated in Figure 7-17. Note that the ordering of surface elements on the projected image plane is the reverse of the ground ordering. In an actual image, a topographic feature will appear to be "laid over" on its side toward the near range (Figure 7-18). Radar layover tends to occur mainly when steep foreslopes are encountered in the near range (large β, small θ).

Figure 7-16 Radar foreshortening of foreslope *AB*, which is projected to the image plane as A_1B_1, relative to backslope *BC*, which is projected as B_1C_1. (Adapted from Ford et al. 1980.)

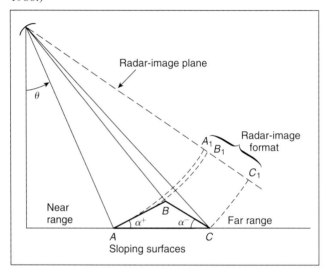

Radar Return Strength and Image Tone

The strength of radar return and, hence, image tone is primarily influenced by the following ground and radar system properties.

Ground Properties
    Surface slope (macroscale relief)
    Surface roughness (microscale relief)
    Complex dielectric constant
    Feature orientation

Radar System Properties
    Operating wavelength
    Antenna depression angle
    Polarization
    Antenna look direction

How these properties individually and in combination influence radar return intensity are discussed in the following sections.

Figure 7-17 Radar layover of foreslope *AB* projected to the image plane as B_1A_1 where the look angle θ is smaller than the foreslope angle α^+. (Adapted from Ford et al. 1980.)

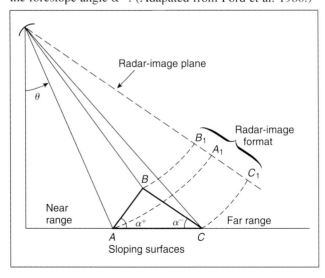

Surface Slope

Very significant in the interpretation of radar images is the relationship between the antenna depression angle of the incident beam and the **surface slope** of macroscale features. The foreslopes of topographic features (slopes facing the antenna) are responsible for strong echoes, with the greatest amount of reflection occurring when the local slope is perpendicular to the radar beam. This condition is known as **normal incidence**, and it means that the incidence and

Figure 7-18 Seasat L-band (23.5-cm) SAR image of the San Francisco volcanic field, Arizona. Note the foreshortening and layover of major topographic features. Layover occurs on *Seasat* images whenever a foreslope angle exceeds about 20°. Look direction is indicated by the arrow.

reflection paths are the same. Normal incidence occurs whenever the sum of the depression angle and the foreslope angle equals 90° ($\beta + \alpha^+ = 90°$). Normal incidence can obliterate any information relating to the microrelief of the foreslope (e.g., vegetation cover), a condition known as **foreslope brightening**. This produces a very bright area on the image. As previously discussed, the foreslope will be shortened to zero length under the condition of normal incidence.

Conversely, the backslopes of topographic features (surfaces that slope away from the antenna) produce weaker echoes—and, therefore, darker image tones—when illuminated. The strongest returns occur when the backslope angle is smaller than the depression angle ($\alpha^- < \beta$). A shadow no-return results whenever the backslope angle is larger than the depression angle ($\alpha^- > \beta$) (Figure 7-14). The differential returns from the foreslopes and backslopes emphasize topography, which can be especially useful for mapping landforms and geologic structures such as faults (Figures 7-1 and 7-15).

Surface Roughness

One of the most important considerations in determining radar return intensity is the interrelationship between the roughness of an object's surface *(microrelief)* and a radar system's wavelength and depression angle. Unlike macroscale relief, which is measured in meters, **surface roughness**, or **microscale relief**, is measured in centimeters (i.e., in wavelength-sized units). This type of roughness is represented by small textural features such as leaves, twigs, cobbles, and pebbles. Although there is no absolute measure,

the average vertical dimension of microrelief is a statistical approximation of a material's surface roughness. Wood- and metal-dowel template and photogrammetric methods for collecting and statistically analyzing information about the roughness of nonvegetated surfaces are described by Schaber et al. (1980).

Two major categories of radar reflection can take place according to surface roughness—**specular** and **diffuse**. As illustrated in Figure 7-19, a **radar-smooth surface** reflects the incident pulse in a single direction away from the antenna. The pulse will be reflected at an angle equal to and

Figure 7-19 Specular and diffuse reflection. Note that pebbles (*A*) are "smooth" compared to wavelength (λ) and the cobbles (*B*) are "rough" compared to wavelength. (Courtesy Goodyear Aerospace Corp.)

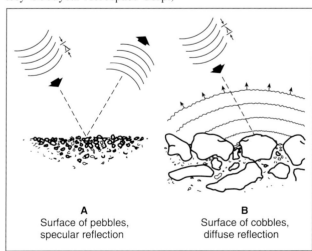

A
Surface of pebbles, specular reflection

B
Surface of cobbles, diffuse reflection

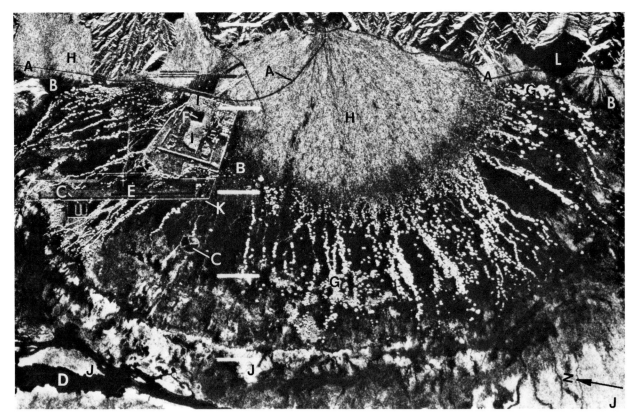

Figure 7-20 X-band (3-cm) SAR image of the Furnace Creek Ranch area, Death Valley, California. Variations in tone are due to differences in surface roughness relative to wavelength. Specular targets include (*A*) paved highway, (*B*) sand zone, (*C*) calm water, (*D*) dry floodplain deposits, (*E*) paved runway, and (*F*) short grass. Diffuse targets include (*G*) stands of honey mesquite, (*H*) cobbles and boulders, (*I*) date palm grove, and (*J*) jagged pinnacles of silty rock salt. A metal fence is shown at *K* and a radar shadow at *L*. Resolution in range and azimuth is about 3 m. Image bars are electronically produced range marks to indicate increments of distance. Scale is about 1:40,000. (Courtesy U.S. Geological Survey.)

at a direction opposite that at which it strikes the surface (i.e., the **Fresnel-reflection direction**). This process is called **specular reflection**, and the surface is said to be a **specular**, or **mirror**, **reflector**. Terrain features that are specular reflectors include playas, pavement surfaces, and calm bodies of water. They are responsible for no or very weak returns (dark image tones), except in the case of normal incidence (Figure 7-20).

A **radar-rough surface** (e.g., cobbles and boulders, lava flows, jagged sea ice, or certain types of vegetation) is composed of numerous small facets that scatter the energy of an incident pulse in many different directions, with some portion of the pulse being reflected back toward the antenna (Figure 7-19). When many of the facets are perpendicular (or nearly so) to the direction of the pulse propagation, regardless of the incidence angle relative to the datum plane, a large proportion of the signal is reflected back toward the antenna. This process is called **diffuse reflection**, and the surface is referred to as a **diffuse reflector**, or **Lambertian surface**. Such surfaces are responsible for bright image

tones; until saturation of the system occurs, the rougher the surface, the brighter the tone (Figure 7-20).

Maximum surface roughness is represented by an **isotropic scatterer**, which is responsible for an extremely strong echo, virtually independent of depression angle and wavelength. Isotropic scatterers include certain forms of vegetation, boulder fields, and rough-surfaced sea ice. In Figure 7-20, the silty rock salt (rough facies) unit represents an isotropic scattering surface.

A given surface, however, may be rough for some microwave wavelengths and smooth for others; or for the same wavelength, a given surface may be either smooth or rough at different depression angles (Figure 7-13). Long microwave wavelengths are not backscattered by relatively fine textured surfaces. Thus, the longer the wavelength, the coarser the surface must be for backscattering to occur. Also, at a given wavelength, a surface becomes smoother as the depression angle decreases, or, conversely, it becomes rougher as the depression angle increases.

The relationship of wavelength and depression angle

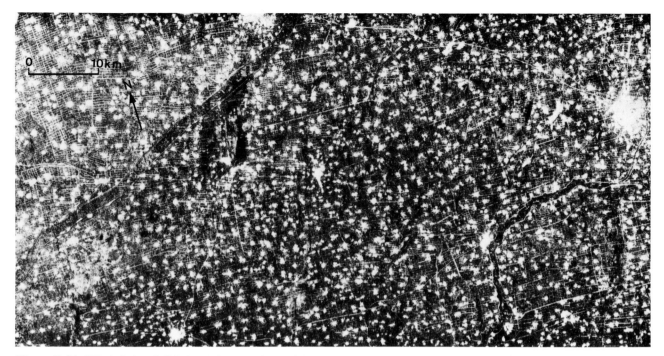

Figure 7-21 SIR-A L-band (23.5-cm) image of Hebei-Shandong, People's Republic of China. The multitude of bright spots are small villages (density of about one per square kilometer). The bright spots coalesce into several large cities. (Courtesy John P. Ford, Jet Propulsion Laboratory.)

to surface roughness can be described by the **Rayleigh criterion**, which considers a surface to be smooth if

$$h < \frac{\lambda}{8 \sin \beta} = \frac{\lambda}{8 \cos \theta},$$ (7-7)

where: h = average vertical height of microrelief,
λ = operating wavelength,
β = antenna depression angle to the local surface, and
θ = antenna look angle to the local surface.

By solving for h, the Rayleigh criterion predicts the approximate boundary or breakpoint between radar-smooth and radar-rough surfaces for a given wavelength and illumination angle (β or θ). Accordingly, surfaces with a value of h less than the calculated value should produce dark image tones, and surfaces with a value of h greater than the calculated value should be responsible for bright image tones. Observe from Equation 7-7 that (1) at any illumination angle, a given surface becomes rougher as wavelength decreases, and (2) independent of wavelength, a given surface becomes smoother as depression angle decreases or look angle increases (i.e., as range increases).

The Rayleigh criterion does not consider that there can be a category of surface roughness intermediate between definitely rough and definitely smooth. Peake and Oliver (1971) modified the Rayleigh criterion to define the upper and lower values of h for a surface of **intermediate roughness**. The **smooth criterion** (h_s) considers a surface to be *definitely* radar smooth if

$$h_s < \frac{\lambda}{25 \sin \beta} = \frac{\lambda}{25 \cos \theta}.$$ (7-8)

The **rough criterion** (h_r) considers a surface to be *definitely* radar rough if

$$h_r > \frac{\lambda}{4.4 \sin \beta} = \frac{\lambda}{4.4 \cos \theta}.$$ (7-9)

Surfaces with a roughness falling between the two calculated boundaries (h_s and h_r) will produce an intermediate radar return and, hence, gray image signatures.

Specular features, however, can be responsible for strong returns if their smooth surfaces are configured to form **corner reflectors**. When oriented perpendicular to the radar beam, a major part of the incident energy is returned to the antenna regardless of wavelength and depression angle. This is called the **cardinal-point effect**. Corner reflectors are a common occurrence in an urban environment where vertical walls intersect the horizontal ground. Consequently, when illuminated from the proper azimuth, cultural structures usually stand out from their surroundings on radar images (Figure 7-21).

Complex Dielectric Constant

The **complex dielectric constant** (ϵ) is dependent upon the electrical properties of a material and is a measure of a material's ability to conduct or reflect microwave en-

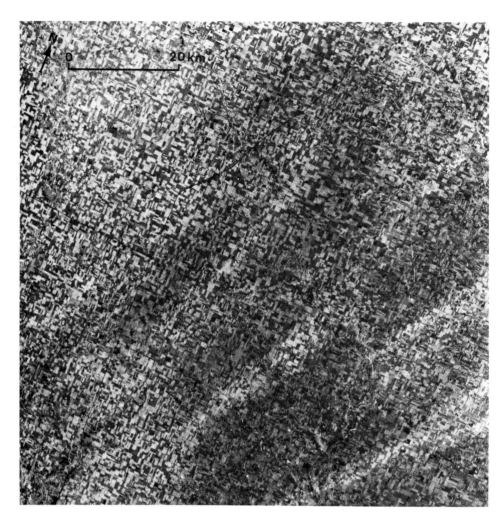

Figure 7-22 *Seasat* L-band (23.5 cm) SAR image of central Iowa. The bright areas are caused by rain-soaked soil and vegetation, which exhibit high reflectivity (elevated ϵ values). Rainfall occurred about 10 hours before the image was generated. (Courtesy John P. Ford, Jet Propulsion Laboratory.)

ergy; as ϵ increases, reflectivity increases, whereas conductivity or penetration decreases. Generally, the dielectric constant of most naturally occurring materials, when dry, ranges from about 3 to 8 at radar wavelengths. Because of this small range, radar generally is considered to be largely insensitive to the electrical properties of a surface material in the absence of water.

When moist, however, a material's effective dielectric constant may approach 80 (i.e., the dielectric constant of liquid water in the same wavelength region). Thus, ground targets produce significantly stronger returns when they are damp (Figure 7-22). The dielectric of a terrain material increases in an approximate linear relationship to increasing moisture content.

For free water and ice, the dielectric constant is not of primary relevance in determining backscatter intensity. Rather, the controlling property is the physical state of their surfaces; calm water and smooth ice are specular reflectors, and rough water and rough ice are diffuse reflectors (Figures 7-23 and 7-24).

Metallic objects, such as bridges, ships, oil platforms, fences, railroad tracks, power lines, and towers, act as antennas, strongly reflecting the incident energy. Such objects often appear as white targets, "blooming" to a size on the image greater than would be indicated by their physical sizes. Echo strength from metal objects can be further intensified by corner reflector configurations (Figure 7-25).

Polarization

Polarization is a fundamental system parameter that influences backscatter strength. Important considerations are the polarization vector of the transmitted signal and the depolarization effects of the terrain. In general, radar echoes from surface features are strong in like polarization (HH or VV) and weaker in cross-polarization (HV or VH). **Like-polarized backscatter** results from **single reflections** along a feature's surface. This is known as **surface scattering**, and it causes little or no depolarization (Figure 7-12). **Cross-polarized backscatter** results from either multiple reflections within a diffuse volume such as a vegetation canopy or from multiple bounce interactions associated with certain "solid" features, such as rough rock surfaces. Generally, as surface roughness increases, depolarization increases. This is known as **volume scattering**, and it causes depolarization (Figure 7-13).

A vegetation canopy provides an excellent example for illustrating surface and volume scattering at two different wavelengths (Figure 7-26). The canopy is not penetrated at short wavelengths (e.g., X-band), and surface scattering will occur along the top of the canopy. However, at long wave-

Figure 7-23 *Seasat* L-band (23.5-cm) SAR image of the southern California coast. Although the image appears to show clouds over the Pacific Ocean, the radar "sees" through the clouds and records only reflections received from the ocean surface. The light and dark areas on the ocean surface represent differences in roughness caused by local surface winds. (Courtesy John P. Ford, Jet Propulsion Laboratory.)

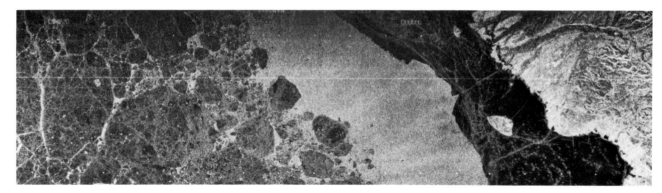

Figure 7-24 *Seasat* L-band (23.5-cm) SAR image showing a portion of the Beaufort Sea ice pack west of Banks Island, Canada (right). The image, obtained at 1:55 A.M. on July 11, 1978, contains a wealth of ice information. The dark zone next to Banks Island is an area of shore-fast ice composed mostly of smooth first-year sea ice. Rough pressure ridges are seen within the shore-fast zone, and to the west is an area of open water, called a shore lead. At the western edge of the lead is a marginal ice zone composed of a mixture of open water and rounded, multiyear ice floes and some first-year ice. Further west is the main polar pack, made up of large floes as much as 20 km in diameter, surrounded by new leads. A random pattern of pressure ridges is visible within the floes. The very bright areas within the floes indicate areas of intensive surface roughness, called rubble fields. (Courtesy Jet Propulsion Laboratory.)

Figure 7-25 SIR-A L-band (23.5-cm) image of a portion of the Arabian Sea showing bright-point targets that include wellhead towers, oil platforms, and ships. The star-shaped patterns result from corner reflections from metal surfaces when the azimuth angle is about 90°. Medium tones indicate rough water surfaces.

lengths (e.g., L-band), the incident beam enters the canopy where the leaves, branches, and trunk act in concert to reflect the beam along multiple paths, causing depolarization.

In like-polarized radar systems (HH or VV), strong echoes occur when the surface scatterers are oriented in the same direction as the polarization of the incident beam. For example, VV polarization is the most sensitive to plant stocks and tree trunks because of their vertical orientation. HH polarization is more sensitive to physical and cultural surfaces that are configured in a horizontally dominant manner. In addition, L-band radars with HH polarization provide an enhanced ability to penetrate mantles of dry sand (described in a later section).

Regarding changes in the complex dielectric constant due to moisture variation, HH polarization is the least sensitive, whereas HV and VV are the most sensitive. Thus, HV and VV polarized returns may contain information concerning soil and snowpack moisture levels.

When interpreting radar images with different polarizations, it is important to remember the following rules:

1. The brighter the tone on HH and VV images, the greater the contribution from surface scattering.

2. The brighter the tone on HV and VH images, the greater the contribution from volume scattering. Because the cross-polarized return is usually weaker than the like-polarized return, the power gain of the cross-polarized receiver is often increased. Therefore, care must be exercised when comparing tonal signatures for the same features on a like- and cross-polarized image pair.

Antenna Look Direction and Feature Orientation

Antenna look direction with respect to the orientation, or strike, of surface features being illuminated by the radar beam plays a major role in determining echo strength and, hence, feature detectability versus background. The intersection of antenna look direction and a feature's strike forms the **orientation**, or **azimuth**, **angle** (Φ), which has limits of 0° and 90°. All else remaining equal, maximum backscatter occurs when radar illumination is at a right angle to the feature's trend ($\Phi = 90°$), with minimum backscatter oc-

Figure 7-26 Surface and volume scattering from a vegetation canopy. Note that volume scattering causes depolarization of the incident radar beam. (Adapted from a NASA drawing.)

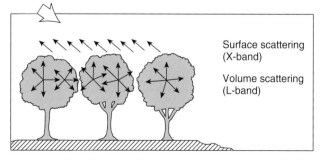

Surface scattering (X-band)

Volume scattering (L-band)

Figure 7-27 Motorola X-band (3.2 cm) RAR images incorporating orthogonal look directions of the same area in Nigeria: (A) east look, (B) south look. Compare fractures indicated by arrows in both images. (Courtesy Ron Gelnett, MARS Associates, Inc.)

curring when radar illumination is parallel to the trend ($\Phi = 0°$).

Images of multiple-look directions, if available, can increase data content appreciably. For example, Gelnett (1978) states that four orthogonal looks are necessary to obtain the maximum amount of data available from radar images:

Four orthogonal looks = 100 percent data content,
Two opposite looks = 90 percent data content,
Two orthogonal looks = 80 percent data content,
One look = 70 percent data content.

Figure 7-27 illustrates two orthogonal looks of a highly fractured and faulted landscape in Nigeria. Note how some fracture sets are more apparent in one look than in the other. Although the increase in data content is shown in this example, the reader must understand that (1) it required flying the survey area twice, effectively doubling the cost of the data collection and data reduction costs of the survey, and (2) the data lost to radar shadows in one direction were not recovered in the second orthogonal look (Gelnett 1978). Recovery of data lost in radar shadow areas is possible, however, with opposite-look images, as is illustrated in Figure 7-28.

Bryan (1979) found that different orientation angles formed between cultural targets and antenna look direction could produce dramatic differences in image gray tone. For example, large areas within the Los Angeles region were found to be portrayed in significantly darker tones on L-band SAR images than adjacent areas having similar land cover. The most pronounced tonal change for common targets occurred when the orientation angle between the antenna look direction and the trend of the streets and the accompanying walls of the structures was less than about 10° or 15°. Consequently, Bryan (1979) acknowledges that an *a priori* knowledge of this orientation is necessary to ensure accurate interpretation of radar images. This consideration is especially important for satellite systems because they have fixed

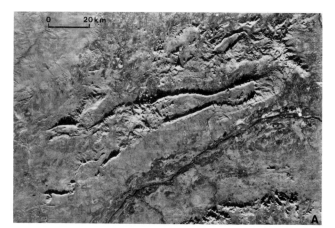

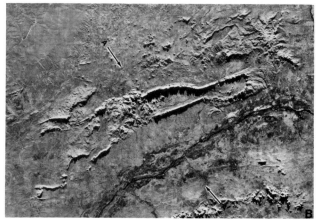

Figure 7-28 Motorola X-band (3.2-cm) RAR mosaics (opposite looks) reduced from original scale of 1:250,000. Shown is an 18,500-km^2 area in east-central Nigeria: (A) north look, (B) south look, and (A-A') trace of a major fault that is not apparent on the north-look mosaic. The prominent structure is an anticline in Cretaceous sandstone. (Courtesy Ron Gelnett, MARS Associates, Inc.)

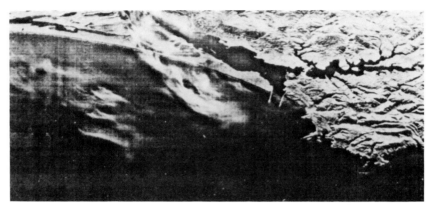

Figure 7-29 K_a-band (0.86-cm) RAR image of a coastal area showing strong backscatter from clouds. (Courtesy National Aeronautics and Space Administration.)

Figure 7-30 Comparison of an X-band (3-cm) SAR image (*left*) and an airphoto (0.5–0.7-μm) (*right*) for cloud-penetration differences. (Courtesy Goodyear Aerospace Corp.)

azimuth angles that may not be optimal with respect to feature orientations. With aircraft radars, flight lines may be changed to compensate for the orientation of features that are of particular importance.

Atmospheric and Surface Penetration

When dealing with imaging radars, two types of penetration must be considered—**atmospheric** and **surface**. Atmospheric scattering effects on microwaves are the most pronounced for the shortest wavelengths. K_a-band radar (no

longer operational) had some cloud-penetration capability, but microwave radiation at this wavelength was backscattered by precipitation and thick cloud formations (Figure 7-29). X-band radar will penetrate haze, dust, fog, most clouds, and moderate precipitation (Figure 7-30). Echoes returned from atmospheric components have bright signatures on radar images (Figure 7-29). C-, L-, and P-band radars are defined as *all weather* because they can operate through any atmospheric condition, including heavy precipitation.

Long-wavelength radars are especially effective in tropical wet environments because of the persistent cloud cover and frequent rainstorms. In numerous tropical countries, radar images have provided the first cloud-free views of vast uncharted regions. Such views, for example, are bound to become increasingly important for monitoring the

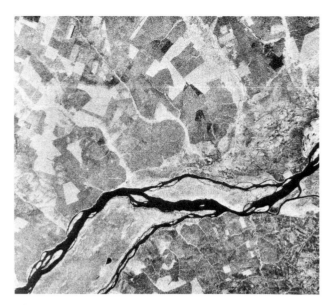

Figure 7-31 SIR-A L-band (23.5-cm) image of a rural area in southern Brazil. The bright areas depict tropical forest stands, whereas the darker areas are mainly deforested pastures of large fazendos or ranches. Also shown is the Paraná River. (Courtesy John P. Ford, Jet Propulsion Laboratory.)

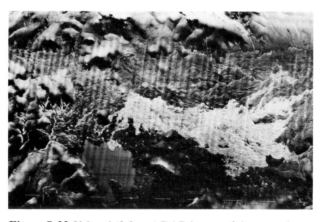

Figure 7-32 X-band (3.2-cm) RAR image of the central part of the Seward Peninsula, Alaska. The image was generated during the late winter when about 1 m of dry snow covered the entire region. The bright area in the middle of the image is a snow-covered lava flow. (Courtesy P. Jan Cannon, Geology/Remote Sensing Consultant.)

deforestation of the world's tropical rainforests (Figure 7-31). With the dual advantage of atmospheric penetration and active illumination, radar is also well suited for applications in the high latitudes—areas characterized by weak solar illumination and adverse weather conditions (Figure 7-24).

The depth of microwave penetration into a surface medium is strongly dependent upon wavelength (λ) and the complex dielectric constant (ϵ) of the surface material at the time of the overflight: (1) as λ increases, penetration in-

creases and (2) as ϵ increases, penetration decreases and reflectivity increases. With appreciable moisture content (determining effective ϵ), the penetration depth is practically negligible at all wavelengths used in imaging radars. Surface penetration can be further enhanced by HH polarization and relatively large look angles or, conversely, relatively small depression angles.

Penetration of Snow and Surficial Materials

Cannon (1980) was one of the first scientists to report on the surface penetration capability of an imaging radar. In an X-band (3.2-cm wavelength) RAR image of the central part of the Seward Peninsula, Alaska, he was able to identify landforms that were covered by about 1 m of fresh, dry snow (Figure 7-32). This unique capability has become known as **subsurface radar imaging**.

More recently, it has been documented that subsurface imaging through dry surficial materials occurred with the two Space Shuttle imaging radars (SIR-A and SIR-B) in the eastern Sahara of Egypt and Sudan (McCauley et al. 1982, 1986) and in north-central Saudi Arabia (Berlin et al. 1985, 1986) and with *Seasat* in the Mojave Desert of California (Blom et al. 1984). In the eastern Sahara, SIR signals penetrated sand sheets and drift sand to a maximum depth of about 4 m, revealing previously unknown paleodrainage systems and geologic structures (Figure 7-33). In the Mojave Desert, as much as 2 m of dry alluvium was penetrated by the *Seasat* signals to reveal buried igneous dikes.

In Saudi Arabia, the most dramatic example of SIR-A and SIR-B subsurface imaging occurred along the southern margin of the Al Labbah Plateau, an area being covered by eolian sand originating from the adjacent An Nafud sand sea (Figure 7-34). Distinct tonal signatures uniquely defined (1) areas where the SIR signals penetrated a relatively thin sand sheet to be diffusely reflected from the buried, "radar-rough" carbonate rocks of the Aruma Formation and (2) a thick sand hill and associated sand-shadow deposit, where SIR subsurface imaging did not occur (Figure 7-34). Depth measurements from more than 80 test holes show that subsurface imaging occurred through a sand layer whose maximum measured thickness is 1.24 m. The maximum thickness of sand where subsurface imaging did not occur is thought to be about 3.1 m. A ground view of the penetrated sand sheet is shown in Figure 7-35.

The conditions under which subsurface radar imaging can occur in surficial deposits must be met *simultaneously* at the time of the overpass. The conditions are as follows:

1. The cover material must be extremely dry (<1 percent moisture content), which is responsible for small ϵ values.

2. The cover material must be fine-grained. Scattering losses do not become prohibitive for subsurface penetration when individual grain sizes are smaller than one tenth the wavelength; scattering losses become appreciable when grain sizes become larger than one fifth the wavelength.

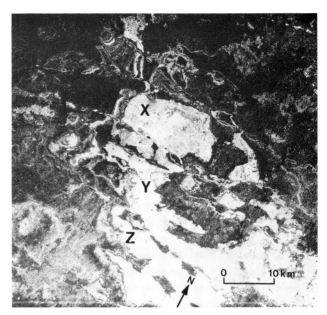

Figure 7-33 *Landsat* MSS band 7 (0.8–1.1-μm) image (*left*) and SIR-A L-band (23.5 cm) image (*right*) of the southern part of the Limestone Plateau in Egypt. The plateau's surface at X, Y, and Z is covered by a veneer of unconsolidated sediment (dark on MSS image); light areas on the MSS image represent Paleocene to Eocene limestone bedrock. The SIR-A signals penetrated the unconsolidated sediment, producing bright responses from the buried bedrock, including localities X, Y, and Z. At locality X, the bedrock was found to be covered by about 1.1 m of drift sand and fine-pebble alluvium. (Courtesy Gerald G. Schaber, U.S. Geological Survey.)

3. The cover material must be essentially free of clay minerals. Water-bearing minerals can severely attenuate the radar signals and thus restrict penetration depths.

4. The subsurface must be rough enough to generate backscatter. If a subsurface is radar smooth, the signals will be specularly reflected away from the antenna, and there will be no image evidence of a buried substrate.

5. The cover material must be thin enough to enable the radar signals to penetrate to a rough substrate and be reflected back through the overburden. If the cover material is excessively thick, the microwaves will be converted to heat energy somewhere along the two-way travel path.

Given these conditions, about 10 percent of the earth's land surface is amenable to subsurface radar imaging.

Penetration of Vegetation Canopies

Microwaves employed in imaging radars can penetrate vegetation canopies to varying extents, depending upon sys-

tem parameters, which include wavelength and polarization plus vegetation conditions such as the geometric structure and thickness of the canopy, leaf size, and the water content of the leaves, which controls the complex dielectric constant. Depth of canopy penetration varies directly with wavelength. For example, X-band incident signals are reflected largely from the canopy's outer surface, whereas L-band signals can enter the canopy where volume scattering occurs (Figures 7-26, 7-31, and 7-36). For most types of vegetation, HH polarization is best for canopy penetration. In addition, penetration depth is inversely related to canopy thickness and the moisture content of the canopy.

One of the most dramatic examples illustrating the ability of L-band radar to penetrate through a vegetation canopy occurred in the mangrove jungle of southern Bangladesh (Sundarbans Forest) during the second Space Shuttle radar mission (SIR-B) in October 1984 (Imhoff et al. 1986). In this study, the L-band (23.5 cm) radar penetrated the 12.5-m-tall and closed mangrove canopy at three illumination angles ($\theta = 26°$, 45°, and 58°; $\beta = 64°$, 45°, and 32°, respectively). Shown in Plate 13 is a color-coded "subsurface" image of the Sundarbans Forest derived from the SIR-B 45° digital image. The colors represent the following field-confirmed conditions beneath the closed mangrove canopy: yellow = areas where the soil surface was completely covered with water; light blue = areas with partial flooding; and brown = areas where the soil surface was free of inundation.

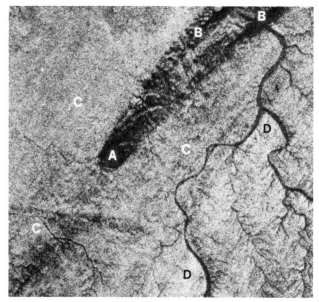

Figure 7-35 Overview of SIR-A penetrated sand sheet on the Al Labbah Plateau, Saudi Arabia. Excavated pit is 82 cm deep, and vegetation cover is less than 1 percent.

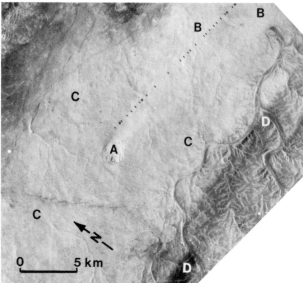

Figure 7-34 SIR-A L-band (23.5-cm) image (*top*) of a portion of the Al Labbah Plateau, Saudi Arabia. Annotated features are (*A*) a sand hill called Anbat, (*B*) the Irq al Ubaytir sand shadow, (*C*) a widespread area where subsurface imaging occurred through a relatively thin sand sheet, and (*D*) rubbly carbonate rock exposures of the Upper Cretaceous Aruma Formation, which also underlie the sand sheet. A *Landsat* TM band 3 image (*bottom*) (0.63–0.69-μm) is included for comparison.

Figure 7-36 Motorola X-band (3.2-cm) RAR image of banana plantations along the Motagua River, Guatemala. The canopy of banana plants was responsible for an exceptionally strong echo. (Courtesy Ron Gelnett, MARS Associates, Inc.)

Interpreting Radar Images

With a proper understanding of SLR operation and the surface properties that control microwave reflectivity, the recognition elements that have been developed for airphoto analysis are applicable to the interpretation of radar images (see Chapter 3). Several of these recognition elements are discussed in the following section.

The tones of a radar image (1) are a qualitative measure of microwave backscatter strength and (2) are a product of many radar system and ground parameters. The radar returns from each resolved ground feature are encoded on photographic film in tones of gray ranging from black to white. The stronger the return, the brighter the tone until saturation occurs. For many ground objects, tonal signatures in radar images are reversed from those observed in photographs representing the visible and near IR spectral regions (Figure 7-37).

TABLE 7-2 Surface States and Representative SLAR Tonal Ranges

Surface State	Reflection Characteristics	Tonal Range
Topographic		
Level terrain	Specular reflection if smooth; weak to no return	Dark tones to black
Foreslope	Relative strong return; maximum return with normal incidence	Medium to light tones
Backslope	Relatively weak return at grazing incidence; no return when slope is not illuminated (radar shadow)	Medium to dark tones Black
Geologic		
Smooth surfaces	Specular reflection when terrain in relatively flat; weak to no return	Dark tones to black
Rough surfaces	Diffuse reflection; moderate to strong returns	Medium to light tones
Smooth and rough surfaces	Reflection influenced by topographic slope	Lighter tones produced by foreslope effect
	Reflection influenced by moisture content (influence of the dielectric constant of water, if water does not produce a smooth surface)	Lighter tones produced by increasing moisture content
Natural corner reflectors	Maximum return with proper orientation angle	Very light tones
Hydrology and Sea Ice		
Smooth water and ice surfaces	Specular reflection; weak to no return	Dark tones to black
Rough water and ice surfaces	Diffuse reflection; moderate to strong returns	Medium to light tones
Vegetation		
Dense woodland	Diffuse reflection; moderate to strong returns; return strength dependent on variables such as density, deciduous versus coniferous, absence or presence of water	Medium to light tones
Grassy areas	Specular and diffuse reflection; moderate to weak returns; return strength dependent on variables such as microrelief, presence or absence of water, and wavelength	Medium to dark tones
Agricultural crops	Variable strong to weak return; return strength dependent on variables such as density, microrelief, presence or absence of water, and wavelength	Light to dark tones
	Reflection influenced by moisture content (influence of the dielectric constant of water)	Lighter tones produced by increasing moisture content
Cultural Targets		
Streets, highways, runways	Specular reflection; weak or no return	Dark tones to black
Railroads, ships, bridges, transmission lines, and other metal objects	Maximum return with proper orientation angle	Very light tones
Corner reflectors produced by vertical walls intersecting horizontal ground plane	Maximum return with proper orientation angle	Very light tones

Adapted from Barr (1968).

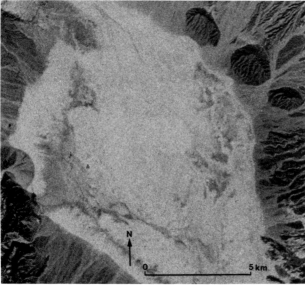

Figure 7-37 X-band (3-cm) SAR image (*top*) and a Skylab visible band photograph (0.4–0.7-μm) (*bottom*) of Cottonball Basin, Death Valley, California. Analysis of the radar image revealed the existence of two suspect fault traces in the evaporite deposits that are less than 2,000 years old (between arrows). The traces are defined in the image because the radar system was able to differentiate surface roughness variations at the centimeter scale. The traces are not recognizable in the photograph because tonal variations are primarily a record of spectral reflectivity differences attributable to changes in surface chemistry and not to small-scale changes in surface roughness. (Berlin et al. 1980.)

Radar echoes are described as being strong, moderate, weak, and no return. Figure 7-38 illustrates variations in radar return strength for a multiple-target terrain strip. Table 7-2 summarizes general surface states and a description of associated echo strengths and image tones.

Radar shadows represent no energy returns and are depicted as black areas in an image (Figure 7-8). Shadows are very important in radar image interpretation because they

indicate the presence of positive and negative terrain. Radar shadows can substantially aid in the interpretation of topographic relief, relative rock resistance (and hence general lithology), and geologic structure (Figures 7-15 and 7-28). The juxtaposition of strong returns from foreslopes and no or weak returns from backslopes has proven to be extremely valuable for showing topographic features that are heavily vegetated, such as those in the Amazon rainforest (Figure 7-39). As with airphotos, most interpreters orient a radar image so that the shadows fall toward them (i.e., *near range at top*). If this is not done, there is apt to be **topography inversion**, or a **pseudoscopic view**, whereby hills appear as valleys and valleys appear as hills (Figure 7-15).

The visual impression of coarseness or smoothness caused by the frequency of tonal changes within an individual surface feature represents image texture. Although a feature's overall tone indicates its ability to reflect microwaves, image texture is a qualitative measure of reflection variability within that particular feature. Image texture is defined as being smooth, grainy, or speckled in order of increasing coarseness.

Uniformity of tone—that is, a smooth image texture—is usually diagnostic of a homogeneous surface. Grainy and speckled textures are commonly associated with radar-rough surfaces because they are responsible for varying degrees of diffuse reflection over a short horizontal distance (e.g., local brightness variance caused by vegetation density differences). Smooth, grainy, and speckled image textures are illustrated in Figure 7-11.

Differentiating natural and cultural features is usually possible because many cultural features have regular geometric shapes (Figures 7-4 and 7-20). Although natural features usually have irregular shapes, many can be identified by their perimeter shapes alone. Examples include lava flows (Figure 7-12), hydrologic features (Figure 7-11), and different types of sea ice (Figure 7-24). The shapes of radar shadows enable the interpreter to infer information about the height, shape, size, and spatial form of terrain features (Figure 7-27).

As with aerial photographs, the larger an object is, the more accurately its shape and size are defined on a radar image. However, metal objects often bloom to a size on a radar image greater than would be indicated by their actual sizes (Figure 7-25). For RAR images, objects that appear as discrete items in near range may appear as a combined—and thus enlarged—feature in far range because of the decay in azimuth resolution (Figure 7-6).

Relief Displacement and Radar Stereo

Radar images differ from vertical airphotos in terms of relief displacement. For a vertical airphoto, relief displacement for features protruding above the datum plane is radially outward from the principal point and radially inward for features below the datum plane (Figure 4-3). For a radar image, however, relief displacement for vertical features above the datum plane is toward the nadir line (i.e., toward near range) and away from the nadir line for features below the datum plane (i.e., toward far range). Thus, the displacement direction is only in the across-track dimension; there is no displacement in the along-track dimension.

Because side-looking radars produce displacements as a function of relief in the across-track direction, stereo radar coverage can be obtained by imaging the same ground feature twice (1) from two different altitudes and printing the images at the same scale, (2) from the same altitude with two different depression angles, or (3) from opposite-look directions. The first two configurations yield **same-side stereo**, and the third configuration produces **opposite-side stereo**. The latter configuration is the least desirable because image shadows fall in opposing directions.

Same-side, same-height stereo coverage usually produces the most satisfactory stereopairs. The larger the difference in the depression angles for this type of stereopair, the greater the vertical exaggeration. A red-yellow SAR stereogram, incorporating two same-side, same-height radar strips, is shown in Plate 13. It can be viewed with a conventional lens or pocket stereoscope to obtain a three-dimensional effect.

Radar Mosaics

Radar strip images are often assembled into **radar mosaics** when there is a need to map large areas. High-quality radar mosaics incorporate ground-range images having shadows falling in the same direction (i.e., images obtained with the same aircraft heading) (Figure 7-40). Image strips are often acquired with 60 percent sidelap to provide stereoscopic enhancement. Because mosaics are often second- or third-generation photographic products, their resolution is somewhat degraded. In addition, shadow-length differences occur across mosaic junction lines between adjoining strips. These differences in shadow length can result in misinterpretation of apparent terrain relief where no differences actually exist.

Radar mosaics have been especially useful for mapping cloud-shrouded terrain. For example, in 1967 a radar mosaic was produced for the persistently cloud-covered Darien Province of eastern Panama as an aid to mapping the

---

Figure 7-38 Radar reflections from a multiple-target terrain strip. (Adapted from Sabins 1987.)

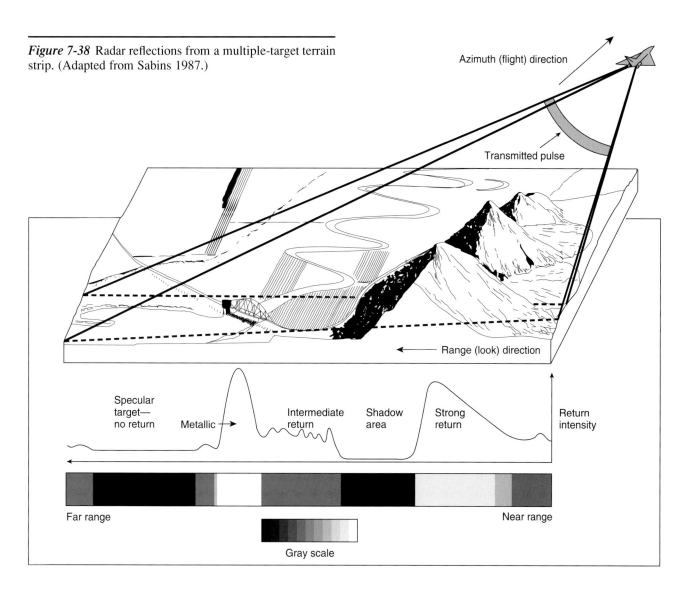

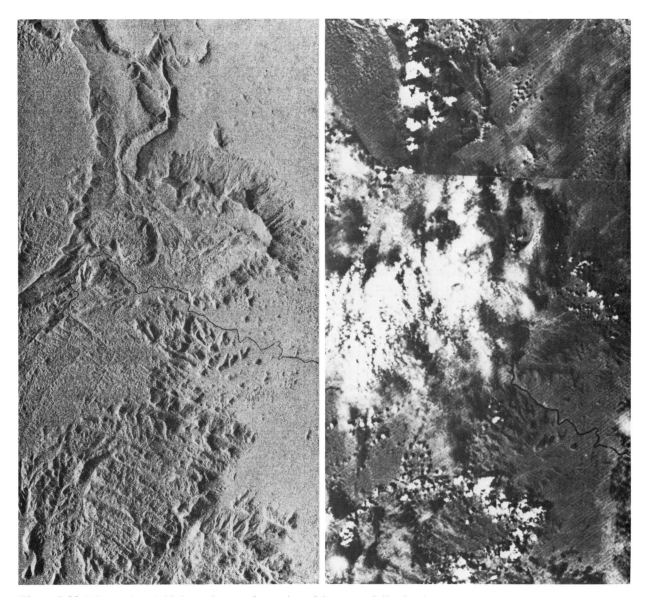

Figure 7-39 SIR-A L-band (23.5-cm) image of a portion of the perennially cloud-covered Amazon rainforest in Venezuela. Despite a closed vegetation canopy, the topography is well displayed because the slope effect locally modulated the backscatter to produce light-toned foreslopes and dark-toned backslopes. The accompanying *Landsat 2* MSS band 7 mosaic (0.8–1.1-μm) shows very little of the topography seen in the SIR-A image. (Courtesy John P. Ford, Jet Propulsion Laboratory.)

region's structural geology (Figure 7-41). Since then, radar mosaics have been produced for many cloud-covered countries (total or partial coverage), including Bolivia, Brazil, Colombia, Indonesia, Japan, Guatemala, New Guinea, Nicaragua, Nigeria, Philippines, Togo, and Venezuela. Radar mosaics employing opposite-look directions (i.e., two sets of radar mosaics) have been compiled for Nigeria and Togo (Figure 7-28).

Since 1980, the U.S. Geological Survey (USGS) has been systematically acquiring X-band SAR image data of the continental United States, Puerto Rico, and the Virgin Islands. The radar acquisition program is in response to congressional legislation to form a public domain remote

sensing data base that complements that of other airborne systems (e.g., NHAP and HAPP photography; Chapter 5). The radar images are acquired in blocks that conform to the USGS 1:250,000-scale topographic map series. The individual image strips are collected with 60 percent sidelap to provide a stereoscopic capability; the mosaics are assembled using the 1° × 2° map quadrangles for format and horizontal control. The Williams, Arizona, radar mosaic is shown in Figure 7-42.

USGS radar products that are available for purchase include strip images, mosaics, and digital data, in the form of computer-compatible tapes (CCTs), for selected areas. The strip images are available at scales of either 1:250,000

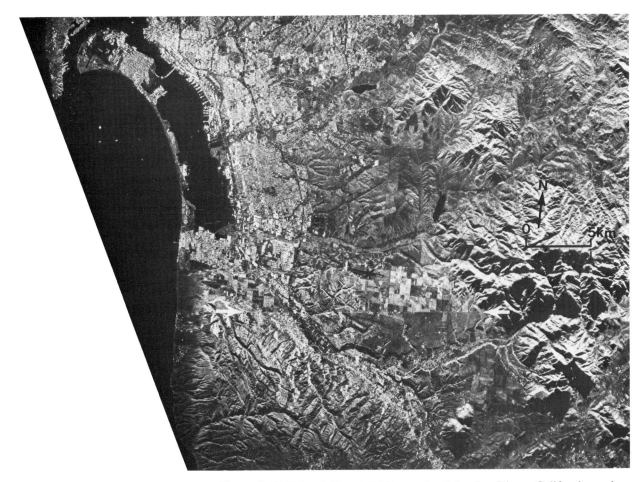

Figure 7-40 X-band (3-cm) SAR mosaic of the San Diego, California, region. The radar images were obtained through partial cloud cover from an altitude of 12 km. Resolution in range and azimuth is 15 m. A portion of the U.S.–Mexico border is discernible (as land-use differences) between the white arrows. (Courtesy Goodyear Aerospace Corp. and Aero Service Division of Western Geophysical of America.)

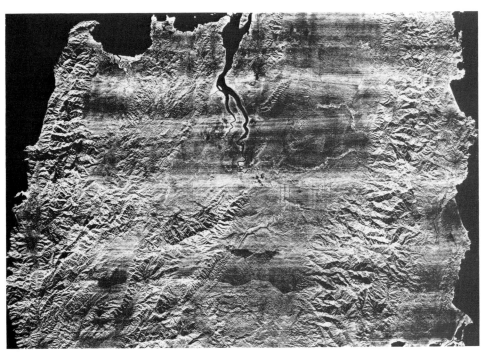

Figure 7-41 Radar mosaic (0.86 cm) of the Darien Province of eastern Panama. Coverage (8 image strips) was obtained in less than 4 hours and represents a land area covering about 20,000 km². Total cloud cover was present during most of the mission.

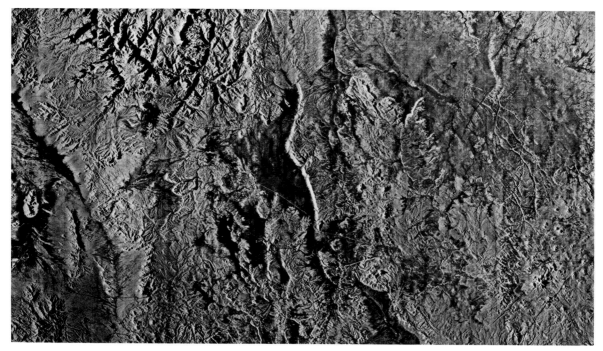

Figure 7-42 Radar mosaic (3 cm) of the Williams, Arizona, 1° × 2° quadrangle, reduced from an original scale of 1:250,000.

or 1:400,000. These products are normally the best medium for interpretation because they have the highest photographic resolution of the available products, and adjoining strips can be viewed stereoscopically. The mosaics are available at scales of 1:250,000 and 1:100,000. The latter products are enlarged from one quarter (NE, SE, SW, or NW) of the 1:250,000-scale mosaics. Both the strip images and mosaics are available as paper prints, film positives, and film negatives. Additional information regarding the USGS radar mosaic program and procedures for placing orders can be obtained from the EROS Data Center, U.S. Geological Survey, User Services Section, Sioux Falls, SD 57198.

NASA Satellite Imaging Radar Systems

Since 1978, NASA has launched three satellite-based imaging radar systems; one was carried by *Seasat* and two were carried aboard the Space Shuttle (SIR-A and SIR-B). All three were experimental missions. Oceanographic studies using SAR images (e.g., monitoring currents, eddy fields, plus surface and internal waves) were the main experiment objectives of the *Seasat* mission. The SIR-A and SIR-B missions were designed to evaluate the utility of SAR image data for applications in geology, hydrology, the vegetation sciences, and cartography. The three SARs collected a substantial amount of imagery which has been placed in the public domain. Table 7-3 summarizes the important *Seasat* and SIR radar parameters.

The first of these systems, the *Seasat* SAR, was successfully placed into earth orbit on June 26, 1978, and op-

erated flawlessly until October 9, 1978, when a massive short circuit in the power system prematurely ended its anticipated year-long mission. However, during its 98-day lifespan, a significant amount of imagery was acquired of both land and water surfaces, which provided a unique view of the earth's surface (Figures 7-11, 7-22, 7-23, and 7-24).

Because the *Seasat* SAR was designed primarily for oceanographic studies, large depression angles—or, conversely, small look angles—were required (Table 7-3). Consequently, layover occurs in mountainous terrain whenever the look angle (17° and 23° for near and far range, respectively) is smaller than a feature's foreslope angle (Figures 7-17 and 7-18). Despite this limitation, a variety of land information can be interpreted from the *Seasat* images.

Inquiries pertaining to *Seasat* radar image coverage and costs can be made to the NOAA Satellite Data Services Division, Room 100, Princeton Executive Center, Washington, DC 20233. Standard products for optically and digitally correlated data include paper prints plus film negatives and positives. Digitally processed scenes are also available on CCTs.

SIR-A and SIR-B possessed many of the characteristics of the *Seasat* SAR system (Table 7-3). However, the principal difference between the two radar systems was that the SIRs incorporated smaller depression angles (Table 7-3). In areas of rugged relief, the smaller depression angles greatly reduced foreshortening and layover effects that characterize many *Seasat* images.

SIR-A and SIR-B acquired images of a wide range of different features (Figures 7-21, 7-25, 7-31, 7-33, 7-34, 7-39, and Plate 13). The images are finding application in a number of earth science disciplines. SIR images in various formats are available from the National Space Science Data

TABLE 7-3 Parameters of *Seasat*, SIR-A, and SIR-B

Parameter	*Seasat* (1978)	SIR-A (1981)	SIR-B (1984)
Wavelength	23.5 cm (L-band)	23.5 cm (L-band)	23.5 cm (L-band)
Polarization	HH	HH	HH
Aperture type	Synthetic	Synthetic	Synthetic
Signal correlation	Optical and digital	Optical	Optical and digital
Azimuth resolution	25 and 50 m	40 m	30 m
Range resolution	25 and 50 m	40 m	17–58 m
Swath width	100 km	50 km	20–50 km
Depression angles (NR to FR)	73°–67°	43°–37°	75°–35° (variable)
Look angles (NR to FR)	17°–23°	47°–53°	15°–55° (variable)
Latitude coverage	10° N–75° N	41° N–36° S	60° N–60° S
Altitude	795 km	259 km	225 and 360 km
Mission duration	98 days	2.5 days	8.3 days
Image coverage	100 million km^2	10 million km^2	6.5 million km^2

Center, World Data Center-A, Code 601, NASA Goddard Space Flight Center, Greenbelt, MD 20771.

The next step in the Space Shuttle radar program is SIR-C, which will be an X-, C-, and L-band system. The C- and L-band channels will have the capability to transmit and receive any combination of horizontally and vertically polarized signals (HH, VV, HV, VH). The X-band channel will be VV-polarized. Present plans call for two SIR-C missions to occur in different seasons in 1992 and 1993.

A modified version of SIR-C will be flown aboard the NASA Earth Observation System (EOS) polar platform that is tentatively scheduled for deployment in the late 1990s. The dedicated EOS-SAR is designed to monitor global deforestation and its potential impact on global warming; soil, snow, and canopy moisture and flood inundation and their relationship to the global hydrologic cycle; and sea-ice properties and their impact on the polar heat flow.

Passive Microwave Systems

Although the technology for the *passive* detection of microwave radiation has been employed in radio astronomy since the mid-1940s, the remote sensing of the earth's atmosphere and its surfaces in the microwave band of the electromagnetic spectrum did not commence on a large scale until the late 1960s. **Passive microwave radiometers** are the general class of remote sensors that detect the low level of natural radiation continuously emitted by the earth's surface and atmosphere at frequencies between about 1 and 200 GHz or, conversely, at wavelengths ranging from about 30 to 0.15 cm, respectively (Figure 1-5).*

Passive microwave radiometers that produce images are called **imaging microwave radiometers**, or **passive microwave imagers**. Most are multichannel, digital systems that operate from earth-orbiting satellites. Passive microwave scanners function in a manner similar to thermal IR scanners, except that the collection mirror is replaced by a "scanning" antenna and the thermal detectors are replaced by ultrasensitive radio receivers. As with imaging radars, passive microwave systems operate at discrete frequencies or wavelengths. A detailed discussion of passive microwave radiometry is given by Ulaby and Carver (1983).

As discussed in Chapter 6 and illustrated in Figure 6-16, objects with nominal temperatures in the neighborhood of 300 K have radiant power peaks at or close to 9.7 µm in the thermal IR region, which fall off rapidly toward the visible region and less rapidly toward the microwave region. Consequently, these objects emit small amounts of microwave radiation. It is for this reason that passive microwave systems must have large instantaneous fields of view in order to record sufficient radiation. The ground resolution from orbital altitudes can reach tens of kilometers.

The **microwave radiance** measured by these systems is often expressed as a **brightness temperature**. A feature's microwave brightness temperature is the product of its kinetic temperature and emissivity, which varies according to observation angle, polarization, wavelength, and surface roughness. Brightness temperatures are determined by comparing the incoming signal to the signal from an internal radiation source of known brightness temperature. The output is a calibrated image of microwave brightness temperature.

Passive remote sensing in the microwave band is traditionally done at frequencies near 1.4, 6, 10, 18, 21, 37, 55, and 90 GHz, and systems using 157 and 183 GHz receivers are being planned (Murphy 1987).* These frequencies have been intentionally selected to fall among the "peaks and

---

*Frequency has traditionally been used for designating passive microwave radiation regions.

*Applicable wavelengths are approximately 21, 5, 3, 1.7, 1.4, 0.81, 0.54, 0.33, 0.19, and 0.16 cm.

TABLE 7-4 Parameters Measured with Passive Microwave Sensors

Observed Medium	Frequency of Observation (GHz)									
	1.4	6	10	18	21	37	50–60	90	160	183
Soil moisture	1	3								
Snow		2	1	1		1		2		
Precipitation										
Ocean			2	1	3	2				
Land				2		1		1		2
Sea surface temperature	1	2	2	2	3					
Sea ice										
Extent				1		1		3		
Type	3	2	1			1		2		
Wind speed (sea surface)			1	2	3	3				
Water vapor										
Total (over ocean)				1	1	2				
Profile					2	3	2	3	2	1
Cloud water (over ocean)					2	1		2		
Temperature profile					3	3	1	3		

Key: 1 = necessary, 2 = important, 3 = helpful.
Adapted from Murphy (1987).

valleys'' of atmospheric attenuation curves (e.g., water vapor and clouds). Measurements at frequencies near the absorption peaks are most useful for monitoring atmospheric phenomena such as water vapor; frequencies in the valleys are least affected by the atmosphere (including clouds) and are ideally suited for monitoring surface features and conditions (e.g., sea temperature, snow cover, and soil moisture) (Murphy 1987). Some of the important parameters which can be measured at these frequencies are listed in Table 7-4.

Imaging Radiometers in the U.S. Space Program

Several imaging microwave radiometers have operated from earth orbit; most are described in detail by Murphy (1987). Channel sensitivities typically fall within atmospheric windows, enabling microwave emission from the earth's surface to reach the sensors with little attenuation. Among the earliest of the imaging systems was the **Electronically Scanned Microwave Radiometer (ESMR)**. Two single-channel ESMRs have been flown, one aboard *Nimbus 5* (1972) operated at 19.3 GHz and a second on *Nimbus 6* (1975) operated at 37 GHz.

The successor to the ESMR was the **Scanning Mul-**

tichannel Microwave Radiometer (SMMR), which was flown on *Seasat* (1978) and *Nimbus 7* (1978); the *Nimbus 7* instrument operated for about 10 yrs. The SMMR operated at five frequencies (6.6, 10.7, 18, 21, and 37 GHz), and the data were processed into images with a resolution of 50 km or larger. SMMR images (CCTs and film products) are available from NOAA Satellite Data Services Division, Room 100, Princeton Executive Center, Washington, DC 20233.

In June 1987, the U.S. Air Force's Defense Meteorological Satellite Program (DMSP) launched its ''F8'' satellite, which carries the **Special Sensor Microwave/Imager (SSM/I)**. The SSM/I is a four-frequency system that measures atmospheric and ocean surface brightness temperatures at 19.3, 22.2, 37.0, and 85.5 GHz. Resolution varies from 12.5 km at the highest frequency to 25 km at lower frequencies. Coverage is global except poleward of 87° N and S latitude. SSM/I images are available to the public through the National Snow and Ice Data Center (NSIDC), CIRES, Campus Box 449, University of Colorado, Boulder, CO 80309.

An SSM/I brightness temperature image (85.5 GHz) of the ice and water surfaces of the North Polar region for November 2, 1987, is shown in Plate 13. The North Pole is in the center of the image, with the area not visible to the sensor indicated by the black circle. Land areas have been intentionally printed in black. Brightness temperatures are displayed in relative ranges from 145 K to 273 K. The coldest areas are dark blue and the colors range through magenta and on to green, with yellow indicating the warmest areas.

New-generation multifrequency passive microwave radiometers, along with a complement of other remote sensing instruments, are scheduled to be carried by the NASA *Nimbus 8* and EOS polar-orbiting satellites. *Nimbus 8* will likely be launched in the early 1990s, whereas the first EOS launch will occur later in the decade.

Sonar Systems

Active remote sensors that use **sound waves** or **acoustical energy** propagated through water are known as **sonar systems**; sonar is an acronym derived from <u>so</u>und <u>n</u>avigation <u>an</u>d <u>r</u>anging. Compared to electromagnetic radiation, which travels at the speed of light (300,000 km/sec in a vacuum), sound waves are extremely slow. In air and seawater, sound-wave velocities are about 300 m/sec and 1,530 m/sec, respectively.

There are two types of sonar systems. Sonars that measure water depths and, hence, changes in bottom topography are called **fathometers**, **echo-sounding profilers**, or **bathymetric sonars**, whereas the systems that generate images of bottom topography and bottom roughness are known as **sidescan imaging sonars**, or simply **imaging sonars**; sonar images are commonly called **sonographs**. Both systems can be operated in concert from a survey ship. The routine use of civilian sonars has largely occurred over the past two decades, and data from both systems are providing completely new views of the earth's water-submerged landscape.

Echo sounders make depth measurements by bouncing sound waves off submerged features, making it relatively easy to produce **topographic profiles** of underwater terrain. The early sounders utilized single beams, but the newest systems use multiple beams, in which a large array of beams is used to measure bottom depths across a swath that is nearly as wide as the water is deep (i.e., water depth is determined for each beam). These systems are known as **swath** or **multibeam echo sounders** (Pittenger 1989). Data from the multibeam systems are generally used to produce **bathymetric charts** or **maps**.

Sidescan Imaging Sonars

Figure 7-43 illustrates the operational parameters of a sidescan imaging sonar, which is similar in many respects to a real-aperture imaging radar system (Figures 7-2 and 7-38). Basically, a sidescan sonar consists of three major components: (1) the **transducers**, (2) the **cable link**, and (3) the **recording system**. The transducers are devices that convert electrical energy into acoustical energy for transmission and backscattered acoustical energy into electrical energy. They are housed in a torpedo-shaped vehicle called the **towfish**, which is towed from a cable below and normally behind the survey ship at a predetermined height off the bottom. The cable also acts as a data transmission link between the transducers and the recording systems, which are located on the ship. The recording system contains most of the sonar's electronics as well as printers for generating hardcopy images in real-time or analog-to-digital (A/D) converters and recorders for storing the image data on magnetic tapes, which can be played back to produce images.

The transducers transmit acoustical pulses through the water to both sides and at right angles to the track of the towfish (Figure 7-43). The pulse is in the form of a fan-shaped beam, very narrow in the horizontal plane and very wide, by comparison, in the vertical plane. **Near-surface sonars** are configured to operate from long ranges and provide regional-scale images of the deep sea floor at a moderate resolution. These systems have imaging swath widths that can reach 30 km per side for a 5-km depth range. Given a 30-km swath, an acoustical pulse lasting 4 sec would be transmitted every 40 sec at a frequency of 6.8 kHz.* **Near-**

For near-surface sonars, a low frequency is desirable because sound waves are absorbed in seawater, and the rate of absorption rapidly increases with increasing frequency of the acoustical energy.

Figure 7-43 Representation of a sidescan imaging sonar system.

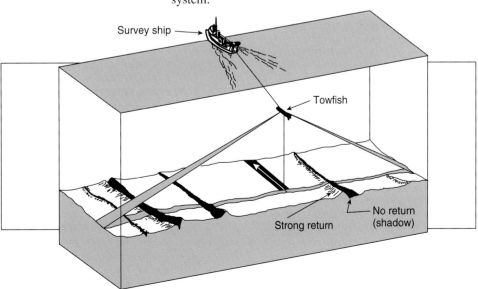

Survey ship

Towfish

Strong return

No return (shadow)

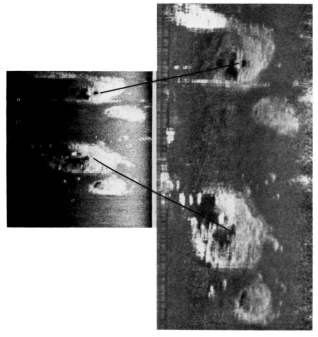

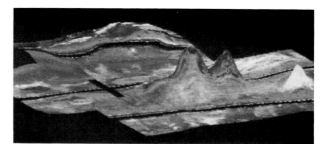

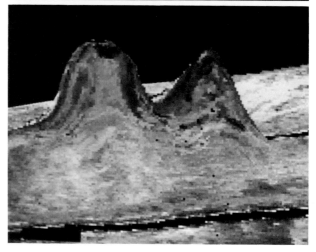

Figure 7-45 Perspective-view sonographs of a double-peak seamount in the Pacific Ocean. (Courtesy Fa Dwan, Texas A&M University.)

Figure 7-44 "Raw" (*left*) and computer-corrected (*right*) sonar images of volcanic terrain west of southern California at about 33° N latitude and 123° W longitude. The large volcanoes are about the size of Mount St. Helens (see frontispiece). The left and right edges are the far- and near-range locations, respectively. The right-hand image incorporates (1) a slant-range to ground-range correction (across-track dimension), (2) a correction for ship velocity variation (along-track dimension), and (3) a shading correction which compensates for the drop-off in returned signal strength as a function of range (across-track dimension). (Courtesy Pat S. Chavez, Jr., U.S. Geological Survey.)

bottom sonars are operated close to the floor (e.g., from 10- to 150-m altitudes) and are designed for high-resolution surveys in shallow water. Imaging ranges are usually less than 1.5 km per side. For a 500-m swath, a typical pulse rate would be 1.5 pulses per second, with the duration of each pulse being 0.1 msec (millisecond); the operating frequency would be about 100 kHz.

Each pulse of acoustical energy encounters the floor and is reflected in varying degrees from irregularities at successively increasing distances from the towfish track. An echo train is intercepted by the transducers, where it is converted to an electrical signal and relayed to the ship, via the tow cable, for direct image construction or digital recording on magnetic tape for subsequent image generation. The electrical signal varies in amplitude proportional to the intensity of acoustical backscatter, which is primarily a function of topography and roughness. As the ship moves forward, the transmission-reception process is repeated, which ultimately produces two continuous strips of image data (**port and starboard looks**) separated by a narrow black area directly beneath the towfish (Figure 7-43).

A paper printer is used for most **analog sonars**, which enables sonographs to be produced in real time. The display of the returning signals is in the form of a series of closely spaced, intensity-modulated lines on a chemically treated paper. The paper is darkened in *inverse* proportion to the strength of the acoustical backscatter. Thus, light tones represent weak signals; white represents acoustic shadow zones behind reflectors; and dark tones indicate strong sonic returns. This tonal record is analogous to a photographic negative.

The tonal record is reversed for **digital sonars**, wherein light image tones represent strong sonic returns and dark tones indicate weak or no returns (Figure 7-44). Data from digital sonars are amenable to computer processing techniques, enabling both geometric and radiometric distortions to be corrected before the images are produced with film-writing equipment (Figure 7-44). In addition, digital image data can be combined with bathymetric digital data by special computer programs to produce **perspective-view (off-nadir) sonographs** of bottom features (Figure 7-45).

Interpreting Sonographs

The intensity of acoustical echoes is a function of topography as well as the roughness properties of the bottom. The tonal descriptions that follow are for digitally processed sonographs, and many of the signatures are illustrated in Figures 7-46 and 7-47. The slopes of topographic features

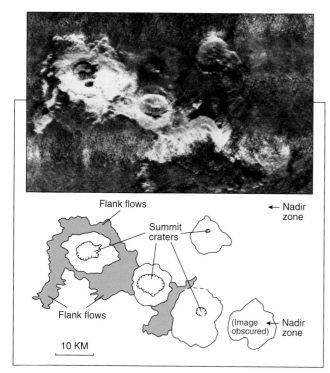

Figure 7-46 Sonograph mosaic (four images, two nadir zones) and line-drawing interpretation of the Taney Seamounts, located about 250 km west of San Francisco. The cratered seamounts rise about 2 km from a water depth of about 4 km. Note (1) how the slopes modulated the backscatter to produce light-toned foreslopes and dark-toned backslopes and (2) the large collapse summit craters and extensive lava flows. (Courtesy James V. Gardner, U.S. Geological Survey.)

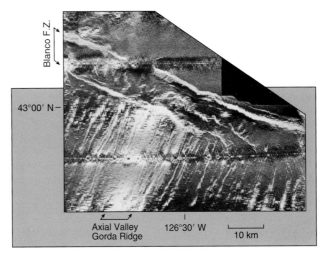

Figure 7-47 Sonograph mosaic of the junction between the Gorda Ridge and Blanco Fracture Zone. Note the tonal contrast between the sediment-covered sea floor (dark) and basalt outcrops on the sea floor (bright). Basalt ridges trending 20° in southern half of the image constitute basement fabric formed along the Gorda Ridge. Fracture-zone ridges trend 112° across the mosaic. Sediment-covered floor of the Cascadia Basin is north of the Blanco Fracture Zone. Black represents regions for which there are no sonar data. (Courtesy James V. Gardner, U.S. Geological Survey.)

U.S. Sonar Surveys

Since 1979, the USGS and the United Kingdom's Institute of Oceanographic Sciences (IOS) have been conducting sonar imaging surveys in the water bodies bordering the United States. The **GLORIA** (for Geological Long Range Inclined Asdic) imaging sonars, which were designed and built by IOS, have been used for these deep-sea surveys. They may be towed at 8 knots or more to image a swath 60 km across (port and starboard looks), giving a maximum coverage in excess of 20,000 km^2 per working day. GLORIA's resolution is constant in range or across track but varies in azimuth or along track. For a 15-km swath width (port or starboard look), range resolution is 20 m, whereas azimuth resolution varies from 80 m at near range to 700 m at far range. The sonar images presented in this chapter were produced during the IOS-USGS surveys in the Pacific Ocean.

Since 1983, when the United States proclaimed an Exclusive Economic Zone (EEZ) that extends to 370 km (200 nautical miles) from its shoreline and the shoreline of its island territories, the USGS has been conducting a reconnaissance-scale mapping program of the entire EEZ, called EEZ-SCAN, using GLORIA sonar images as the principal mapping tool.* The first phase of EEZ-SCAN, completed in 1984, provides continuous overlapping GLORIA images of

facing the transducers (i.e., foreslopes) are better reflectors than slopes facing away from the transducers (i.e., backslopes). Hence, foreslopes are normally depicted in lighter image tones than backslopes. Whenever the backslope angle is larger than the depression angle, protruding topographic features produce acoustic shadow zones. Here transmitted waves are not intercepted, and hence no echoes are recorded; acoustic shadows are depicted as black voids.

Depending largely upon their roughness characteristics, various materials on the ocean floor are responsible for backscatter of different intensities. Smooth materials, such as sand and mud, reflect the acoustical beam specularly; hence, these materials have weak returns and are depicted in dark image tones. Rough materials, including cobbles and boulders, are better reflectors of acoustical energy and will scatter much of the incident beam back to the transducers. Therefore, they are recorded in lighter image tones.

Belderson et al. (1972) provide an important reference to imaging sonar and interpretation principles. Articles describing sonar applications, with illustrative images, are found in a variety of scientific journals, including *Deep-Sea Research, Geology, Geomarine Letters, Marine and Petroleum Geology, Marine Geology, Marine Geotechnology, Oceanology,* and *Oceanological Acta.*

*The proclaimed Exclusive Economic Zones give the United States sole jurisdiction over their vast living and nonliving resources.

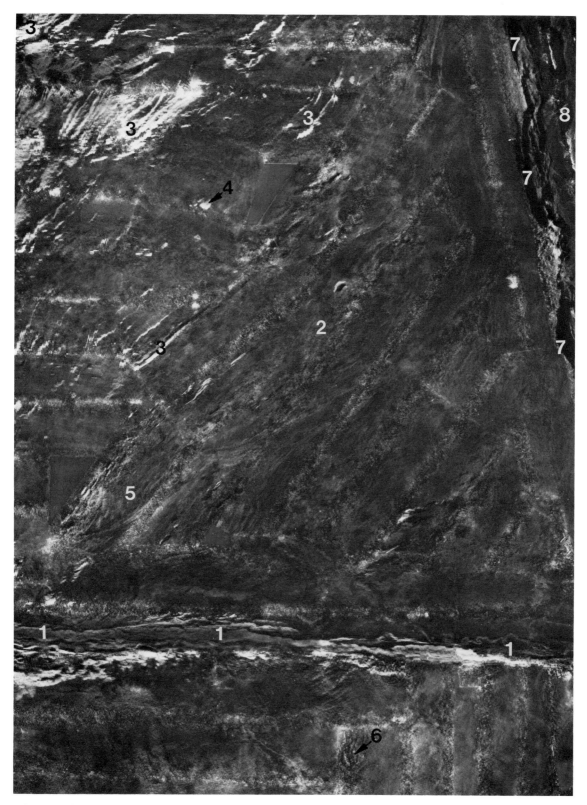

Figure 7-48 GLORIA sonar mosaic showing a 2° × 2° area of the Pacific Ocean floor off the coast of Eureka, California, reduced from an original scale of 1:500,000. Annotations are as follows: (1) Mendocino Fracture Zone, (2) Gorda Deep Sea Fan, (3) basement ridges, (4) volcano, (5) large canyon or channel, (6) slump, (7) tectonic front (subduction zone), and (8) continental margin. Water depth ranges from about 1,000 m (*upper right*) to 3,100 m (*left center*). (Courtesy James V. Gardner, U.S. Geological Survey.)

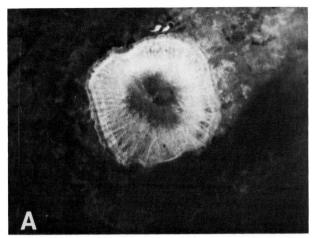

Figure 7-49 Three film chips of Meteor Crater, Arizona, representing visible, thermal infrared, and microwave wavelengths.

the entire EEZ off the western United States, an area of some 850,000 km². The complete data set in the form of 2° × 2° computer-enhanced sonar mosaics, together with geologic interpretations at a scale of 1:500,000, has been published in an atlas format (EEZ-SCAN 84 Scientific Staff 1986). One of the 2° sonar mosaics is presented in Figure 7-48. Atlases for the Atlantic and Gulf of Mexico EEZs are currently in preparation.

The GLORIA images have revealed a plethora of geologic information. In the western United States EEZ, for example, midocean ridges, transform faults, volcanic provinces, and sediment fans are seen in detail. Also, many features never before mapped in this region have been located; examples include large canyons, seamounts, slumps, meandering channels, and calderas (EEZ-SCAN 84 Scientific Staff 1988).

CD-ROM (compact disc–read-only memory) discs containing the GLORIA digital images for the EEZs are being made available to the public once they are properly compiled and formatted. Information regarding the GLORIA CD-ROM program and associated software packages can be obtained from the NOAA National Geophysical Data Center, 325 Broadway/Code E-GC3, Boulder, CO 80303.

Questions

1. Briefly compare and contrast radar and sonar sensors plus radar and passive microwave sensors.
2. The terrain for a study area has low topographic relief. Will the expression of topography be best expressed in the near- or far-range portion of a radar image? Explain.
3. Describe the causation of radar shadows and how the following parameters influence shadow length: (a) depression angle, (b) look angle, (c) relative relief, and (d) slant range (distance).
4. What is the primary difference between real-aperture and synthetic-aperture radar systems?
5. A SLAR system with a pulse length of 0.1×10^{-6} sec is used to obtain images of features at depression angles of 10° (far range) and 29° (near range). Use Equation 7-3 to calculate the ground-range resolution elements.
6. For a real-aperture, X-band (3-cm wavelength) SLAR system with an antenna length of 3.9 m, use Equation 7-4 to calculate the azimuth resolution elements from slant-range distances of 15 km and 25 km.
7. If the depression angle to the center of a 3-cm wavelength radar image is 60°, use Equations 7-8 and 7-9 to calculate the rough and smooth criteria, and complete the following table:

Roughness Category		Predicted Image Tone
Rough	> _____ cm	_____
Intermediate	= _____ cm	_____
Smooth	< _____ cm	_____

8. Examine the three film chips of Meteor Crater, Arizona, in Figure 7-49 and identify the airphoto (0.5–0.7-μm), the thermal infrared image (8–14-μm), and the X-band radar image (3 cm). Explain your reasoning.
   a. film chip A = _____
   b. film chip B = _____
   c. film chip C = _____

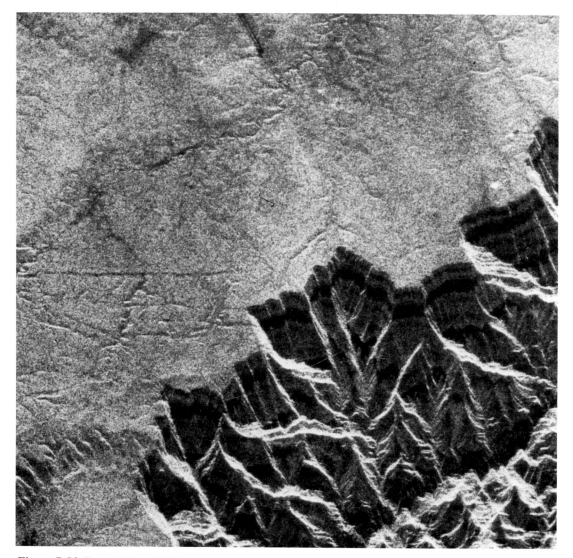

Figure 7-50 *Seasat* L-band (23.5-cm) SAR image of the South Rim, Grand Canyon, Arizona.

9. Examine the *Seasat* SAR image of the Grand Canyon, Arizona, in Figure 7-50, and answer the following questions:

a. Does the top of the image represent near range or far range? Explain your reasoning.

b. Two parallel traces on the South Rim appear to intersect the canyon. What might these represent? What cultural feature appears to be associated with one of the traces?

c. The image clearly shows cliff (dark bands) and slope (lighter bands) topography within the canyon. The cliff-forming formations, primarily sandstone and limestone, have a slope range from about 75° to 90°. The slope-forming formations, principally shales, have a slope range from about 28° to 49°. Given an antenna depression angle of 70° at the South Rim, why are the cliff faces dark toned and the slope faces lighter toned?

d. Do you see any evidence of foreshortening and layover? Explain your reasoning.

Bibliography and Suggested Readings

Allan, T. D., ed. 1983. *Satellite Microwave Remote Sensing.* New York: Halsted Press.

Barr, D. J. 1968. Use of Side-Looking Airborne Radar (SLAR) Imagery for Engineering Soils Studies. Technical Report 46-RT. Fort Belvoir, Va.: U.S. Army Engineer Topographic Laboratories.

Berlin, G. L., G. G. Schaber, and K. C. Horstman. 1980. Possible Fault Detection in Cottonball Basin, California: An Application of Radar Remote Sensing. *Remote Sensing of Environment* 10:33–42.

Berlin, G. L., G. G. Schaber, R. C. Kozak, and P. S. Chavez, Jr. 1982. Cliff and Slope Topography of Part of the Grand Canyon, Arizona as Characterized on a Seasat Radar Image. *Remote Sensing of Environment* 12:81–85.

Berlin, G. L., M. A. Tarabzouni, K. M. Sheikho, and A. H. Al-Naser. 1985. SIR-A and Landsat MSS Observations of Eolian Sand Deposits on the Al Labbah Plateau, Saudi Arabia. *Pro-*

ceedings Nineteenth International Symposium on Remote Sensing of Environment. Ann Arbor, Mich.: Environmental Research Institute of Michigan, pp. 311–21.

Berlin, G. L., M. A. Tarabzouni, A. H. Al-Naser, K. M. Sheikho, and R. W. Larson. 1986. SIR-B Subsurface Imaging of a Sand-Buried Landscape: Al Labbah Plateau, Saudi Arabia. *IEEE Transactions on Geoscience and Remote Sensing* GE-24:595–602.

Belderson, R. H., N. H. Kenyon, and A. H. Stride. 1972. *Sonographs of the Sea Floor.* Amsterdam: Elsevier.

Blom, R. G., R. E. Crippen, and C. Elachi. 1984. Detection of Subsurface Features in Seasat Radar Images of Means Valley, Mojave Desert, California. *Geology* 12:346–49.

Bryan, M. L. 1979. The Effect of Radar Azimuth Angle on Cultural Data. *Photogrammetric Engineering and Remote Sensing* 45:1097–1107.

Cannon, P. J. 1980. Applications of Radar Imagery to Arctic and Subarctic Problems. *Radar Geology: An Assessment.* JPL Publication 80-61. Pasadena, Calif.: Jet Propulsion Laboratory, pp. 265–74.

Chavez, P. S., Jr. 1986. Processing Techniques for Digital Sonar Images from GLORIA. *Photogrammetric Engineering and Remote Sensing* 52:1133–45.

EEZ-SCAN 84 Scientific Staff. 1986. *Atlas of the Exclusive Economic Zone, Western Conterminous United States.* Miscellaneous Investigations Series I-1972. Reston, Va.: U.S. Geological Survey.

———. 1988. Physiography of the Western United States Exclusive Economic Zone. *Geology* 16:131–34.

Elachi, C. 1987. Radar Images of the Earth from Space. *Scientific American* 247:54–61.

Ford, J. P., R. G. Blom, M. L. Bryan, M. I. Daily, T. H. Dixon, C. Elachi, and E. C. Xenos. 1980. *Seasat Views North America, the Caribbean, and Western Europe with Imaging Radar.* JPL Publication 80-61. Pasadena, Calif.: Jet Propulsion Laboratory.

Gelnett, R. H. 1978. Importance of Look Direction and Depression Angles in Geologic Applications of SLAR. Technical Report TR-04823. Phoenix, Ariz.: MARS Associates, Inc.

Imhoff, M., M. Story, C. Vermillion, F. Khan, and F. Polcyn. 1986. Forest Canopy Characterization and Vegetation Penetration Assessment with Spaceborne Radar. *IEEE Transactions on Geoscience and Remote Sensing* GE-24:535–42.

Jensen, H., L. C. Graham, L. J. Porcello, and N. Leith. 1977. Side-Looking Airborne Radar. *Scientific American* 237:84–95.

Leberl, F. W. 1989. *Radargrammetric Image Processing.* Norwood, Mass.: Artech House Books.

Matthews, R. E., ed. 1975. *Active Microwave Workshop Report.* NASA Special Publication-376. Washington, D.C.: U.S. Government Printing Office.

McCauley, J. F., G. G. Schaber, C. S. Breed, M. J. Grolier, C. V. Haynes, B. Issawi, C. Elachi, and R. Blom. 1982. Subsurface Valleys and Geoarchaeology of the Eastern Sahara Revealed by Shuttle Radar. *Science* 218:1004–19.

McCauley, J. F., C. S. Breed, G. G. Schaber, W. P. McHugh, B. Issawi, C. V. Haynes, M. R. Grolier, and A. El Kilani. 1986. Paleodrainages of the Eastern Sahara—The Radar Rivers Revisited. *IEEE Transactions on Geoscience and Remote Sensing* GE-24:624–48.

Moore, R. K., L. J. Chastant, L. Porcello, and J. Stevenson. 1983. Imaging Radar Systems. In *Manual of Remote Sensing,* edited by R. N. Colwell, 2d ed., 429–74. Falls Church, Va.: American Society for Photogrammetry and Remote Sensing.

Murphy, R. Chairman. 1987. *HMMR High-Resolution Multifrequency Microwave Radiometer Instrument Panel Report.* Vol. IIe. Washington, D.C.: National Aeronautics and Space Administration.

Peake, W. H., and T. L. Oliver. 1971. The Response of Terrestrial Surfaces at Microwave Frequencies. Report APAL-TR-70-301. Columbus, Ohio: Ohio State University.

Pittenger, R. F. 1989. Exploring and Mapping the Seafloor. *National Geographic* 177:61A, January.

Sabins, F. F., Jr. 1987. *Remote Sensing Principles and Interpretation,* 2d ed. New York: W. H. Freeman & Co.

Schaber, G. G., G. L. Berlin, and W. E. Brown. 1976. Variations in Surface Roughness within Death Valley, California: Geologic Evaluation of 25-cm-Wavelength Radar Images. *Geological Society of America Bulletin* 87:29–41.

Schaber, G. G., G. L. Berlin, and R. J. Pike. 1980. Terrain Analysis Procedures for Modeling Radar Backscatter. *Radar Geology: An Assessment.* JPL Publication 80-61. Pasadena, Calif.: Jet Propulsion Laboratory, pp. 168–99.

Simonett, D. S., and R. W. Davis. 1983. Image Analysis-Active Microwave. In *Manual of Remote Sensing,* edited by R. N. Colwell, 2d ed., 1125–81. Falls Church, Va.: American Society for Photogrammetry and Remote Sensing.

Trevett, J. W. 1986. *Imaging Radar for Resource Surveys.* London: Chapman and Hall.

Tsang, L., J. A. Kong, and R. T. Shin. 1985. *Theory of Microwave Remote Sensing.* New York: John Wiley & Sons.

Ulaby, F. T., and K. R. Carver. 1983. Passive Microwave Radiometry. In *Manual of Remote Sensing,* edited by R. N. Colwell, 2d ed., 475–516. Falls Church, Va.: American Society for Photogrammetry and Remote Sensing.

Ulaby, F. T., R. K. Moore, and A. K. Fung. 1981. *Microwave Remote Sensing: Active and Passive.* Vol. I: Microwave Remote Sensing Fundamentals and Radiometry. Reading, Mass.: Addison-Wesley Publishing Co.

———. 1982. *Microwave Remote Sensing: Active and Passive.* Vol. II: Radar Remote Sensing and Surface Scattering and Emission Theory. Reading, Mass.: Addison-Wesley Publishing Co.

———. 1986. *Microwave Remote Sensing: Active and Passive.* Vol. III: From Theory to Applications. Reading, Mass.: Addison-Wesley Publishing Co.

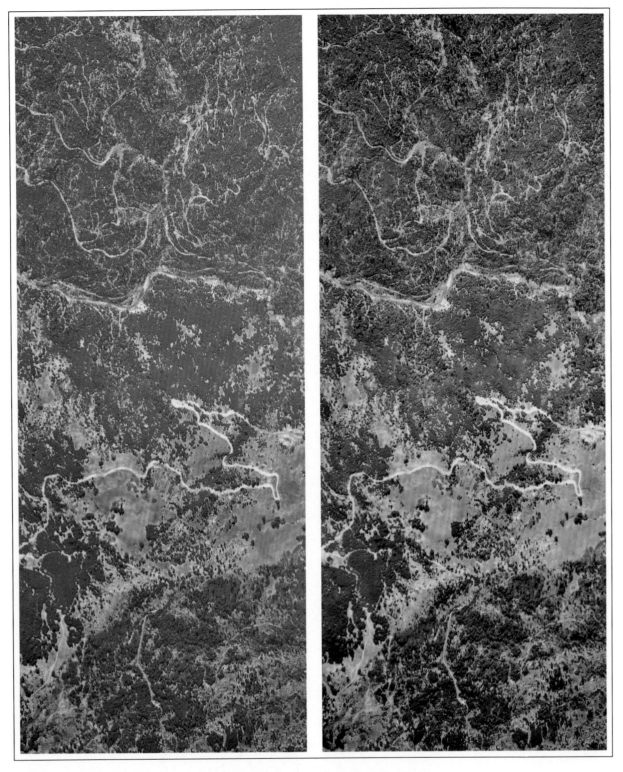

Plate 1   High-altitude normal color airphoto of redwood stands and open grass areas in Redwood Creek Basin, California, with (*left*) and without (*right*) atmospheric haze. The haze was removed from the original transparency, after it was digitized, by a special computer operation described in Chapter 15. (Courtesy Pat S. Chavez, Jr., U.S. Geological Survey.)

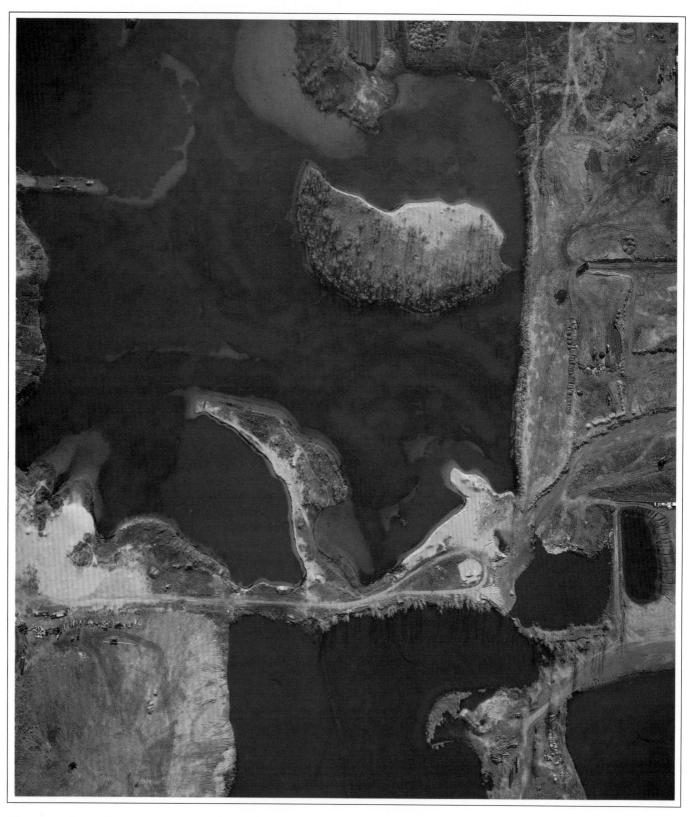

Plate 2   Low-altitude normal color airphoto of an active landfill operation in the midwestern United States. Exposure was made during the early winter. (Courtesy National Aeronautics and Space Administration.)

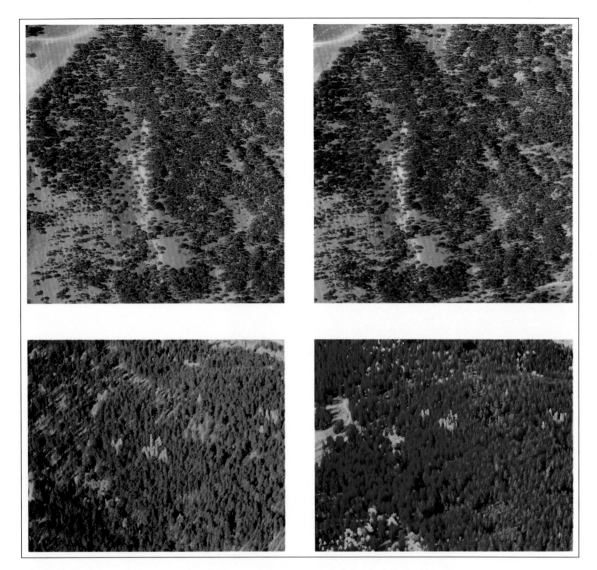

Plate 3   *Top:* Normal color stereogram illustrating damage caused by the mountain pine beetle on ponderosa pine in the Black Hills, South Dakota. Dying trees have yellow crowns and are clearly discernible in stereo. Scale is about 1:6,200. (Courtesy U.S. Forest Service.) *Bottom:* Normal color and color infrared oblique airphotos (Autumn) of beetle damage to ponderosa pine in an adjacent area of the Black Hills. In the normal color photo, dying ponderosa pines are light yellow; aspen foliage is medium yellow; and defoliated pines are gray. In the color infrared photo, there is no separation between the dying ponderosa pine and the aspen, as both appear white; defoliated pines appear blue. (Courtesy Robert C. Heller, formerly of the U.S. Forest Service.)

Plate 4   Normal color and color infrared airphotos of two reservoirs in Contra Costa County, California. Scale is about 1:34,000. Panchromatic and black-and-white infrared airphotos of the same area appear in Figure 13-14. (Courtesy U.S. Forest Service and the University of California, Berkeley.)

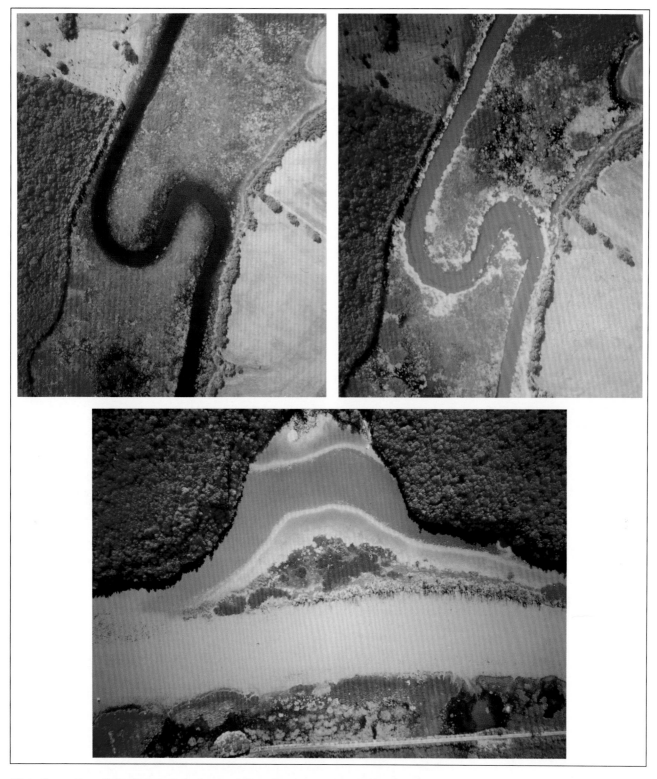

Plate 5   *Top:* Color infrared airphotos of East Creek (southern Lake Champlain region) during a low-water year (*left*) and a high-water year (*right*). The low-water exposure shows extensive stands of wild rice along both sides of the sharp meander, while the high-water exposure shows inundated areas plus relatively extensive stands of cattail and virtually no wild rice. *Bottom:* Color infrared airphoto of water chestnut infestations (light and dark pink) in the Red Rock Bay area of the Lake Champlain Narrows. The dense mats of water chestnut inhibit boat traffic and the use of the narrows for other recreational purposes. (Courtesy Donald J. Bogucki, State University of New York, Plattsburgh.)

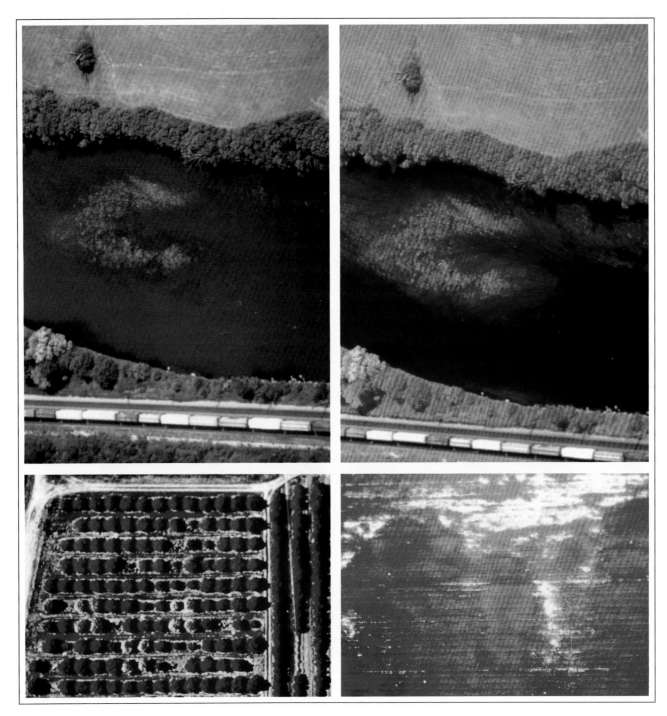

Plate 6 *Top:* Normal color and color infrared airphotos of an algae bloom. (Courtesy U.S. Geological Survey.) *Bottom left:* Color infrared airphoto of a citrus grove; trees with pinkish-white crowns are suffering from a nutrient deficiency. *Bottom right:* Color infrared airphoto of cotton affected by soil salinity: red = healthy cotton, low salinity; pink = unhealthy cotton, moderate salinity; white = bare soil with salt accumulations. (Courtesy U.S. Department of Agriculture.)

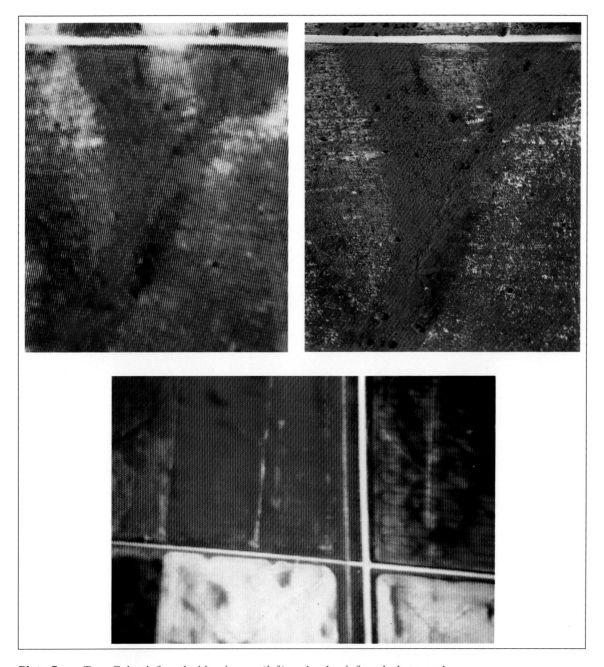

Plate 7 *Top:* Color infrared video image (*left*) and color infrared photograph (*right*) of a rangeland area near Roma, Texas, December 1985. Scale is about 1:20,000. Note in both simultaneously acquired pictures how the riparian and drought-stressed vegetation on the upland areas have similar red and brown colors, respectively. The comparison also illustrates the poorer resolution of the video image versus the photograph; note that individual trees can be distinguished in the photograph but generally cannot be distinguished in the video image. (Courtesy James H. Everitt and David E. Escobar, Agricultural Research Service, U.S. Department of Agriculture.) *Bottom:* Color infrared video image of an agricultural area near Crookston, Minnesota, August 1984. The image was taken by the Biovision color infrared video camera shown in Figure 6-2. Red hues denote a potato field and roadside vegetation; bare, dry soil appears white; and dark areas indicate high levels of soil moisture. (Courtesy Marvin E. Bauer, Remote Sensing Laboratory, University of Minnesota.)

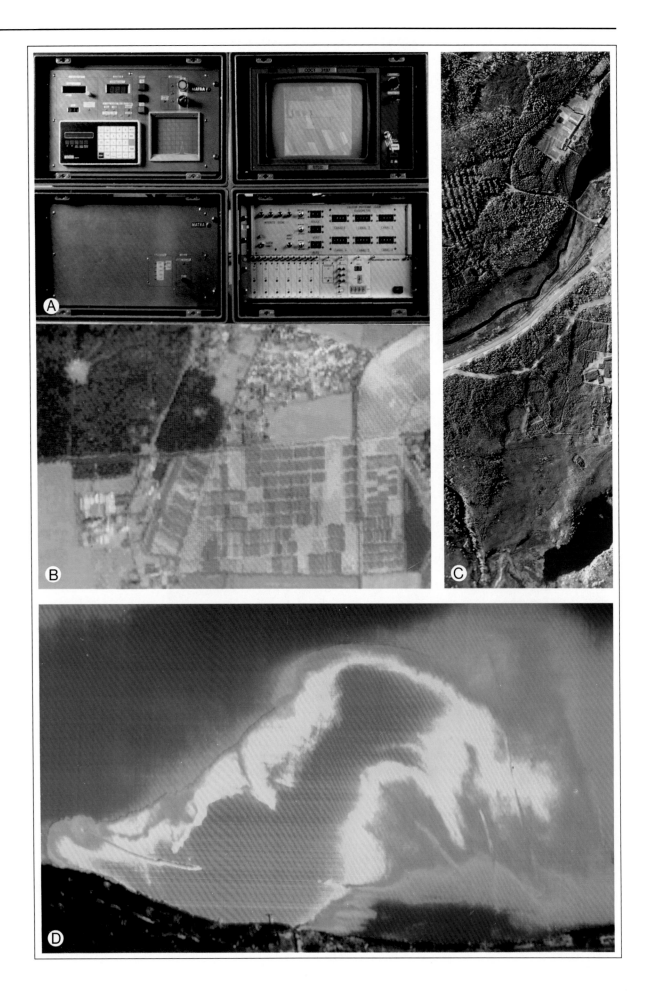

Plate 8 (*Left*)   *(A)* Electronics unit, operator's console, and quick-look TV-display unit of the Matra Thematic Scanner. *(B)* False color image of a rural area near Grignon, France, produced in flight by the Matra Thematic Scanner; the image incorporates three spectral channels: (1) 0.48–0.57 μm = blue, (2) 0.58–0.69 μm = green, and (3) 10.5–12.5 μm = red. (Courtesy B. Deffes and Jean Seligmann, Matra Optique, Bois-d Arcy, France.) *(C)* Pushbroom scanner image of an area near Petawawa, Ontario, produced by the Multispectral Electro-optical Imaging Scanner (MEIS). The false color image incorporates three spectral channels (center wavelengths): (1) 0.445 μm = blue, (2) 0.590 μm = green, and (3) 0.871 μm = red; ground resolution is 1.2 m. (Courtesy G.P. Jackson, MacDonald Dettwiler and Associates, Ltd., and the Canada Centre for Remote Sensing.) *(D)* Color-coded thermal infrared image of a warm water plume. The image was acquired in August 1970 at 7:20 A.M. local time. Each 1°C temperature interval is depicted in one of six colors. The sequential temperature relationship from warm to cool is red, yellow, green, cyan, blue, and magenta. This calibrated image was produced by the Daedalus DIG-ICOLOR™ process. (Courtesy George England, Daedalus Enterprises, Inc. and the University of Wisconsin.)

Plate 9 (*Below*)   *Landsat 3* Multispectral Scanner (MSS) color infrared composite image of a portion of New Zealand's South Island. Color assignments are as follows: band 4 (0.5–0.6 μm) = blue, band 5 (0.6–0.7 μm) = green, and band 7 (0.8–1.1 μm) = red. The image measures 185 × 185 km. Compare with Figure 6-39. (Courtesy U.S. Geological Survey.)

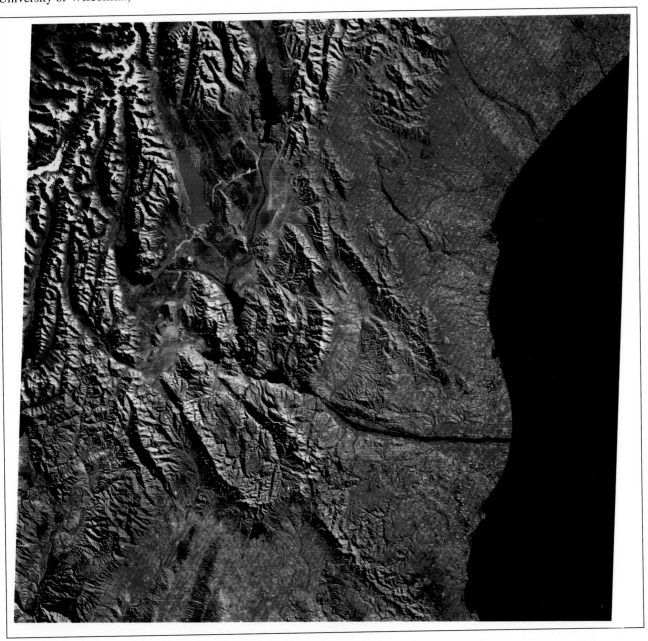

Plate 10 *Landsat 4* Thematic Mapper (TM) color composite subscene images of the San Francisco Bay region. The images used to make each color composite incorporate linear contrast stretches see Chapter 15. (Courtesy U.S. Geological Survey.)

(A) Natural color image—band/color assignments are as follows: band 1 = blue, band 2 = green, and band 3 = red. Compare with Figure 6-41.

(B) Color infrared image—band/color assignments are as follows: band 2 = blue, band 3 = green, and band 4 = red. Compare with Figure 6-41.

(C) False color image—band/color assignments are as follows: band 1 = blue, band 4 = green, and band 5 = red. Compare with Figure 6-41.

TM Band Sensitivities	
Band 1 =	0.45–0.52 μm
Band 2 =	0.52–0.60 μm
Band 3 =	0.63–0.69 μm
Band 4 =	0.76–0.90 μm
Band 5 =	1.55–1.75 μm

0   10 km

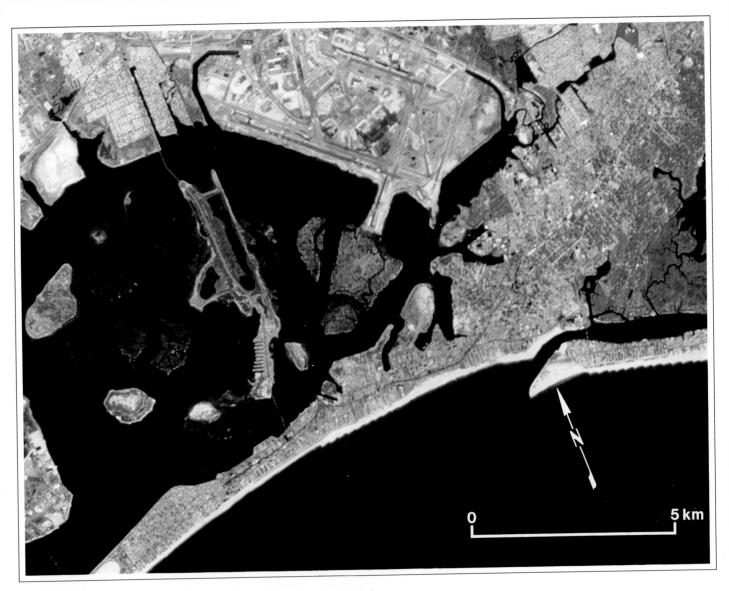

Plate 11   *SPOT 1* High Resolution Visible (HRV) color infrared composite sub-scene image of a portion of Long Island, New York, showing John F. Kennedy International Airport and Jamaica Bay. Band/color assignments are as follows: band 1 (0.50–0.59 μm) = blue, band 2 (0.61–0.68 μm) = green, and band 3 (0.79–0.89 μm) = red. The images used to make the composite incorporate linear contrast sketches; see Chapter 15. Compare with Figure 6-44. (Copyright © 1988 CNES, provided courtesy of SPOT Image Corporation, Reston, Virginia.)

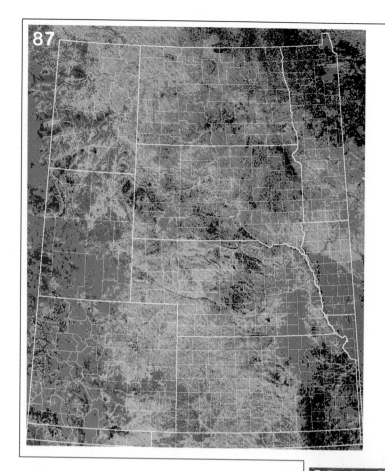

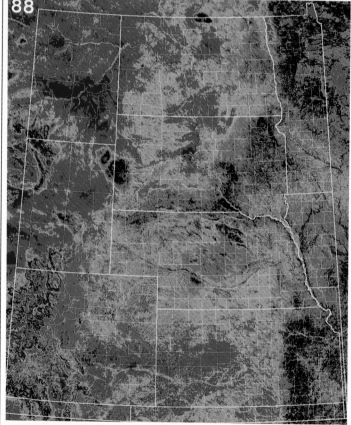

Plate 12   Normalized Difference Vegetation Index (NDVI) image mosaics derived from *NOAA 9* Advanced Very High Resolution Radiometer (AVHRR) scenes [bands 1 (0.58–0.68 μm) and 2 (0.725–1.10 μm)] of the northern Great Plains for early June 1987 (normal year) (*top*) and early June 1988 (drought year) (*bottom*). Resolution is approximately 1.1 km. The NDVI technique and color-coding scheme are described in Chapter 6. (Courtesy William C. Draeger, EROS Data Center, U.S. Geological Survey.)

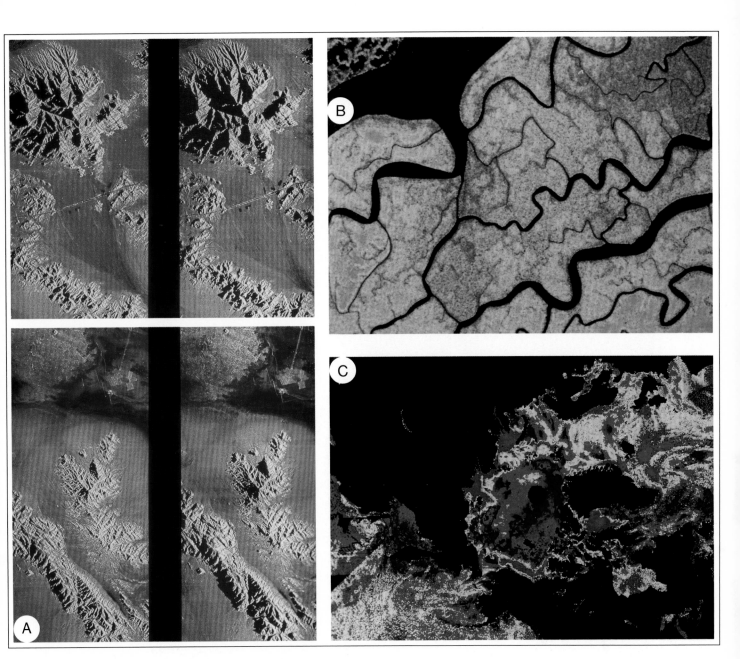

Plate 13 *(A)* Red-yellow X-band (3 cm) stereo-gram of the Amboy region, Mojave Desert, California. (Courtesy Goodyear Aerospace Corporation and Aero Service Division of Western Geophysical of America.) *(B)* Shuttle-Imaging Radar (SIR-B) L-band (23.5 cm) color coded image of the Sundarbans Forest, Bangladesh. Here the heavy cloud cover and the closed mangrove canopy were penetrated by the radar beam; colors identify the following subcanopy areas: yellow = flooded, blue = partially flooded, and brown = nonflooded. (Courtesy Marc Imhoff, National Aeronautics and Space Administration.) *(C)* Special Sensor Microwave/Imager (SSM/I) passive microwave image (85.5 GHz) for the North Polar region. The color scheme, representing brightness temperatures, is described in Chapter 7; land areas have been intentionally printed in black. (Courtesy Claire Hanson, National Snow and Ice Data Center, University of Colorado.)

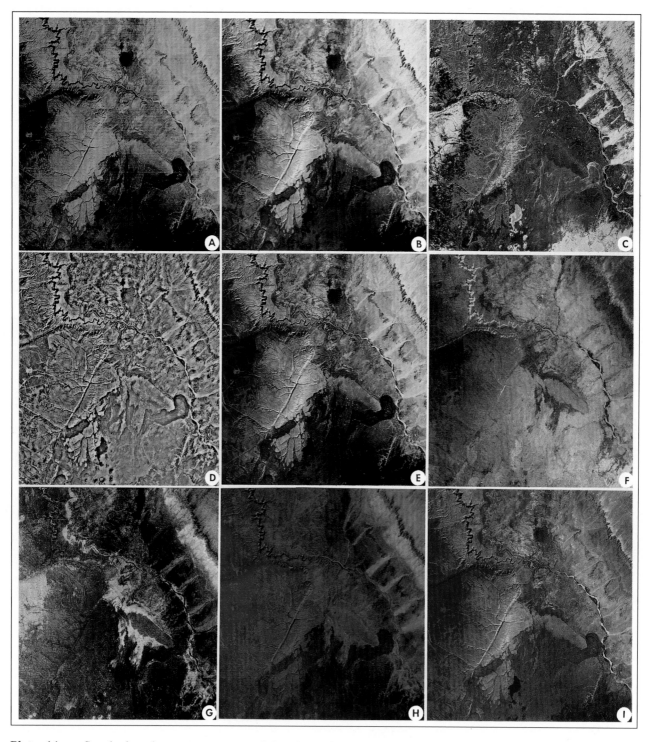

Plate 14 Standard and computer-processed Landsat Multispectral Scanner (MSS) color composite images of a volcanic-sedimentary area in northern Arizona. *(A)* MSS 4, 5, 7 standard color infrared image (band 4 = blue, band 5 = green, band 7 = red); *(B)* linear contrast stretched MSS 4, 5, 7 color infrared image; *(C)* sinusoidal contrast stretched MSS 4, 5, 7 false color image; *(D)* 31 × 31 high-pass filtered MSS 4, 5, 7 false color image; *(E)* edge-enhanced MSS 4, 5, 7 color infrared image; *(F)* triple-ratio false color image (4/5 = blue, 5/6 = green, 6/7 = red); *(G)* triple-ratio false color image (5/4 = blue, 6/4 = green, 7/5 = red); *(H)* hybrid-ratio false color image (5/4 = blue, 5 = green, 7/5 = red); *(I)* simulated natural color image ("band 3" = blue, band 4 = green, band 5 = red). See Chapters 6 and 15. (Courtesy U.S. Geological Survey.)

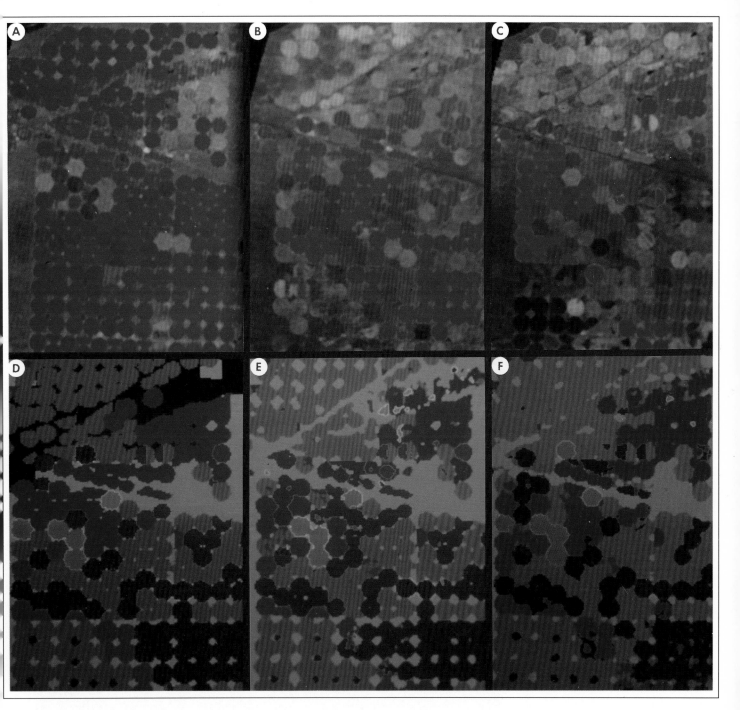

Plate 15 Crop type discrimination for an area near Clarke, Oregon, using a 12-band multitemporal Landsat Multispectral Scanner (MSS) digital data set (three 4-band MSS scenes combined). *(A)* MSS 4, 5, 7 color infrared image acquired June 3, 1979; *(B)* MSS 4, 5, 7 color infrared image acquired July 18, 1979; *(C)* MSS 4, 5, 7 color infrared image acquired September 10, 1979; *(D)* ground-reference data representing six major crops; *(E)* six-channel canonical transformed minimum distance to mean classification used for comparison to the ground-reference data; *(F)* maximum likelihood classification used for comparison to the ground-referenced data. The color scheme for images *E* and *F* is as follows: brown = wheat, red = alfalfa, green = potatoes, light blue = corn, dark blue = soybeans, and yellow = rangeland. See Chapter 15. (Courtesy Thomas M. Holm, Technicolor Government Services, Inc., EROS Data Center, U.S. Geological Survey.)

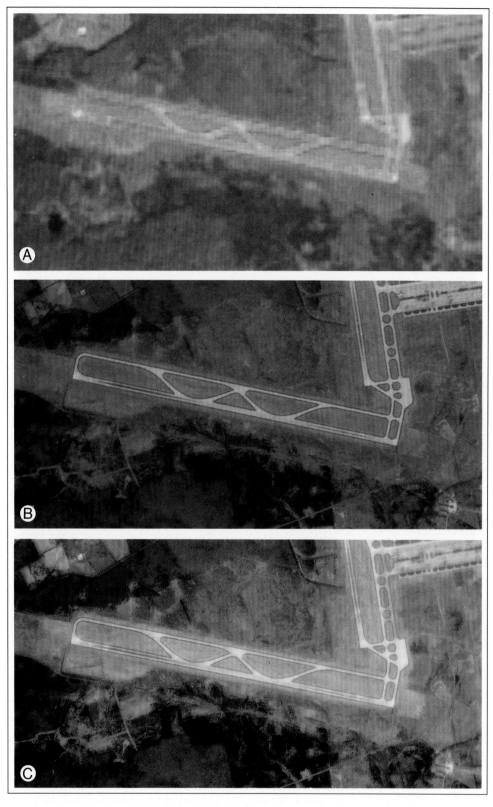

Plate 16 Separate and digitally merged Landsat Thematic Mapper (TM) images and digitized National High Altitude Photography (NHAP) photograph of a portion of Dulles International Airport and vicinity, Virginia. *(A)* TM bands 2, 3, 4 color infrared image at 30-m resolution; *(B)* digitized NHAP aerial photograph at a resolution of about 4 m; *(C)* digital merging of images *A* and *B*, with spectral resolution provided by the TM component and spatial resolution provided by the NHAP component. See Chapter 15. (Courtesy Pat S. Chavez, Jr., U.S. Geological Survey.)

GIS and Land Use and Land Cover Mapping

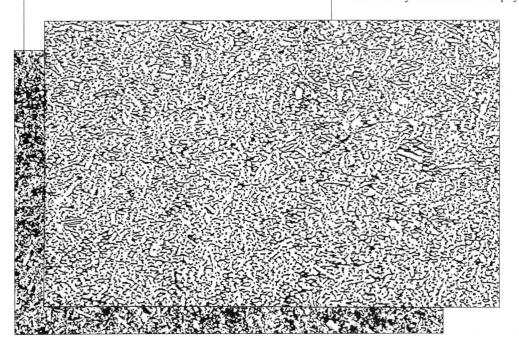

Chapter 8

2. **Inventory and Data Handling:** The collecting, collating, analyzing, and reporting of information on land and natural resources and associated socioeconomic conditions.

3. **Decision Making:** Consideration of alternatives, evaluation of impacts of proposed actions, and resolution of conflicts.

4. **Action:** Converting plans to action.

This chapter is largely devoted to the second phase, that is, to the role computerized **geographic information systems (GIS)** can play in inventory and data handling activities and **land use and land cover mapping**—an important and necessary element in all comprehensive GISs.

Overview of Geographic Information Systems

Planning organizations need vast amounts of accurate and timely information on physical resources and related socioeconomic factors to help guide their management and planning decisions. This ideally requires the organization and storage of what is known and the provision for rapid information retrieval in forms acceptable to an array of users. Over the past several years, a number of different computer-based systems have been developed to help meet this need.

A GIS is designed to accept, organize, statistically analyze, and display diverse types of spatial data that are digitally referenced to a common coordinate system of a particular projection and scale (Figure 8-1). Each variable is archived in a computer-compatible digital format as a geographically referenced layer or plane called a **data base**. Data bases can represent many different kinds of areal information; representative examples include terrain descriptors, soil and lithology types, climate, land use, land cover, population density, land ownership, and digital image radiance data. When digitally registered to each other, the data sets of n layers compose the GIS **data bank** related to a given problem (Figure 8-2).

If each data set is visualized as an independent overlay for a given base map, then two or more data observations can be analyzed for a single location—a technique known as **overlay** or **composite analysis**. The computer search for

Planning Phases

The process of environmental and natural-resources planning and management may be arbitrarily divided into four chronological phases:

1. **Awareness and Organization:** The recognition that a problem exists and that detailed studies based on specific objectives will be required for successful planning.

such formatted data is analogous to that of passing a needle through each registered map overlay (Figure 8-2).

From the foregoing discussion, it is apparent that a geographic information system has a dual data-handling responsibility (i.e., it must handle **positional** or **map data** as well as **attribute** or **descriptive data**). Map data are explicit locational identifiers (i.e., Cartesian coordinates) associated with the spatial entities of points, lines, and areas or polygons (Figure 8-3). Attributes can represent both qualitative data (e.g., soil type) and quantitative data (e.g., soil texture, soil salinity, soil porosity) keyed to each spatial entity.

GIS Capabilities

A state-of-the-art GIS should be capable of the following:

1. Accepting data inputs in one or more formats (e.g., analog map and overlay information, tabulations, digital-image data).

2. Storing and maintaining information with the necessary spatial relationships.

3. Manipulating data (search and retrieval, computations, etc.) in a timely manner.

4. Some level of modeling that takes into account data interrelationships and cause-and-effect responses of the appropriate factors.

5. Presenting data outputs in a variety of ways (e.g., tabulations, video displays, and computer-generated maps).

An ideal GIS would be designed to serve diverse users; furthermore, it would be capable of continuously being updated as new data became available. In view of the great amount of information it may use, it is not surprising that a GIS relies heavily on high-speed digital computers, a comprehensive and powerful program or software package, and a variety of peripheral input-output devices (Figure 8-4).

Technical Elements of a Digital GIS

A GIS is built around a framework of five basic **technical elements**: (1) **encoding**, (2) **data input**, (3) **data management**, (4) **manipulative operations**, and (5) **output products** (Figure 8-5). Unless otherwise specified, the following discussion of these elements is largely summarized from the works of Greenlee (1979), Short (1982), and Marble et al. (1983).

Encoding

Two position indexing systems can be used to encode spatial entities that are portrayed as points, lines, or polygons: (1) **grid-cell**, or **raster**, **coding**, and (2) **polygon**, or **vector**, **coding**. Grid-cell coding is conceptually a matrix system superimposed over the geography such that the attribute information can be collected by a systematic array of grid squares or cells. Normally, the information category most dominant for each cell is encoded (Figure 8-6). Because cell size largely determines class accuracy, two methods have emerged to assist in preserving data integrity: (1) decreasing cell size or (2) listing the relative amounts of each

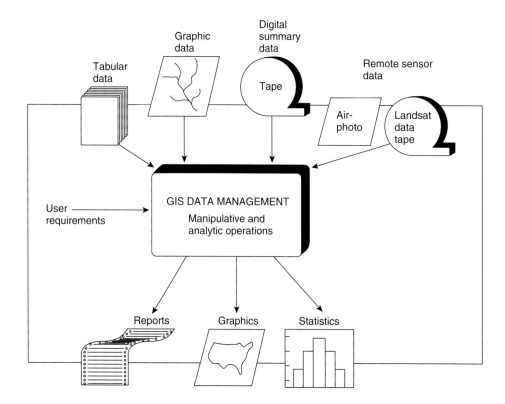

Figure 8-1 Schematic diagram showing a generalized GIS for data management. (Adapted from Short 1982.)

data type falling within a cell. Grid cells are functionally identical to the picture elements, or pixels, that compose a digital image (see Chapter 15).

With polygon coding, the perimeter of each areal unit containing the desired attribute data is digitally encoded and stored. One type of polygon indexing is **topological coding**, whereby arcs are formed by connecting nodes, and polygons are formed by connecting arcs (Figure 8-7). Polygon coding more accurately defines boundaries and requires less computer storage space than does the grid-coding structure.

Data Input

Analog information (e.g., hardcopy map data) is converted to the digital domain by the digitization process for GIS input. Methods of data capture are by **manual**, or **hand-tracing**, **digitizing** (e.g., tablet or table digitizers) and **automatic digitizing** (e.g., drum or laser-beam scanners). An integral part of both methods is a connected display system that enables an operator to edit the data for erroneous values and to recapture omissions. Data already in digital form (e.g., satellite images) usually have to be reformatted and scaled to match the geometry of the GIS reference map projection.

Data Management

Because of the large volume and variety of data, plus the wide range of potential applications, data management is extremely important for the successful and efficient operation of a GIS. The management system consists of a series of computer programs to perform all **data entry**, **storage**, **retrieval**, and **maintenance tasks**.

Manipulative Operations

GISs are capable of performing two kinds of automated analysis: (1) **surface analysis** and (2) **overlay analysis**. Surface analysis applies to intravariable relationships that exist within one data plane. For example, soil categories can be grouped together, analyzed, and labeled according to agricultural value (McFarland 1982). Area measurements could also be made for both the original and interpretive data bases. Most surface analysis produces new variables that can be applied to other surface or overlay analysis procedures.

A GIS is capable of performing automated overlay analysis that applies to intervariable relationships created by overlaying or stacking two or more data planes. Thus, location is held constant and several variables are simultaneously evaluated. Because raster formatting spatially prearranges data, vector-formatted information is often converted to a grid-cell structure for overlay analysis processing to improve data handling efficiency.

One of the most common uses of overlay analysis is to derive statistical data and special maps describing shared characteristics. For example, given the data bases of land

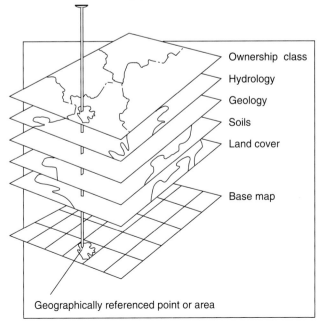

Figure 8-2 A computerized GIS can be visualized as a base map accompanied by several registered overlays. For any point or area on the base map, resource data can be analyzed by means of computer programs. (Courtesy Project LIST, Texas A & M University.)

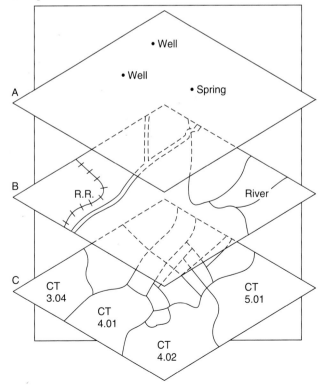

Figure 8-3 Schematic representation of geographic phenomena: (*A*) point phenomena, (*B*) linear phenomena, and (*C*) polygon, or area, phenomena, in this case census tracts. (Adapted from Greenlee 1979.)

use and land cover and of topographic slope, one could ascertain how many hectares of agricultural land exceed a particular slope. An interpretive map could also be produced to show where these conditions were met. This new data set could be integrated with other data bases (e.g., soil type, land ownership) for additional analysis (McFarland 1982).

Quantitative interpretation by overlay analysis is accomplished by establishing a numerical index for each qualitative variable. For example, with vegetation type, soil type, and topography data bases, an interpretation of erosional properties could be calculated by using quantitative indices for these three variables as they relate to erosion potential (Figure 8-8). Overlay analysis also allows data variables to be weighted according to their relative importance. In the previous example, topography may be determined to be more important than either vegetation type or soil type in determining erosion potential.

By using quantitative interpretation and weighting techniques, data bank variables can be used with two types of prediction models. For example, **evaluative models** can be developed to assess environmental characteristics (e.g., wildlife habitat, forest fire potential, groundwater contamination, accessibility to transportation systems), and **allocative models** can be developed to indicate areas best suited for specific land uses (e.g., urban development, transportation routing, irrigated agriculture development).

Johnson and Loveland (1980) developed a GIS allocative model to evaluate several types of geobased data for determining *land irrigation suitabilities* in the Stanfield, Oregon, area. The variables of existing land cover, soil characteristics, and topographic slope (Figures 8-9, 8-10, and 8-11) were used to create a composite irrigability map that defined the physical capability of the land to support irrigation development (Figure 8-12). Two additional variables, horizontal and vertical distance from water (Figures 8-13 and 8-14), were then evaluated in terms of development costs to determine their effect on the location of irrigation development (pumping costs increase as distance from water increases, and the greater the pump lift, the greater the cost of providing water). The land-factor irrigability map (Figure 8-12) was then digitally merged with the distance-from-water data (economic factors) to modify irrigability potential (Figure 8-15). This revised analysis predicted the actual location of irrigation development.

Output Products

A GIS can retrieve and display data in graphic or tabular form, or both. Most systems are capable of producing hardcopy charts, scatter diagrams, tables, and maps in various forms and sizes. In addition, all systems have a TV monitor on which graphic or tabular information for segments of a data base or multiple data bases can be displayed

Figure 8-4 Schematic representation of a GIS. In addition to commercial GISs, most digital image processing systems (IPS) can be modified to store and manipulate geographically encoded data (see Chapter 15). (Courtesy Project LIST, Texas A & M University.)

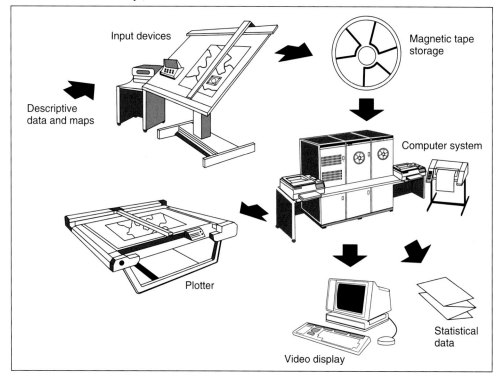

Sources of GIS Information

With the recent and burgeoning growth of GIS technology and the community of GIS users, several periodicals devoted entirely to GIS topics are now published on a regular schedule. These include the *International Journal of Geographic Information Systems, Geo Info Systems*, and *GIS World*. In addition, many established journals now carry frequent GIS articles and news. These include the major remote sensing journals (Table 1-5) plus *Environmental Management, Professional Surveyor, Journal of Forestry, Journal of Soil and Water Conservation*, and *Journal of the American Planning Association*.

Several recently published books devoted entirely to GIS topics have been authored or edited by Burrough (1986), Aronoff (1989), Marble et al. (1989), and Rhind and Mounsey (1990). Guptill (1988) presents a comprehensive guide for evaluating different geographic information systems.

(Figure 8-4). This represents **interactive analysis** because the retrieval and display of data are in near real time (see Chapter 15).

Figure 8-5 Flow diagram of steps in a typical GIS (Adapted from Short 1982).

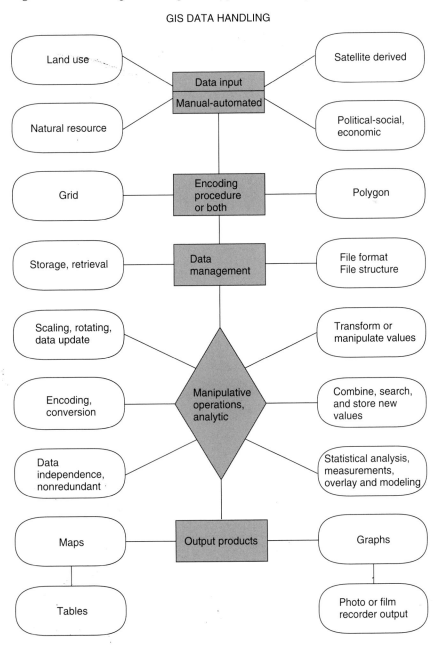

Land Use and Land Cover Defined

Although the terms **land use** and **land cover** have been used interchangeably, it is important to remember that the two expressions are not necessarily synonymous. Land use encompasses several different aspects of people's relationship to the environment (e.g., activity, ownership, land quality). By comparison, land cover is represented by the natural and artificial compositions covering the earth's surface at a certain location. For example, the land cover for a given area might be classified as deciduous forest when the land use is that of a wildlife refuge or mining operation (Figure 8-16). The distinction between the two terms is particularly important when such information is interpreted from aerial photographs and electro-optical images.

Classification Systems

Classification systems for describing land use and land cover have been the subject of extensive research and symposia in North America and Europe for more than three decades. During the last decade, the major emphasis has been on the development of classification schemes that incorporate information derived from remotely sensed data. Although no system will probably ever be developed that is universally acceptable, several classification schemes are being used at the national, regional, and local levels (Jensen et al. 1983).

The expandable system devised by the U.S. Geological Survey (USGS) (Anderson et al. 1976) represents a national classification scheme that has achieved widespread acceptance and is being used in a number of operational mapping programs. With two levels of detail, it can be used with remotely sensed data at various resolutions and scales.

It has also served as a framework for more detailed regional and local classification systems because it is sufficiently inclusive (Table 8-1). Much of the remainder of the chapter is devoted to that national system, herein referred to as the **USGS system**.

Figure 8-7 Topological elements of a polygon map. (Adapted from Mitchell et al. 1977.)

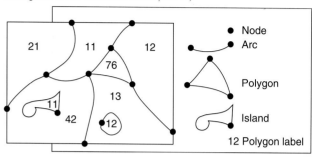

Figure 8-8 Overlay analysis using a grid model. (Adapted from Short 1982.)

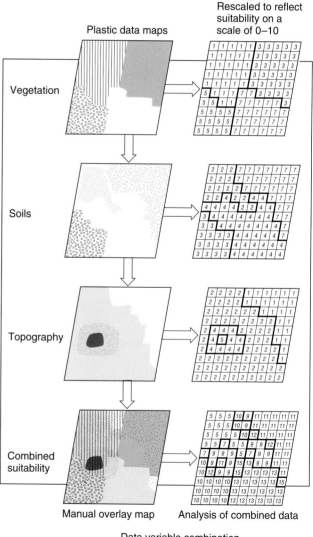

Figure 8-6 Grid-cell coding. (Adapted from Greenlee 1979.)

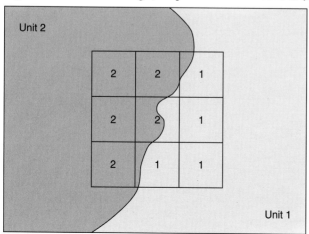

TABLE 8-1   Land Use and Land Cover Classification System for Use with Remote Sensor Data[a]

Level I (and Map Color)	Level II
1 Urban or built-up land (red)	11 Residential
	12 Commercial and services
	13 Industrial
	14 Transportation, communications, and utilities
	15 Industrial and commercial complexes
	16 Mixed urban or built-up land
	17 Other urban or built-up land
2 Agricultural land (light brown)	21 Cropland and pasture
	22 Orchards, groves, vineyards, nurseries, and ornamental horticultural areas
	23 Confined feeding operations
	24 Other agricultural land
3 Rangeland (light orange)	31 Herbaceous rangeland
	32 Shrub and brush rangeland
	33 Mixed rangeland
4 Forest land (green)	41 Deciduous forest land
	42 Evergreen forest land
	43 Mixed forest land
5 Water (dark blue)	51 Streams and canals
	52 Lakes
	53 Reservoirs
	54 Bays and estuaries
6 Wetland (light blue)	61 Forest wetland
	62 Nonforested wetland
7 Barren land (gray)	71 Dry salt flats
	72 Beaches
	73 Sandy areas other than beaches
	74 Bare, exposed rock
	75 Strip mines, quarries, and gravel pits
	76 Transitional areas
	77 Mixed barren land
8 Tundra (green-gray)	81 Shrub and brush tundra
	82 Herbaceous tundra
	83 Bare ground tundra
	84 Wet tundra
	85 Mixed tundra
9 Perennial snow or ice (white)	91 Perennial snowfields
	92 Glaciers

[a]*Level I is based primarily on surface cover; level II is derived from both cover and use. Color codes are based on recommendations of the International Geographical Union.*

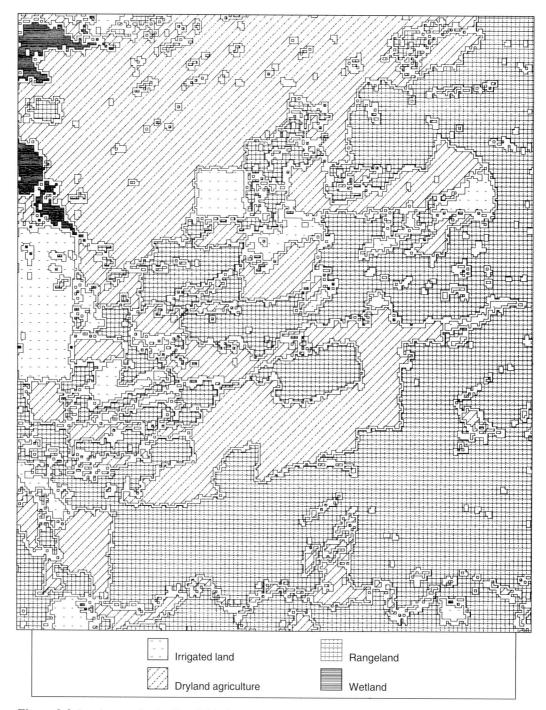

| | Irrigated land | | Rangeland |
| | Dryland agriculture | | Wetland |

Figure 8-9 Land cover in the Stanfield, Oregon, area based on a digital classification of a 1972 Landsat Multispectral Scanner (MSS) image. The assumption of the classification scheme is that both dryland agriculture and rangeland constitute potentially irrigable land areas, whereas wetlands are not considered capable of irrigation development. (Courtesy Thomas R. Loveland, Technicolor Government Services, EROS Data Center.)

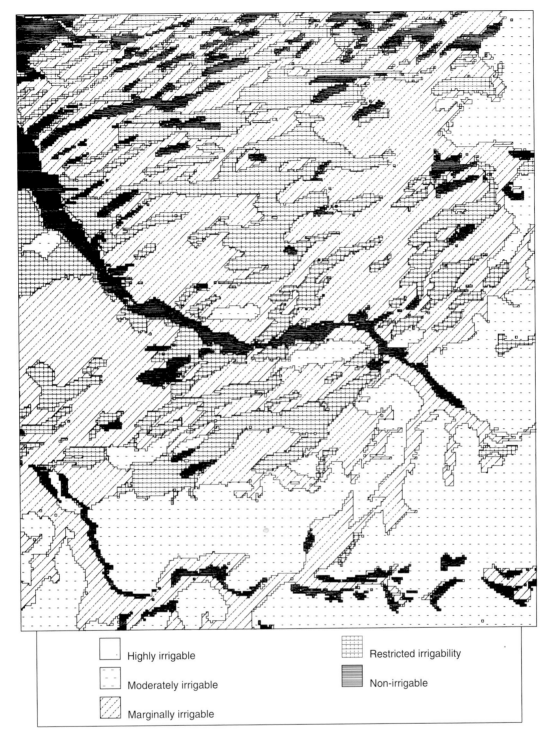

Figure 8-10 Soil potential for irrigation based on an evaluation of soil characteristics (e.g., thickness, texture, salinity). (Courtesy Thomas R. Loveland, Technicolor Government Services, EROS Data Center.)

Legend:
- Highly irrigable
- Moderately irrigable
- Marginally irrigable
- Restricted irrigability
- Non-irrigable

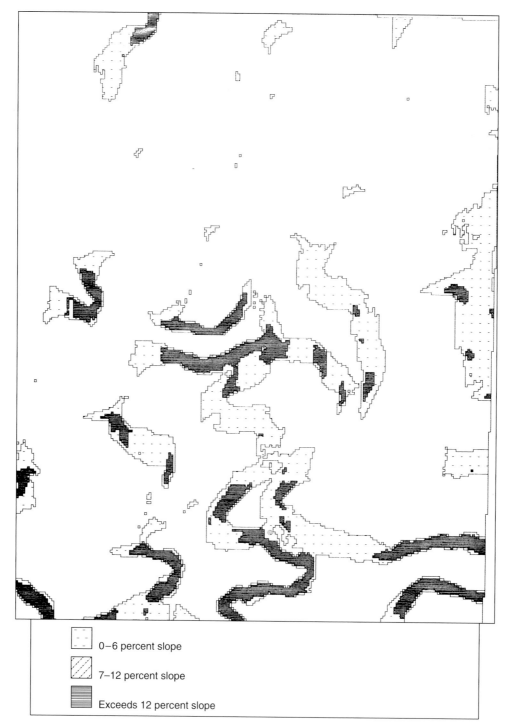

0–6 percent slope

7–12 percent slope

Exceeds 12 percent slope

Figure 8-11 Percent slope calculated from digital terrain data: slopes greater than 12 percent—unirrigable by any present irrigation system; 7–12 percent slope—irrigable by center-pivot irrigation systems; and 0–6 percent slope—irrigable by most conventional irrigation methods. (Courtesy Thomas R. Loveland, Technicolor Government Services, EROS Data Center.)

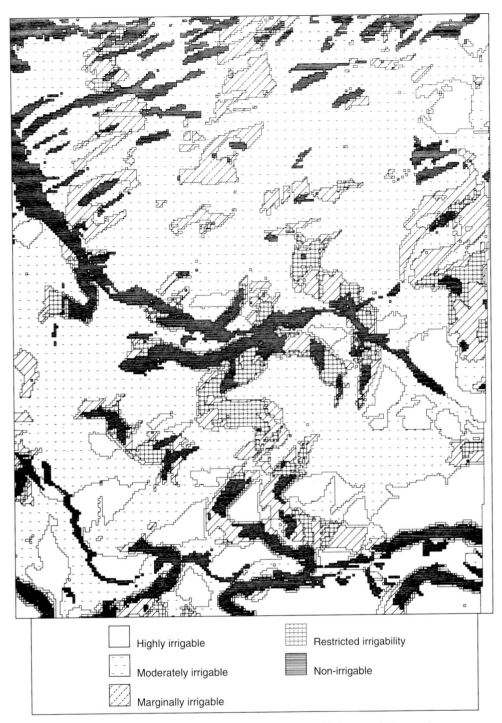

	Highly irrigable		Restricted irrigability
	Moderately irrigable		Non-irrigable
	Marginally irrigable		

Figure 8-12 Composite irrigability based on land cover, soil characteristics, and percent slope. (Courtesy Thomas R. Loveland, Technicolor Government Services, EROS Data Center.)

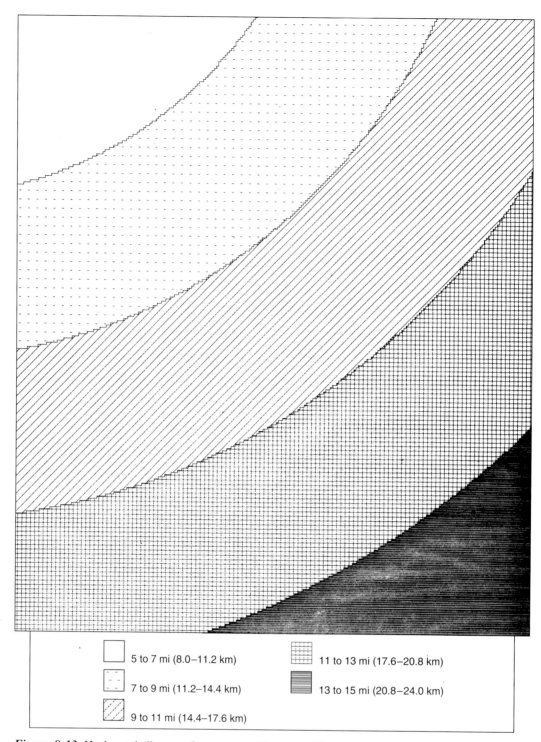

5 to 7 mi (8.0–11.2 km)

7 to 9 mi (11.2–14.4 km)

9 to 11 mi (14.4–17.6 km)

11 to 13 mi (17.6–20.8 km)

13 to 15 mi (20.8–24.0 km)

Figure 8-13 Horizontal distance from water. (Courtesy Thomas R. Loveland, Technicolor Government Services, EROS Data Center.)

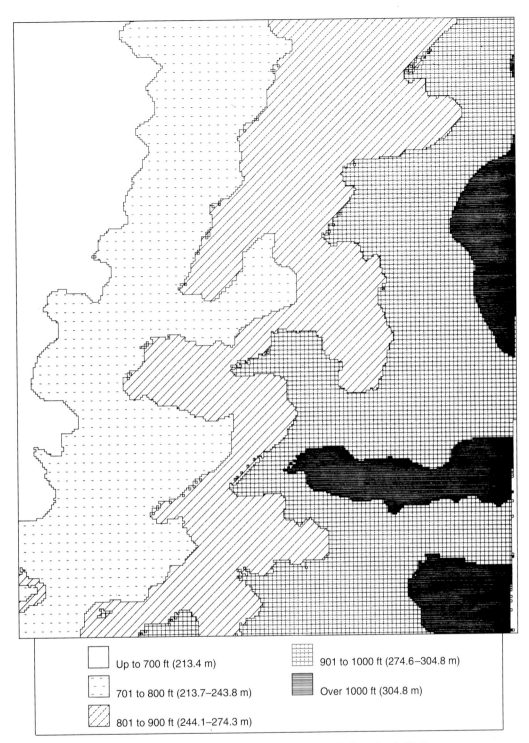

Legend:
- Up to 700 ft (213.4 m)
- 701 to 800 ft (213.7–243.8 m)
- 801 to 900 ft (244.1–274.3 m)
- 901 to 1000 ft (274.6–304.8 m)
- Over 1000 ft (304.8 m)

Figure 8-14 Vertical distance from water. (Courtesy Thomas R. Loveland, Technicolor Government Services, EROS Data Center.)

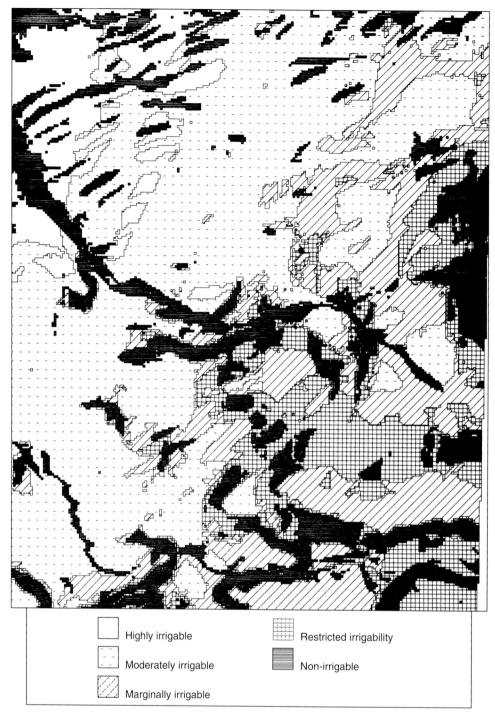

Highly irrigable

Moderately irrigable

Marginally irrigable

Restricted irrigability

Non-irrigable

Figure 8-15 Composite irrigability including economic considerations (distance from water). Compare with Figure 8-12. (Courtesy Thomas R. Loveland, Technicolor Government Services, EROS Data Center.)

Figure 8-16 This quiltlike pattern of trees and small clearings connected by roads is an oil field in McKean County, Pennsylvania. Two pumping stations are circled. The difference between older and newer drillings is shown by the regrowth of vegetation. Scale is 1:20,000. (Courtesy U.S. Department of Agriculture.)

Features of the USGS System

The **USGS hierarchical system** (i.e., ordered classes) incorporates the features of several existing classification systems that are amenable to data derived from remote sensors, including images and photographs from satellites and high-altitude aircraft. The system attempts to meet the need for current overview assessments of land use and land cover on a basis that is uniform in categorization at the first and second levels of detail. It is intentionally left open-ended so that various levels of government, for example, may have flexibility in developing more detailed classifications at the third and fourth levels. Such an approach permits various agencies to meet their particular needs for land-resource management and planning and at the same time remain compatible with the national system (Table 8-1). For detailed descriptions at the 9 level I and 37 level II categories, the reader is referred to Anderson et al. (1976). Compilation specifications are documented by Loelkes (1977) and Loelkes et al. (1983).

The types of land use and land cover categorization developed in the USGS classification system can be related to systems for classification of land capability, vulnerability to certain management practices, potential for any particular

activity, or land value, either intrinsic or speculative. The functions that lands fill will usually be associated with certain types of land cover. Thus, the image interpreter attempts to identify land cover patterns and shapes as a means of deriving information about land use (Figure 8-17).

Classification Criteria

According to the USGS, a land use and land cover classification system that can effectively employ orbital and high-altitude remote sensor data should meet the following criteria:

1. Interpretation accuracy in the identification of land use and land cover categories from remote sensor data should be 85 percent or greater.

2. The accuracy of interpretation for the several categories should be about equal.

3. Repeatable or repetitive results should be obtainable from one interpreter to another and from one sensing time to another.

4. The classification system should be applicable over extensive areas.

5. The categorization should permit vegetation and other types of land cover to be used as surrogates for activity.

6. The classification system should be suitable for use with remote sensor data obtained at different times of the year.

7. Effective use of subcategories that can be obtained from ground surveys or from the use of larger-scale or enhanced remote sensor data should be possible.

8. Aggregation of categories must be possible.

9. Comparison with future land use data should be possible.

10. Multiple uses of land should be recognized when possible.

For land use and land cover data needed for planning and management, accuracy of interpretation at the generalized first and second levels is satisfactory when the interpreter makes the correct interpretation 85 to 90 percent of the time. Except for urban and built-up areas, this can often be achieved with satellite images. For regulation of land use or for tax assessment, for example, greater accuracy may be required, and greater accuracy will normally imply higher costs.

The problem of classifying *multiple uses* occurring on a single parcel of land is not easily solved. Multiple uses may occur simultaneously, as in the instance of agricultural land or forest land being used for recreational activities, hunting, or camping. Uses may also occur alternately, as would be the case with a major reservoir that provided flood

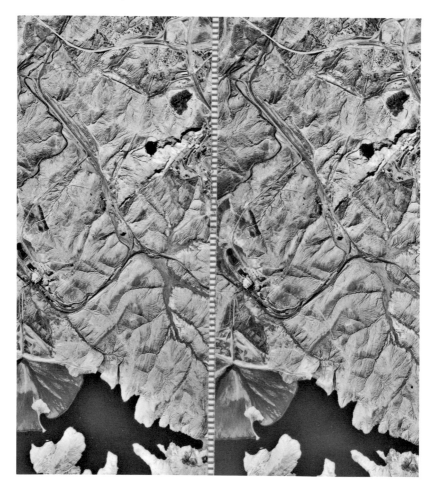

Figure 8-17 Stereogram of a copper-mining area in Tennessee. Most of the vegetation has been killed by smelter gases. (Courtesy Tennessee Valley Authority.)

control during spring runoff and generated power during summer peak-demand periods. All these activities would not be detected on a single aerial image; thus the selected categorization may be necessarily based on the dominant or apparent use on the date of the sensor image (Figure 8-18).

Vertical arrangements of land uses above or below terrain surfaces produce added complexities for image interpreters. Mineral deposits under croplands or forests, electrical transmission lines crossing pastures, garages underground or on roofs of buildings, and subways beneath urban areas all exemplify situations which must be resolved by individual users and compilers of land use data.

Minimum Areas and Image Resolution

The minimum area that can be classified as to land use and land cover depends on (1) the scale and resolution of the original sensor image or data source, (2) the scale of data compilation or image interpretation, and (3) the final scale of the land use information or map. It is difficult to delineate and symbolize any map area smaller than 0.25 to 0.5 cm on a side. Actual ground area represented by such parcels depends on the scale of the map.

A wide variety of remote sensing image data can be used for the different levels of land use and land cover classifications. Each sensor provides a degree of image resolution that is dependent upon flight altitude and effective focal length (or scale). For example, assuming a focal length of 15 cm, the flight altitudes and image scales in Table 8-2 would be appropriate when manual (nonautomated) interpretation is anticipated.

The foregoing recommendations are approximate guidelines and not absolutes. With future technological advances, such tabulations will need to be revised. In fact, the entire classification system may undergo a complete metamorphosis as greater dependence is placed on automatic data analysis and automatic image interpretation (see Chapter 15).

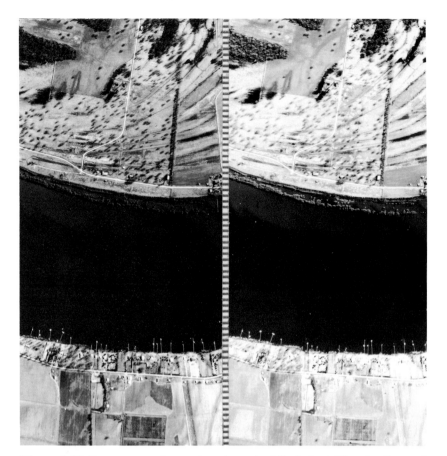

Figure 8-18 Stereogram of a section along the Mississippi River in Concordia Parish, Louisiana. Land use is predominantly agricultural, but the housing and boat docks along one shoreline would likely be classed as urban-residential or urban-recreational. Scale is about 1:24,000. (Courtesy U.S. Geological Survey.)

TABLE 8-2	Appropriate Flight Altitudes and Image Scales for Manual Interpretation of Land Use and Land Cover	
Classification Level	Sensor Platform or Altitudes	Approximate Range of Image Scales
I	Earth satellites	1:250,00 to 1:3,000,000
II	9,000–22,000 m	1:60,000 to 1:125,000
III	3,000–9,000 m	1:20,000 to 1:60,000
IV	1,200–3,000 m	1:8,000 to 1:20,000

Expanded Classifications

One of the primary virtues of a hierarchical classification system is that it is structured for developing categories at more detailed levels. This feature also permits subsequent aggregation or disaggregation of land units without difficulty. The USGS system is aimed at complete standardization at levels I and II only. Users of the system are encouraged to develop their own subcategories for levels III and IV. Table 8-3 shows how residential land might be subdi-

vided in a level III classification. This particular breakdown employs criteria of density, type, height, and phase of construction as the discriminating factors among classes.

A level III land use and land cover classification system has been developed for two types of applications in Connecticut (Table 8-4). Once the entire state is mapped at this level, the resulting land use and land cover data will be correlated with water use and evapotranspiration coefficients to produce information (1) for the Connecticut Water Use Information System, and (2) for developing water-balance models for the state's drainage basins (National Mapping Division 1982). The primary source materials for the level III interpretations are 1:80,000-scale and 1:12,000-scale black-and-white aerial photographs.

TABLE 8-3   Examples of Subdivisions in a Level III Land Use and Land Cover Classification

Level I	Level II	Level III
1 Urban or built-up land	11 Residential	111 Single-family units (<5 units/0.5 ha)
		112 Single-family units (>5 units/0.5 ha)
		113 Multifamily units (single-story)
		114 Multifamily units (multiple-story)
		115 Mobile-home units
		116 Residential estates (lot size >2 ha)
		117 Residential under construction
		118 Other residential

TABLE 8-4   Proposed Connecticut Land Use and Land Cover Classification System

Level I	Level II	Level III
1 Urban or built-up land	11 Residential	111 Rural
		112 Low density
		113 Medium density
		114 High density
	12 Commercial and services	121 Low impervious cover
		122 Medium impervious cover
		123 Medium impervious cover, mostly buildings
		124 High impervious cover
		125 High impervious cover, mostly buildings
	13 Industrial	131 Electric-power generating stations
		132 Other industrial
	14 Transportation, communications, and utilities	141 Limited access highways
		142 Railway facilities
		143 Airports
		144 Port facilities
		145 Oil and gas storage facilities
		146 Water treatment facilities
		147 Sewage treatment facilities
		148 Waste disposal sites
		149 Other transportation, communications, and utilities
	17 Other urban or	171 Golf courses

Level I	Level II	Level III
	built-up land	172 Other urban or built-up land
2 Agricultural land	21 Cropland and pasture	210 Cropland and pasture
	22 Orchards, groves, vineyards, nurseries, and ornamental horticultural areas	221 Orchards 222 Greenhouses 223 Other groves, vineyards, nurseries, and ornamental horticultural areas
	23 Confined feeding operations	231 Dairy confined feeding operations 232 Poultry confined feeding operations 233 Other confined feeding operations
	24 Other agricultural land	240 Other agricultural land
3 Rangeland	32 Shrub and brush rangeland	321 Eastern brushland
4 Forest land	41 Deciduous forest land	411 Deciduous, 10–50 percent crown cover 412 Deciduous, greater than 50 percent crown cover
	42 Evergreen forest land	421 Evergreen, 10–50 percent crown cover 422 Evergreen, greater than 50 percent crown cover
	43 Mixed forest land	431 Mixed, 10–50 percent crown cover 432 Mixed, greater than 50 percent crown cover
5 Water	51 Streams and canals	510 Streams and canals
	52 Lakes	520 Lakes
	53 Reservoirs	530 Reservoirs
	54 Bays and estuaries	540 Bays and estuaries
6 Wetland	61 Forested wetland	611 Deciduous forested wetland 612 Evergreen forested wetland 613 Mixed forested wetland
	62 Nonforested wetland	621 Freshwater nonforested wetland 622 Brackish and saltwater nonforested wetland
7 Barren land	72 Beaches	720 Beaches
	73 Sandy areas other than beaches	730 Sandy areas other than beaches
	75 Strip mines, quarries, and gravel pits	751 Sand and gravel pits 752 Other strip mines and quarries
	76 Transitional areas	760 Transitional areas

Courtesy Richard E. Witmer, U.S. Geological Survey.

USGS Mapping and Data Compilation Program

Since 1974 the USGS has been engaged in a mapping and data compilation program to provide *nationwide* land use–land cover and associated polygon maps at scales of 1:250,000 and 1:100,000 and area statistical data. The associated maps portray certain types of natural and administrative areas that can be graphically and statistically compared to the land use and land cover information. The maps presently provided by the USGS program consist of the following:

1. Land use and land cover.

2. Political units (state and county boundaries).

3. Hydrographic units (watershed boundaries corresponding to USGS-defined drainage basins).

4. Census county subdivisions (boundaries for census tracts within Standard Metropolitan Statistical Areas and for minor civil divisions outside of those areas).

5. Federal land ownership (inventory of ownership for 28 agencies).

6. State land ownership (optional, data provided by co-operating state).

The land use–land cover maps are compiled to show level II categories of the USGS classification system (Table 8-1). A primary source material for land use–land cover compilation is 1:58,000-scale color infrared photographs produced by the National High-Altitude Photography Program (see Chapter 5). Compilation specifications (e.g., minimum sizes of mapping units, minimum densities of appropriate objects, and line weight for delineating polygons) are documented by Loelkes (1977). The specifications ensure the replicability and coherence of the data sets derived from using the USGS classification system.

As part of the USGS compilation program, the land use–land cover maps and associated maps are digitized, and the resultant digital data bases are handled by a computerized GIS—the Geographic Information Retrieval and Analysis System, or GIRAS (Mitchell et al. 1977). GIRAS has the capability to input, manipulate, analyze, and output the digital spatial data. The general system flow of GIRAS is presented in Figure 8-19. Digital data base tapes and a variety of graphical and statistical products can be produced by GIRAS. Kleckner (1981) describes several current capabilities of GIRAS:

One ability is to plot and replicate the land use and land cover and associated map data originally compiled from remotely sensed data. The entire map, selected map parts, or a combination of two or more whole maps or map parts can be plotted using high speed computer-driven plotters. This makes possible the ready comparison of the distribution patterns of different categories. The user can also plot out selected categories of land use and land cover. Changes in map scale and projection can also be accomplished, facilitating the overlay of maps in GIRAS with other maps such as soil maps.

Another application of the GIRAS is computer-assisted cartography leading to the actual color printing of land use and land cover maps. Many manual techniques are involved in preparing a map for lithographic printing. The USGS is now exploring ways to eliminate many of these manual techniques by using digital data and computer techniques.

Another capability involves calculating the area of land use and land cover types within specified units such as counties, census county subdivisions, and drainage basins. Land use and land cover statistical data can then be compared to other data sets.

Additional information on the status of the USGS mapping and data compilation program and the availability of current data may be obtained from the Office of Geographic Research, National Mapping Division, 521 National Center, U.S. Geological Survey, Reston, VA 22092.

Land Use and Land Cover Changes

In analyzing the development of any area, being able to identify and map land use and land cover *changes* over time is one of the most valuable indicators of rural, urban, and industrial growth. The changes can also indicate trends that can be used to predict future land use patterns. Sequential remote sensing data can play an important role because a trained interpreter can identify and map changes in land use and land cover at two or more distinct times. An example of changing land use and land cover pattern owing to the displacement of agricultural land to residential growth is shown in Figure 8-20.

For many purposes, the USGS level II categories are adequate for monitoring and assessing land use and land cover changes in urban and regional environments (Table 8-1 and Figure 8-20). Level II categories can usually be interpreted with reliability from high-altitude aircraft photographs alone (Table 8-2), but problems with identifications can arise when more detailed breakdowns are desired. Verifications may require supplemental sources of information, including ground surveys and interviews with local residents.

Figure 8-19 General system flow of GIRAS. (Adapted from Mitchell et al. 1977.)

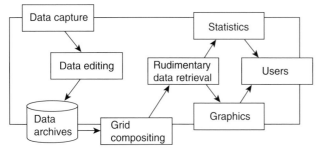

Figure 8-20 Two airphotos of a part of Tempe, Arizona, in 1970 (*top*) and 1979 (*bottom*). Each photograph is overlaid with a polygon map delineating level II land use and land cover categories (see Table 8-1). (Milazzo 1980.)

Figure 8-21 Panchromatic airphoto of an urban area along the east coast of southern Florida.

Questions

1. Upon reviewing Figure 8-21, Table 8-1, and Anderson et al. (1976), identify the land use and land cover types at levels I and II. Also, devise an appropriate level III category for each unit.

Unit	Level I Code	Level II Code	Level III Name
A	_____	_____	_____
B	_____	_____	_____
C	_____	_____	_____
D	_____	_____	_____
E	_____	_____	_____
F	_____	_____	_____
G	_____	_____	_____
H	_____	_____	_____

2. Three single-family housing areas are shown at locations 1–3 in Figure 8-21. Devise a level III scheme that could be used for differentiating them.

Housing Area	Level III Category
1	_____
2	_____
3	_____

3. Examine Figure 8-22 in stereo and describe the land use and land cover changes that have occurred in a time interval of 7 years.

Figure 8-22 Stereograms of an area photographed at a 7-year interval.

Bibliography and Suggested Readings

Anderson, J. R., E. E. Hardy, J. T. Roach, and W. E. Witmer. 1976. *A Land Use and Land Cover Classification System for Use with Remote Sensor Data.* USGS Professional Paper 964. Reston, Va.: U.S. Geological Survey.

Aronoff, S. 1989. *Geographic Information Systems: A Management Perspective.* Ottawa: WDL Publications.

Burrough, P. A. 1986. *Principles of Geographical Information Systems for Land Resources Assessment.* London: Clarendon Press.

Greenlee, D. D. 1979. *Reference Notes: Spatial Analysis Concepts.* Sioux Falls, S. Dak.: U.S. Geological Survey.

Guptill, S. C., ed. 1988. *A Process for Evaluating Geographic Information Systems.* USGS Open-File Report 88-105. Reston, Va.: U.S. Geological Survey.

Jensen, J. R., M. L. Bryan, S. Z. Friedman, F. M. Henderson, R. K. Holz, D. Lindgren, D. L. Toll, R. A. Welch, and J. R. Wray. 1983. Urban/Suburban Land Use Analysis. In *Manual of Remote Sensing,* edited by R. N. Colwell, 2d ed., 1571–666. Falls Church, Va.: American Society for Photogrammetry and Remote Sensing.

Johnson, G. E., and T. R. Loveland. 1980. The Columbia River and Tributaries Irrigation Withdrawals Analysis Project: Feasibility Analysis and Future Plans. In *Symposium, Identifying Irrigated Lands Using Remote Sensing Techniques,* 37–47. Omaha, Nebr.: Missouri River Basin Commission.

Kleckner, R. L. 1981. A National Program of Land Use and Land Cover Mapping and Data Compilation. In *Planning Future Land Uses,* 7–13. Madison, Wis.: ASA, CSSA, SSSA.

Lindgren, D. T. 1985. *Land Use Planning and Remote Sensing.* Dordrecht: Martinus Nijhoff.

Loelkes, G. L., Jr. 1977. *Specifications for Land Use and Land Cover and Associated Maps.* USGS Open-File Report 77-555. Reston, Va.: U.S. Geological Survey.

Loelkes, G. L., Jr., G. E. Howard, Jr., E. L. Schwertz, Jr., P. D. Lampert, and S. W. Miller. 1983. *Land Use/Land Cover and Environmental Photointerpretation Keys.* USGS Bulletin 1600. Reston, Va.: U.S. Geological Survey.

Marble, D. F., H. W. Calkins, and D. J. Peuquet, 1989. *Basic Readings in Geographic Information Systems,* 2d ed. Williamsville, N.Y.: SPAD Systems Limited.

Marble, D. F., D. J. Peuquet, A. R. Boyle, N. Bryant, H. W. Calkins, T. Johnson, and A. Zobrist. 1983. Geographic Information Systems and Remote Sensing. In *Manual of Remote Sensing,* edited by R. N. Colwell, 2d ed., 923–58. Falls Church, Va.: American Society for Photogrammetry and Remote Sensing.

McFarland, W. D. 1982. Geographic Data Bases for Natural Resources. In *Remote Sensing for Resources Management,* 41–50. Ankeny, Ia.: Soil Conservation Society of America.

Milazzo, V. A. 1980. *A Review and Evaluation of Alternatives for Updating U.S. Geological Survey Land Use and Land Cover Maps.* USGS Circular 826. Reston, Va.: U.S. Geological Survey.

Mitchell, W. B., S. C. Guptill, K. E. Anderson, R. G. Fegas, and C. A. Hallam. 1977. *GIRAS: A Geographic Information Retrieval and Analysis System for Handling Land Use and Land Cover Data.* USGS Professional Paper 1059. Reston, Va.: U.S. Geological Survey.

National Mapping Division. 1982. *Research, Investigations, and Technical Developments, National Mapping Program.* USGS Open-File Report 82-236. Reston, Va.: U.S. Geological Survey.

Rhind, D., and H. Mounsey. 1990. *Understanding Geographic Information Systems.* London: Taylor & Francis.

Short, N. M. 1982. *The Landsat Tutorial Workbook.* NASA Publication 1078. Washington, D.C.: U.S. Government Printing Office.

Witmer, R. E. 1978. U.S. Geological Survey Land Use and Land Cover Classification System. *Journal of Forestry* 76:661–66.

Chapter 9

Prehistoric and Historic Archaeology

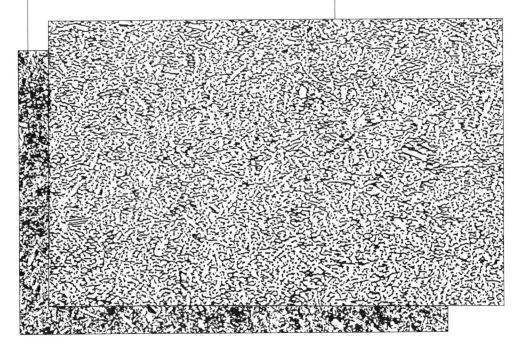

A New Perspective

Remote sensing photographs and images are a vital tool of archaeologists, historical geographers, and others who are concerned with the discovery, evaluation, and preservation of prehistoric and historic sites. Although many such sites are still being discovered by accident (e.g., during surveys for a new highway alignment), aerial imagery provides a systematic means of searching out features that may have gone unnoticed for centuries (Figure

9-1). Furthermore, the remote sensor data itself constitutes historical data; once acquired, it becomes a historical document of conditions that existed at a certain time and place.

In arid climates, the *detection* and *delineation* of archaeological sites may be fairly simple, especially where vegetation is sparse or absent. By contrast, detection may be extremely difficult in high-rainfall regions such as tropical forests. Potential sites may not only be obscured by vegetation, but their ground scale alone may render them virtually invisible to ground observation. The greater range and vertical perspective afforded by aerial imagery thus provides a new dimension in the search for and delineation of archaeological sites. Even when such sites are not clearly discernible, probable locales for detailed ground exploration may be predicted by image analysis.

Site detection may be accomplished by visual aerial reconnaissance or by use of various remote sensors such as aerial cameras, thermal infrared scanners, and side-looking airborne radar. This chapter describes techniques that might be used with any type of imagery, but emphasis is placed on conventional photography because of its superiority in terms of scale and resolution (Figure 9-2).

Early Developments

Aerial reconnaissance and photographic interpretation have been tools of the archaeologist for many years, and entire books have been devoted to the chronology of site discoveries from the air (Deuel 1969). Several early findings were based on visual sightings of unusual ground markings by World War I pilots in the Near East. Such discoveries were later followed up with extensive airphoto coverage, and much of the pioneering work in **photoarchaeology** was conducted in Europe, the Mediterranean region, and North Africa (Figure 9-3).

A contribution of special interest to historical geographers was the work of John Bradford, an English archaeologist who was interested in reconstructing the rural landscape of the Romans. He found that aerial photographs revealed the Roman system of dividing conquered territory into squares of 20 × 20 *actus* (710 × 710 m). In many instances, the land subdivisions were still intact. Small-scale vertical airphotos made it possible to see more of the landscape in a single view and from a different perspective, thus revealing patterns of fields in areas where they had been previously unrecognized (Bradford 1957).

Figure 9-1 Oblique view of Indian habitation site (Second Canyon Ruin) in Pima County, Arizona. The village covers an area of about 150 × 300 m; the ruin was first discovered in 1969. Excavation revealed that the site was occupied during two different periods by Hohokam and Salado Indians (circa A.D. 700 to 1000 and A.D. 1250 to 1350). (Courtesy Arizona Department of Transportation and the Arizona State Museum.)

In 1922 the Englishman O.G.S. Crawford demonstrated the utility of patterns of various **ground markings** to delineate probable sites in the United Kingdom. His successes are exemplified by the fact that he discovered more Celtic, Roman, and ''henge'' sites in 1 year than had previously been found during 100 years of ground reconnaissance. This work firmly established aerial techniques for archaeological exploration in England (Crawford 1953).

In southern Peru, mysterious markings in the desert soil had been observed over a period of many years. The markings are scattered over an area of about 16 × 64 km and were formed by the piling up of varnished rocks from the desert floor, which caused the lighter-colored soils underneath to be exposed. From aerial photographs, archaeologists were able to discern patterns from the markings. Rectangles, trapezoids, and centers from which lines radiated were observed, as were giant effigies of birds and spiders. Due to the great size of the figures, their identities could not be determined from the ground. It is believed that the markings were made by Nazca Indians, but the significance of the figures has not been fully explained.

In spite of the foregoing examples of notable successes during the 1920s and 1930s, photoarchaeological techniques were still somewhat limited during this period. This restraint was at least partially due to shortages in trained professional personnel and the reluctant acceptance of a new methodology by individuals and institutions that were in a position to support and use the results of significant discoveries. It is, therefore, not surprising that many early site discoveries were made by amateur archaeologists, pilots, and aerial observers.

North American Explorations

Early aerial explorations in North America include coverage of the Cahokia Indian mounds in western Illinois and both visual reconnaissance and photographic flights by Charles and Anne Lindbergh over the American Southwest and the Yucatán peninsula of Mexico (Figure 9-4). Early flights over tropical areas produced only limited success because of insufficient advance planning and dense, masking ground cover. Nevertheless, a few important sites were located, and the attendant publicity generated induced several foundations and universities to send archaeological expeditions into Mexico and South America. Since 1934, most of the site discoveries in these regions have been made through aerial reconnaissance or photographic interpretation.

The following listing, though quite incomplete, provides *examples of sites* in the United States that have been discovered, delineated, or mapped through remote sensing techniques:

1922—Mapping of Cahokia Indian mounds in Illinois.

1930—Lindbergh flights over pueblo ruins of Chaco Canyon, New Mexico.

1932—Photographic evidence of effigy sites (giant Indian intaglios) on bluffs above the Colorado River near Blythe, California, and delineation of Hohokam Indian irrigation canals in central Arizona.

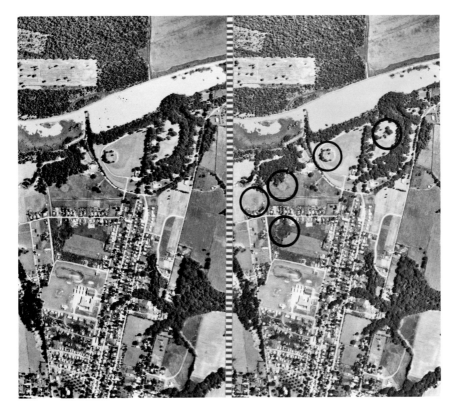

Figure 9-2 Marksville prehistoric Indian site in Avoyelles Parish, Louisiana, as pictured in 1968. Circled are some of the known and probable mound sites dating from the Burial Mound I and II periods (1000 B.C. to A.D. 700). Scale is about 1:20,000. (Courtesy U.S. Department of Agriculture.)

1948—Mapping of Zuni and Hopi Indian pueblos in Arizona and New Mexico.

1953—Detailed site mapping of concentric banks of earthworks (part of mound complex) near Poverty Point, Louisiana.

1959—Discovery of village sites in the Alaskan tundra.

1965—Discovery and delineation of fortified village sites along the Missouri River in South Dakota.

1969—Discovery of prehistoric Indian agricultural plots
1977　(with thermal infrared images) in northern Arizona.
1990

1970—Discovery and mapping of ancient roadways and Indian pueblos in the vicinity of Chaco Canyon, New Mexico.

In spite of technological advances, the applications of aerial archaeology have been limited in the United States until the past few years. This may be a reflection of our greater interest in the present than in our ancestry, but there are probably other reasons also. Among these are (1) the relatively short period of established human settlement, (2) the feeling that many, if not most, sites have already been described, and (3) the attitude that more impressive sites are likely to be found outside United States boundaries (Figures 9-5 and 9-6).

Site Detection Principles

The archaeologist who relies on ground reconnaissance for the detection of archaeological sites is limited to sites that are (1) small enough to be comprehended on the ground from visible remains, (2) accessible within practical and economical limits, (3) still visible in spite of modern-day cultivation and construction, and (4) recognizable, even though the erosional effects of nature may have been operating over a long time.

Fortunately, aerial discovery techniques are not as severely limited by the foregoing conditions. Remains of past landscapes that are too large to be comprehended from the ground, or which may have been incorporated into the present landscape and thus have gone unrecognized, are often detectable on some form of aerial image. And the advantage of a greater range and vertical perspective, as depicted on an aerial view, helps our understanding of the patterns of things that are seen but not understood on the ground (Figures 9-7 and 9-8).

For example, subtle suggestions of buried landscapes are sometimes revealed on conventional aerial photographs by shadow patterns, variations in soil coloration, or differences in the height, density, or color of the plants that grow above the buried features. Moreover, because some remote

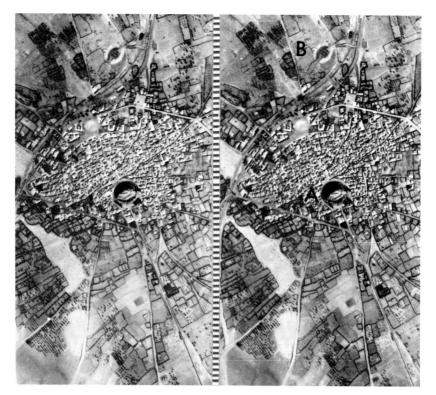

Figure 9-3 Ancient Roman ruins in North Africa. The coliseum at *A* dominates the contemporary mud and brick Arab homes and also serves as a focal point for roads that radiate outward from the village. This is a prime example of the influence of a past culture on a more recent settlement pattern; the present village has retained many of the spatial features of a typical Roman city. Much of the original coliseum structure has remained intact. Just outside the village, at *B,* is evidence of an abandoned amphitheater, with trees growing on the floor of the ancient structure. (Courtesy Henry Svehlak, from the Frank Beatty airphoto collection of North Africa.)

sensors operate outside the visible portion of the electromagnetic spectrum, they provide the archaeologist with another eye into the past. For example, thermal infrared images may denote potential sites if buried or faintly expressed surface features of archaeological significance have measurable radiant temperature differences from their backgrounds.

Shadow Marks

Shadow marks are site indicators produced by the sun's rays falling obliquely on minor terrain configurations or irregularities. Such surface irregularities may have been caused by soil accumulations on the ground, by mounds of earth that resulted from older structures, or by buried remains of archaeological value. Old earthworks, unnoticed banks and ditches, and other characteristics of previous landscapes may sometimes be discernible from shadow marks on aerial photographs—even though such features are virtually invisible to a ground observer (Figure 9-9).

Shadow marks denote variations in surface relief through contrasting tones of shadows, normal photographic tones, and highlighted areas. Because an oblique sun angle is ordinarily required for good shadow marks to be produced, photographic flights should be planned for early morning or late afternoon. The lower the relief, the more oblique the sun angle must be for details to be discerned. Detection of very faint relief may also be aided by the presence of a light snow cover on the terrain.

The direction of the sun's rays and the altitude of the sun at the time of photography are important in that they produce diagnostic shadow marks. When the sun's rays are parallel to a bank or ditch, the tone on the photograph may be perfectly uniform and nonrevealing. For the best contrast between light and shadow, the sun's rays should form an angle as close as possible to 90° with suspected linear relief features. In planning exploratory flights, it may therefore be necessary to photograph an area at different times of day or during different seasons. For oblique photography, the diagnostic effectiveness of shadows is improved when the camera tends to face into a low sun.

Additional problems in obtaining good shadow marks are caused by shadows of obscuring objects, such as hills, trees, and buildings. The masking of minor shadow marks by vegetation may be somewhat alleviated with dormant-season photography (i.e., photography when deciduous

Figure 9-4 Large Indian pueblo village (*A*) near Taos, New Mexico, as photographed in 1962. The remains of an abandoned habitation site can be seen at *B*. Scale is about 1:20,000. (Courtesy U.S. Department of Agriculture.)

plants are leafless). In all instances, clear air and a minimum of atmospheric haze provide the best results (Figure 9-10).

Shadow marks have proven especially valuable for revealing old field systems in England. During the Celtic era, it was the practice to define ownership boundaries with ditches. A ditch was dug along the boundary by the two landowners, and a ridge of earth was thrown up on both sides. Remnants of these double hedgebanks, with a ditch in between, are often revealed by shadow marks.

Large earthworks of low relief may also be revealed from marks. It was this technique that led to the delineation of an extensive Indian village site on a presently cultivated floodplain at Poverty Point, Louisiana. The site contains the remains of six concentric banks and ditches, each one more than 1 km in diameter. Cultivation and erosion had reduced the broad, low banks to a height of about 1.2 m, and they remained unrecognized for many years because of the sheer size of the earthworks (Figure 9-11).

Soil Marks

Soil marks are variations in the natural color, texture, and moisture of the soil; these variations may result from ditches and depressions, excavations, or earth fills. In many instances, the soil profile has been so severely disturbed that the original subsoil has become the present surface soil (Figure 9-12). The contrasts in photographic tones may be striking and quite definitive, even though such marks are rarely apparent to ground observers. Soil marks may permit the archaeologist to distinguish layers of past human occupation and to detect architectural patterns, ditches, canals, or other human alterations of previous landscapes.

The type of subsoil present is an important factor in the production of soil marks. Light-colored subsoils in combination with dark surface soils provide excellent marks; for example, where a chalky subsoil contrasts strongly with a brownish surface soil, outstanding soil marks are rendered. In some regions of western Europe, definitive soil marks are closely associated with the distribution of loess; conversely, limestone subsoils appear to be poor for forming soil marks. The mark-producing capability of several soil types in West Germany has been summarized by Martin (1971).

Soil marks are most easily detected after the first plowing of a field that has gone uncultivated for a long time. Weathering tends to emphasize soil marks in fields that have been plowed and left fallow for several years. Soil marks may reappear annually upon plowing and gradually become less distinct. Eventually, as the surface soil becomes nearly uniform, the marks may disappear. Tractors that plow as deep as 40 to 50 cm can destroy soil marks that have persisted for hundreds, or even thousands, of years. Harrowing, drilling, and ridging practices are particularly destructive to soil marks.

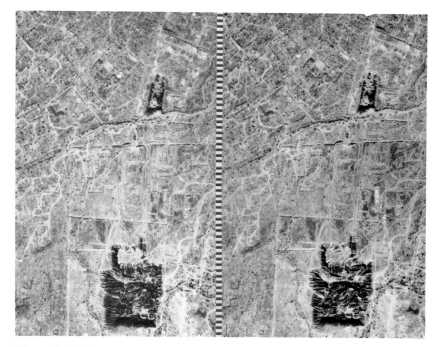

Figure 9-5 Ruins of pre-Inca pyramid, ancient walls, and related structures in Peru. Scale is approximately 1:6,000.

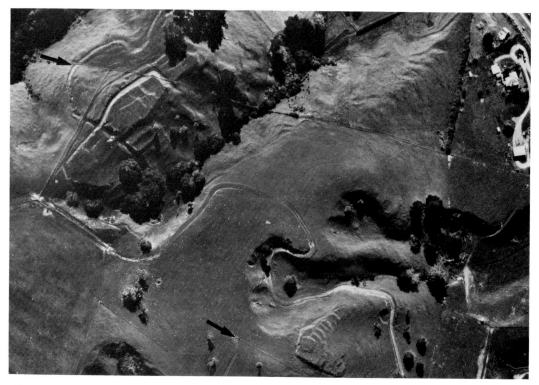

Figure 9-6 Remnants of Maori fortifications can be seen at the upper left and lower center of this vertical view taken near Maketu, New Zealand. Maketu was once the headquarters of the Arawa tribe; the ancestors of this tribe arrived from Hawaiki about A.D. 1350. (Courtesy New Zealand Aerial Mapping, Ltd.)

Figure 9-7 Intaglio Indian effigy, possibly representing mythical figures, on a bluff above the Colorado River near Needles, California. (Bureau of Reclamation, photography by Heilman.)

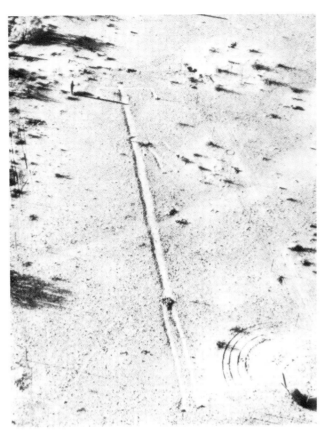

Figure 9-8 Giant effigy figure in the desert, on the present Gila River Indian Reservation, Arizona. Note how the figures are enhanced by their shadow marks. (Courtesy Arizona State Museum.)

The best time for photographing soil marks is soon after plowing and following a heavy rainfall. At such times, distinctive marks may be produced by the differential drying rates of contrasting soil types or soil mixtures. When *moisture variations* constitute a major diagnostic factor, infrared films are preferred. However, where *soil color differences* are significant, panchromatic films may be equally suitable or even superior for delineating soil marks (Figure 9-13).

Although many significant discoveries revealed by soil marks have been in the more arid regions, important discoveries also have occurred in humid areas. In certain sections of England, such as the Fen Basin and chalk regions, soil marks have revealed numerous remains of previous landscapes. Similar markings have also served to outline buried ditches enclosing ancient Roman fields.

Plant or Crop Marks

Cultivated crops and native plants (e.g., grasses) may reveal the existence of buried landscapes by variations in

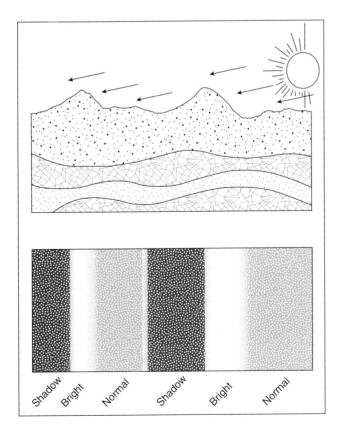

Figure 9-9 (*Left*) Profile of terrain model (*top*) showing the formation of shadow marks as a result of a low sun angle. The lower sketch represents the same terrain as seen in a vertical view.

Figure 9-10 Shadow marks outlining an ancient medieval village site in England.
(Cambridge University Collection, copyright reserved.)

Figure 9-11 Poverty Point, an Indian village habitation site in West Carroll Parish, Louisiana, as pictured in 1969. Dating from the close of the Late Archaic Period, the principal remains consist of a concentric octagonal figure about 1,200 m across and composed of six rows of earth ridges. Site construction and occupation are estimated at circa 800 to 600 B.C. Scale is about 1:20,000. (Courtesy U.S. Department of Agriculture.)

their color, density, or height. These variations, which may indicate differences in plant-root penetration, can result from the remains of such features as ditches, pits, or buried wall fragments. **Plant** or **crop marks** may reappear year after year, even though the causal buried remains may lie well below all cultivation levels, and they may continue to show up long after all traces of soil marks have vanished.

Plant or crop marks may be classed as either *positive* or *negative*. **Positive marks** result where growth is stimulated by filled-in ditches, whereas **negative marks** (inhibited growth) can result from buried foundations and walls. Positive marks are the more common type. They are affected by the width and depth of the original excavation in the subsoil and are most pronounced when the excavation was large and deep. The minimum width of excavation required for the production of a plant mark is perhaps about 1 m. However, a plant or crop mark will seldom be as wide as the ditch or other feature beneath it.

For positive plant marks to be produced, the subsoil must be well drained. Thus, during dry periods, plants growing in deeper soil will be the only ones to flourish. Marks may not be present in areas with loose subsoils, because the roots of crops may extend as far down in undisturbed subsoils as in the loose silt of old excavations. The best subsoils for crop or plant marks are compact gravels, chalk, or silt. Limestone and sandy subsoils, along with loose gravels and

Figure 9-12 Profile of terrain model (*top*) showing the formation of soil marks. The lower sketch represents the same terrain as seen in a vertical view.

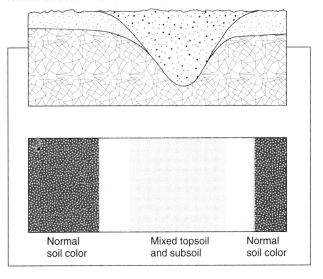

Normal soil color   Mixed topsoil and subsoil   Normal soil color

Figure 9-13 Oblique view of distinctive soil marks in an English field. (Cambridge University Collection, copyright reserved.)

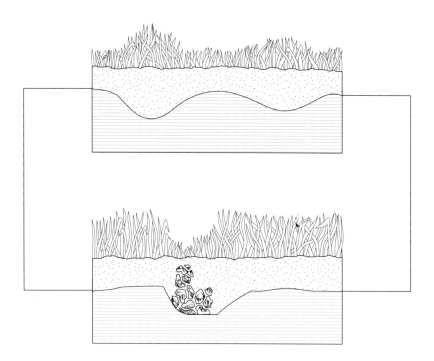

Figure 9-14 Diagrammetric representation of positive crop marks (*top*) and negative crop marks (*bottom*).

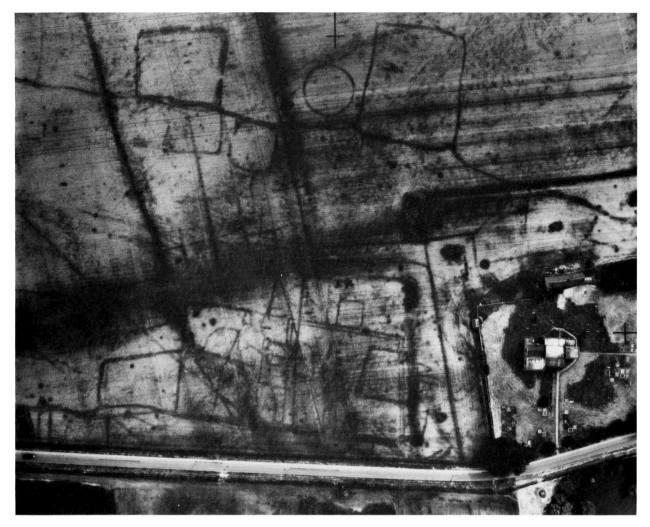

Figure 9-15 Vertical photograph of crop marks in a field in southern England. (Cambridge University Collection, copyright reserved.)

clays, are generally unsuitable. Negative marks are usually independent of the subsoil, because buried foundations, walls, and roads almost always have an adverse effect on the crops growing above them (Figures 9-14 and 9-15).

In the production of crop marks, the type of plant cover is almost as important as the surface soil. Cereal crops are the best medium by which buried remains are revealed, but clover, sugar beets, and grass also give good results. In very dry weather, almost any type of vegetation may produce distinctive tonal signs; crop marks appear gradually, with the contrast and amount of detail steadily increasing. During periods of wet weather, color differences quickly vanish, but variations in plant height and density tend to remain. Also, wet-season plant marks may be visible for a longer period, because crops such as grains ripen more slowly and are harvested later.

Marks that result from plant *density* differences are best recorded on *vertical* photographs, whereas *oblique* views may be superior for detection of faint marks based on plant *color* or *height* differences. The oblique photographs are best obtained in midmorning or midafternoon during the drier summer months, the time most suitable for detection of faint tonal differences and minor plant height variations.

Plant or crop marks have revealed evidence of past Roman landscapes in Great Britain. Numerous Roman military remains such as camps, forts, battlefields, and roads have been discovered by plant marks as registered on aerial photographs.

Site Evaluations

For known historic or prehistoric sites, and particularly important for those that have not been excavated, aerial images can help determine the spatial extent, orientation, and significance of ancient structures (Figures 9-16, 9-17, and 9-18). From evaluations of aerial images it may be possible for archaeologists to draw inferences about past cultures, environmental zones, and the levels of technology that prevailed at a given time and place.

The concept of spatial extent is exemplified by such

Figure 9-16 Fortress of Napoleonic era (circa 1800) at Alderney, a channel island off the Normandy Coast of France. This 1945 view also depicts more recent coastal defenses, since the fortifications were renovated and employed by the German Army during World War II. (Courtesy Alberta Center for Remote Sensing.)

findings as giant Indian pictographs or clusters of earth mounds that were sometimes built as ceremonial burial grounds, effigy sites, or religious temples. Large earthworks of low relief (e.g., the concentric banks of Poverty Point, Louisiana) may be incomprehensible without the perspective afforded by an aerial view.

Where exploratory flights reveal only minor indications of habitation sites, it may be feasible to make inferences about other features or structures that are not discernible but which could have been logically associated with such sites. For instance, the discovery of an ancient Indian ball court (Figure 9-19) might lead an archaeologist directly to an entire buried village.

For those features whose spatial extent is already known, aerial images and precise photogrammetric maps can be valuable in assessment of the significance, orientation, or original purpose of unusual structures. Stonehenge (Figure 9-20) and the Big Horn medicine wheel (Figure 9-21) are somewhat analogous structures of this type; both apparently functioned as early astronomical observatories.

The Big Horn medicine wheel reveals an especially interesting case study that has been documented by Eddy (1974). Constructed around A.D. 1700 at an elevation of nearly 3,000 m, it was first discovered by Europeans in the late 1800s. At that time, the local Indians interviewed were aware that the wheel existed, but none appeared to know its precise location or purpose. Because Indians tended to equate the word *medicine* with magic or the supernatural,

the medicine wheel was long assumed to have served a religious or ceremonial function. It remained for astronomer J. A. Eddy, who worked with detailed photographs and field maps, to explain the orientation, solstitial alignments, and related functions of this primitive outdoor observatory.

Photogrammetric Site Mapping

In many instances, remote sensing procedures are more efficient than traditional methods for estimating excavation costs, site mapping, and making detailed studies of large structures and relationships. *Photogrammetric mapping*, a standard technique of engineers for many years, is now employed in the preparation of **archaeological site plans** (Figures 9-22 and 9-23).

Applications of data derived through photogrammetry are readily apparent. Precisely compiled site maps are invaluable for ruin reconstruction and stabilization. Areas of structural weakness are more easily recognized and the volume of materials to be removed during excavation or restoration can be quickly estimated. The site maps and aerial photographs themselves provide a permanent record of the site, its dimensions, and its spatial characteristics. Finally, when photogrammetric data are computer digitized, auto-

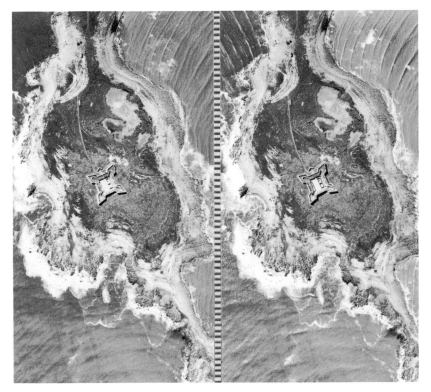

Figure 9-17 Fort Prince of Wales at the mouth of the Churchill River, Hudson Bay, Canada. The massive stone fortification was completed in 1772, some 40 years after construction was initiated. Scale is 1:15,840. (Courtesy Manitoba Center for Remote Sensing.)

mated plotters can quickly reproduce site plans or other graphic information in the form of precision-drafted maps, overlays, or TV monitor displays.

Sequential Photography

The step-by-step excavation or restoration of a historical site can be recorded by means of **sequential photography**. Such photography may be obtained through conventional aerial techniques or through the use of ground-controlled camera platforms such as large **tripods, bipods**, or **captive balloons**.

For smaller sites and "spot" excavations, the ground-based tripod or bipod platform works well. Regular hand cameras or aerial cameras that use 70-mm film can be rigged to provide vertical photographs—either as single exposures or in stereoscopic pairs. When the selected cameras are manually operated, the photographer must have access to the camera platform to advance and change the film and possibly to cock and release the shutter. Therefore, such setups prove most efficient when the height of the camera above the site is less than 10 m.

For larger sites and where the camera must be suspended 10 to 300 m above ground, tethered balloons can be used in conjunction with remotely controlled cameras for sequential photography. The design and suspension of several camera mounts for use with tethered balloons have been outlined by Whittlesey (1970). Sequential exposures obtained at various levels of excavation provide permanent records for the study of previous landscapes at successive periods of occupation. Used in this context, such imagery would effectively serve as "time-lapse" photography of past cultures.

Obtaining Aerial Images

Numerous trials of photographic and nonphotographic sensors have been conducted in attempts to determine their suitability for archaeological exploration. Because the major objective of each investigator tends to differ, however, a listing of rigid specifications cannot be set forth. On the other hand, there are general recommendations that can be made regarding imagery of ancient sites; those made here will refer to conventional photography unless otherwise stated.

As a general rule, vertical photographs are preferred for reconnaissance flights and for the detailed photogrammetric mapping of known sites. Oblique photographs may be specified for detection of plant or crop marks, and they also provide important records during site excavation or restoration (Figure 9-24).

For the detection and evaluation of most archaeolog-

Figure 9-18 Fort Loudoun, a partially restored log palisade on the Little Tennessee River near Vonore, Tennessee, as seen in 1965. Completed in 1757, the fort was built to protect England's claims in the South and was the first British settlement in what is now Tennessee. It was surrendered in 1760 as a result of a siege by Cherokee Indians. Scale is about 1:4,800. (Courtesy Tennessee Valley Authority.)

ical sites, dry seasons of the year are preferred over wet periods, because the loss or retention of moisture by various soils provides more striking tonal contrasts during dry seasons. Density and condition of covering vegetation are additional seasonal factors for consideration. Growing-season photography is required for the detection and evaluation of plant or crop marks, for example.

Reconnaissance flights over humid regions are likely to be more successful when masking deciduous plants are leafless. Also, soil marks are most readily discernible after plowing but before the establishment of an agricultural crop. In summary, the photographic season must be selected on the basis of specific project objectives; there is no single period of the year that is best for all forms of archaeological exploration.

Time of Day

The time of day specification is largely governed by the desired sun angle on the date of photography. For any given latitude and day of the year, the sun's declination can be determined in advance from a solar ephemeris or from special charts available from aerial film manufacturers.

The detection of shadow marks, as noted previously, may require a very low sun angle. When shadow marks are the objective, early morning or late afternoon photography may be required, especially at the lower latitudes. For most vertical reconnaissance photography, however, a high sun angle is desired, for it minimizes shadows and provides maximum illumination of terrain features. Adequate sunlight is especially important for producing correctly balanced color photographs.

Choice of Film

Early archaeological investigations used panchromatic film largely because this was the least expensive and most common type of film available. Panchromatic airphotos, especially those made with a minus-blue filter, are still considered very useful and may be regarded as fairly standard for preliminary reconnaissance surveys.

Since the early 1960s, the use of normal color, color infrared, and black-and-white infrared films has rapidly expanded. For example, Strandberg (1967) conducted a multifilm experiment along the Missouri River in South Dakota, where extensive ground surveys and excavations had been previously completed. Examination of the various photographs revealed all the known archaeological sites, but a normal color photograph showed what turned out to be a wholly unexpected aboriginal Indian village (Figure 9-25). The village was strikingly evident on the color photo because of the film's ability to capture subtle color variations.

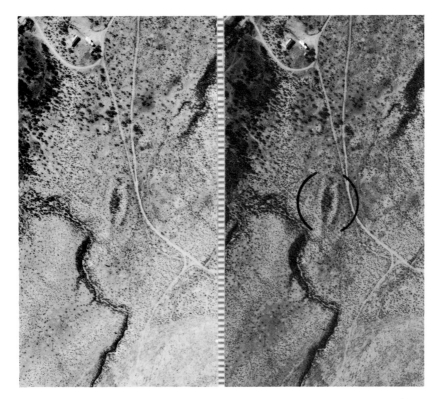

Figure 9-19 Prehistoric Indian ball court (unexcavated) in Pima County, Arizona. The estimated time of use by the Hohokam Indians (Santa Cruz-Scaton phases) was A.D. 800 to 1000. Scale is approximately 1:6,000. (Courtesy Arizona Department of Transportation.)

More recently, color infrared airphotos have revealed a network of previously unknown footpaths in the mountainous Tilaran region of northwestern Costa Rica that were used by local inhabitants more than 1,000 years ago (Sheets and Sever 1988). The paths tend to run straight across hills and valleys, linking such archaeological sites as villages, cemeteries, and sources of building stone. Rigaud and Hersé (1986) report on the superiority of black-and-white infrared photographs for depicting a Gallic farm in France that cannot be detected on simultaneously acquired panchromatic photographs (Figure 9-26).

Photographic Scale

In consideration of photographic scale, we must distinguish between reconnaissance overflights covering large regions and detailed photography of known or excavated sites. In the first instance, scales of 1:10,000 to 1:20,000 have been successfully employed in several countries, though on the basis of research reports, scales in the range of 1:3,000 to 1:10,000 appear to be preferred. The potential of very small scale imagery obtained from earth-orbiting satellites has not been fully evaluated from the site discovery viewpoint, but certain linear features such as ancient roadways have been discerned on such imagery.

For photographing known, exposed, or excavated sites, various investigations have used photographic scales of from 1:500 to 1:5,000. Here, the scale specified is governed by the physical extent of the site and degree of detail required. Extremely large scales are usually limited to sites being excavated or photogrammetrically mapped (e.g., burial mounds, fortifications, small villages, or pueblos).

Thermal Infrared Images

Although images from a number of nonphotographic sensors are currently being evaluated for archaeological applications, thermal infrared images hold considerable promise for use in searching for new sites or for surveying areas surrounding known sites for additional discoveries. Thermal infrared images can be valuable because of their ability to display subtle radiant temperature differences (day and/or night) that may be related to buried or faintly visible surface features, such as walls, or cultural modification of soils and other surficial materials. As an example, Thomas Sever, a NASA archaeologist, has demonstrated the archaeological promise of thermal infrared remote sensing at Chaco Canyon, a famous Anasazi complex in northwestern New Mexico (Figure 9-22), where he found buried walls, prehistoric roadways, and agricultural fields (McAleer 1988).

Interpretation of thermal infrared images by Schaber and Gumerman (1969) and Berlin et al. (1977, 1990) has led to the discovery of several previously unknown, prehistoric Sinagua agricultural sites in northern Arizona. One of the

Figure 9-20 Stonehenge, an ancient megalithic monument on the Salisbury Plain, Wiltshire, England. Solstitial alignments indicate that the circular feature was constructed as an astronomical observatory. (Cambridge University Collection, copyright reserved.)

Figure 9-21 Oblique view of the Big Horn medicine wheel in northern Wyoming. Diameter is about 25 m. Built about A.D. 1700, the rock feature is believed to have served as a primitive astronomical observatory. (Courtesy Roger M. Williams, district ranger, U.S. Forest Service.)

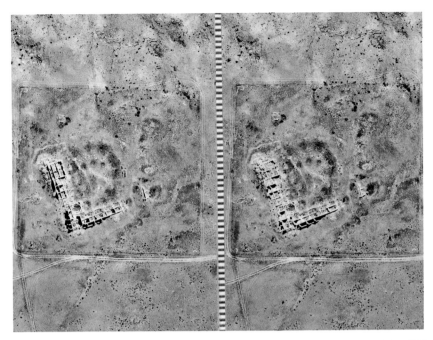

Figure 9-22 Pueblo Pintado, Chaco Canyon, New Mexico, photographed in 1973 at a scale of 1:3,000. The top of the photograph faces east. Several similar Anasazi pueblo sites, dating from about A.D. 900 to 1150, are situated in this part of northwestern New Mexico. (Courtesy Koogle and Pouls Engineering and the Chaco Center, National Park Service and University of New Mexico.)

Figure 9-23 Planimetry and topography of Pueblo Pintado. Contour interval is 0.3 m. (Courtesy Koogle and Pouls Engineering and the Chaco Center, National Park Service and University of New Mexico.)

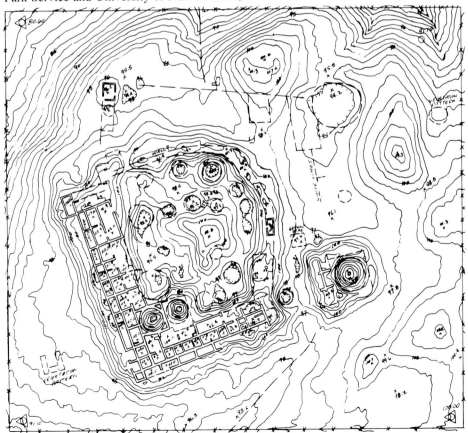

Figure 9-24 Locale of Big Hidatsa Indian village near Stanton, North Dakota, as it appeared in 1967. The circular depressions mark the sites of earth lodges that were located within a stockade. The village was occupied when Lewis and Clark explored the Missouri River (1804–1806) and continued in use until about 1845. (Courtesy North Dakota Highway Department.)

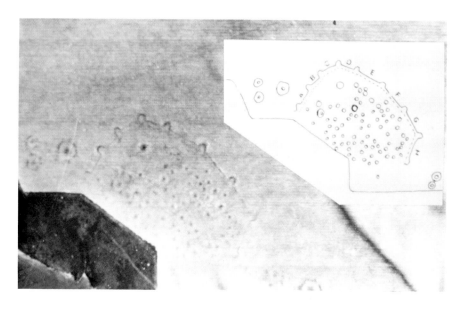

Figure 9-25 Vertical airphoto and sketch map of a fortified Indian village site along the Missouri River, about 35 km south of Pierre, South Dakota (original in color). The solid outer line on the sketch map marks the location of the moat; the dashed inner line represents the location of the palisade. The smaller circles indicate locations of older living structures that were occupied at the time when the village was actively defended. The larger, double circles show locations of more recent Native American earthen lodges. Residual traces of these features can be seen in the photograph. The moat and palisade were probably built about A.D. 1200 to 1400. Scale is approximately 1:10,000. (Courtesy Carl H. Strandberg, Remote Sensing Consultant.)

Figure 9-26 Black-and-white infrared photograph (*top*) of a beet field showing the remains of a Gallic farm in France (circa late first century B.C. to the first century A.D.) that is not visible on a simultaneously acquired panchromatic photograph (*bottom*). The buried walls of the structure are revealed in the infrared photograph because of moisture-enhanced crop marks. (Courtesy M. Hersé, Centre National de la Recherche Scientifique.)

---

field systems is shown in Figure 9-27. It is composed of a series of alternating narrow ridges of black volcanic ash separated by slight troughs or swales of exposed soil. In the image, which was generated in the afternoon, the ridges are represented by light tones indicative of warm radiant temperatures (low thermal inertia), whereas the swales are portrayed in darker tones that are indicative of cooler radiant temperatures (high thermal inertia). The Sinagua collected the volcanic ash into thicker accumulations to act as an effective mulch for dry-farming enhancement. This reconfiguration of the volcanic ash was responsible for the unique temperature signatures that were captured by the thermal infrared scanner. Pollen remains recovered from the soil lying beneath the ash reveal that the plots were used for growing corn (Figure 9-27). Identification of pot shards or fragments of known age indicates that the farming activity occurred between A.D. 1064 and 1250.

Site Prediction

Many archaeological remains are of insufficient size to fall within the resolution of airborne and especially space-borne remote sensors, making it impossible for a *direct site discovery*. However, the location of archaeological sites can often be inferred with remote sensing techniques by defining areas where human activity likely occurred. This *indirect* method of detection is known as **site prediction**.

Various determinant forces led ancient peoples to select particular sites for their hunting camps, villages, forts, and burial grounds. These forces include the following: (1) a reliable source of water, (2) a means of finding game and edible wild plants and/or the capability of growing crops, (3) an acceptable environment (i.e., a favorable climate) in relation to other alternatives known to be available, (4) shelter from the elements and protection from enemies, and (5) land or water routes for transporting essentials to a selected habitation site. It will be recognized that many of these *controlling elements* of human environments can be identified or inferred from the study of remote sensing data (Avery and Lyons 1981). Features from arid and semiarid environments that can often be identified on aerial and orbital photographs and images and that can be used in site prediction procedures include the following: (1) sand blowouts in dune fields, (2) lava flows (seeps and springs often occur along their edges), (3) playas, (4) fossil stream channels, and (5) stands of phreatophytic vegetation (indicating shallow groundwater).

Site predictions have been made in a number of regions utilizing remote sensing data in conjunction with a knowledge of the form, preferences, and habits of the culture being studied and a particular region's natural environment (Lyons and Avery 1977). As an example, interpretation of panchromatic airphotos of central New Mexico revealed a large "peninsula" in a region where some of the earliest American inhabitants hunted large game animals. This feature represented an area where such game might have been easily trapped; this delineation thereby restricted the ground search to a specific area. Ground reconnaissance subsequently revealed several sites that are believed to be the remains of hunting camps dating back 9,000 years or more.

Spaceborne radar systems, such as the Shuttle Imaging Radars (SIR-A and SIR-B), offer tremendous potential for site prediction investigations in arid regions because of their ability to provide subsurface observations through dry surficial materials. For example, the ground-penetration capability of SIR-A and SIR-B revealed a network of previously unknown fossil river systems whose presence is now obscured by eolian sand deposits of Quaternary age in the eastern Sahara in Egypt and Sudan (Figure 9-28). During subsequent ground surveys and excavations at several of these paleorivers, thousands of Stone Age artifacts, such as hand axes and flakes, were discovered. These artifacts date from 150,000 to 500,000 years ago—a time when human groups were widely present in a subhumid riparian environment along the edges of these ancient valleys (McHugh et al. 1988).

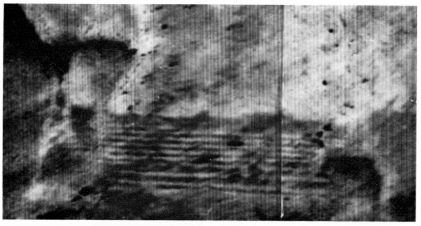

Figure 9-27 Simultaneously acquired thermal infrared image (8–14-μm) (*top*) and panchromatic photographs (0.5–0.7-μm) (*bottom*) of a prehistoric Sinagua agricultural site in north-central Arizona. The field system measures approximately 67 × 265 m. (Berlin et al. 1977.)

Figure 9-28 Shuttle Imaging Radar (SIR-A) image (23.5 cm) of the Sahara in southern Sudan showing buried fossil river systems. Background image is a computer-enhanced Landsat Multispectral Scanner (MSS) band 7 image (0.8–1.1-μm). (Courtesy Gerald G. Schaber, U.S. Geological Survey.)

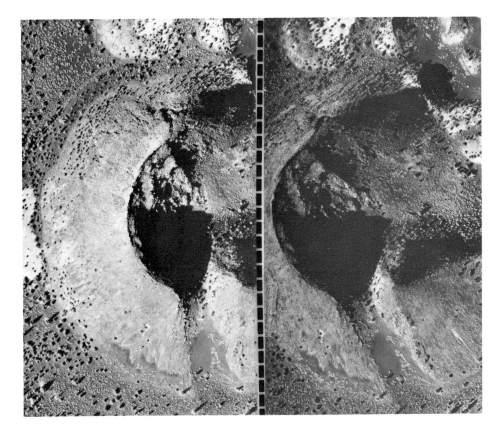

Figure 9-29 Stereogram of a breached cinder cone in north-central Arizona. Scale is about 1:18,000.

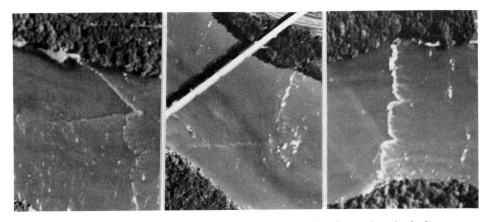

Figure 9-30 Portions of three panchromatic photographs showing archaeological rock structures in the Potomac River near Washington, D.C. (Courtesy Carl H. Strandberg, Remote Sensing Consultant.)

Questions

1. Examine Figure 9-29 with a lens stereoscope, and answer the following questions.

    a. What evidence is there to suggest the presence of an archaeological site?

    b. What do you think the site represents?

    c. If another photo mission was to be conducted using panchromatic film from the same platform altitude, what factor might be taken into account that would enable one to search for additional sites on the processed photographs?

2. Examine Figure 9-30, and answer the following questions.

    a. What do the underwater stone structures represent?

    b. Why were the underwater structures detected by the panchromatic film?

    c. Why would properly filtered black-and-white infrared film (0.7 to 0.9 μm) be a poor choice for detecting these underwater features?

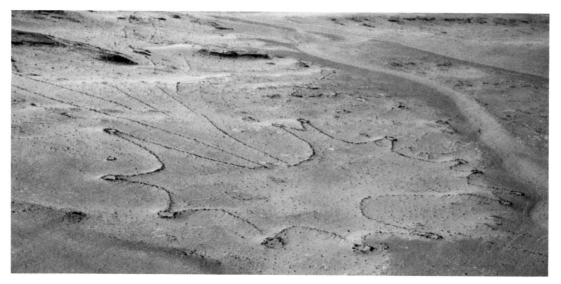

Figure 9-31 Oblique airphoto of a prehistoric MAR-TU "stone-kite" structure in north-central Saudi Arabia. The small circular features are about 1.5 m in diameter. The structural remains of a MAR-TU village are shown in Figure 3-4.

3. Examine Figure 9-31, and answer the following questions.

   a. Given that the "stone-kite" structures were used by the MAR-TU for perhaps four millennia (6000 to 2000 B.C.) when the area received more precipitation (savanna climate) and brush was attached to the low block walls, explain the function of such a structure.

   b. Even though the block walls have a color similar to the surrounding bedrock outcrops and surficial deposits, why is the structure clearly depicted on the airphoto?

Bibliography and Suggested Readings

Asfaw, B., C. Ebinger, D. Harding, T. White, and G. Wolde-Gabriel. 1990. Space-Based Imagery in Paleoanthropological Research: An Ethiopian Example. *National Geographic Research* 6:418–34.

Avery, T. E., and T. R. Lyons. 1981. *Remote Sensing: Aerial and Terrestrial Photography for Archaeologists.* Washington, D.C.: U.S. Government Printing Office.

Berlin, G. L., J. R. Ambler, R. H. Hevly, and G. G. Schaber. 1977. Identification of a Sinagua Agricultural Field by Aerial Thermography, Soil Chemistry, Pollen/Plant Analysis, and Archaeology. *American Antiquity* 42:588–600.

Berlin, G. L., D. E. Salas, and P. R. Geib. 1990. A Prehistoric Sinagua Agricultural Site in the Ashfall Zone of Sunset Crater, Arizona. *Journal of Field Archaeology* 17:1–16.

Bradford, J. 1957. *Ancient Landscapes.* London: G. Bell and Sons.

Crawford, O.G.S. 1953. *Archaeology in the Field.* New York: Praeger.

Deuel, L. 1969. *Flights into Yesterday: The Story of Aerial Archaeology.* New York: St. Martin's Press.

Ebert, J. I. and T. R. Lyons, author/editors. 1983. Archaeology, Anthropology, and Cultural Resources Management. In *Manual of Remote Sensing,* edited by R. N. Colwell, 2d ed., 1233–304. Falls Church, Va: American Society for Photogrammetry and Remote Sensing.

Eddy, J. A. 1974. Astronomical Alignment of the Big Horn Medicine Wheel. *Science* 184:1035–43.

Gumerman, G. J., and T. R. Lyons. 1971. Archeological Methodology and Remote Sensing. *Science* 172:126–32.

Gumerman, G. J., and J. A. Neely. 1972. An Archeological Survey of the Tehucan Valley, Mexico: A Test of Color Infrared Photography. *American Antiquity* 37:520–27.

Limp, W. F. 1989. *The Use of Multispectral Digital Imagery in Archeological Investigations.* Fayetteville: Arkansas Archeological Survey.

Lyons, T. R., and T. E. Avery. 1977. *Remote Sensing: A Handbook for Archaeologists and Cultural Resource Managers.* Washington, D.C.: U.S. Government Printing Office.

Martin, A. M. 1971. Archeological Sites—Soils and Climate. *Photogrammetric Engineering* 37:353–57.

McAleer, N. 1988. Pixel Archaeology. *Discover* August: 72–77.

McHugh, W. P., J. F. McCauley, C. V. Haynes, C. S. Breed, and G. G. Schaber. 1988. Paleorivers and Geoarchaeology in the Southern Egyptian Sahara. *Geoarchaeology* 3:1–40.

Rigaud, P., and Hersé, M. 1986. Télédection Photographique Infrarouge à Partir de Ballons Captifs. *C.R. Academie des Sciences Paris.* 19:1703–08.

Schaber, G. G., and G. J. Gumerman. 1969. Infrared Scanning Imagery: An Archeological Application. *Science* 164:712–13.

Sheets, P., and T. Sever. 1988. High-Tech Wizardry. *Archaeology* November/December: 28–35.

Strandberg, C. H. 1967. Photoarchaeology. *Photogrammetric Engineering* 33:1152–57.

Strandberg, C. H., and R. Tomlinson. 1970. Analysis of Ancient Fish Traps. *Photogrammetric Engineering* 36:865–73.

Whittlesey, J. H. 1970. Tethered Balloon for Archeological Photos. *Photogrammetric Engineering* 36:181–86.

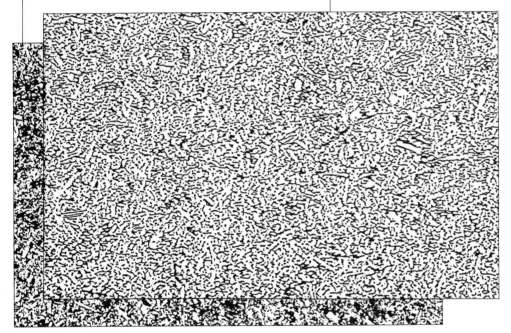

Chapter 10

Agriculture and Soils

The principal objectives of modern agriculture are to cultivate the soil for increasing crop yields while protecting the land from erosion or misuse. One of the truly critical problems facing the world today is that of increasing agricultural production to feed a continually expanding population; it has been estimated that two-thirds of the world's people have diets that are nutritionally deficient. By the year 2000, it is projected that the amount of arable land per capita will be only about 0.2 ha, as compared with about 0.5 ha in 1950.

Increases in crop yields will not solve the basic problem of unrestricted population growth. However, there is a definite need for international cooperation in the planning and administration of world agricultural programs. An important requirement for developing nations is a periodic **agricultural census** and **inventory**. Such information permits the monitoring of current production and the prediction of crop yields, so that serious food shortages or distribution problems can be recognized in time and corrective action can be taken. Also, because some areas of the world have untapped agricultural resources, current inventories can assist in the search for unused but potentially arable lands.

The Role of Remote Sensing

Because of the size, remoteness, diversity, variability, and vulnerability of the world's crops, *accurate information* on production during and following the crop season is limited to the crops grown in the more advanced countries. Even for these, the need for increased accuracy, timeliness, and detail of crop information is continually growing.

One promising way of meeting the current and future needs for crop information is remote sensing. Conventional, medium-scale aerial photographs have been used for decades in some regions for identification of major crops and monitoring of crop-area allotments. The use of more sophisticated techniques (e.g., high-altitude color infrared photography, multispectral scanning, earth-satellite imaging) offers the potential of macroscopic agricultural surveys on a synoptic basis, along with detailed observations of selected croplands.

Remote sensing applied to a global program of assessment of unused but potentially arable land resources can hasten the attainment of a better balance between food requirements and food production for the world. Surveys of agricultural land from space altitudes permit the identifica-

World Food Problems

The earth's agricultural base is vast, encompassing about 4 billion hectares (ha) of arable land, pasture, and rangelands. Annual (as opposed to perennial) crops occupy 95 percent of the cultivated lands. The principal crops grown are summarized in Table 10-1. It can be seen that grains and oilseeds are planted on 80 percent of the total cultivated area. Major types of farming in the United States are shown in Figure 10-1.

TABLE 10-1 Generalized Breakdown of World Crops in Major Producing Countries

Agricultural Crop	Percentage of Total Cultivated Area
Grains, including wheat, rice, corn, millet, sorghum, barley, oats, rye	73
Oilseeds, including soybean, peanut, sunflower	7
Roots and tubers, including potatoes, yams, cassava	5
Pulses, including peas, beans, lentils	4
Fibers, including cotton, flax, jute	4
Fruits and vegetables	3.5
Sugarcane and sugar beets	2
Other crops (e.g., coffee, tea, tobacco)	1.5
Total	100[a]

[a]Total approximate due to rounding of values.
From U.S. National Research Council and U.S. Department of Agriculture.

tion of present use of the land and show population settlement patterns and transportation networks. Such surveys also permit the identification of land characteristics, such as major soil types, drainage, and topographical relief patterns, as a basis for evaluating the best potential use of the land. The following listing includes some of the *more common applications* of remote sensing in agricultural surveys:

Assessment of potentially arable lands.

Classification of agricultural lands.

Identification and distribution of cultivated crops.

Crop areas and predicted crop yields.

Periodic changes in crops or farming patterns.

Detection of crop damage.

Evaluation of plant diseases and plant vigor.

Surveys and mapping of agricultural soils.

Analysis of terrain and soil forming processes.

Evaluation of wind and water erosion.

Figure 10-1 Major types of farming in the conterminous United States. Specialized agricultural production areas are largely an indication of the effects of climate, soil type, and site. Proximity to markets is also a contributing factor in the location of fruit, vegetable, and dairy industries. (Courtesy U.S. Department of Agriculture.)

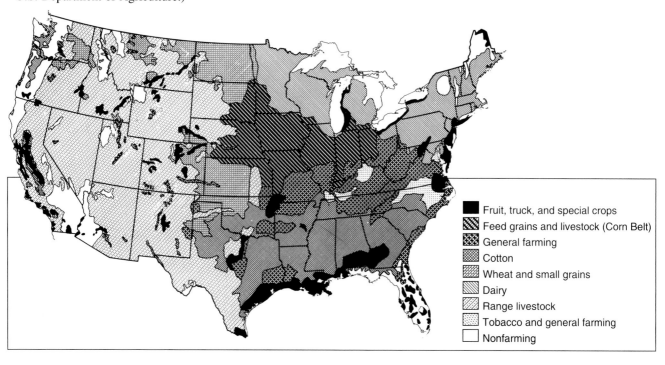

- Fruit, truck, and special crops
- Feed grains and livestock (Corn Belt)
- General farming
- Cotton
- Wheat and small grains
- Dairy
- Range livestock
- Tobacco and general farming
- Nonfarming

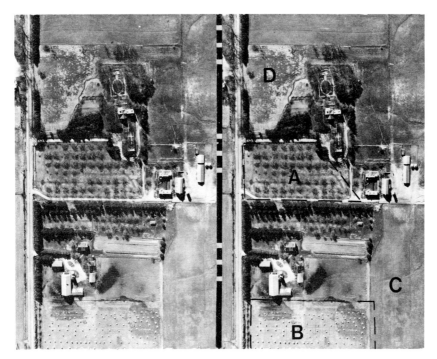

Figure 10-2 Agricultural land classes easily recognized on this Wisconsin farm include (*A*) an orchard; (*B*) shocks of grain; (*C*) field used for annual row crops; and (*D*) pasture. Scale is about 1:7,200.

Land and Crop Classifications

The *first step* in the classification of agricultural land is that of learning to recognize broad categories that are easily separable on conventional photographs (Figure 10-2). In many regions of the world, the following six categories can be identified: seasonal row crops; continuous-cover crops; improved pasturelands; fallow or abandoned fields, including unimproved grazing lands; orchards; and vineyards. It is also usually possible to differentiate between irrigated and dry farming areas.

The *next step* is to compile a listing of all crops that occur in each type within the land area of interest; this can be accomplished with the assistance of local agricultural scientists. Farming practices in a given region are fairly stable, and completely foreign crops are rarely introduced on a large-scale basis. Therefore, crop identification procedures can be developed with little concern that familiar crops will suddenly be replaced by entirely new plant species.

Once a crop listing has been made, it is recommended that the interpreter learn to identify each *on the ground.* Ground identifications should be recorded directly on a set of current aerial photographs for reference purposes. For the nonagriculturalist, it will also be helpful to obtain ground or aerial oblique photography of each crop at periodic intervals during the planting, growing, and harvesting seasons. Some interpreters will then wish to prepare **selective photographic keys** based on the combined vertical and oblique views of each crop.

Crop Calendars

An intimate knowledge of the progressive development of each crop is essential if one is to make reliable identifications on aerial photographs at different times. This type of data can be summarized in the form of a **crop calendar**, that is, a detailed listing of the specific crops grown in an area, along with rotational cycles. The calendar determines which crops must be identified and the times of the year when they are visible and subject to discrimination. It may also be important to document the months during which a field changes from bare soil to crop A and then to crop B or back to bare soil.

Crop calendars can be used in conjunction with existing aerial photographs to determine (1) whether there is a *single date* when certain crops can be distinguished from other crops or (2) what *combinations of dates* during the growing season will provide maximum crop discrimination. In many instances, unique spectral signatures will exist at one stage during the season, so that several individual crops can be separated. If a single date cannot be specified for a given kind of imagery, sequential photo coverage may be required. In this approach, crops are discriminated on the basis of changing patterns (e.g., bare soil to continuous-cover crop to bare soil) at particular dates throughout the year.

Table 10-2 describes monthly growth cycles for three major crops in the midwestern United States. It can be seen that one does not have to discriminate these crops when all

Month	Winter Wheat	Corn	Soybeans
January	Frozen or snow Vegetation brown	Plowed, pasture or corn stalks	Plowed or stubble
February	Frozen or snow Vegetation brown	Plowed, pasture or corn stalks	Plowed or stubble
March	Ground with vegetation brown	Plowed, pasture or corn stalks	Plowed or stubble
April	Becoming green to short green	Plowed, pasture or disked stalks	Plowed or stubble
May	Green, medium to tall	Planting, May 5 to June 20	Planting, May 10 to June 30
June	Yellow, harvest, June 20 to August 5	Short, green	Planting
July	Harvest or stubble	Green, ground covered	Dark ground, short green in rows
August	August 5—stubble or plowed	Green, full height	Green, ground essentially covered
September	Planting, September 10 to November 1	Drying starts, green to light brown	Drying starts, harvest September 10 to November 1
October	Planting, dark soil and short green in drill rows	Harvest, September 25 to December 5	Harvest
November	Dark soil and short green in drill rows	Harvest or corn stubble	Plowed or stubble
December	Frozen or snow Vegetation green to brown	Corn stubble, some fields cut for silo filling	Plowed or stubble

From NASA and U.S. Department of Agriculture.

have green foliage. For example, fields where winter wheat is to be sown are bare in September at a time when corn is tall and either green or light brown. A similar state of affairs will occur at different times in different latitudes. Thus, the differences in planting, maturity, and harvesting dates for various crops can aid in their identification on aerial photographs.

Single-Date Photography

The reliability of crop identifications on **single-date photography** can be improved by observance of the following rules:

1. Schedule aerial coverage during the month when the most important crops are distinctly separable.

2. When a given crop exhibits no unique spectral signature during the growing season, obtain aerial coverage during the time when the fewest other similar crops are present.

3. Use the critical bare-soil months, or optimum crop discrimination periods, to predict the occurrence of the next crop in the rotational cycle.

Where interpretation is based solely on conventional panchromatic photographs, crop identification is extremely difficult (Figure 10-3). During early phases of the growing season, spring-planted crops are almost identical in tone and general appearance. After harvesting begins in late summer, crop differentiation is again difficult, especially for small grains. In a study of crop identification in northern Illinois, the optimum conditions for recognition of crops on panchromatic photographs were found to occur between July 15 and July 30 (Goodman 1959). During this brief period, cul-

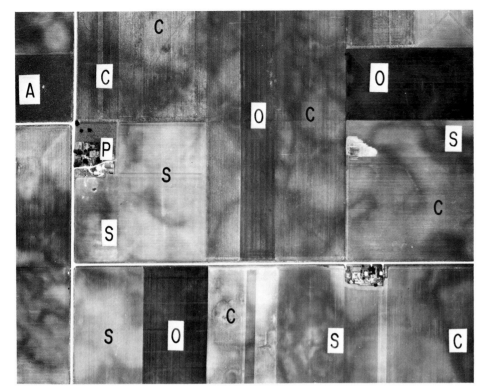

Figure 10-3 Panchromatic photograph taken July 15 in McLean County, Illinois. Crops shown are alfalfa (*A*), corn (*C*), oats (*O*), pasture (*P*), and soybeans (*S*). Scale is about 1:7,200. (Courtesy University of Illinois.)

tivated crops, including alfalfa, wheat, corn, barley, oats, and soybeans, were identified.

Photographic tone and texture are the most important factors to be considered in recognition of individual crops on black-and-white prints. Local variations in farm practices, methods of plowing, and harvesting techniques have proved to be of limited value in the identification process. Tones may range from nearly black, in the case of oats and alfalfa fields, to almost white, as exhibited by stands of ripe wheat. Corn and soybeans are intermediate in tone.

After corn and soybeans have begun to mature, they may be separated on the basis of texture and differences in height. The mottled texture (light and dark spots) seen on many photographs of agricultural land is usually due to differences in soil moisture. The drier portions of fields, that is, higher elevations, tend to show up in light tones on panchromatic prints.

Irrigated or flooded crops such as rice are easily recognized by the presence of low, wavy terraces that show up as irregular lines on panchromatic photographs (Figure 10-4). Pasturelands can be detected by the presence of stock ponds or well-trodden lanes leading to and from barns or across roads that bisect fenced lands. Detailed studies of farmsteads and ranches on large-scale prints may also reveal the presence of dairy barns, horse stables, tent-shaped hog houses, and similar structures for animals (Figure 10-5).

Color and color infrared photographs, along with multispectral scanner images, appear to offer the greatest relia-

bility for single-date crop discrimination. Limited studies employing radar images indicate that crops such as corn and sugar beets are easily separated. However, where several similar crops (e.g., grains) occur in the same area, the accuracy of identification from single-date photography will rarely exceed 55 to 65 percent.

Multidate Photography

Optimum conditions for crop identification are found when photography is available in more than one spectral band and on more than one date during the crop's rotational cycle (i.e., **multidate photography**). Under such circumstances, crop identification accuracy can exceed 80 percent.

As an example, a group of skilled interpreters were asked to classify 125 fields in the Mesa, Arizona, area into seven crop categories (Figure 10-6). Overall accuracy of identification on high-altitude, multidate color infrared photography was 81 percent (Lauer 1971). Multidate photography has also been employed for a semi-operational inventory of wheat, barley, and alfalfa on approximately 200,000 ha of cropland in Maricopa County, Arizona. Accuracy achieved by three skilled interpreters using 1:120,000-scale multidate color photography approached 90 percent for the entire county.

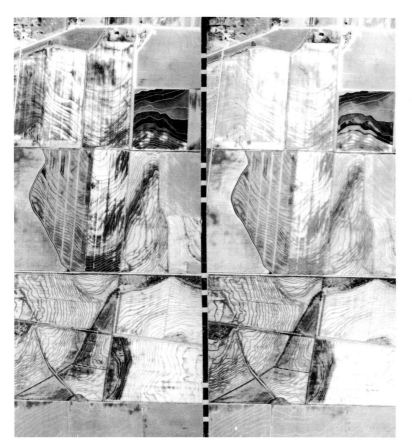

Figure 10-4 Rice cultivation near Pine Bluff, Arkansas. Scale is 1:24,000. (Courtesy U.S. Department of Agriculture.)

Crop Area and Yield Estimates

Information on *crop areas* and *crop yields* and *forecasts* during the growing season are of vital interest to agriculture. Techniques of area measurement are discussed in Chapter 4. It is worthy of mention that one of the principal reasons for the existence of the Agricultural Stabilization and Conservation Service in the U.S. Department of Agriculture is that agency's responsibility for monitoring crop areas.

Maintaining up-to-date checks of each farmer's annual planting allotment would be virtually impossible today without some form of aerial reconnaissance. Accordingly, almost all sizable agricultural areas of the conterminous United States are rephotographed for the Agricultural Stabilization and Conservation Service at scheduled intervals of about 3 to 7 years. Photographic enlargements, rectified to an exact scale by ground checks, are used to determine each owner's field area planted to price-supported crops such as cotton, wheat, peanuts, and tobacco. Although a few citizens have professed to resent this ''spy method'' of crop monitoring, it remains the most efficient technique for the detection of overplanted areas and the maintenance of equitable allotments for a majority of the nation's farmers.

Predictions of crop yield are derived from the product of field area and sample-based estimates of yield per unit area. Random ground samples are best measured as each field nears maturity, so that yield differences due to density, vigor, or disease incidence are taken into account. Where fields appear to be of uniform health and density on aerial photographs, a minimum of ground plots will be required. For certain types of cropland (e.g., orchards and vineyards), it *may* be feasible to estimate yields solely from high-quality color or color infrared photographs.

Orchards and Vineyards

As a rule, orchards are characterized by uniformly spaced rows of trees that give the appearance of a grid pattern (Plate 6). Orchards planted on level terrain (as are pecan and citrus orchards) are usually laid out in squares so that the same spacing exists between rows as between individual trees in the same row. On rolling to hilly terrain, tree rows may follow old cultivation terraces or land contours (as in peach and apple orchards). On the latter type, the sinuous lines of trees, when viewed on small-scale photographs, tend somewhat to resemble fingerprints.

Figure 10-5 Cattle farm in eastern Tennessee. Cattle are visible grazing in improved pasture (*A*); note also the large stock pond (*B*). Scale is about 1:5,000.

Figure 10-6 Agricultural test site showing ground identifications of irrigated crop types near Mesa, Arizona. Most of the crops (depicted in mid-March) were alfalfa and barley; several of the bare fields were planted to cotton about 1 month later. Land area is approximately 42 km². (Courtesy National Aeronautics and Space Administration.)

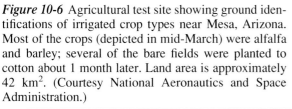

B	= Barley; field green, average height 18 in., inflorescence not emerged
W	= Wheat; field green, average height 16 in., inflorescence not emerged
A_m	= Alfalfa nearing maturity or ready to be cut
A_c	= Alfalfa recently cut and alfalfa pastures
BS_m	= Cultivated fields not yet planted (bare soil relatively moist from recent irrigation)
BS_d	= Cultivated fields not yet planted (bare soil relatively dry)
SB	= Sugar beets
H	= Housing development or other structures
	= Arizona State University experiment farm

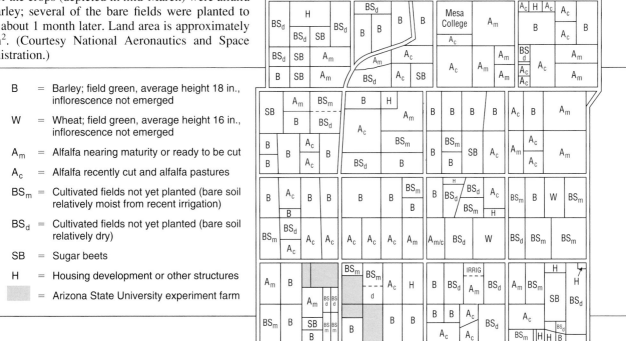

Figure 10-7 Sheep grazing in an improved pasture at South Canterbury, South Island, New Zealand. Scale is about 1:2,800. (Courtesy New Zealand Aerial Mapping, Ltd.)

For most orchards, the key identification characteristics are row spacing, crown size, crown shape, total height, and type of pruning employed (often visible in shadow patterns on large-scale photographs). Whether the plants are deciduous or evergreen is also of assistance for some localities.

Vineyards present a uniformly linear pattern on aerial photographs. Because of the localization of grape cultivation and the wider spacing between individual rows, vineyards are not likely to be confused with corn or other row crops of similar height and texture. Grapes are grown locally over much of the eastern United States, but the largest production area is found in north-central California.

Seasonal Changes in Farm Patterns

Photographic comparisons of the same area taken during the seasons show pronounced differences in the tones of soils, vegetation, and erosional features. **Seasonal contrasts** are particularly significant in midlatitude regions that have humid temperate climates. Where such areas are under intensive cultivation, changes can be detected not only in vegetation and soil moisture but also in the outlines of the fields themselves. These periodic changes, as depicted on panchromatic photographs, may be summarized as follows.

Spring. Field patterns are sharp and distinct due to differences in the state of tillage and crop development. Mottled textures due to differences in soil moisture content are very distinct. High topographic positions, even those only a few centimeters above adjacent lower sites, tend to photograph in light tones. Low topographic positions photograph dark because of large local variations in soil moisture. Recently cultivated fields exhibit very light photo tones, which imply good internal soil drainage.

Summer. Photographs are dominated by dark tones of mature growing crops and heavily foliaged trees. Soil moisture content is normally low; therefore, bare soil tends to photograph light gray. Field patterns are somewhat subdued because of the predominance of green vegetation.

Autumn. Field patterns are relatively distinct because of various stages of crop development and harvesting. Differences in tone resulting from variations in soil moisture content are subdued.

Winter. Photographic tones are drab and dull, with some field patterns indistinct. Mottling due to variations in soil moisture is practically nonexistent, and bare ground tends to photograph in dark tones because soil moisture content is uniformly high. A low angle of illumination causes sharp shadow patterns in wooded areas, producing a distinctive form of flecked texture. Gullies are usually more pronounced in winter than in summer, because the low winter sun casts denser shadows and dormant, leafless vegetation does not mask the surface.

Sequential photographic coverage on an annual cycle provides a basis for detection of complete changes in land use, for example, losses of agricultural lands due to highway construction, strip-mining, urbanization, and other factors. While such losses are not easily arrested, their adverse effect may be reduced by the selective allocation of less productive lands to some of these nonagricultural uses. Earth-satellite imagery holds great promise for making such evaluations periodically.

Large-scale aerial photographs can also be useful for livestock surveys, including counts of animal type, animal distribution, and grazing preferences (Figure 10-7).

Damage Detection

Locating and mapping natural disasters present problems for those in agricultural resources management. Among the major causes of damage to agricultural production are wind and water erosion, floods, fires, insects, and diseases. Knowledge of the extent and degree of damage is especially critical for those who must implement disaster relief. And, later on, crop losses must be assessed and provisions made for crop salvage or restoration.

During floods or extreme droughts, aerial coverage may indicate areas from which people and livestock should be rescued or relocated. In some instances, crop losses may be discernible directly from airphotos, along with damage to farm buildings, fences, corrals, and roads. The weather

accompanying some natural disasters (e.g., hurricanes) may prohibit the acquisition of aerial photographs until the weather has changed.

Plant Disease Detection

Losses due to floods, windstorms, or droughts are minor compared to those resulting from plant diseases. Damage may begin when the crop is planted, continue throughout its growing period, and persist after harvest, when products are transported and placed in storage. Unless diseases are prevented or controlled, there can be no sustained improvement in crop production.

The economic gain from *early disease detection* is that of effectively increasing agricultural productivity. To employ fungicides or herbicides effectively, the agriculturalist must have sufficient advance warning of disease incidence, severity, and rate of spread. This need for immediate diagnosis and control has resulted in numerous research studies aimed at previsual detection of losses in plant vigor.

Most successful experiments in the detection of plant vigor losses have used color infrared photographs (Plate 6). The reason for this appears to be that many diseases result in a decreased reflectance of plant foliage in the near IR portion of the spectrum. The explanation of this phenomenon, as detailed by the National Research Council (1970), is as follows:

The spongy mesophyll tissue of a healthy leaf, which is turgid, distended by water, and full of air spaces, is a very efficient reflector of any radiant energy and therefore of the near-infrared wavelengths. These pass the intervening palisade parenchyma tissue (which absorbs blue and red and reflects green from the visible). When its water relations are disturbed and the plant starts to lose vigor, the mesophyll collapses, and as a result there may be great loss in the reflectance of near-

Figure 10-8 Soil survey maps prepared by Soil Conservation Service techniques (*left*) and by automated classification from multispectral scanner data (*right*). This portion of the Little Cottonwood Canyon (Utah) measures about 1 × 1.5 km. Classifications are (1) rockland and shallow soils, (2) deep gravelly and cobbly soils and rockland, (3) deep gravelly and cobbly soils, (5) deep gravelly and cobbly soils with dark surfaces and clayey subsoils, (6) deep gravelly and cobbly soils in park areas. (Courtesy Earth Information Services, McDonnell-Douglas Corp.)

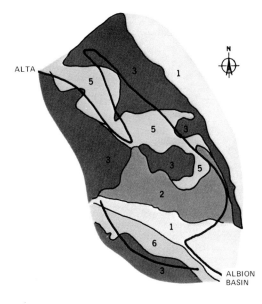
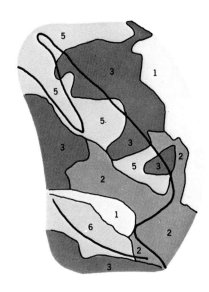

infrared energy from the leaves almost immediately after the damaging agent has struck a plant. Furthermore, this change may occur long before there is any detectable change in reflectance from the visible part of the spectrum, since no change has yet occurred in the quantity or quality of chlorophyll in the palisade parenchyma cells. To detect this change photographically, a film sensitive to these near-infrared wavelengths is used.

Plant diseases that are susceptible to detection through infrared photography include stem rusts of wheat and oats, potato blight, leaf spot of sugar beets, bacterial blight of field beans, and "young tree decline" of citrus trees. As a general rule, aerial photographs have proven most successful when scales were 1:8,000 and larger.

Soil Surveys

Soil surveys constitute essential information for sustained agricultural production. For all practical purposes, soil characteristics determine the type of crop that can be grown and the production potential of that crop. As outlined by the National Research Council (1970):

Information on soils is particularly important in an area in which new land is being brought into production. In many areas, pressing agricultural development has taken place without the essential data on soils needed for assessing the potential for successful development. There are many examples of failures in draining or irrigating land not suitable for drainage and irrigation. In other situations, land suitable for forest was cleared, put into agricultural production, and then found to be too erodible for crop production.

Until recently, soil surveyors have relied on conventional panchromatic photographs, in conjunction with extensive fieldwork, to delineate soil boundaries. However, more promising and reliable results are now being obtained with color photography and multispectral scanner imagery (Figure 10-8). Photo scales of 1:6,000 or larger are usually preferred, and aerial flights are ideally scheduled soon after agricultural fields have been plowed.

Soil-Forming Processes

Terrain elements are the features, attributes, and materials that make up a landscape. The more important factors to be considered in the evolution of landscapes and associated soils are *topography, drainage patterns, local erosion, natural vegetation,* and *the works of humans.* Topography is the result of the interaction of erosional and depositional agents, the nature of the rocks and soils, the structure of the earth's crust, and the climatic regime. The topographic surface is, in effect, a synthesis of all environmental elements into a single expression. As such, it plays an important role

in soil surveys, because it provides a key for deducing the soil-forming processes at work in a given region (Figure 10-9).

A number of attempts have been made to classify drainage patterns into specific regional groupings. When this can be done, much can be inferred with regard to soil type, geologic structure, amount and intensity of local precipitation, and land tenure history. However, stream patterns are almost infinitely variable, and the various types often grade into one another so that no single pattern appears to predominate. In some instances, a large drainage system may display several subtypes of drainage simultaneously. For example, the gross drainage pattern of a region may be dendritic, while associated lesser stream patterns may be pinnate. This situation is quite common in areas of deep loess deposits.

Wind and Water Erosion

Features produced by **wind and water erosion** are important aids in photo interpretation because they are diagnostic of surface soil textures, soil profiles, and soil moisture. Specific implications of each type are discussed in the following paragraphs.

Wind Erosion

Evidences of wind erosion include blowouts, which are smoothly rounded and irregularly shaped depressions; sand streaks, which are light-toned but poorly defined parallel streaks; and sand blotches, which are light-toned and poorly defined patches. Evaluation of such features depends on a knowledge of prevailing wind direction, wind velocity, and the general climatic regime. Climate is important because it provides some indication of probable soil moisture. Any surface unprotected by vegetation and not continuously moist may be eroded by the wind. Both local and regional topographic configurations should be kept in mind during the evaluation of eolian action, because mountains, hills, or other features may channel air movements in such a way that erosion is severe in one locality and insignificant in another.

Plowed fields, beaches, and floodplains are examples of surfaces especially susceptible to wind erosion (Figure 10-10). In general, the finer the grain size, the greater the distance surface material is transported. As a result, a blowout with evidence of immediate deposition downwind implies relatively coarse-grained material, whereas a blowout without such evidence implies fine-grained material.

Many small erosional forms resulting from wind action are difficult to identify on airphotos. As a rule only the larger blowouts are readily picked out. Evidence of deposition is more easily detected, because resulting dunes or sheets present distinctive shapes or light-toned streaks and blotches. These are of considerable significance in regional land use studies. In any given locality, wind-deposited materials tend to be of uniform size, resulting in homogeneous soils. This, in turn, implies that agricultural conditions in any one locality will be approximately uniform, provided slope, vegetation, and moisture conditions are similar.

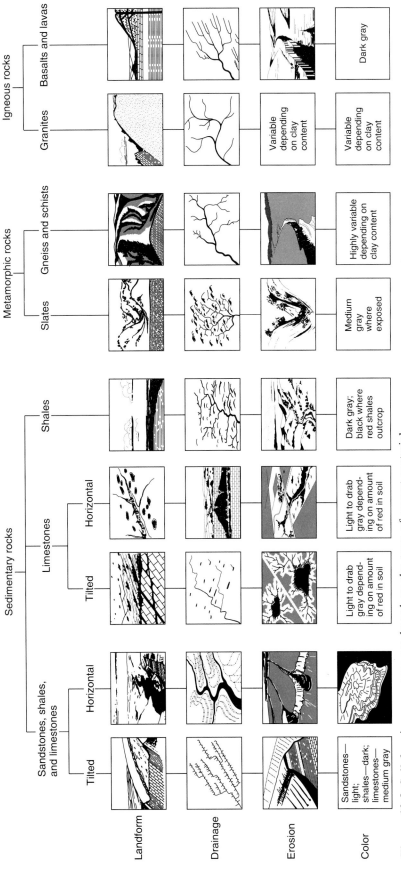

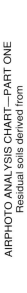

Figure 10-9 Soil-forming processes are based on the type of parent material, landform, drainage pattern, and local erosion (next three pages). (Courtesy Purdue University and U.S. Department of Commerce.)

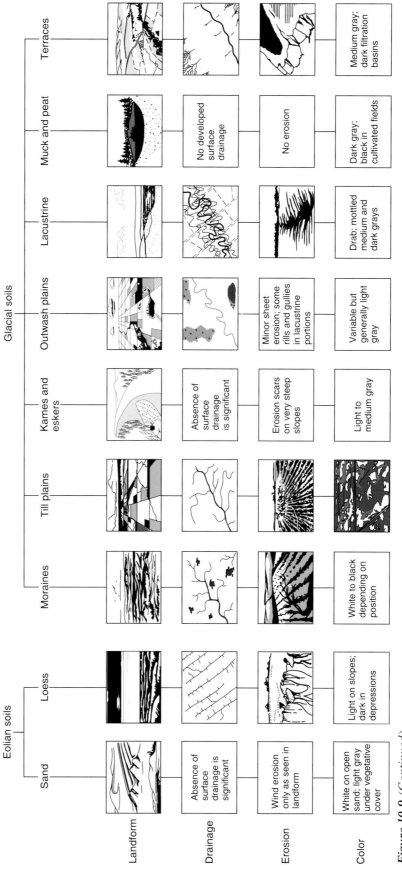

Figure 10-9 (Continued).

AIRPHOTO ANALYSIS CHART—PART THREE

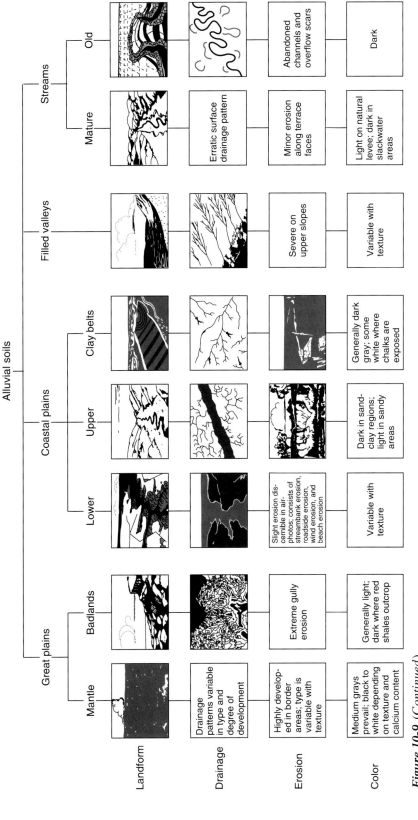

Figure 10-9 (Continued).

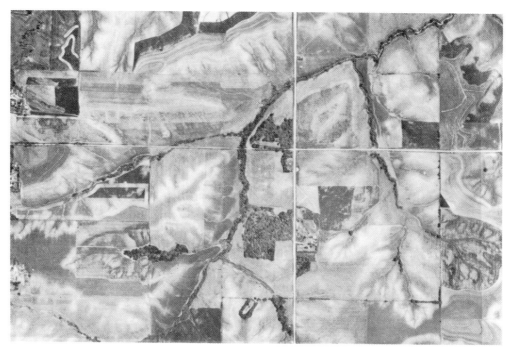

Figure 10-10 Cultivated portion of a dissected loess plain in Harrison County, Iowa. The fine-grained eolian soils, deposited from glacial outwash areas to the north, are easily eroded by wind and water. Corn is the leading crop in this rolling plains area, with livestock providing the bulk of the farm income. Scale is about 1:20,000. (Courtesy U.S. Department of Agriculture.)

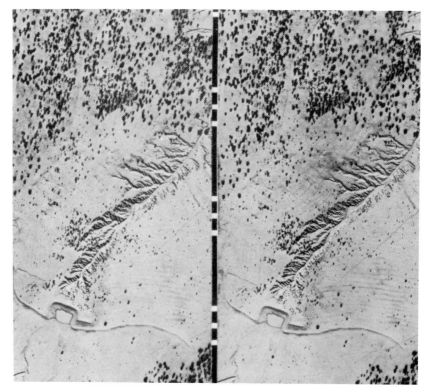

Figure 10-11 Dendritic drainage pattern formed in soft sediment near the Rio Grande in New Mexico. Scale is about 1:6,000. (Courtesy Abrams Aerial Survey Corp.)

Figure 10-12 French longlot patterns in Assumption Parish, Louisiana. This pattern, found in several European countries, was brought over by early colonists, who depended on river transportation; each landowner thus had river frontage. Roads and dwellings are concentrated on artificial levees on either side of the river. The principal crops grown here are sugarcane, cotton, and rice. (Courtesy U.S. Department of Agriculture.)

Water Erosion

Moving water is the major active agent in the development of the earth's surface configuration. Despite its awesome power in the form of floods and tidal waves, moving water is delicately responsive to variations in environment, and modest changes in the material being eroded or the climatic regime can profoundly modify the surface expressions produced. Therefore, the landscape patterns produced through the action of moving water are of great importance to the photo interpreter (Figure 10-11). In addition, the interpreter should have a basic knowledge of the interrelations between climate, surface materials, surface configuration, and vegetation (Figure 10-12). The relative importance of various factors influencing runoff varies according to specific environmental conditions that occur in a given area. Surface runoff is governed by the following general considerations:

1. The amount and intensity of rainfall determine the degree of runoff. A heavy rainfall of short duration may produce more runoff than the same amount over a longer period of time.

2. The amount of runoff is dependent upon the moisture in the soil before rainfall. A given rainfall on wet soil will produce more runoff than the same rainfall on dry soil. A proportion of the incident water will be stored by the dry soil, whereas the wet soil has less available storage capacity.

3. A noncohesive soil is eroded more readily than a cohesive soil.

4. The greater the permeability of a soil, the less the surface runoff.

5. In general, the greater the density of vegetation, the less the runoff for a given quantity of incident water.

6. The steeper the slope, the greater the surface runoff.

Figure 10-13 Panchromatic airphoto of a semiarid region in western Texas.

Figure 10-14 Panchromatic airphoto showing late summer patterns of cropland near Sturgis, South Dakota.

Questions

1. Explain the cause of the dark patches in the otherwise lighter-toned surfaces in Figure 10-3.
2. Explain the cause of the stock pond's tonal signature in the minus-blue stereogram shown in Figure 10-5. Could the pond have been better detected with another type of film? Explain.
3. Explain the tonal disparity in Figure 10-13.
4. Examine the cropland patterns in Figure 10-14, and answer the following questions:
   a. What do the light tones indicate?
   b. What do the darker tones indicate?
   c. What do the straight field patterns tell you about the local topography?

Bibliography and Suggested Readings

Bauer, M. E. 1975. The Role of Remote Sensing in Determining the Distribution and Yield of Crops. *Advances in Agronomy* 27:271–304.

Berg, A. ed. 1981. *Application of Remote Sensing to Agricultural Production Forecasting.* Rotterdam: A. A. Balkema.

Condit, H. R. 1970. The Spectral Reflectance of American Soils. *Photogrammetric Engineering* 36:955–66.

Godby, E. A., and J. Otterman. 1978. *The Contribution of Space Observations to Global Food Information Systems.* Oxford: Pergamon Press.

Goodman, M. S. 1959. A Technique for the Identificaton of Farm Crops on Aerial Photographs. *Photogrammetric Engineering* 25:131–37.

Hay, C. M. 1974. Agricultural Techniques with Orbital and High-Altitude Imagery. *Photogrammetric Engineering* 40:1283–93.

Lauer, D. T. 1971. Testing Multiband and Multidate Photography for Crop Identifications. In *Proceedings International Workshop on Earth Resource Survey Systems,* 33–45. Washington, D.C.: U.S. Government Printing Office.

Meyer, M. P., and L. Calpouzos. 1968. Detection of Crop Diseases. *Photogrammetric Engineering* 34:554–57.

Myers, V. I., and others. 1983. Crops and Soils. In *Manual of Remote Sensing,* edited by R. N. Colwell, 2d ed., 1715–813. Falls Church, Va.: American Society for Photogrammetry and Remote Sensing.

National Research Council. 1970. *Remote Sensing with Special Reference to Agriculture and Forestry.* Washington, D.C.: National Academy of Sciences.

Paine, D. P. 1981. *Aerial Photography and Image Interpretation for Resource Management.* New York: John Wiley & Sons.

Parry, J. T., W. R. Cowan, and J. A. Heiginbottom. 1969. Soil Studies Using Aerial Photographs. *Photogrammetric Engineering* 35:44–56.

Philipson, W. R., and T. Lang. 1982. An Airphoto Key for Major Tropical Crops. *Photogrammetric Engineering* 48:223–33.

Poulton, C. E., and others. 1983. Range Resources: Inventory, Evaluation and Monitoring. In *Manual of Remote Sensing,* edited by R. N. Colwell, 2d ed., 1427–78. Falls Church, Va.: American Society for Photogrammetry and Remote Sensing.

Soil Conservation Service. 1966. *Aerial Photo-Interpretation in Classifying and Mapping Soils.* Agricultural Handbook 294. Washington, D.C.: U.S. Government Printing Office.

Steiner, D. 1970. Time Dimension for Crop Surveys from Space. *Photogrammetric Engineering* 36:187–94.

White, L. P. 1977. *Aerial Photography and Remote Sensing for Soil Surveys.* Oxford: Clarendon Press.

Chapter 11

Forestry Applications

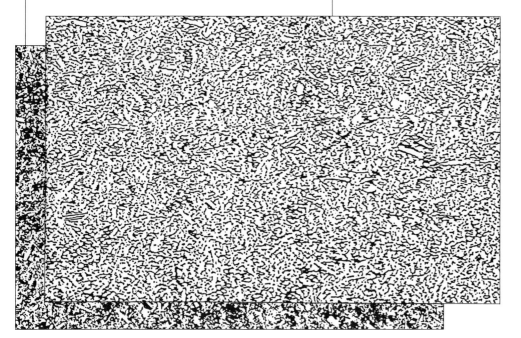

(Plate 3). In addition to these applications, aerial photographs have proved valuable in range and wildlife habitat management, in outdoor recreation surveys, and in estimations of the volumes of standing trees.

In this chapter, emphasis is placed on the recognition and classification of vegetative types, identification of plant species on large-scale photographs, forest inventory techniques, and detection of plant vigor. Although photo interpretation can make the land manager's job easier, there are limitations that must be recognized. Accurate measurements of such items as tree diameter or quantity of forage are possible only on the ground. Aerial photographs are therefore used to complement, improve, or reduce fieldwork rather than take its place.

Vegetation Occurrence and Distribution

The *occurrence* and *distribution* of native vegetative cover in a given locality are governed by such elements as (1) annual or seasonal rainfall, (2) latitude, (3) elevation above sea level, (4) length of the growing season, (5) solar radiation and temperature regimes, (6) soil type and drainage conditions, (7) topographic aspect and slope, (8) prevailing winds, (9) salt spray, and (10) air pollutants. Within any given climatic zone, the *distribution* of available moisture can be as critical as the total amount. Temperature *extremes* are also controlling factors, because such extremes influence evaporation-transpiration ratios in various plant communities.

In some parts of temperate North America, notably the U.S. Southwest, land elevation and precipitation are the two principal factors that determine the distribution of vegetative types. The controlling nature of these two elements in Arizona, for example, is illustrated in Table 11-1. This stratification is quite general, and local variations occur because of differences in soils, slope, and aspect.

A sound knowledge of plant ecology and the factors controlling the natural evolution of plant communities are of inestimable value to interpreters of vegetative features. Armed with such background information, one may often *predict* the kinds of native vegetation that will be encountered under specified environmental conditions.

Overview of Forestry Applications

Foresters and range managers use aerial photographs for preparing forest-type maps, locating access roads and property boundaries, determining bearings and distances, and measuring areas. Skilled interpreters may also be adept at recognizing individual plant species and at appraising fire, insect, or disease damage by means of special photography

Vegetation Zone or Cover Type	Range of Land Elevation (Meters)	Range of Annual Precipitation (Millimeters)
Sonoran desert	450–1,200	75–380
Chaparral	1,050–1,500	300–430
Pinyon and juniper	1,350–2,250	380–480
Ponderosa pine	1,800–2,550	500–660
Aspen and Douglas-fir	2,400–2,850	580–740
Spruce and fir	2,550–3,450	680–890
Timberline	3,450–3,600	700+

Classifying Vegetation

The *simplest* classification method is one that merely discriminates *vegetated* from *nonvegetated* lands, followed by a subdivision of plant associations into *productive* or *nonproductive* sites. As an alternative, vegetated areas might be classed as one of the following basic **ecological formations**: desert scrub, grassland, chaparral, woodland, forest, or tundra. Such primary stratifications can sometimes be made from earth-satellite images.

Major problems one encounters in devising any rational classification system are (1) defining vegetative types so that the classes are mutually exclusive and (2) making allowance for the handling of transition zones, that is, areas where one plant community gradually changes to a different cover type. Wherever possible, it is desirable to adopt a standardized, hierarchical classification system that will be applicable across diverse geographic and political boundaries (see Chapter 8).

As an example of one type of classification approach, the following system has been proposed by the Food and Agriculture Organization of the United Nations (Lanly 1973). It is designed for an area classification of existing land use that could be employed for varied forest inventory projects, particularly those in *tropical countries*. The *first step* is the separation of land and water areas, *followed* by a breakdown of the total land area into these categories:

Forest Area

1. Natural forests

    a. Broad-leaved, excluding mangroves
    b. Coniferous
    c. Mixed broad-leaved and coniferous
    d. Pure bamboo
    e. Mangrove
    f. Coastal and riverine palms
    g. Temporarily unstocked

2. Planted forests (items *a* through *g* as applicable)

Other Wooded Area

1. Savanna: open woodlands
2. Heath: stunted and scrub forest
3. Trees in lines: windbreaks and shelterbelts
4. Other areas

Nonforest Area

1. Agricultural land

    a. Crops and improved pastures
    b. Plantations

2. Other lands

    a. Barren
    b. Natural rangelands and grasslands
    c. Swamp
    d. Heath, tundra
    e. Urban, industrial, and communication
    f. Other areas

As with any classification system, the foregoing types must be clearly defined in rigorous terms. Otherwise, the most elemental discriminations (e.g., what *is* forestland?) can result in inconsistent image interpretations.

Identifying Cover Types

As outlined earlier, interpreters of vegetation should be *well versed* in plant ecology and the various factors that influence the distribution of native trees, shrubs, forbs, and grasses. *Field experience* in the region of interest is also a prime requisite, because many cover types must be deduced or inferred from associated factors instead of being recognized directly from their photographic images.

The *inferential approach* to **cover-type identifications** becomes more and more important as image scales and resolution qualities are reduced. Range managers may rely exclusively on this technique where they must evaluate the grazing potential for lands obscured by dense forest canopies.

The degree to which cover types and plant species can be recognized depends on the quality, scale, and season of photography, the type of film used, and the interpreter's background and ability. The shape, texture, and tone or color of plant foliage as seen on vertical photographs can also be influenced by stand age or topographic site. Furthermore, such images may be distorted by time of day, sun angle, atmospheric haze, clouds, or inconsistent processing of transparencies and prints. In spite of insistence on rigid specifications, it is often impossible to obtain uniform photographs of extensive landholdings. Nevertheless, experienced interpreters *can* reliably distinguish cover types in diverse vegetative regions when photographic flights are carefully planned to minimize the foregoing limitations.

The first step in cover-type recognition is to determine which types should and should not be expected in a given locality. It will also be helpful for the interpreter to become familiar with the most common plant and environmental associations of those types most likely to be found. Much of this kind of information can be derived from generalized cover-type maps (Figure 11-1 and Table 11-2) and by ground or aircraft checks of the project area in advance of photo interpretation. And for limited regions, vegetative photo interpretation keys will be available.

The chief diagnostic features the interpreter uses in recognizing vegetative cover types are photographic texture (smoothness or coarseness of images), tonal contrast or color, relative sizes of crown images at a given photo scale, and topographic location or site. Most of these characteristics constitute rather weak clues when observed singly, but together they may comprise the final link in the chain of "identification by elimination." Several important cover types occurring in the United States and Canada are illustrated in Figures 11-2 through 11-7.

Recognizing Individual Species

The *recognition* of an *individual species* on aerial photographs is most easily accomplished when that species occurs naturally in pure, even-aged stands. Under such circumstances, the cover type and the plant species are synonymous. Therefore, reliable delineations *may* be made on medium-scale photographs. As a rule, however, individual plants can be identified only on large-scale photographs. The listing in Table 11-3, based on a synopsis of several research reports, illustrates the relationship between photo scales and expected levels of plant recognition.

Figure 11-1 Forest regions of the United States, excluding Hawaii.

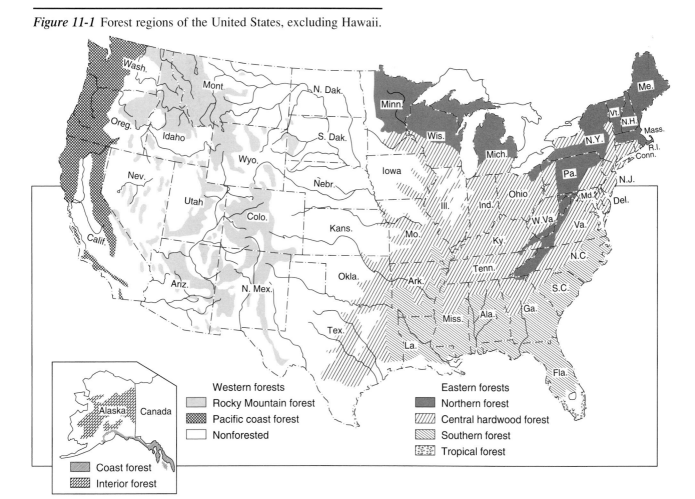

TABLE 11-2 Principal Trees of the Forest Regions of the United States[a]

Rocky Mountain Forest

Northern Portion (Northern Idaho and Western Montana)
Lodgepole pine
Douglas-fir
Western larch
Engelmann spruce
Ponderosa pine
Western white pine
Western redcedar
Grand and alpine firs
Western and mountain hemlocks
Whitebark pine
Balsam poplar

Eastern Oregon, Central Idaho, and Eastern Washington
Ponderosa pine
Douglas-fir
Lodgepole pine
Western larch
Engelmann spruce
Western redcedar
Western hemlock
White, grand, and alpine firs
Western white pine
Oaks and junipers (in Oregon)

Central Montana, Wyoming, and South Dakota
Lodgepole pine
Douglas-fir
Ponderosa pine
Engelmann spruce
Alpine fir
Limber pine
Aspen and cottonwoods
Rocky Mountain juniper
White spruce

Central Portion (Colorado, Utah, and Nevada)
Lodgepole pine
Engelmann and blue spruces
Alpine and white firs
Douglas-fir
Ponderosa pine
Aspen and cottonwoods
Pinyons
Rocky Mountain and Utah junipers
Bristlecone and limber pines
Mountain-mahogany

Southern Portion (New Mexico and Arizona)
Ponderosa pine
Douglas-fir
White, alpine, and corkbark firs
Engelmann and blue spruces
Pinyons
One-seed, alligator, and Rocky Mountain junipers
Aspen and cottonwoods
Limber, Mexican white, and Arizona pines
Oaks, walnut, sycamore, alder, boxelder
Arizona cypress

Pacific Coast Forest

Northern Portion (Western Washington and Western Oregon)
Douglas-fir
Western hemlock
Grand, noble, and Pacific silver firs
Western redcedar
Sitka and Engelmann spruces
Western white pine
Port Orford cedar and Alaska cedar
Western and alpine larches
Lodgepole pine
Mountain hemlock
Oaks, ashes, maples, birches, alders, cottonwoods, madrone

Southern Portion (California)
Ponderosa and Jeffrey pines
Sugar pine
Redwood and giant sequoia
White, red, grand, and Shasta red firs
California incense-cedar
Douglas-fir
Lodgepole pine
Knobcone and Digger pines
Bigcone spruce
Monterey and Gowen cypresses
Sierra and California junipers
Singleleaf pinyon
Oaks, buckeye, California laurel, alder, madrone

Southern Forest

Pine Lands
Shortleaf, loblolly, longleaf, slash, and sand pines
Southern red, black, post, laurel, cherrybark, and willow oaks
Sweetgum
Winged, American, and cedar elms
Black, red, sand, and pignut hickories
Eastern and southern redcedars
Basswoods

Alluvial Bottoms and Swamps
Sweetgum and tupelos
Water, laurel, live, overcup, Texas, and swamp white oaks
Southern cypress
Pecan, water and swamp hickories
Beech
River birch
Ashes

Alluvial Bottoms and Swamps
Red and silver maples
Cottonwoods and willows
Sycamore
Hackberry
Honeylocust
Holly
Redbay and sweetbay
Southern magnolia
Pond and spruce pines
Atlantic white-cedar

Central Hardwood Forest

Northern Portion
White, black, northern red, scarlet, bur, chestnut, and chinquapin oaks
Shagbark, mockernut, pignut, and butternut hickories
White, blue, green, and red ashes
American, rock, and slippery elms
Red, sugar, and silver maples
Beech
Pitch, shortleaf, and Virginia pines
Yellow-poplar
Sycamore
Chestnut
Black walnut
Cottonwoods
Hackberry
Black cherry
Basswoods
Ohio buckeye
Eastern redcedar

Southern Portion
White, post, southern red, blackjack, Shumard, chestnut, swamp chestnut, and pin oaks
Sweetgum and tupelos
Mackernut, pignut, southern shagbark, and shellbark hickories
Shortleaf and Virginia ("scrub") pines
White, blue, and red ashes
Yellow-poplar
Black locust
Elms
Sycamore
Black walnut
Silver and red maples
Beech
Dogwood
Persimmon
Cottonwoods and willows
Eastern redcedar
Osage-orange

Texas Portion
Post, southern red, and blackjack oaks
Eastern redcedar, Ashe juniper

Florida and Texas Forest—Tropical

Mangrove, false mangrove
Royal and thatch palms; palmettos
Florida yew
Wild figs
Seagrapes ("pigeon plum")
Blolly
Bahama lysiloma ("wild tamarind")
Wild-dilly
Gumbo-limbo
Poisontree
Inkwood
Button-mangrove
False-mastic ("wild olive")
Fishpoison-tree ("Jamaica dogwood")

Northern Forest

Northern Portion
Red, black, and white spruces
Balsam fir
Eastern white, red ("Norway"), jack, and pitch pines
Hemlock
Sugar and red maples
Beech
Northern red, white, black, and scarlet oaks
Yellow, paper, sweet, and gray birches
Quaking and bigtooth aspens
Basswoods
Black cherry
American, rock, and slippery elms
White and black ashes
Shagbark and pignut hickories
Butternut
Northern white-cedar
Tamarack

Southern Portion (Appalachian Region)
White, northern red, chestnut, black, and scarlet oaks
Chestnut
Hemlock
Eastern white, shortleaf, pitch, and Virginia ("scrub") pines
Sweet, yellow, and river birches
Basswood
Sugar and red maples
Beech
Red spruce
Fraser fir
Yellow-poplar
Cucumber magnolia
Black walnut and butternut
Black cherry
Pignut, mockernut, and red hickories
Black locust
Tupelos ("black gums")
Buckeye

Alaska—Forest

Coast Forest
Western hemlock (important)
Sitka spruce (important)
Western redcedar
Alaska cedar
Mountain hemlock
Lodgepole pine
Black cottonwood
Red and Sitka alders
Willows

Interior Forest
White (important) and black spruces
Alaska paper (important) and Kenai birches
Black cottonwood
Balsam poplar
Aspen
Willows
Tamarack

[a]*The order indicates the relative importance or abundance of the trees.*

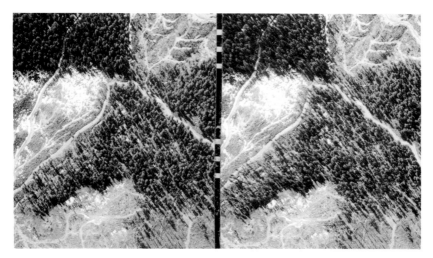

Figure 11-2 Panchromatic stereogram of a recently logged stand of Douglas-fir in Lewis County, Washington. Scale is about 1:6,000. (Courtesy Northern Pacific Railroad.)

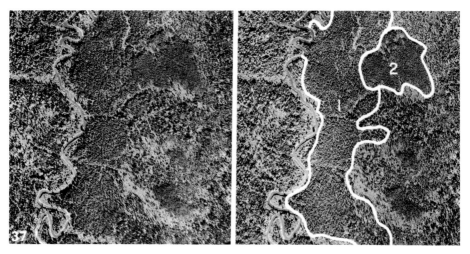

Figure 11-3 Summer panchromatic stereogram of (1) balsam fir and (2) black spruce in Ontario. Scale is about 1:16,000. (Courtesy V. Zsilinszky, Ontario Department of Lands and Forests.)

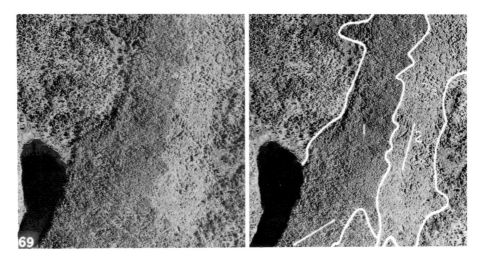

Figure 11-4 Summer panchromatic stereogram of (1) aspen and white birch and (2) young beech stand in Ontario. Scale is about 1:16,000. (Courtesy V. Zsilinszky, Ontario Department of Lands and Forests.)

Figure 11-5 Summer panchromatic stereogram of red pine plantation (dark crowns) in the lower peninsula of Michigan. Scale is about 1:20,000. (Courtesy Abrams Aerial Survey Corp.)

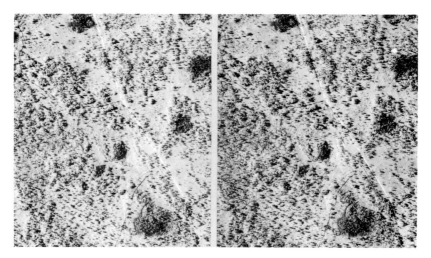

Figure 11-6 Autumn modified infrared stereogram (0.6–0.9-μm) of longleaf and slash pines (light-toned crowns) in the Georgia coastal plain. The dark, water-filled depressions are "ponds" of southern cypress. Scale is about 1:16,000.

Figure 11-7 Panchromatic stereogram of a cable-logged area near Springfield, Louisiana. Principal cover types pictured are bottomland hardwoods and cypress-tupelo gum. Scale is about 1:6,000.

TABLE 11-3 Levels of Plant Recognition to Be Expected at Selected Image Scales

Type or Scale	General Level of Plant Discrimination
Earth-satellite images	Separation of extensive masses of evergreen versus deciduous forests
1:25,000–1:100,000	Recognition of broad vegetative types, largely by inferential processes
1:10,000–1:25,000	Direct identification of major cover types and species occurring in pure stands
1:2,500–1:10,000	Identification of individual trees and large shrubs
1:500–1:2,500	Identification of individual range plants and grassland types

Photographic identification of individual plants requires that interpreters become familiar with a large number of species *on the ground*. For example, there are more than a thousand species of woody plants that occur naturally in the United States; professional foresters and range managers rarely know more than a third of this number. In Australia there are more than 800 species of eucalyptus trees; many of these species are difficult to separate even when they are within arm's reach.

Species Identification Characteristics

As a minimum, the interpreter should be familiar with the **branching characteristics**, **crown shapes**, and **spatial distribution patterns** of important species in the locality. Mature trees in sparsely stocked stands can often be recognized by the configuration of their crown shadows falling on level ground (Figure 11-8). A familiarity with tree crowns as seen from above (Figure 11-9) can be of invaluable assistance when large-scale photographs are being interpreted.

Species identification can usually be aided by the use of black-and-white infrared or color infrared photographs (Figure 2-32 and Plate 4). In a study of Rocky Mountain rangelands, color infrared photographs at scales of 1:800 to 1:1,500 proved superior to conventional color photographs for the identification of shrubs (Driscoll and Coleman 1974). One experiment showed that 7 of 11 shrub species could be identified 83 percent of the time: the diagnostic image characteristics employed by interpreters were as follows:

1. **Plant height**

   a. ≧ 1.5 m
   b. ≦ 1.5 m

2. **Shadow**

   a. Distinct
   b. Indistinct

3. **Crown margin**

   a. Smooth
   b. Wavy
   c. Irregular
   d. Broken

4. **Crown shape**

   a. Indistinct
   b. Round
   c. Oblong

5. **Foliage pattern**

   a. Continuous
   b. Clumpy
   c. Irregular

6. **Texture**

   a. Fine
   b. Medium
   c. Coarse
   d. Stippled
   e. Mottled
   f. Hazy

7. **Color**: Numerically coded according to standardized color charts

Because range plants progress through distinctive growth stages each year, the date of photography is of utmost importance for the identification of various species. Most studies have indicated that the time of near-maximum foliage development is the best time of year for detection of the kind and amount of forage through remote sensing.

Keys for Species Recognition

Photo interpretation keys are useful aids in the recognition of plant species, especially when such keys are illustrated with high-quality stereograms. Vegetation keys for U.S. and Canadian species are most easily constructed for northern and western forests where conifers predominate. In these regions, there are relatively few species to be considered and crown patterns are fairly distinctive for each important group.

A sample **elimination key** for identification of northern conifers is reproduced in Table 11-4. In its original form, this tree species key was supplemented by descriptive materials and several illustrations. Selected examples of other tree species that may be identified on panchromatic stereo-grams are shown in Figures 11-10 through 11-13. A forest-type map based on a classification scheme developed by the Society of American Foresters is presented in Figure 11-14.

Tree Volume Estimates

Where large-scale, stereoscopic photo coverage is available, it may be feasible to make direct estimates of in-dividual **tree volumes**, by species or species groups. Sam-ple-strip coverage obtained with 70-mm format cameras is often specified, at scales ranging from 1:1,000 to 1:5,000

(Figure 11-15). The determination of the *exact* scale is ide-ally accomplished by employment of a radar altimeter in the photographic aircraft.

Panchromatic photographs have been used success-fully for individual tree evaluations in northern boreal for-ests; in several photographic experiments, the fall season (i.e., after deciduous trees are leafless but before snowfall) has been specified. This timing is regarded as ideal for spe-cies recognition. In other regions, of course, different film-season combinations may be preferred. Regardless of the film emulsion specified, the use of positive transparencies will generally yield more reliable measurements than the interpretation of photographic prints.

A common approach to volume determination is to substitute photographic measures of crown diameter (or crown area) and total tree height for the usual field tallies of stem diameter at breast height (dbh) and merchantable

Figure 11-8 Silhouettes of 24 forest trees. When tree shadows fall on level ground, they often permit identification of individual species.

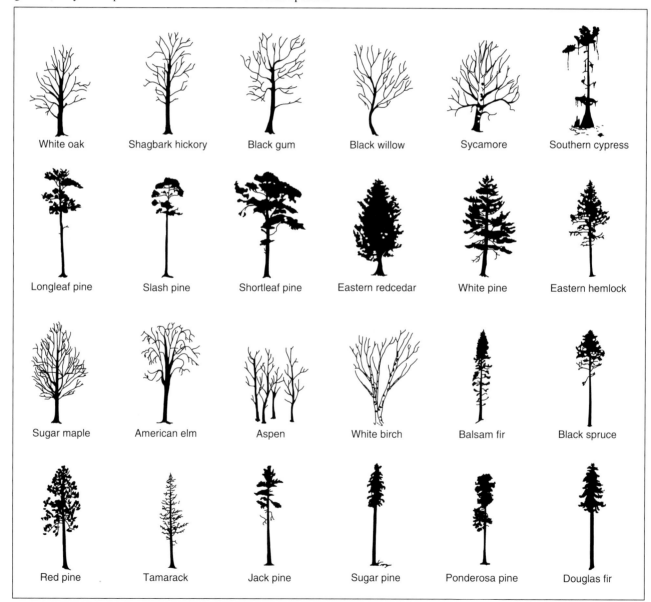

height, respectively (Figure 11-16). Regression equations are then developed for each species or species group for use in volume estimation. For example, the generalized linear equation $\hat{Y} = a + bX$ may be employed if a "combined variable" (the product of crown and height measurements) is substituted as a value of X. Thus, the general equation becomes:

$$\hat{Y} = a + b(cd^2h), \text{ or } \hat{Y} = a + b(ca\ h), \qquad (11\text{-}1)$$

where: $\hat{Y}$ = tree volume, determined from a sub-sample of ground measurements to establish the regression coefficients a and b,
cd = tree crown diameter,
ca = tree crown area, and
h = total tree height.

Crown Diameters or Areas

The value of crown measurements in equations predicting tree volume depends on the relationship that exists between crown dimensions and corresponding stem diameters or basal areas. High correlations of these variables often can be established for even-aged conifers that have not been subjected to undue suppression or stand competition; the relationship is usually linear for trees in the middle diameter or age classes.

The photographic determination of **crown diameter** is simply a linear measure, but measurements can be difficult because of the small sizes of tree images, the effects of crown shadows, and noncircular crowns (Figure 11-17). Various linear scales, magnifiers, and "crown wedges" are available for photographic measurements. Careful interpreters can measure to within ± 0.1 mm, so accuracy is dependent on photo scale, film resolution, and the ability of the interpreter.

Measurements of **crown area** offer an alternative to crown diameter evaluations. Area determinations can be made with finely graduated dot grids or can be calculated from stereoplotter coordinates of points along the crown perimeters. As more sophisticated interpretation equipment becomes available, crown areas may entirely supplant crown diameters in tree volume equations.

Tree Volume Tables

Aerial **tree volume tables** are often compiled from volume prediction equations based on crown and height measurements. However, because more interpretation time is required for the measurement of heights than crowns, several single-entry volume tables (based on crown diameter or crown area alone) have been proposed. The approach is valid where tree heights are fairly uniform within specified crown

---

Figure 11-9 Sketches of overhead views of tree crowns for several boreal species. (Courtesy U.S. Forest Service Remote Sensing Project, Berkeley, California.)

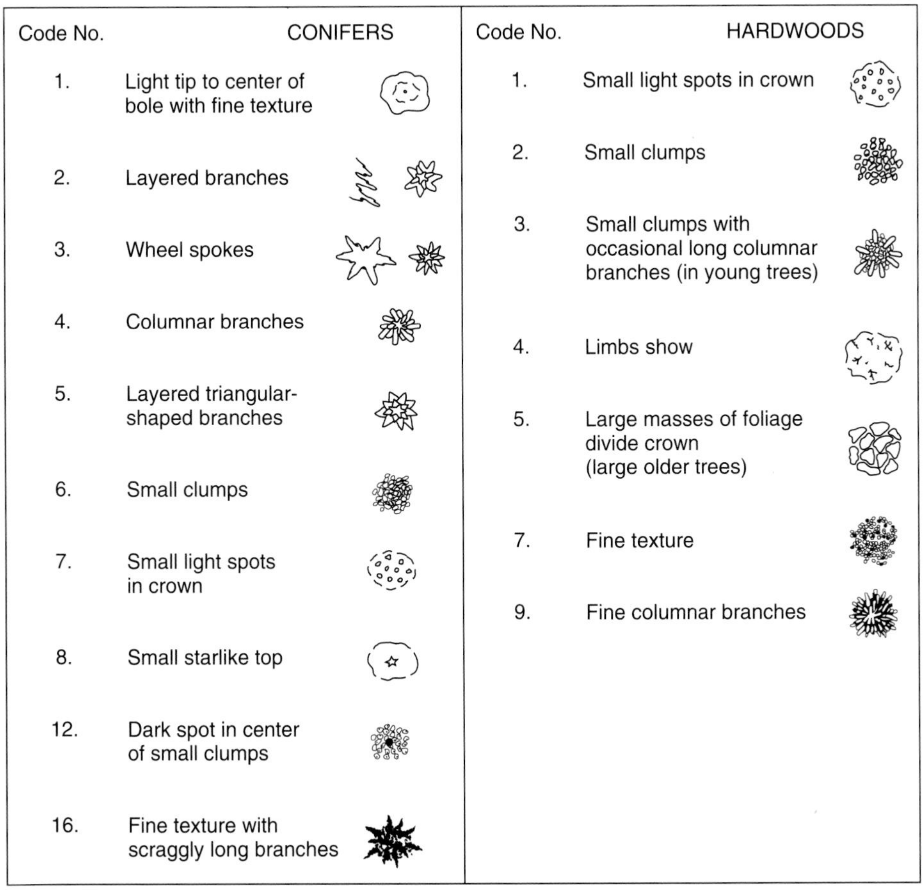

Code No.	CONIFERS		Code No.	HARDWOODS	
1.	Light tip to center of bole with fine texture		1.	Small light spots in crown	
2.	Layered branches		2.	Small clumps	
3.	Wheel spokes		3.	Small clumps with occasional long columnar branches (in young trees)	
4.	Columnar branches		4.	Limbs show	
5.	Layered triangular-shaped branches		5.	Large masses of foliage divide crown (large older trees)	
6.	Small clumps		7.	Fine texture	
7.	Small light spots in crown		9.	Fine columnar branches	
8.	Small starlike top				
12.	Dark spot in center of small clumps				
16.	Fine texture with scraggly long branches				

TABLE 11-4 Identification Key for the Northern Conifers[a]

1. Crowns small, or if large, then definitely cone-shaped		
	2. Crown broadly conical, usually rounded tip, branches not prominent	Cedar
	2. Crowns have a pointed top, or coarse branching, or both	
	Crowns narrow, often cylindrical, trees frequently grow in swamps	Swamp-type black spruce
	Crowns conical, deciduous, very light toned in fall, usually associated with black spruce	Tamarack
	Crowns narrowly conical, very symmetrical, top pointed, branches less prominent than in white spruce	Balsam fir
	Crowns narrowly conical, top often appears obtuse on photograph (except northern white spruce), branches more prominent than in balsam fir	White spruce, black spruce (except swamp type)
	Crowns irregular, with pointed top, has thinner foliage and smoother texture than spruce and balsam fir	Jack pine
1. Crowns large and spreading, not narrowly conical, top often well defined		
	3. Crowns very dense, irregular or broadly conical	
	4. Individual branches very prominent, crown usually irregular	White pine
	4. Individual branches rarely very prominent, crown usually conical	Eastern hemlock
	3. Crowns open, oval (circular in plan view)	Red pine

[a]*From Canada Department of Forestry.*

classes. Table 11-5 was formulated from existing *tarif tables** by the following steps:

1. A tree volume equation was produced from optical dendrometer measurements of 58 standing trees; on the basis of this data, a tarif access table was derived.

2. An existing tarif table was selected for the area from which sample trees were drawn.

3. A crown diameter–stem diameter relationship, based on 600 tree measurements, was established for the sample area.

4. The crown diameter–stem diameter relationship was used to convert the selected tarif table to a single-entry aerial volume table based solely on crown diameter.

Although Table 11-5 is applicable only to a limited area in northern Arizona, the *method* of constructing the single-entry table may be useful in other areas where crown and stem diameters are closely correlated.

Stand Volume Tables

Where only small-scale aerial photographs are available to interpreters, emphasis is on measurement of **stand variables** rather than individual tree variables. Aerial **stand**

The term tarif *is of Arabic origin and simply means tabulated information.* Tarif tables *are standardized local volume tables which list various volumetric parameters by stem diameter at breast height (dbh).*

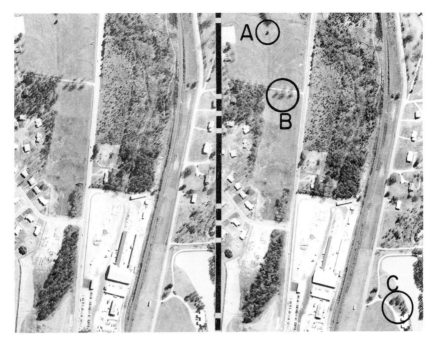

Figure 11-10 Winter stereogram taken near Charlotte, North Carolina, showing distinctive tree shadows of eastern redcedar (*A*), oaks devoid of foliage (*B*), and shortleaf pines (*C*). Compare with drawings in Figure 11-8. Scale is about 1:16,000.

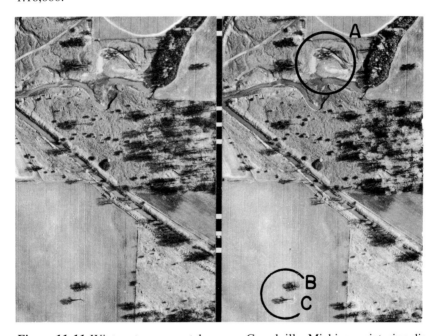

Figure 11-11 Winter stereogram taken near Grandville, Michigan, picturing distinctive tree shadows of American elm (*A, C*) and oak (*B*). Compare with drawings in Figure 11-8. Scale is about 1:6,000.

volume tables are multiple-entry tables that are usually based on assessments of two or three photographic characteristics of the dominant-codominant crown canopy—average stand height, average crown diameter (or crown area), and percentage of crown closure. These tables may be derived by multiple regression analysis; photographic measurements of the independent variables are made by several skilled interpreters, and a volume prediction equation is developed.

Crown Closure

Crown closure, also referred to as **crown cover** and **canopy closure**, or **density**, is defined by photo interpreters as the percentage of a forest area occupied by the vertical projections of tree crowns. The concept is primarily applied to even-aged stands or to the dominant-codominant canopy level of uneven-aged stands. In this context, the maximum value possible is 100 percent.

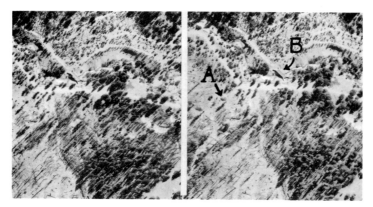

Figure 11-12 Stereogram from the Plumas National Forest in California, illustrating the shadow pattern of a sugar pine (A). Compare with drawing in Figure 11-8. At B is an abandoned bridge with roadway approaches washed out by severe erosion. Scale is about 1:5,000.

Figure 11-13 Stereo-triplet on 70-mm panchromatic film from the Superior National Forest in Minnesota. Species circled are: (*1*) balsam fir, (*2*) quaking aspen, (*3*) paper birch, (*4*) red maple, (*5*) white spruce, (*6*) red pine, and (*7*) white pine. Scale is about 1:1,600. (Courtesy U.S. Forest Service Remote Sensing Project, Berkeley, California.)

In theory, crown closure contributes to the prediction of stand volume, because such estimates are approximate indicators of stand density (e.g., the number of stems per hectare). Because basal areas and numbers of trees cannot be determined directly from small-scale photographs, crown closure is sometimes substituted for these variables in volume prediction equations. Photographic estimates of crown closure are normally used, because reliable ground evaluations are much more difficult to obtain (Figures 11-18 and 11-19).

At photo scales of 1:15,000 and smaller, estimates of crown closure are usually made by ocular judgment, and stands are grouped into 10 percent classes. Ocular estimates are easiest in stands of low density; they become progressively more difficult as closure percentages increase. Minor stand openings are difficult to see on small-scale photographs, and they are often shrouded by tree shadows. These

factors can lead to overestimates of crown closure, particularly in dense stands. And, if ocular estimates are erratic, the variable of crown closure may contribute very little to the prediction of stand volume.

With high-resolution photographs at scales of 1:5,000 to 1:15,000, it may be feasible to derive crown closure estimates with the aid of finely subdivided dot grids. Here, the proportion of the total number of dots that fall on tree crowns provides the estimate of crown closure. This estimation technique has the virtue of producing a reasonable degree of consistency among various photo interpreters; it is therefore recommended wherever applicable.

A modification of the foregoing technique involves the copying of aerial photographs onto 35-mm slides or microfilm. The images are then enlarged by conventional projection or by use of a microfilm reader, and dot counts and crown closure estimates can be made.

Estimating Stand Volumes

Once an appropriate aerial volume table has been selected (or constructed), there are several procedures that can be employed in the derivation of **stand volumes**. One approach is as follows:

1. Outline tract boundaries on the photographs, using the effective area of every other print in each flight line. This assures stereoscopic coverage of the area on a minimum number of photographs and avoids duplication of measurements by the interpreter.

2. Delineate important cover types. Except where type lines define stands of relatively uniform density and total height, they should be further broken down into homogeneous units so that measures of height, crown closure, and crown diameter will apply to the entire unit. Generally, it is unnecessary to recognize stands smaller than 2 to 5 ha.

3. Determine the area of each condition class with dot grids or a planimeter. This determination can sometimes be made on contact prints.

4. By stereoscopic examination, measure the variables for entry into the aerial volume table. From the table, obtain the average volume per hectare for each condition class.

Figure 11-14 Simplified forest cover-type map compiled from aerial photographs.

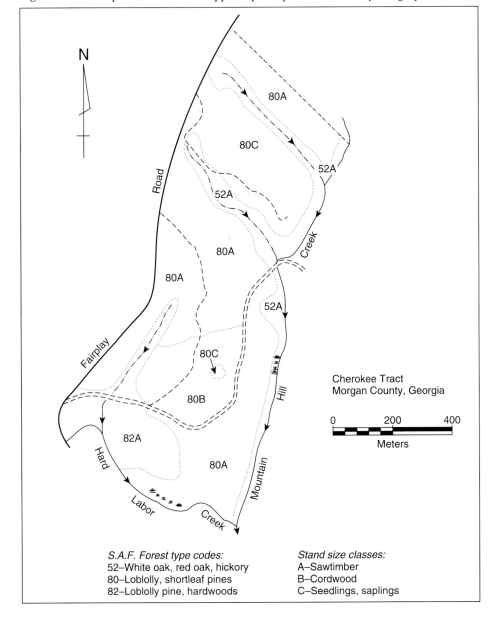

S.A.F. Forest type codes:
52–White oak, red oak, hickory
80–Loblolly, shortleaf pines
82–Loblolly pine, hardwoods

Stand size classes:
A–Sawtimber
B–Cordwood
C–Seedlings, saplings

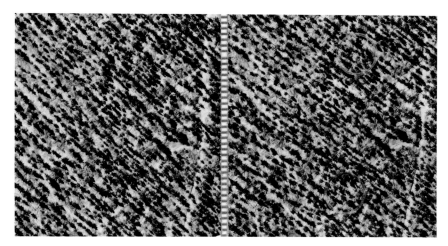

Figure 11-15 Panchromatic stereogram of a thinned radiata pine plantation near Rotorua, New Zealand. Tree counts, heights, and crown measurements can be made on such photographs. Scale is about 1:3,000. (Courtesy New Zealand Forest Service and New Zealand Aerial Mapping, Ltd.)

5. Multiply volumes per hectare from the table by condition class areas to determine gross volume for each class.

6. Add class volumes for the total gross volume on the tract.

Volume Adjustment from Field Checks

Aerial volume tables and volume prediction equations are not generally reliable enough for purely photographic estimates, and some allowance must be made for differences between gross volume estimates and actual net volumes on the ground. Therefore, a portion of the stands (or condition classes) that are interpreted should be checked in the field. If field volumes average 60 m³/ha as compared with 80 m³/ha for the photo estimates, the adjustment ratio would be 60/80, or 0.75. When the field checks are representative of the total area interpreted, the ratio can be applied to photo volume estimates to yield adjusted net volume. It is desirable to compute such ratios by forest types, because deciduous, broad-leaved trees are likely to require larger adjustments than conifers.

The accuracy of aerial volume estimates depends not only upon the volume tables used but also on the ability of interpreters who make the essential photographic assessments. Because subjective photo estimates often vary widely among individuals, it is advisable to have two or more interpreters assess each of the essential variables.

Photo Stratification for Ground Cruising

A **photo-controlled ground cruise** combines the features of aerial and ground estimating, offering a means of obtaining timber volumes with maximum efficiency. Photographs are used for area determination, for allocation of field sample units by forest type and stand size classes, and for designing of the pattern of fieldwork. Tree volumes, growth, cull percentages, form class, and other data are ob-

| TABLE 11-5 | Tree Volume Table for Young-Growth Ponderosa Pine |

Crown Diameter (Meters)	Merchantable Volume (Cubic Meters to a 10-cm Top Diameter)
2.5	0.0453
3.0	0.0821
3.5	0.1246
4.0	0.1813
4.5	0.2492
5.0	0.3086
5.5	0.3823
6.0	0.4673
6.5	0.5523
7.0	0.6457
7.5	0.7562
8.0	0.8638
8.5	0.9771

Adapted from Hitchcock (1974).

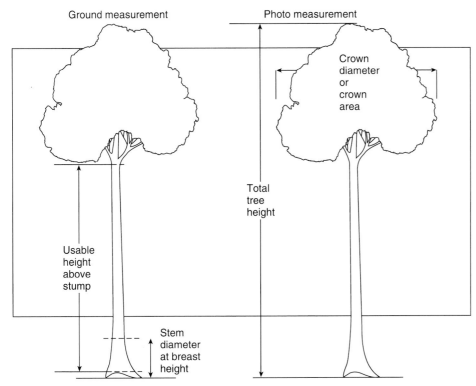

Figure 11-16 Comparison of ground and photographic measurements in the determination of individual tree volumes.

Figure 11-17 The shapes of tree crowns as seen from above can make the measurement of crown diameter difficult.

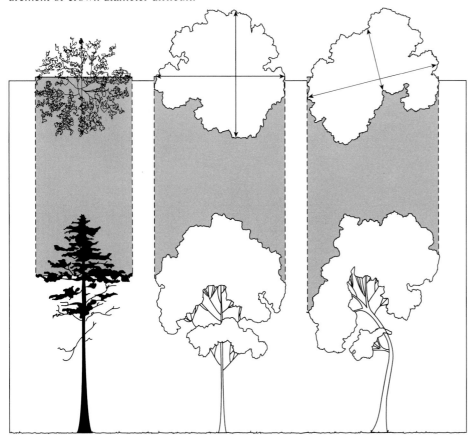

Figure 11-18 A ground view of crown closure as seen by a canopy camera. (Courtesy U.S. Forest Service.)

tained on the ground by conventional methods. A photo-controlled cruise may increase the efficiency and reduce the total cost of an inventory on tracts as small as 50 ha.

The approach to an inventory of this kind is largely dependent on the types of strata recognized and the method of allocating field sample units. The total number of field plots to be measured is determined by cost considerations or by the statistical precision required. Once the number has been determined, the individual sample units are commonly distributed among various photo classifications by the technique of stratified random sampling.

If type boundaries have been accurately delineated and stands are homogeneous within the recognized classes, field plots can sometimes be taken along routes of easy travel without much bias being introduced. Usually, however, some kind of coordinate system is designated as a sampling frame; then a random selection of sample units is made within each stratum. Field measurements are taken by conventional procedures. Cumulative tally sheets or point-sampling may be employed to speed up the tree tally. After the volume per hectare for each stratum has been determined by field sampling, the values are multiplied by the appropriate stand areas. The result is the total volume on the tract, by cover types.

Special Uses of Aerial Photographs

Foresters and range managers have long used aerial photographs in various activities related to the prevention and control of wildfires. The potential fire danger in a given locality can be predicted by the intensive analysis of seasonal changes in plant cover. These ''forest fuels'' are readily mapped by special photographic flights timed at known periods of critical fire danger.

Presuppression activities include aerial photo searches for reliable sources of water during expected drought. Advance photographic coverage also provides information on existing fire lines and makes it feasible to lay out new lines and access routes before the occurrence of wildfires. Woodlands subject to heavy use by campers, hunters, and fishermen can be regularly monitored by means of up-to-date aerial photographs. The detection of wildfires with thermal infrared images was outlined in Chapter 6.

Special-purpose photography may also be used to advantage by foresters in estimation of timber volumes re-

Figure 11-19 Crown closure estimates are difficult when foliage and shadows obscure canopy openings (*above*) or when deciduous trees are leafless (*below*). Scale is about 1:1,000. (Courtesy Canada Department of Forestry.)

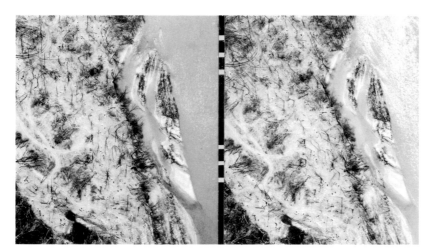

Figure 11-20 Site of a timber-harvesting operation near Newcomerstown, Ohio. Individual tree stumps and residual tops are discernible. It is also evident that the river flat has been subjected to periodic flooding in the past. Scale is about 1:3,000. (Courtesy Abrams Aerial Survey Corp.)

moved during harvesting operations or in assessment of logging and wind damage to residual stands of timber. In Figure 11-20, for example, individual stumps may be counted and merchantable logs that were removed may be estimated by measurement of photo distances between paired stumps and undisturbed treetops. The shadows reveal that stumps were cut rather high—probably an indication of the predominance of swell-butted bottomland hardwood species. The stand was rather heavily cut, and subsequent flooding of the river flat will probably result in severe soil erosion. An example of wind damage to a timber stand is shown in Figure 11-21.

Inventories of Floating Roundwood

Cut roundwood being rafted down rivers, towed in booms, or stored in ponds can be inventoried with fair accuracy from large-scale aerial photographs. One technique for counting floating pulpwood sticks is based on a tally of in-

Figure 11-21 Timber blow-down on the Kaibab National Forest, in Arizona. Most downed trees are ponderosa pines. Scale is about 1:5,000. (Courtesy U.S. Forest Service.)

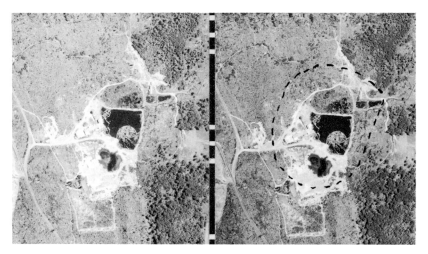

Figure 11-22 Sawmill storage pond for logs in Lewis County, Washington. Average log length and a reliable log count can be determined from such photography. Scale is about 1:6,000. (Courtesy Northern Pacific Railroad.)

dividual bolts on sample "plots" that are randomly located in storage areas. The wood-storage perimeters are then delineated, and areas are determined with a dot grid or planimeter to provide expansion factors for the plot estimates. Photographic resolution and image size are major factors affecting the accuracy of such pulpwood stick counts.

Photography flown especially for inventories of floating roundwood should be taken when water areas are calm and when floating timber is spread out in a single layer. Where roundwood is piled high in several layers or covered by snow and ice, reliable counts are virtually impossible. The extremes in seasonal photographic coverage of floating roundwood are illustrated in Figures 11-22 and 11-23.

Detecting Plant Vigor and Stress

Stress symptoms in vegetation result from a loss of vigor, which indicates an abnormal growing condition. Among the causal agents for loss of plant vigor are diseases,

insects, soil moisture deficiencies, soil salinity, decreases in soil fertility, air pollutants, and so on. Because the symptoms of plant stress tend to be similar regardless of the causal agent, the agent itself usually must be determined by ground examination.

The primary role of remote sensing research is (1) to find the best combination of films, filters, and image scales for detection of damaged plants and (2) to ascertain whether plants under stress can be detected *before* visual symptoms of decline are apparent (i.e., *previsual detection*). The most notable successes in previsual detection have been with infrared-sensitive films.

When plant foliage suddenly changes over extensive areas, conventional color or color infrared films provide an effective sensor at scales of 1:4,000 to 1:8,000. Skilled interpreters can then delineate the afflicted trees or shrubs, so that control or salvage operations can be planned. Losses from epidemics can also be quickly determined. However, this technique is limited, because the vegetation that has be-

Figure 11-23 Storage of floating roundwood along river banks in Aroostock County, Maine. A boom of wood is also being towed across the river. The mantle of snow and ice on the wood prohibits a reliable inventory. Scale is about 1:20,000. (Courtesy U.S. Department of Agriculture.)

come visually detectable is already dying or dead. Control measures that can *save* such vegetation may thus be dependent on much earlier (previsual) detection—an achievement that has met with some success with black-and-white infrared and color infrared films.

An attempt to generalize some of the research findings in plant stress detection resulted in the compilation of Table 11-6. It should be emphasized that this tabulation merely provides a *few examples* of reported research and that the results are necessarily condensed and generalized. For a more complete summary, readers are referred to the references at the end of the chapter. An excellent guidebook to forest damage assessment has been prepared by Murtha (1972).

TABLE 11-6 Selected Examples of Plant Stress Detection

Causal Agent	Primary Species Affected	Imagery and Scale	Results or Detection Accuracy	Reporting Scientists
Balsam woolly aphid	*Abies amabilis*	Color infrared; 1:1,000	83% accuracy	Murtha and Harris (1970)
Black Hills beetle	*Pinus ponderosa*	Conventional color and color infrared; 1:7,920 and smaller	80%–90% acuracy at 1:7,920 scale	Heller (1971) (see Plate 3)
Pine butterfly	*Pinus ponderosa*	Color infrared; 1:127,000	70% of area, as compared to visual ground survey	Ciesla (1974)
Douglas fir beetle	*Pseudotsuga menziesii*	Conventional color and pan; 1:5,000–1:10,000	High accuracy; no numerical results	Wear, Pope, and Orr (1966)
Smog (air pollution)	*Pinus ponderosa* and *Pinus jeffreyi*	Conventional color; 1:8,000	80%–90% accuracy	Wert, Miller, and Larsh (1970)
SO_2 fumes	Most forest vegetation and some shrubs	600–700 nm; satellite imagery	Delineation of damage zones	Murtha (1973)
Dutch elm disease	*Ulmus americana*	Color infrared; 1:8,000	90%–100% acuracy	Meyer and French (1967)

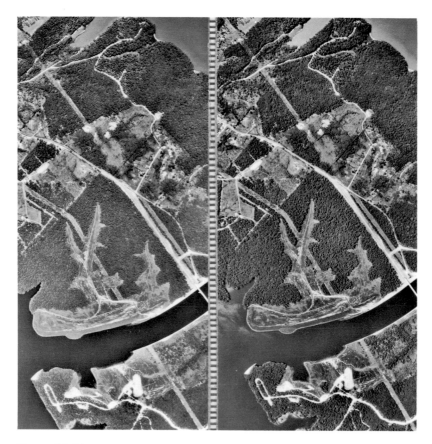

Figure 11-24 Land-between-the-lakes area (Kentucky) where recreational facilities are highly developed. Scale is about 1:24,000. (Courtesy Tennessee Valley Authority.)

Land Capability for Outdoor Recreation

Existing techniques are readily available for the physical inventory of recreational sites; the *real problem* is that of establishing inventory criteria, that is, standards of what should be considered a recreational resource. The approach used for the Canada Land Inventory is quoted here.*

Seven classes of land are differentiated on the basis of the intensity of outdoor recreational use, or the quantity of outdoor recreation, which may be generated and sustained per unit area of land per annum, under perfect market conditions.

"Quantity" may be measured by visitor days, a visitor day being any reasonable portion of a twenty-four-hour period during which an individual person uses a unit of land for recreation.

"Perfect market conditions" implies uniform demand and accessibility for all areas, which means that location relative to population centres and to present access do not affect the classification.

Canada Land Inventory Report No. 6, Land Capability Classification for Outdoor Recreation, *Lands Directorate of the Department of the Environment, Ottawa, Canada, 1970.*

Intensive and dispersed activities are recognized. Intensive activities are those in which relatively large numbers of people may be accommodated per unit area, while dispersed activities are those which normally require a relatively larger area per person.

Some important factors concerning the classification are:

1. *The purpose of the inventory is to provide a reliable assessment of the quality, quantity and distribution of the natural recreation resources within the settled parts of Canada.*
2. *The inventory is of an essentially reconnaissance nature, based on interpretation of aerial photographs, field checks, and available records, and the maps should be interpreted accordingly.*
3. *The inventory classification is designed in accordance with present popular preferences in nonurban outdoor recreation. Urban areas (generally over 1,000 population with permanent urban character), as well as some nonurban industrial areas, are not classified.*
4. *Land is ranked according to its natural capability under existing conditions, whether in natural or modified state; but no assumptions are made concerning its capability given further major artificial modifications.*
5. *Sound recreation land management and develop-*

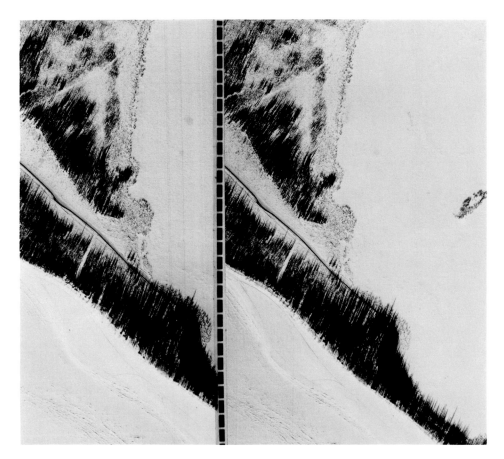

Figure 11-25 Panchromatic stereogram of a winter scene in Alaska.

ment practices are assumed for all areas in practical relation to the natural capability of each area.

6. *Water bodies are not directly classified. Their recreational values accrue to the adjoining shoreland or land unit.*

7. *Opportunities for recreation afforded by the presence in an area of wildlife and sport fish are indicated in instances where reliable information was available, but the ranking does not reflect the biological productivity of the area. Wildlife capability is indicated in a companion series of maps.*

Recreational Surveys

The objective of a **recreational survey** is to locate potential sites and transfer these areas to a base map of suitable scale. With up-to-date photographs in the hands of skilled interpreters, a preliminary recreational survey can be accomplished with a minimum of fieldwork.

Most types of photographic films are suitable, although color emulsions are generally preferred. Exposures should be planned during the dormant season when deciduous trees are leafless or during the season when the greatest numbers of people would be likely to use potential features.

Photographic scales of 1:5,000 to 1:12,000 have been successfully employed; if large regions must be covered by a preliminary survey, the smallest scale that can be reliably interpreted should be chosen to avoid the handling and stereoscopic study of excessive numbers of airphotos (Figure 11-24).

The photo interpretation phase will require the identification and delineation of such features as

Natural vegetation	Existing structures
Land use patterns	Historical features
Scenic terrain features	Access roads
Water resources	Paths or trails
Beaches and inlets	Soils and drainage
Potential docks or ramps	Topography

The more promising sites are then checked on the ground for verification of the interpreter's assessments of current land use, present ownership, site availability, and potentially undesirable features (e.g., polluted water, excessive noise, industrial fumes, or lack of suitable access).

After the elimination of those areas that are unavailable or undesirable, a final report is prepared, summarizing and ranking the recreational potential of each site recommended. The report should be accompanied by both ground and aerial photographs that have been annotated to emphasize salient features, needed improvements, and possible trouble spots.

Figure 11-26 Stereogram on 70-mm panchromatic film of balsam fir trees near Ely, Minnesota. Scale is about 1:1,600; compare with Figure 11-13. (Courtesy U.S. Forest Service Remote Sensing Project, Berkeley, California.)

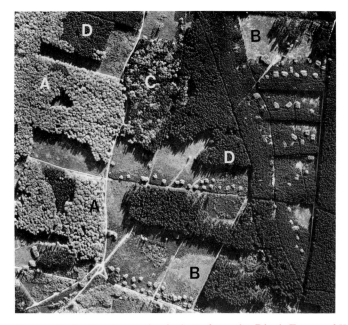

Figure 11-27 Panchromatic airphoto from the Black Forest of West Germany.

Questions

1. Using either a dot grid or polar planimeter (Chapter 4), calculate the surface area of the three booms of floating roundwood shown in Figure 11-23, given a photo scale of 1:20,000.

2. Using a lens stereoscope, study the silhouettes of the tree shadows shown in Figure 11-25. Compare the silhouettes to those shown in Figure 11-8, and identify the type of tree.

3. Given the same scale, film type, and time of year, why do the balsam fir trees appear different in Figure 11-26 as compared to Figure 11-13?

4. Examine Figure 11-27, and answer the following questions:

    a. What time of year was the exposure made? Explain.

    b. Using the U.S. Geological Survey's level II land use and land cover categories (Table 8-1), identify areas A through D.

Unit	Level II Code	Level II Name
A	_____	_____
B	_____	_____
C	_____	_____
D	_____	_____

Bibliography and Suggested Readings

Aldred, A. H., and J. K. Hall. 1975. Application of Large-Scale Photography to a Forest Inventory. *Forestry Chronicle* 51(1):1–7.

Aldrich, R. C. 1971. Space Photos for Land Use and Forestry. *Photogrammetric Engineering* 37:389–401.

Bonner, G. M. 1968. A Comparison of Photo and Ground Measurements of Canopy Density. *Forestry Chronicle* 44(3):12–16.

Brown, H. E., and D. P. Worley. 1965. The Canopy Camera in Forestry. *Journal of Forestry* 63:674–80.

Ciesla, W. M. 1974. Forest Insect Damage from High-Altitude Color-IR Photos. *Photogrammetric Engineering* 40:683–89.

Department of Agriculture. 1978. *Forester's Guide to Aerial-Photo Interpretation.* Agricultural Handbook 308. Washington, D.C.: U.S. Government Printing Office.

Driscoll, R. S., and M. D. Coleman. 1974. Color for Shrubs. *Photogrammetric Engineering.* 40:451–59.

Heller, R. C. 1971. Detection and Characterization of Stress Symptoms in Forest Vegetation. In *Proceedings, International Workshop on Earth Resource Survey Systems.* Washington, D.C.: U.S. Government Printing Office.

Heller, R. C., and others. 1983. Forest Resource Assessments. In *Manual of Remote Sensing,* edited by R. N. Colwell, 2d ed., 2229–324. Falls Church, Va.: American Society for Photogrammetry and Remote Sensing.

Heller, R. C., G. E. Doverspike, and R. C. Aldrich. 1964. *Identification of Tree Species on Large-Scale Panchromatic and Color Aerial Photographs.* Agricultural Handbook 261. Washington, D.C.: U.S. Government Printing Office.

Hitchcock, H. C., III. 1974. Constructing an Aerial Volume Table from Existing Tarif Tables. *Journal of Forestry* 72:148–49.

Hudson, W. D. 1991. Photo Interpretation of Montane Forests in the Dominican Republic. *Photogrammetric Engineering and Remote Sensing* 57:79–84.

Kucher, A. W., and I. S. Zonneveld, eds. 1988. *Vegetation Mapping.* Dordrecht: Kluwer Academic Publishers.

Lanly, J. P. 1973. *Manual of Forest Inventory, With Special Reference to Mixed Tropical Forests.* Rome: Food and Agriculture Organization of the United Nations.

Meyer, M. P., and D. W. French. 1967. Detection of Diseased Trees. *Photogrammetric Engineering* 33:1035–40.

Murtha, P. A. 1972. *A Guide to Air Photo Interpretation of Forest Damage in Canada.* Publication 1292. Ottawa: Canadian Forestry Service.

Murtha, P. A. 1973. ERTS Records SO_2 Fume Damage to Forests, Wawa, Ontario. *Forestry Chronicle* 49(6):251–52.

Murtha, P. A., and J.W.E. Harris. 1970. Air Photo Interpretation for Balsam Woolly Aphid Damage. *Journal of Remote Sensing* 1(5):3–5.

Myers, N. 1988. Tropical Deforestation and Remote Sensing. *Forest Ecology and Management* 23:215–25.

Null, W. S. 1969. Photographic Interpretation of Canopy Density— A Different Approach. *Journal of Forestry* 67:175–77.

Paine, D. P. 1981. *Aerial Photography and Image Interpretation for Resource Management.* New York: John Wiley & Sons.

Wear, J. F., R. B. Pope, and P. W. Orr. 1966. *Aerial Photographic Techniques for Estimating Damage by Insects in Western Forests.* Portland, Ore.: Pacific Northwest Forest and Range Experiment Station.

Wert, S. L., P. R. Miller, and R. N. Larsh. 1970. Color Photos Detect Smog Injury to Forest Trees. *Journal of Forestry* 68:536–39.

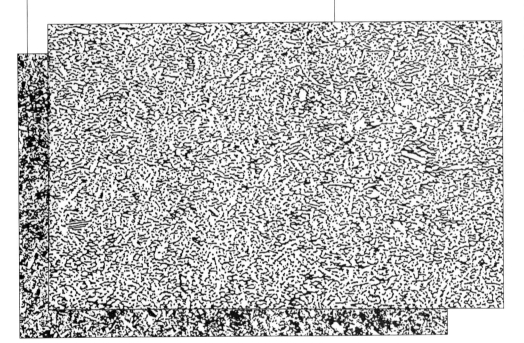

Chapter 12

Geology Applications

Nature of Photogeology

The use of aerial photographs to obtain both qualitative and quantitative geologic information is referred to as **photogeology**. Aerial photographs are widely used today for identifying and mapping landforms, drainage patterns, structural features such as faults and folds, and rock or lithologic units. Knowledge of these surface attributes of a landscape also enables a **photogeologist** to infer or predict subsurface characteristics and relationships. Airphotos are routinely used for the following types of geologic studies: (1) com-

piling topographic and geologic maps, (2) exploring for mineral, hydrocarbon, and groundwater deposits, (3) identifying hazardous features or sites such as active earthquake faults and areas prone to landslides, (4) identifying and mapping landscape changes caused by a natural hazard event such as a hurricane or earthquake, and (5) selecting potential construction sites for critically engineered facilities such as dams and nuclear power plants.

The geologic interpretation of aerial photographs is based on the fundamental **recognition elements** that include shape, size, pattern, shadow, tone or color, texture, association, and site (see Chapter 3). The quantity and quality of geologic information that can be interpreted from aerial photographs is dependent upon the following factors: (1) the type of terrain, vegetation and soil cover, and the stage of the erosional cycle; (2) the type and scale of the photography; (3) whether stereopairs are available and, if so, the amount of vertical exaggeration that is present in the stereographic models; and (4) the training and experience of the interpreter in geology and remote sensing.

Although **oblique airphotos** are often of value to the photogeologist, most studies make use of **vertical airphotos** and accompanying **stereoscopic analysis** (Figures 12-1 and 12-2). The three-dimensional view of the terrain reveals important topographic information that cannot be obtained by viewing single photographs (Figures 3-29 and 12-2).

Most of the photographs used in photogeology are acquired under relatively high solar-illumination angles to ensure that ground detail is not hidden by cast shadows. However, when there is a need to enhance surface irregularities, special-purpose photographs are acquired with low-sun-angle illumination. Whenever feasible, both high- and low-sun-angle photographs are obtained for a given study area (Figure 12-3). The effects of six different angles of artificial illumination on relief enhancement for a plaster topographic model are shown in Figure 12-4. High- and low-sun-angle photographs of dissected alluvial and bedrock terrain are shown in Figure 3-12.

Medium- to large-scale airphotos are best suited to the detailed study of a localized area, whereas small-scale airphotos, such as those obtained from earth orbit, find their greatest utility for regional surveys (Figures 12-5 and 12-6). In photogeology, the **convergence-of-information principle** is often employed, whereby an interpreter starts with small-scale photographs for a synoptic view and gradually focuses upon a local or target area by interpreting successively larger scale photographs. Regardless of scale, however, it is important to remember that ground observations (i.e., **field geology**) have not been replaced by photogeology,

Figure 12-1 (*Left*) Oblique view looking south across the central portion of the Grand Canyon, Arizona. The Bright Angel normal fault is seen at *A-A'*; the upthrown block is to the right. The bright banded unit seen near the top of the canyon is the cliff-forming Coconino sandstone of Permian age. A vegetated projection of the North Rim is observed at *B*. The area surrounding *C* is shown in a vertical view in Figure 12-2. A south-look radar image of the Grand Canyon is shown in Figure 7-15. (Courtesy U.S. Geological Survey.)

Figure 12-2 (*Below*) Stereogram covering part of the central Grand Canyon, Arizona. Annotations are as follows: (*A*) V-shaped inner gorge of the Colorado River cut in the Precambrian Vishnu schist, (*B*) Bright Angel Canyon (fault), (*C*) Sumner Butte, (*D*) Zoroaster Temple, (*E*) Clear Creek, and (*F*) Howlands Butte. Scale is about 1:80,000. (Courtesy U.S. Geological Survey.)

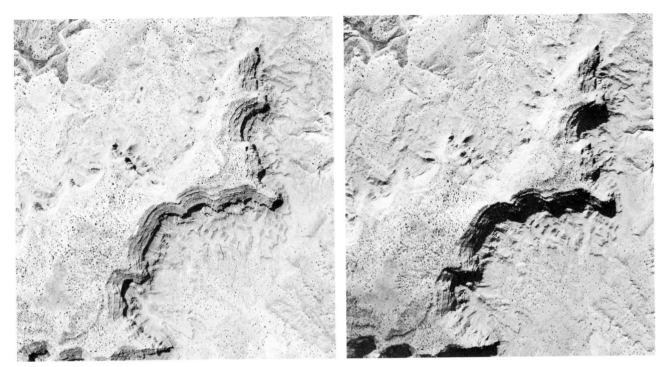

Figure 12-3 High- and low-sun-angle vertical airphotos of a mesa landscape in New Mexico; sedimentary strata are nearly horizontal.

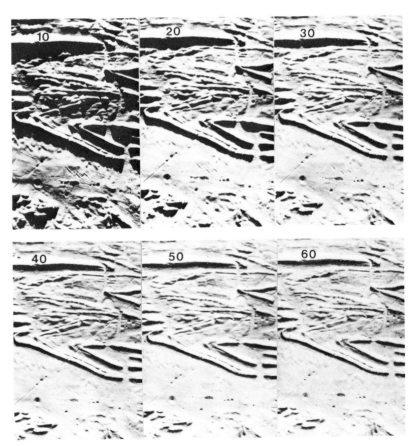

Figure 12-4 Photographs of a plaster topographic model under different angles of artificial illumination. In this series, the greatest detail of topographic relief (when not covered by shadow) is recognizable when the illumination angle is 10°. (Adapted from Hackman 1966.)

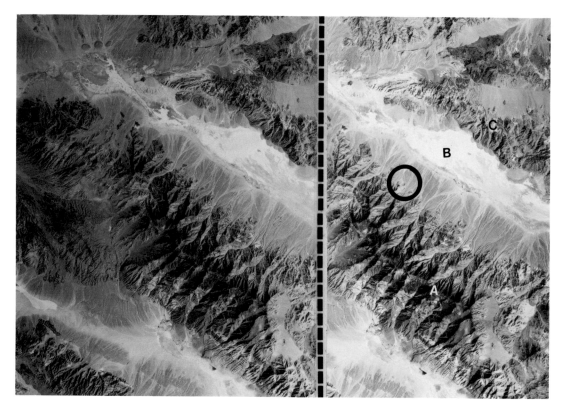

Figure 12-5 Large Format Camera (see Chapter 5) stereogram covering part of the Basin and Range Province of the western United States. Annotations are as follows: (*A*) Panamint Mountains, (*B*) Death Valley saltpan, and (*C*) Black Mountains. In this province, the valleys have been faulted down (structural rather than erosional) and the mountains have been raised. Generally, the faults lie along the western sides of the mountains, and the fault blocks have been tilted to the east. Scale is about 1:500,000. The circled area is shown in a large-scale format in Figure 12-6.

as many types of information are obtainable only by close field inspection.

Detailed analyses of geomorphology, stratigraphy, and structural geology from airphoto interpretation require professional training in geology with considerable field experience. However, nongeologists may develop a proficiency in the recognition of general rock types, distinctive landforms, drainage patterns, and structural features. Consequently, the focus of this chapter is to provide an introduction to the study of these landscape elements through the use of aerial photographs, with an emphasis on panchromatic stereopairs. The computer processing of digital image data for geologic applications is discussed in Chapter 15. Because geologic times are mentioned throughout this chapter, the *geologic time scale* is presented on the inside back cover.

Qualitative Information from Airphotos

The geologic information that can be interpreted from aerial photographs may be grouped into four major categories: (1) **lithology**, (2) **structure**, (3) **drainage**, and (4) **landforms**. The following comments introduce these key considerations, which are discussed later in detail.

The interpreter usually desires to classify general lithologies or rock types with respect to their origin. The rock unit could be formed (1) directly from a molten material (**igneous**), (2) from the particles of preexisting rocks or plant and animal remains (**sedimentary**), or (3) by the action of heat and pressure on previously existing rock (**metamorphic**). When exposed, **bedrock** of these classes often show up with striking clarity on aerial photographs.

Most **surficial deposits** (the unconsolidated sedimentary rocks) are readily distinguishable from consolidated rocks on aerial photographs. Surficial materials that were transported and deposited by water (**alluvial**), ice (**glacial**), wind (**eolian**), and gravity (**mass wasting**) often have distinctive forms. Common features that are readily identifiable on airphotos include alluvial fans, sand bars and spits, river terraces, eskers, drumlins, sand dunes, loess (eolian silt) deposits, and talus or scree deposits.

Structural features denote geologic structures that are formed by **diastrophism**, or the internal deformation forces that can fold (**plastic deformation**) or rupture (**brittle deformation**) any type of rock. Many structural features have unique geometric attributes that are discernable on airphotos where they intersect the earth's surface.

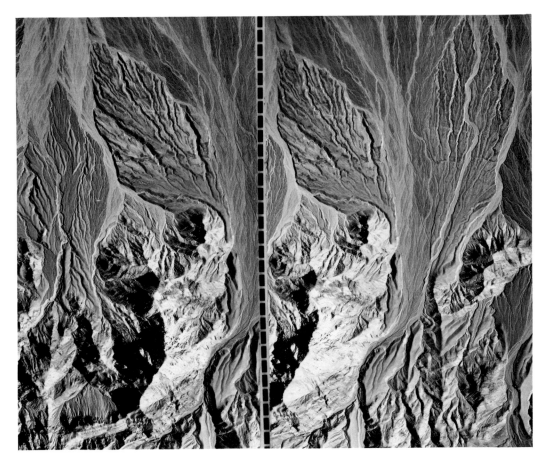

Figure 12-6 Stereogram covering the upper reaches of Death Valley Canyon alluvial fan along the eastern flank of the Panamint Mountains (Death Valley, California). The faint lines running at right angles to the dry washes are normal faults. Drainage pattern is dichotomic. Scale is 1:48,000.

Drainage patterns represent the arrangement and repetition of stream channels that have been formed in response to natural forces acting upon the earth's land surface. The patterns, which are normally visible on airphotos, reflect to varying degrees the lithology and structure of a region. Drainage patterns are one of the most important identifiers of landforms.

Landforms are the distinctive geometric configurations of the earth's land surface (i.e., **terrain** or **landscape features**) that can be classified or identified by compositional, physical, and visual characteristics. Landforms may be defined as **erosional** or **depositional**. Erosional landforms are those created by the agents of erosion, such as running water, glacial ice, and wind. Depositional landforms have the character and shape of the deposits from which they are composed. Erosional landforms may be regarded as **destructional** and depositional landforms as **constructional**.

Quantitative Information from Airphotos

On vertical photographs of *appropriate and known scale,* it is possible to make several types of direct measure-

ments; these are described in Chapter 4. For example, the heights of topographic features can be determined by the measurement of parallax differences on stereopairs. Under special conditions, heights may also be computed from measurement of shadow lengths. Area measurements of geologic surfaces can be made directly from vertical airphotos when the terrain is level to gently rolling; both dot grids and polar or digital planimeters can be used for this purpose. In addition, the lengths of curved or irregular features (e.g., shorelines and river courses) can be determined easily with an opisometer.

It is also possible to make direct measurements of the dip and strike of a planar structure (e.g., bedding plane, fault, or dike) on vertical airphotos of the proper scale (Figure 12-7). **Dip** is the angle of inclination, or the amount a surface is tilted from a horizontal plane; dip angles are measured downward from the horizontal plane and range between 0° (horizontal) and 90° (vertical). The **dip direction** is toward the steepest inclination of the dipping surface and is expressed as a compass bearing or azimuth. **Strike** is the bearing or azimuth of an imaginary line running along the trend of the feature. The dip direction is always measured at right angles to the strike.

When **bedding surfaces** coincide with **topographic surfaces**, the angle of dip (θ) can be calculated with photo measurement data. This is accomplished by first determining the height or elevation difference between any two points (H), one directly down-dip from the other. The H parameter can be derived from measurements of stereoscopic parallax as described in Chapter 4 (Equation 4-6). After the horizontal distance between the same two points is determined (D), the dip angle can be computed by the following trigonometric relationship:

$$\tan \theta = \frac{H}{D}. \qquad (12\text{-}1)$$

If relief in an area is low, the horizontal distance may be scaled directly from a single photograph without significant error in computation of the dip. However, when relief is moderate or high, a correction for the relief displacement of the upper point with respect to the lower point should be made. Ray (1960) describes the use of an overlay procedure from which corrected horizontal distances can be obtained:

The overlay procedure requires first laying out on transparent material a line equal in length to the adjusted photobase. The overlay is then placed over the right photograph of the stereoscopic pair so that the line drawn is coincident with the flight direction and its right end terminates at the photographic center. Radial lines are then drawn on the overlay from the photographic center through all points whose relative positions are to be determined. The procedure is repeated with the overlay positioned over the left photograph, again with the original line coincident with the flight direction and its left end terminating at the photographic center. The intersection of a pair of lines through the same image points is the corrected horizontal position for that point.

In the unique circumstance where the strike is radial from the photographic center, or the surface on which the dip to be measured is near a photographic center point, there is little or no relief displacement in the dip direction. Therefore, no correction in scaling the horizontal distance need be made (Ray 1960).

The strike line generally can be determined with a protractor by inspection of the stereoscopic model and notation of two points of equal altitude on a bed. Where dips are low,

however, tilt in the photographs will affect the direction of the strike. The lower the dip, the greater the effect of tilt on the change in azimuth or bearing of the strike line (Ray 1960).

In areas where outcropping beds of sedimentary rocks are horizontal or nearly horizontal, **stratigraphic thicknesses** can be determined directly by converting to meters or feet the parallax difference between the top and bottom of a bed seen in a stereopair (Equation 4-6); no correction is needed for relief displacement (Ray 1960).

If the beds are inclined, however, the angle of dip must first be determined, and then corrections must be made for relief displacement and for the effect of dip on the stratigraphic thickness. As defined by Ray (1960), the thickness may be determined from the following formula, which incorporates the trigonometric relationships shown in Figure 12-8:

$$T = (H)\cos \theta + (D)\sin \theta, \qquad (12\text{-}2)$$

where: T = stratigraphic thickness,
θ = dip angle determined from Equation 12-1,
H = height difference between some point on the lower contact of the bed and some point, along a line at right angles to the strike line, at the upper contact of the bed; H is determined from the parallax formula (Equation 4-6), and
D = corrected horizontal distance between measured points at lower and upper contacts of the bed; horizontal distances are corrected by use of the overlay procedure described in the previous section.

When dips are steep (Figure 12-9), it is best to relate the horizontal distance between the top and bottom of the bed (D) and the angle of dip (θ) to stratigraphic thickness (T) according to the relation

$$T = (D)\sin \theta. \qquad (12\text{-}3)$$

Figure 12-7 Diagram of strike line and dip angle for gently dipping beds of sedimentary rocks.

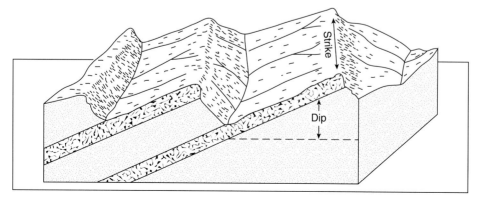

Structural Features

As was previously mentioned in this chapter, internal deformation forces can fold (plastic deformation) or rupture (brittle deformation) the earth's crustal rocks. The lateral forces of **compression** can flex strata and form an alternating series of **anticlines** or arches and **synclines** or troughs (Figure 12-10). The **downfold** whose sides, called **limbs**, or **flanks**, are inclined upward on either side of the concave flexure is the syncline. The **upfold**, whose limbs are inclined downward on either side of the convex flexure, is the anticline. Folding may be **symmetrical** (i.e., one limb is the mirror image of the other), indicating opposing horizontal

Figure 12-8 Diagram of gently dipping beds of sedimentary rocks showing relation of stratigraphic thickness to differential parallax determined at any two points along dip direction and at the formation contacts. (Adapted from Ray 1960.)

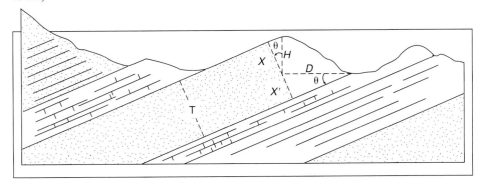

Figure 12-9 Diagram of steeply dipping beds of sedimentary rocks showing relation of stratigraphic thickness to dip angle and horizontal distance between the top and bottom of the bed. (Adapted from Ray 1960.)

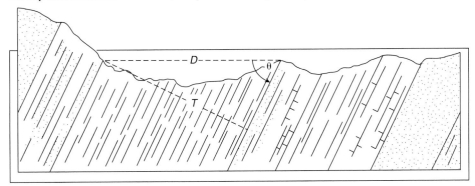

Figure 12-10 Diagram illustrating the geometrical characteristics of folded strata and the terms used to describe the various types of folds.

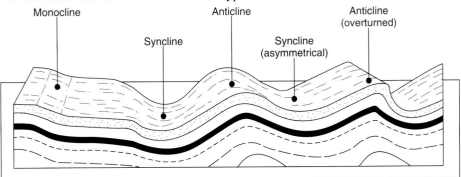

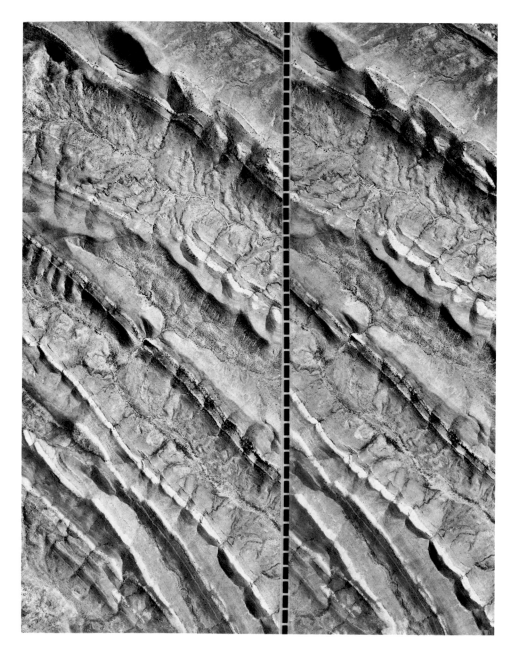

Figure 12-11 Stereogram of eroded anticlinal/synclinal folds in interbedded sedimentary rocks in West Texas. Scale is about 1:26,000.

forces of equal magnitude. If the opposing forces are of unequal magnitude, the resulting folds will be **asymmetrical**; an extreme but not uncommon asymmetrical fold can even be **overturned** (Figure 12-10).

In their initial stages of development, large anticlines form mountains, ridges, and domes, whereas synclines form valleys or basins (Figure 12-10). Through geologic time, however, the crest of an anticline may be removed by erosion, leaving an **anticlinal valley** (Figure 12-11). A **synclinal ridge** is formed where a remnant of resistant rock acts as a cap retarding the erosion of the underlying layers. These two examples both represent a **topographic reversal**. The partial erosion of large folded structures produces distinctive and elegantly expressed landforms (Figures 12-11 and 12-12).

With interbedded sedimentary rocks, the relative ages of the beds can be determined by the folding pattern. The age sequence of anticlinal beds is *oldest to youngest* from the center outward, whereas the age sequence for synclinal beds is *youngest to oldest* from the center outward (Figure 12-12).

The flexing of strata by the lateral forces of tension produce **monoclines**, which are steplike bends in otherwise horizontal or gently dipping beds (Figure 12-10). For sedimentary rocks, limestone bends readily to form monoclines, while sandstone is more brittle and tends to break into faults under similar tensional forces (Figure 12-13).

Brittle deformation produces two major types of **rock fractures**. **Joints** are fractures in a rock mass along which no measurable movement has occurred. Joints are often

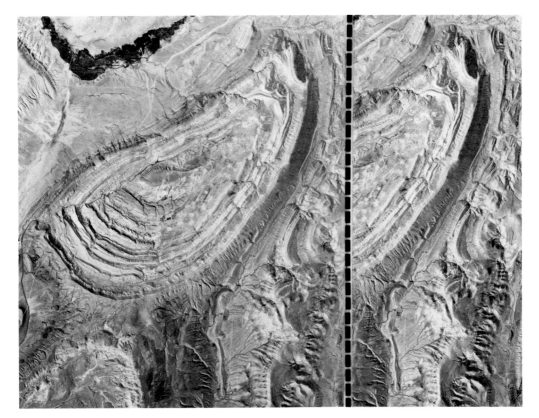

Figure 12-12 Stereogram of Circle Ridge Dome, Freemont County, Wyoming. This eroded asymmetrical anticline has a central valley underlain by Triassic shale and sandstone. The anticlinal valley is surrounded by younger, outward-dipping sandstone ridges separated by weaker shale units, largely of Jurassic age. A localized oil field occupies the central lowland (note well pads). Scale is about 1:58,000.

found as **sets**, where their strike is uniform over an extended area; erosion along joints produces linear depressions that are readily seen on aerial photographs (Figure 12-14). **Faults** are fractures along which measurable displacement of the rocks on either side of the **fault plane** has taken place. Faults may be described and classified on the basis of relative displacements: (1) **dip-slip faults** are those in which movement along the fault plane is predominantly vertical, being parallel to the fault plane's dip; (2) **strike-slip faults** are those in which movement is predominantly horizontal, being parallel to the fault plane's strike; and (3) **oblique-slip faults** are those in which movements include a combination of vertical and horizontal displacements (Figure 12-15).

Two major types of dip-slip faults are recognized. A **reverse fault** separates rock masses where the **hanging wall** appears to have moved up relative to the **footwall** (Figure 12-15); if the dip of the fault plane is small (usually less than 15°), the term **thrust fault** is used. A **normal fault** separates rock masses in which the hanging wall appears to have moved down relative to the footwall (Figure 12-15). Topographic escarpments caused by dip-slip faulting are called **fault scarps**. Normal faults are shown in Figures 12-1, 12-6, and 12-13.

Normal faults that divide an area into a series of elevated and depressed blocks are said to produce **block faulting**. The uplifted blocks may either be horizontal, forming **horsts** with two pronounced fault scarps, or tilted, forming **tilt blocks** with only one pronounced scarp (Figure 12-16); the Sierra Nevada and most of the mountains of the Basin and Range Province are tilt blocks (Figure 12-5). The blocks that are depressed between parallel faults are called **grabens** (Figure 12-16). A graben in limestone strata is shown in Figure 12-13.

Because a strike-slip fault separates blocks that have experienced primarily horizontal displacement, large scarps are not produced by this kind of movement. The fault plane is usually vertical or near vertical. This type of fault is further classified as **right lateral** (**dextral**) or **left lateral** (**sinistral**). If one views a block on the far side of a fault trace that has moved to the right, the fault is right lateral; if the far block has been displaced to the left, the fault is left lateral. The San Andreas, which separates the Pacific and North American plates, and most other strike-slip faults in California have been associated with right-lateral displacements (Figures 12-17 and 12-18). Notable exceptions include the left-lateral Garlock and Big Pine faults (Figure 12-19).

Mapping **lineaments** on remote sensing imagery is an effective procedure for recognizing *possible faults*. O'Leary et al. (1976) define a lineament as a "mappable, simple, or composite linear feature of a surface whose parts are aligned in a rectilinear or slightly curvilinear relationship and which differs distinctly from the pattern of adjacent features and

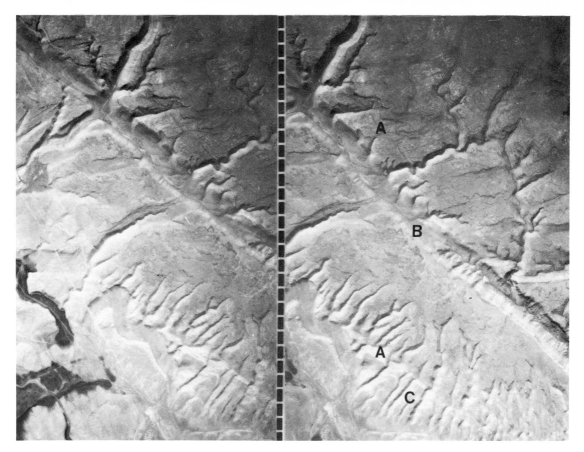

Figure 12-13 Stereogram of structural features in the Permian Kaibab limestone, Coconino County, Arizona. Annotations are as follows: (*A*) normal fault, (*B*) Mesa Butte graben, and (*C*) Additional Hill monocline. Scale is about 1:80,000.

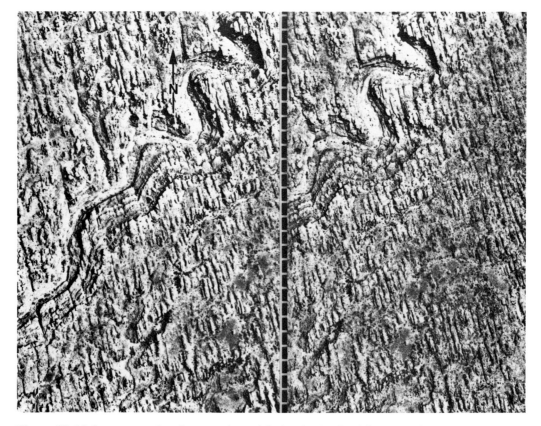

Figure 12-14 Stereogram showing prominent jointing in the flat-lying Jurassic Entrada sandstone near Moab, Utah. Scale is 1:20,000. (Courtesy U.S. Geological Survey.)

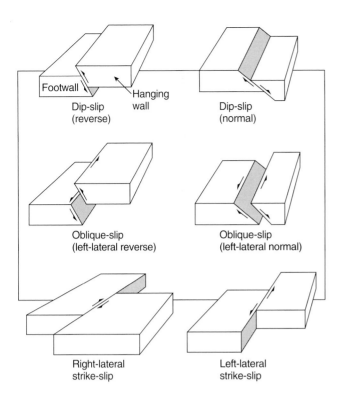

Figure 12-15 Diagram illustrating dip-slip, strike-slip, and oblique-slip faults. Fault planes are shaded; right-lateral reverse and right-lateral normal faults are not shown. (Adapted from Berlin 1980.)

Footwall
Hanging wall
Dip-slip (reverse)
Dip-slip (normal)

Oblique-slip (left-lateral reverse)
Oblique-slip (left-lateral normal)

Right-lateral strike-slip
Left-lateral strike-slip

Figure 12-16 Diagram illustrating the topographic attributes of horsts, grabens, and tilt blocks. (Adapted from Bunnett 1968.)

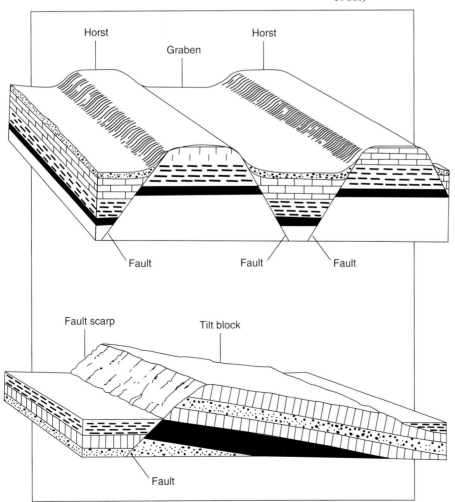

Horst
Graben
Horst
Fault
Fault
Fault

Fault scarp
Tilt block
Fault

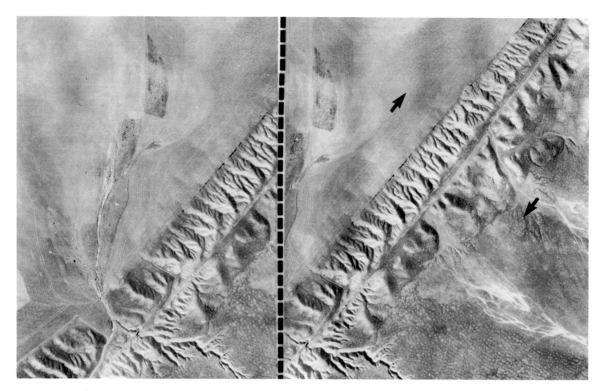

Figure 12-17 Stereogram of the right-lateral San Andreas fault on the Carrizo Plain, California. Note offset drainage and trench in Quaternary deposits along the fault trace. Scale is 1:20,000.

Figure 12-18 Stereogram of the right-lateral Hayward fault (*A-A'*) and the University of California, Berkeley. Memorial Stadium is sited directly on the fault. Scale is 1:60,000.

Figure 12-19 Small-scale (1:120,000) vertical photograph of the left-lateral Garlock fault (*A-A'*) near Searles Valley, California. Cloud shadows obscure ground detail at *B*. Shorelines of Pleistocene Searles Lake are evident at *C*. Exposure was made with a sun elevation angle of 34°. (Courtesy U.S. Geological Survey.)

presumably reflects a subsurface phenomenon.'' Field investigation is normally required to determine the validity of lineaments mapped as inferred or suspected faults.

Clues for locating possible faults on airphotos include the following: (1) deflected and/or straight stream channels (Figure 12-20); (2) straight contacts between erosional and depositional features (Figure 12-19); (3) straight valleys in hard rock areas; (4) linear alignments of natural vegetation and lakes, including sag ponds (Figure 12-21); (5) linear features crossing drainage channels (Figure 12-6); (6) topographic scarps (Figure 12-5); (7) distinct hue or tonal changes on opposite sides of a lineament; (8) lineaments detectable in both rock and adjoining surficial materials; and (9) offsets in drainage channels, topographic features, and lithologies (Figure 12-17).

Drainage Patterns

The type of drainage system prevailing on a given terrain surface is largely controlled by the soil type or surficial deposit, slope, parent material, and underlying structure. Generally speaking, most large surfaces develop diagnostic drainage patterns that are easily recognizable on aerial photographs because of their geometric attributes. These patterns can provide a great deal of information about the surface and subsurface characteristics of a landscape.

The absence of a drainage system also provides information of significance. For example, the lack of a well-defined drainage network might indicate the presence of **porous rock**, such as basaltic lava, where surface water percolates downward through cracks and cavities. In other instances, **soluble rock**, such as limestone, may absorb runoff through sinkholes and underground solution channels.

Twelve common drainage patterns are shown in Figure 12-22; the following descriptions are adapted from Way (1978), Strandberg (1967), and von Bandat (1962).

1. The **dendritic pattern** is the most common of all stream patterns. It is characterized by a random, tree-like branching system in which the tributaries join the gently curving mainstream at acute angles (Figure 12-23). This pattern indicates homogeneous soil or rock materials (i.e., same resistance to erosion) with little or no structural control. It is typified by landforms composed of soft, flat-lying sedimentary rocks, massive crystalline rocks, volcanic tuff, and thick glacial till. Tidal marshes and sandy coastal plains may also develop dendritic drainage.

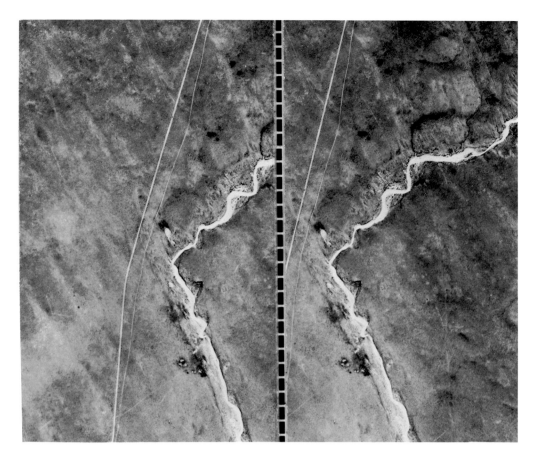

Figure 12-20 Deflected, straight-channel segment of an intermittent stream in New Mexico indicating the location of a possible fault. Scale is 1:20,000.

Figure 12-21 Vertical airphoto of a northwest-trending (photo *left*) strike-slip fault on the floor of Owens Valley, California, marked by sag ponds in alluvial deposits. Scale is 1:37,400.

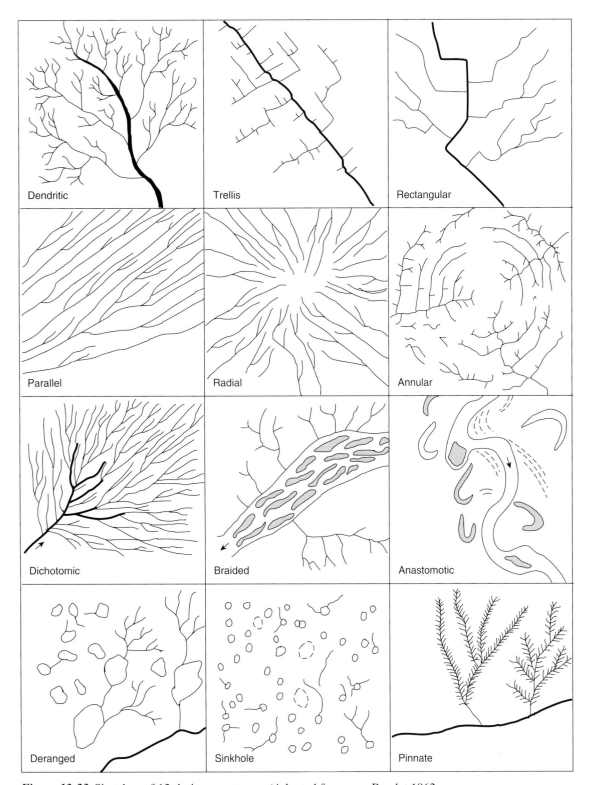

Figure 12-22 Sketches of 12 drainage patterns. (Adapted from von Bandat 1962 and Strandberg 1967.)

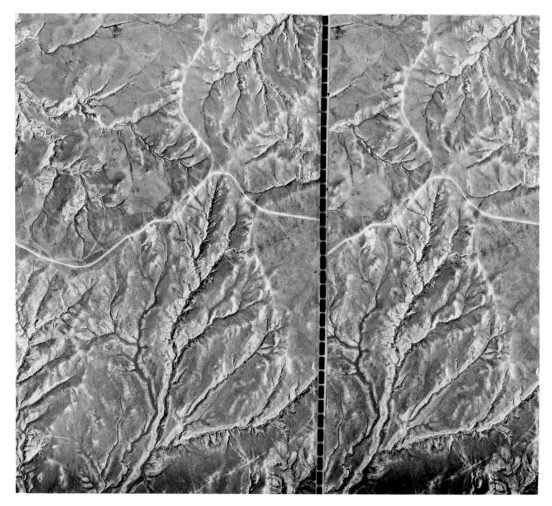

Figure 12-23 Stereogram showing dendritic stream pattern developed on the flat-lying Fort Union Formation (clay shale, siltstone, and sandstone) in eastern Montana. Note how the roads follow stream divides. Scale is about 1:20,000.

Figure 12-24 Stereogram showing trellis stream pattern developed along joints in flat-lying sandstone; location is New Mexico. Scale is 1:20,000.

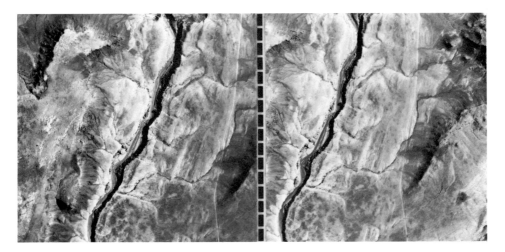

Figure 12-25 Stereogram of a semiarid region in New Mexico showing rectangular drainage to the right of the entrenched mainstream and dendritic drainage to its left. Scale is 1:20,000.

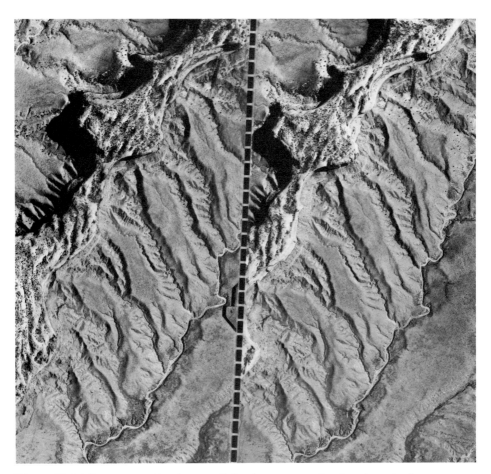

Figure 12-26 Stereogram showing parallel stream pattern developed on sloping, fine-textured alluvium; location is New Mexico. Scale is 1:20,000.

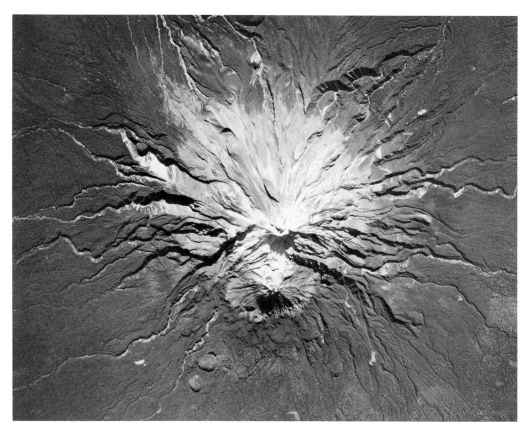

Figure 12-27 Vertical airphoto of Mt. Egmont, New Zealand, showing the centrifugal pattern of radial drainage. This snow-capped composite volcano rises 2,520 m above sea level and was photographed from an altitude of about 9,100 m by New Zealand Aerial Mapping, Ltd.

2. The **trellis pattern** resembles a vine trellis and is a modified dendritic form. It is characterized by straight, parallel primary tributaries and shorter secondary tributaries that join the larger branches at right angles (Figure 12-24). This pattern is structurally controlled, developing along folded and tilted sedimentary strata or along faults and joints (areas of weakness) in hard resistant rocks of granular texture (e.g., granite, slate, and massive sandstone).

3. The **rectangular pattern**, another variation of the dendritic system, consists of tributaries that join the mainstream at approximate right angles. This pattern frequently reflects a regional pattern of intersecting joints or faults and foliations. The "stronger" the stream imprint, the thinner the soil cover. The rectangular pattern is often formed in metamorphic rocks (e.g., slate, schist, and gneiss), resistant sandstone in arid climates, or in sandstone in humid climates where the soil profile is thin (Figure 12-25).

4. The **parallel pattern** consists of steams flowing side by side in the direction of the regional slope; the parallel channels characteristically join a mainstream at about the same angle (Figure 12-26). This pattern develops where the streams are formed on steep slopes of the same fine-textured material or along parallel fractures in hard, resistant rock.

5. The **radial pattern** resembles a spoked wheel and may be either centrifugal or centripetal. With the **centrifugal pattern**, streams flow radially outward and downward from a symmetrical hill, such as a dome or volcano (hub of the wheel higher than the rim) (Figure 12-27). The **centripetal pattern** develops where streams flow radially inward and downward toward a basin or depression such as a dry lakebed or playa (rim of the wheel is higher than the hub) (Figure 12-28). A volcanic cinder cone often has centrifugal drainage on its sides and centripetal drainage inside its crater (Figure 12-29).

6. The **annular pattern** is formed when stream courses adjust to follow a circular path around the base of resistant hills. It may also develop as a modification of the radial pattern when an intruded body has upwarped bedded sedimentary rock of different strengths. Here, the ringlike tributaries follow the less resistant layers of the tilted beds, intersecting the radial channels at approximate right angles (Figure 12-30).

7. The **dichotomic pattern** commonly develops on alluvial fans and deltas (Figures 12-5 and 12-6). The stream courses fan out, distributing the flow from the main channel through a series of branching **distributary channels**. For alluvial fans, runoff disappears

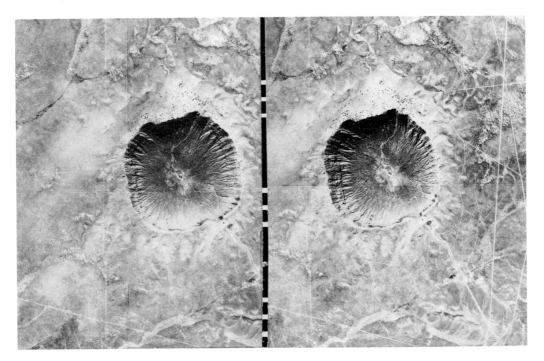

Figure 12-28 Stereogram of Meteor Crater, Arizona, showing the centripetal pattern of radial drainage. This impact feature is about 1,265 m across and 175 m deep. Scale is about 1:40,000.

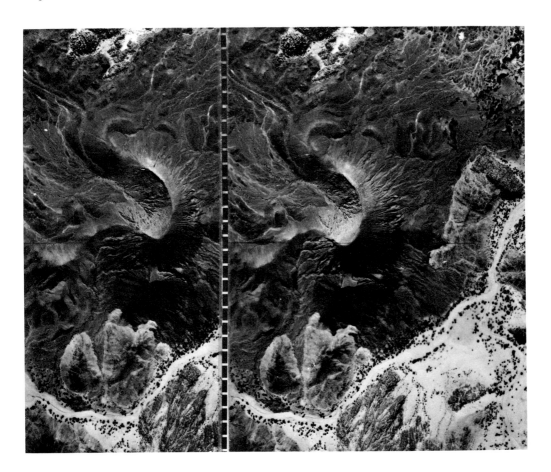

Figure 12-29 Stereogram of a volcanic cinder cone breached by a subsequent basaltic lava flow in Africa. Note centrifugal drainage pattern on the side of the cone and centripetal drainage pattern inside the crater. (Courtesy U.S. Air Force.)

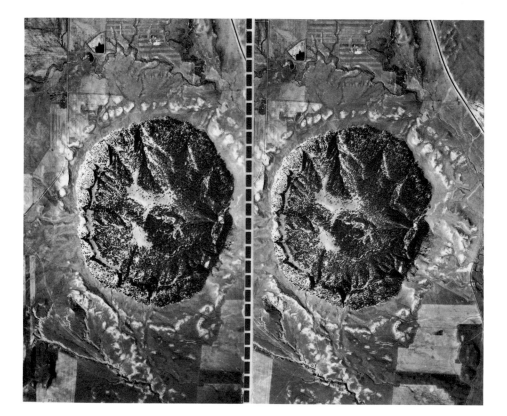

Figure 12-30 Stereogram of Green Mountain near Sundance, Wyoming, showing an annular drainage pattern. Green Mountain is a classic example of a youthful laccolithic dome; it was produced by the intrusion of an igneous mass between the bedding planes of sedimentary rock strata, forming a lenticular mass convex upward. The crystalline core is not yet exposed, but several layers of weak strata have been eroded away, leaving an upturned ridge of serrated sandstone (hogback) encircling its base. The small central depression indicates that another layer of strata is beginning to be removed (Curran et al. 1984). Scale is about 1:40,000. (Courtesy U.S. Geological Survey.)

into coarse, granular sediments (high permeability), with the coarsest materials found at the apexes. When this arrangement of streams forms on the birdfoot type of river delta, it indicates the deposited material consists of fine-grained sediments (low permeability).

8. The **braided pattern** develops on broad floodplains or alluvial terraces and is controlled by the load of the stream (Figure 12-31). It occurs when stream velocity becomes insufficient to carry bed and suspended loads, depositing them in the channels. Braided stream channels are good sources of sand and gravel, and large volumes of water can often be obtained from shallow wells sited along their banks.

9. The **anastomotic pattern** is characteristic of mature floodplain drainage. The meandering of the mainstream produces meander scrolls or loops and interlocking channels along its serpentine course; drainage features include meander scars and oxbow lakes (Figures 3-6 and 12-32).

10. The **deranged**, or **disordered**, **pattern** represents nonintegrated and very irregular drainage systems. The pattern usually indicates a relatively young landform with a level or slightly undulating surface, high water table, and poor drainage. The deranged pattern is characterized by short streams and random swamps, bogs, small lakes, or ponds. Regional streams may meander through the area, but they do not influence the local drainage. This drainage pattern typically develops on glacial till plains and granular moraines.

11. The **sinkhole**, or **swallow-hole**, **pattern** consists of short streams that end in depressions or that disappear and flow underground; these subterranean streams may reemerge at the surface as large springs. The bedrock underlying the areas in which this pattern develops is normally massive limestone, where pits and sinks have formed by chemical solution or by the collapse of caves (i.e., **karst topography**).

12. The **pinnate pattern**, a modification of dendritic drainage, indicates a high silt content of the residual soil and typically forms where loess blankets an area. The drainage follows a featherlike branching pattern composed of many short, parallel gullies and tributaries that intersect mainstreams at slightly acute angles upstream; headwater basins are often pearshaped (Figure 12-33).

Figure 12-31 Stereogram showing a braided stream pattern in New Mexico. Flow was occurring in the mainstream when the exposure was made, but the local tributaries were dry. Scale is 1:20,000.

Drainage Texture

Drainage patterns can be further classified by variations in channel density per unit area (subjectively defined). This is known as **drainage texture**, for which there are three main categories (Figure 12-34); each type is readily observable on aerial photographs.

1. **Fine-textured drainage** has a high drainage density (closely spaced channels) and develops on easily eroded formations where surface runoff is high. This texture may be associated with weak sedimentary strata or soils of low permeability (e.g., shale and clay).

2. **Medium-textured drainage** has a moderate drainage density (moderately spaced channels) and develops on soil and bedrock having a moderate permeability (e.g., thin-bedded sandstone).

3. **Coarse-textured drainage** has a low drainage density (widely spaced channels) and develops on hard, resistant rock formations (e.g., granite, gneiss, and quartzite) and highly permeable materials (e.g., sand and gravel) because little water is available as surface runoff.

Lithologic Analysis

The crust of the earth is composed of various kinds of rock that can be exposed (**outcrops**) or concealed by soil, surficial deposits, and vegetation. The climate and stage of erosion are important influences on rock appearance. Because climate controls the amount of moisture in a region, it directly influences soil formation, the degree of weathering, the rate of erosion, and the amount and type of vegetative cover. As a general rule, major lithologic units are more easily identified on airphotos of arid and semiarid regions where soil and vegetation cover is sparse. However, in humid regions where obscuring agents are strong, it is often possible to identify the principle lithologic units indirectly on airphotos by the criteria of topographic expression, drainage pattern and texture, residual soil color or tone, structural imprints, and the zoning patterns of natural vegetation. The following sections, adapted from Way (1978) and von Bandat (1962), describe the criteria of greatest value for identifying the major sedimentary, igneous, and metamorphic rock units.

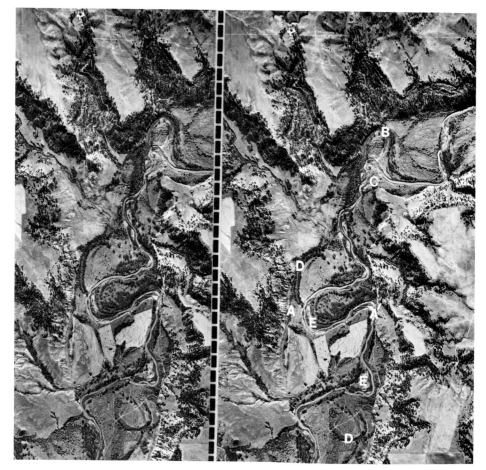

Figure 12-32 Stereogram of meander floodplain of the graded Belle Fourche River (anastomatic drainage pattern), Crook County, Wyoming. Annotated features are as follows: (*A-A'*) floodplain boundary (confining valley walls), (*B*) partially filled oxbow lake, (*C*) abandoned channel, (*D*) filled meander channel, and (*E*) pointbar with meander scars. In stereo, it is possible to see alluvial terraces at different heights above the river. Scale is about 1:28,000. (Courtesy U.S. Geological Survey.)

Figure 12-33 Oblique view of a heavily eroded loess deposit near the Ching-Ho River in the People's Republic of China. The slopes are terraced and intensively cultivated. (Courtesy U.S. Air Force.)

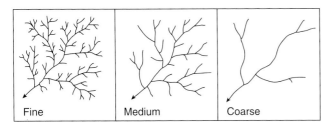

| Fine | Medium | Coarse |

Figure 12-34 Sketches of fine-, medium-, and coarse-textured dendritic drainage patterns. (Adapted from a U.S. Geological Survey drawing.)

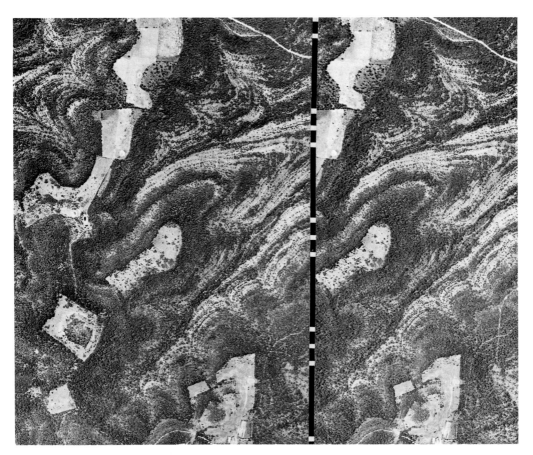

Figure 12-35 Stereogram of flat-lying, interbedded sedimentary rocks in Baxter County, Arkansas. Dark bands supporting heavier vegetation are sandstone; lighter-toned bands are principally shale or limestone beds with less vegetation. The cap rock is sandstone. Scale is 1:20,000.

Flat-Lying Sedimentary Rocks

Sedimentary rocks, primarily **sandstone**, **shale**, and **limestone**, are the most common outcropping formations on the continents. Consequently, they are the principal rock types encountered by the photo interpreter. Sedimentary rocks are originally laid down in horizontal layers (Figure 12-14), but they may later become tilted or folded by the forces of diastrophism (Figures 12-11 and 12-12). This section is devoted to a discussion of flat-lying sedimentary formations that are tilted no more than a few degrees.

Sandstone is an aggregate of cemented sand grains and is a hard, weather-resistant rock. Because of sandstone's rigidity, it is broken and dislocated more easily than other more plastic rocks such as shale; fractures become visible on airphotos when they become widened and deepened by water or wind erosion (Figure 12-14). Sandstone topography

in humid regions is rolling to hilly; large hills are rounded with steep slopes (Figure 12-35). There is little surface erosion because of sandstone's high resistance to weathering and erosion and its relatively high porosity; the drainage pattern tends to be coarse dendritic.

In the mature stage of erosion, sandstone topography in arid and semiarid regions is rugged and angular, with isolated flat-topped plateaus, mesas, and buttes (Figures 12-36 and 12-37). Cliffs normally occur where sandstone overlies weaker sedimentary rock (Figure 7-37). Because of a lack of a thick residual soil cover, the fractures in sandstone often have maximum control over the drainage pattern (e.g., angular, dendritic, trellis, or rectangular). Sandstone outcrops normally photograph in light tones unless they are coated with desert varnish. In the latter case, the ferruginous sandstone will register in dark tones.

Shale is formed from the deposition and compaction

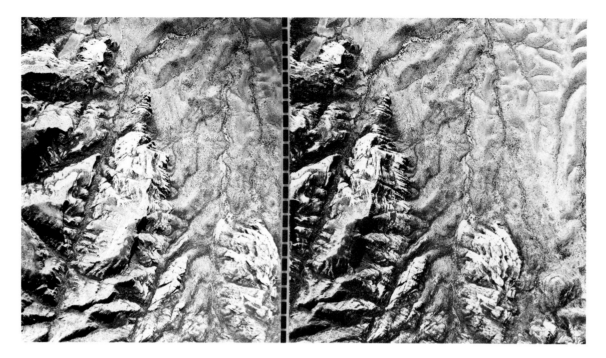

Figure 12-36 Stereogram of rugged sandstone topography in Coconino County, Arizona; note the extensive jointing in the massive, crossbedded sandstone. Less rugged areas are underlain principally by shale. Scale is about 1:52,000.

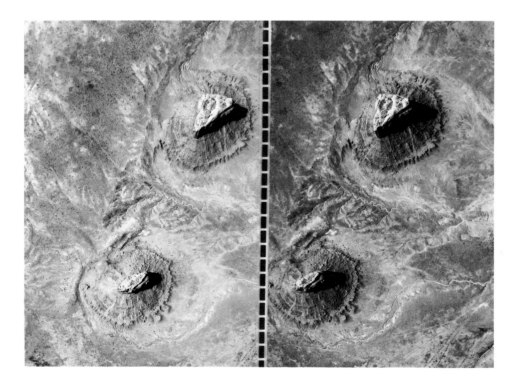

Figure 12-37 Stereogram of two sandstone-capped buttes in New Mexico; note how the sandstone forms vertical walls. The sandstone is underlain by soft, easily eroded shale and shaly siltstone, which forms slopes. These buttes are the erosional remnants of a once-continuous cover of sedimentary rocks. Scale is 1:20,000.

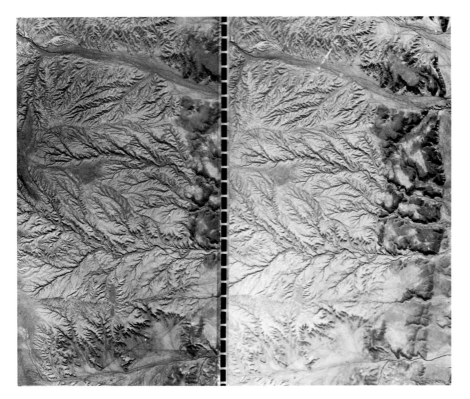

Figure 12-38 Stereogram of dissected shale terrain (i.e., badland topography) in New Mexico. The fine-grained shale produces a closely spaced (fine-texture) pinnate pattern resembling the veins of a cabbage leaf. Scale is 1:20,000.

Figure 12-39 Vertical airphoto showing karst topography with its characteristic sinkhole lakes in the Central Ridge District, Florida. Note the absence of a surface drainage system. This region of the state is famous for its citrus groves. Scale is about 1:20,000.

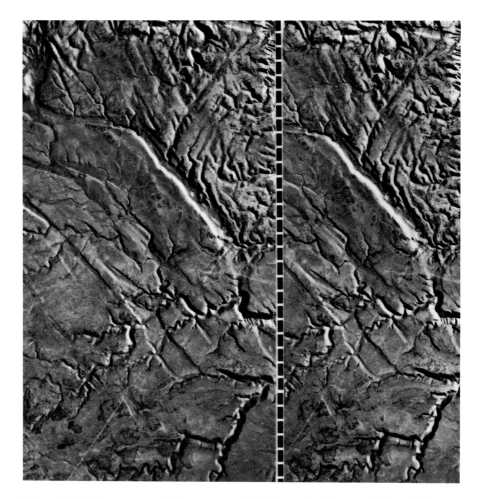

Figure 12-40 Stereogram of limestone terrain in Coconino County, Arizona. The drainage pattern is very angular, following fracture alignments. Scale is about 1:100,000. (Courtesy U.S. Geological Survey.)

of silts and clays. Shale is an impervious rock, but it is weak and easily eroded. In humid regions, smooth rounded hills are characteristic of shale deposits, and photographic tones are mottled because of variations in moisture and organic material. Soft shales exert no control over the drainage system, permitting a medium- to fine-textured dendritic pattern to develop.

Shale terrain in arid regions is called **badland topography**, which is characterized by minutely dissected hills with sharp ridgelines and steep sideslopes, reflecting the soft nature of rock; drainage is fine textured and pinnate (Figure 12-38). The general tonality of shales is light on panchromatic airphotos.

Limestone is formed by the consolidation of calcareous shells of marine animals or by the chemical precipitation of calcium carbonate from seawater. Mature landscapes of limestone in humid regions are undulating to hummocky and are easily recognized by their circular- or oval-shaped sinkholes (**karst topography**) and associated internal drainage (Figure 12-39). Because of solution cavities within the rock and the high permeability of the residual soil, limestone regions are drained internally, leaving little water to be collected in a surface water system. Consequently, few major streams are developed (Figure 12-39).

Because little moisture is available for chemical weathering in arid climates, limestones erode very little. They form caprocks with vertical faces, developing none of the characteristics associated with karst topography. The drainage system is well developed and tends to be very angular, following fracture alignments in the bedrock (Figure 12-40).

Differential erosion of flat-lying, interbedded sedimentary rocks produces **stair-stepped**, or **terracelike, topography**. In sandstone-shale combinations, the more resistant sandstone remains as a caprock with steep sideslopes or vertical cliffs, whereas shale forms more gradual slopes (Figure 12-37). In limestone-shale combinations, limestone occupies the hilltops and uplands and may have solution features in humid regions. Both sandstone and limestone maintain steep escarpments in arid regions (Figure 12-37).

Flat-lying beds are indicated by strong contrast in photographic tone (**banded pattern**), which results primarily from the exposure of the different bedding traces along topographic contours in dry regions and zonal differences in vegetation in humid regions (Figures 3-9 and 12-35). In dry regions, light-toned beds are often sandstone and limestone, whereas the darker-toned beds are shale (Figure 3-9). In humid regions, vegetation tends to be preferentially concen-

Figure 12-41 Stereogram of gently dipping strata in San Juan County, New Mexico. The resistant, cross-bedded sandstone beds stand out as low ridges with steep scarp slopes and gentler back slopes; such asymmetrical ridges are called cuestas. Note that the back slope of the uppermost ridge is parallel to the dip and has a parallel drainage pattern. The valley between the two cuestas is composed of weak and easily eroded shale. Here, the drainage pattern is not determined by the dip of the shale complex; shale, regardless of dip, will always develop a dendritic-pinnate form. A stream divide is clearly discernible in the shale complex. Scale is 1:20,000.

trated along areas underlain by sandstone, which produces dark photographic tones (Figure 12-35).

Tilted Sedimentary Rocks

Horizontal sedimentary strata may become tilted or inclined through folding and faulting (Figures 12-10 and 12-16). The residual landforms of tilted sedimentary strata have distinctly different appearances than those of horizontally bedded strata. Because different sedimentary rocks have different resistances to weathering and erosion, the more resistant of the tilted beds dominate in the landscape as **upland features**, whereas the softer units form **lowland features**.

Interbedded sedimentary rocks that have been faulted into tilt blocks and differentially eroded form a parallel or nearly parallel series of resistant ridges that may be closely spaced or separated by wide valleys. For sandstone-shale combinations in arid regions, the strongest sandstones form sharp-crested ridges, with the thinner beds forming the sharpest crests, whereas the shale units are eroded to valleys or low rounded hills (Figures 12-41, 12-42, and 12-43). Ridge crests tend to be more rounded in humid regions, with forests covering the steep slide slopes.

For gently dipping strata (Figure 12-8), the resistant beds stand out as low ridges with steep scarp slopes and gentler back slopes; this type of **asymmetrical ridge** is known as a **cuesta** (Figures 12-41 and 12-42). For steeply dipping strata (Figure 12-9), the resistant beds form narrow-crested ridges in which the front and back slopes are both steep; this type of **symmetrical ridge** represents a **hogback** (Figure 12-43).

The limbs of folded strata also form distinctive topographic expressions in the youthful and mature stages of erosion. For example, dissected intrusive domes are often encircled by upturned sandstone hogbacks of various heights and shapes that dip away from the intrusive mass (Figure 12-30). When a lower bed is weak (e.g., shale), a lowland is eroded between the hogback and the intrusion. In moderate- to high-relief terrain of folded strata, parallel hogbacks or cuestas with recognizable dip slopes mark the flanks of anticlines and synclines (Figures 12-11 and 12-12).

Intrusive Igneous Rocks

Intrusive igneous rocks, also called **plutonic rocks**, are formed when molten **magma**, the parent material, slowly cools and crystallizes within the earth's crust. Here it as-

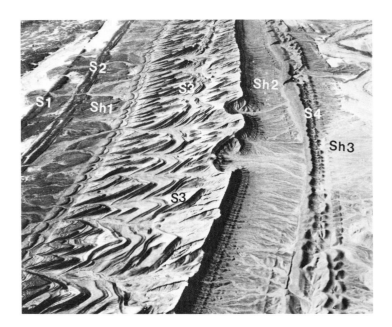

Figure 12-42 Oblique view of a sandstone-shale complex on a structural flank in the Atlas foothills, Mauritania. The gently dipping sandstone beds (*S1–S4*) are interbedded with shales (*Sh1–Sh3*). The sandstones form resistant cuestas, whereas the shale units are eroded to valleys. *S3* is a thick sandstone sequence with thin shale beds; the triangular-shaped sandstone spurs are called flatirons. Most of the sandstones are coated with dark desert varnish. (Courtesy U.S. Air Force.)

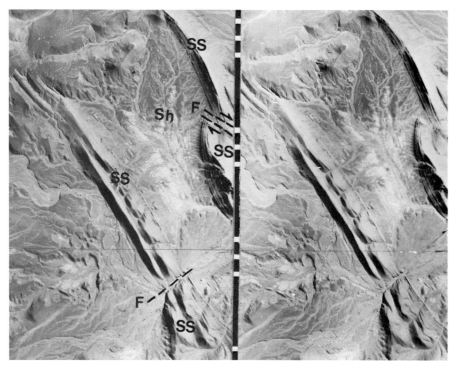

Figure 12-43 Stereogram of tilted and faulted sandstone formations in the northern Sahara Desert. Clearly discernible are steeply dipping (near vertical) sandstone ridges called hogbacks (*SS*) that are dislocated by faults (*F*). The large valley between the hogbacks is a shale complex (*Sh*). Scale is about 1:40,000. (Courtesy U.S. Air Force.)

sumes a variety of forms that may later be exposed at the earth's surface through erosion. The largest and deepest seated intrusive bodies are called **batholiths**, which often occur as roots of mountain systems (e.g., Sierra Nevada, Andes) or as crystalline shields (e.g., Canadian Shield, Deccan Plateau in India, Arabian Shield). **Laccoliths** are less extensive and dome-shaped, which arches up the overlying strata (Figure 12-44). **Stocks** are small intrusive masses, usu-

ally being a few kilometers in diameter (Figure 12-45). The intruding magma will often send projections into the surrounding rocks, forming tabular **dikes** and **sills**. When the mass of magma cuts *across* bedding planes (e.g., along fault planes), it forms wall-like dikes, which may be vertical or inclined. Most dikes stand up as linear ridges when exposed at the surface (Figure 12-46). Much less common are **ring dikes**, which are formed by the intrusion of magma along

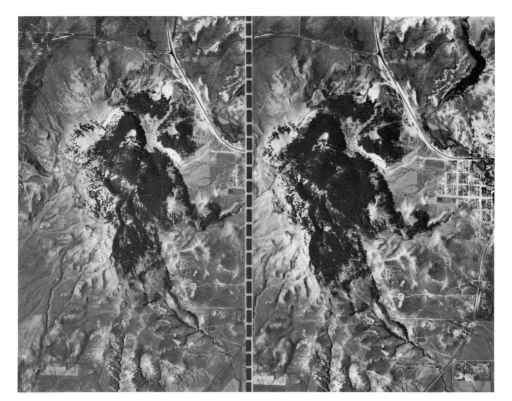

Figure 12-44 Stereogram of Sundance Mountain, Wyoming, a maturely eroded laccolith. The arched-up sedimentary strata have been eroded away, exposing the domed roof of the crystalline core. Note that the dome is eroding as a homogeneous body, not a stratified one (Curran et al. 1984). A nearby laccolith in the youthful stage of erosion is shown in Figure 12-30. Scale is about 1:40,000. (Courtesy U.S. Geological Survey.)

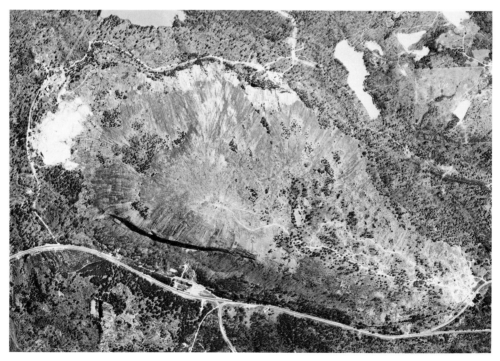

Figure 12-45 Vertical airphoto of Stone Mountain, Georgia, a monadnock formed from a granitic stock. Its smoothly rounded surface is thought to be caused by erosion of large exfoliation shells (Curran et al. 1984). Stone Mountain is about 11 km in circumference and 360 m in height. Scale is about 1:20,000.

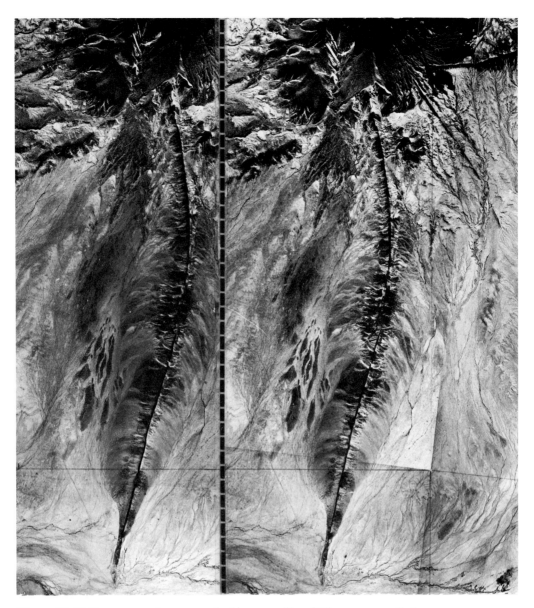

Figure 12-46 Stereogram of a near-vertical igneous dike of Pliocene age radiating from Shiprock, a deeply eroded volcanic neck in San Juan County, New Mexico. The dike material filled a fissure and now forms a wall-like ridge that has been etched into relief by erosion of the surrounding Mancos shale of Cretaceous age. Scale is 1:20,000. (Courtesy U.S. Department of Agriculture.)

the trace of circular fault. Sills form when a tabular sheet of magma is intruded *between* bedding planes; some sills that are tilted by faulting form ridgelike escarpments when exposed by erosion.

Common plutonic rocks include **granite** (most common), **diorite**, **diabase**, and **gabbro**, which develop similar landforms and drainage patterns. As seen in panchromatic airphotos, their major difference, for a common landscape feature, is mainly in tone. **Acidic** igneous rocks, such as granite, are usually light-toned unless coated with desert varnish, whereas **basic** igneous rocks, such as diabase and gabbro, are dark to black (Figures 12-45 and 12-46).

The porosity of granite and other intrusives is very low, making them highly weather resistant in all climatic settings. Because they are massive and homogeneous, their resistance to weathering is uniform when their masses are only moderately fractured. The relief of granitic formations in humid and arid regions typically shows as massive, rounded, domelike hills (Figure 12-47). A dendritic drainage pattern of medium to fine texture is common; the domelike hills cause curvilinear segments (resembling sickle shapes) to develop, and these are important evidence in the identification of granite (Figure 12-47).

The domelike appearance of granitic hills reflects a weathering process known as **exfoliation**. This occurs when thin concentric shells break off from the parent rock mass

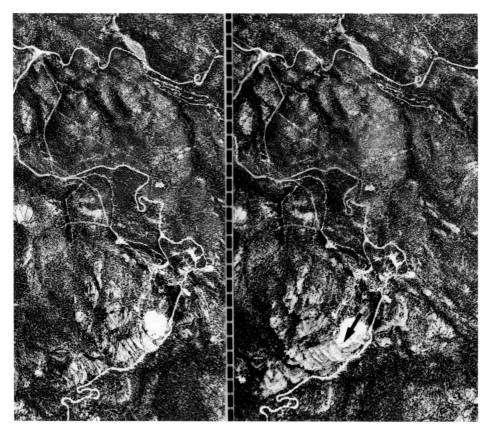

Figure 12-47 Stereogram of granitic topography in the Black Hills, South Dakota. Irregularly spaced joints in several different orientations are detectable in outcrop areas. The occurrence of trees of predominantly one variety (ponderosa pine) is suggestive of bedrock with a uniform composition. The granites are Precambrian in age. Arrow points to the location of the Mount Rushmore National Memorial. Scale is 1:23,600. (Courtesy U.S. Geological Survey.)

as a result of a combination of temperature changes, freeze-thaw action, and perhaps by minor chemical effects (Figures 12-45 and 12-47).

Because granitic rocks solidify from molten material, the cooling magma becomes fractured. This fracturing will be close-spaced on the outer part of the mass, where contraction is more intense, than in its core, where cooling is a very slow process. At the surface, fractured granite will weather into large cuboidal blocks called **woolstacks**; fracture adjustment causes the development of an angular drainage pattern, such as trellis or rectangular (Figure 12-48).

Extrusive Igneous Rocks

Extrusive igneous rocks, also called **volcanic rocks**, are formed by the rapid cooling and solidification of molten material after it breaks through the earth's crust via a **vent** (hole) or a **fissure** (crack). Common extrusive rocks are **andesite, basalt, dacite**, and **rhyolite**. Each extrusive rock has an intrusive equivalent—for example, rhyolite from granite and basalt from diabase. The acidic group of extrusive rocks, such as dacite and rhyolite, are usually light-toned in panchromatic photographs, whereas the basic extrusive rocks, such as basalt, register in dark tones. The massive forms of

extrusive rocks are called **lavas** (nonviolent eruption), whereas the fragmental materials are called **pyroclastics** (violent eruptions); pyroclastics form **breccia** and **tuff** when consolidated.

Volcanoes are mounds or cone-shaped features built by the eruption of molten rock through a relatively small central vent. There are three types of volcanoes: (1) **Cinder cones** are built entirely of pyroclastics of various sizes (Figures 3-28 and 12-29) and are the smallest type of volcano; (2) **lava cones**, also called **lava domes**, or **shield volcanoes**, are built by outpourings of lava; slopes are gentle for low-viscosity (highly fluid) lavas and steep for high-viscosity lavas, and the best examples are seen in the Hawaiian Islands; and (3) **composite cones**, or **stratovolcanoes**, are built of alternating layers of lava and pyroclastics; most of the world's majestic volcanoes are composite cones (Figure 12-27).

Volcanoes are readily identified on airphotos on the basis of their shapes. As they are dissected over time, a radial drainage pattern is developed on their slopes, which can further assist in their identification (Figures 12-27 and 12-29). The texture of the radial pattern is dependent upon climate, with the finest texture found in arid regions.

In the old stage of erosion, some or all of the internal

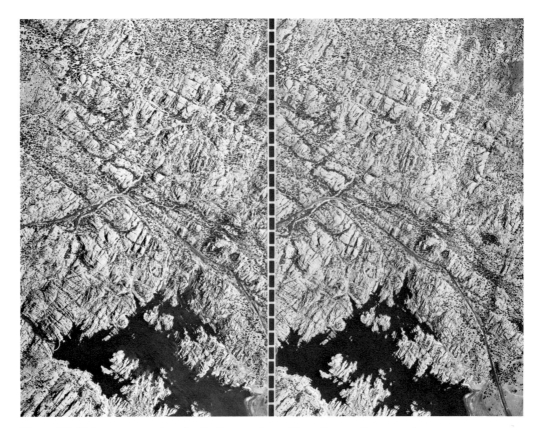

Figure 12-48 Stereogram of profusely fractured granitic rocks near Prescott, Arizona. Note the strong control of drainage by joints and faults. The granites are Precambrian in age. Scale is about 1:40,000.

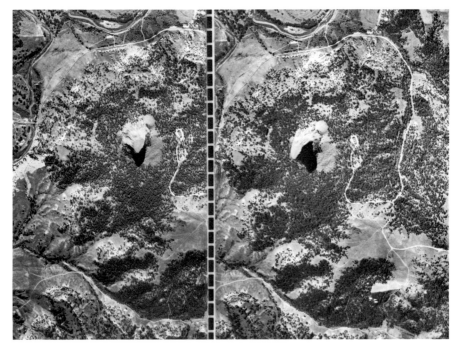

Figure 12-49 Stereogram of Devils Tower in northern Wyoming. This exposed volcanic neck of intrusive igneous rock is extremely steep-sided and has large-scale columnar jointing that was caused by shrinkage during cooling. It rises about 215 m above the surrounding hills. Around its base is an apron of talus. Scale is about 1:28,400. (Courtesy U.S. Geological Survey.)

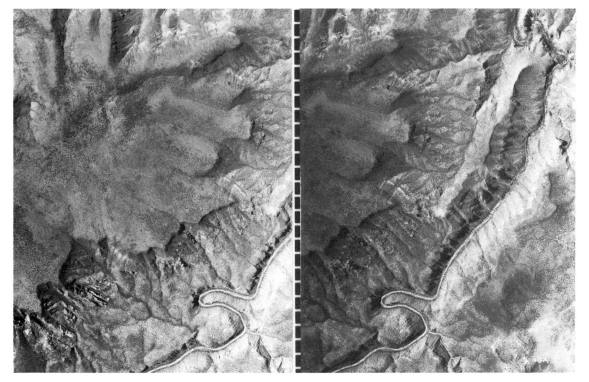

Figure 12-50 Stereogram of a basaltic lava flow overlying sedimentary strata in Arizona. Note the entrenched meander. Scale is about 1:52,000.

features of a volcano may be exposed, including the **neck**, or **plug**, and **radiating dikes** (Figure 12-46). The volcanic neck is a vertical shaft of igneous rock that represents the former feeder conduit. This vent material may be pyroclastic breccia, as is the case for Shiprock in New Mexico (Figure 12-46) or crystalline intrusive rock, as is the case for Devils Tower in Wyoming (Figure 12-49).

The most widely distributed extrusive rock is **basaltic lava**, which originally flowed over preexisting surfaces in the form of thin tongues or sheets, called **lava flows** (Figures 12-29 and 12-50). A lava flow may be flat or hilly, and minor surface irregularities are common. Canyon wall slopes are nearly vertical when breached by rivers, and both stratification and columnar jointing may be encountered (Figure 12-50). Many lava flows lack surface drainage because numerous fractures create high permeability. Basaltic lava flows normally appear in very dark tones on panchromatic airphotos (Figures 12-29 and 12-50).

Metamorphic Rocks

When extreme heat and pressure alter the mineral composition, texture, and structure of preexisting sedimentary and igneous rocks, the resultant materials are known as **metamorphic rocks**. This class of rock is associated with areas that have undergone severe deformation, including the uplifted cores of many mountain ranges and areas where rocks have been strongly folded and faulted. Common metamorphic rocks are **quartzite**, **slate**, **marble**, **gneiss**, and **schist**.

Metamorphism usually makes sedimentary rocks harder and, hence, more resistant to weathering and erosion.

Normally resistant sandstone, for example, becomes even more resistant when changed to quartzite, which forms prominent sharp-crested ridges under all climatic regimes. In spite of its original weakness, shale becomes much stronger when turned into slate. In both humid and arid regions, slate topography is rugged and is characterized by angular drainage, most often rectangular.

Gneiss is a term for a varied series of coarse-grained crystalline rocks with a banded or foliated structure that does not show morphologically. Gneisses originate from both igneous rocks (**orthogneiss**) and sedimentary rocks (**paragneiss**). Orthogneiss is commonly derived from granite and is morphologically similar to it in appearance and landforms. Glaciated regions develop the same topography, except that the topographic highs may be more rounded as a result of glacial smoothing (Figure 12-51). There is a great variety of paragneisses, depending on the sediments from which the rocks came. Paragneiss shows sharp-crested, parallel ridges when it is derived from massive sandstone and is smooth and irregular when it comes from tuffs and shales.

Schists are medium-grained crystalline rocks with a highly foliated or laminated, sometimes wavy, structure that are most frequently derived from sedimentary rocks. At least one mineral, such as mica or chloride, is crystallized into a platty form. Due to the cleavage of the parallel platty components, these rocks split easily along the banded laminations. Because of this friability, schists are easily broken down by weathering processes. Large areas of schist in humid climates develop deep residual soils and rounded hills with steep sideslopes. In arid regions, schist topography appears fairly rugged, with the form of the ridges and valleys being controlled by regional foliation (Figure 12-52).

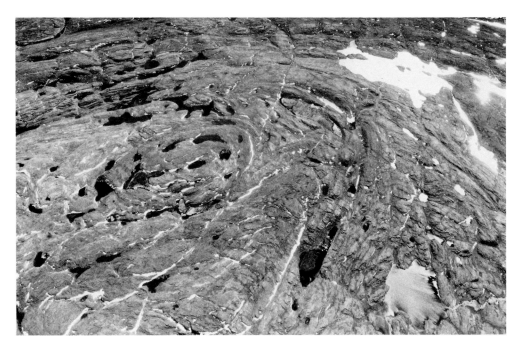

Figure 12-51 Oblique view of glaciated orthogneiss of Precambrian age in the Canadian Shield.

Fluvial Landforms

As defined here, **fluvial landforms** refer to those features formed by stream erosion, transportation, and deposition. **Meander floodplains** are formed by full-maturity streams subject to periodic flooding. During overflow periods, stream deposits on adjacent surfaces result in the formation of a broad valley of low relief; sediments are generally fine grained. These floodplains are characterized by many special features, including meander scrolls or loops, oxbow lakes, meander scars, abandoned channels, levees, and terraces along valley walls (Figures 3-6, 12-32, and 12-53).

Filled valleys are commonly found in arid and semiarid intermontane basins that have accumulated materials washed down from the bounding mountain ranges (Figure 12-5). Valley fill contains alluvium of a wide textural range, with the coarsest material found near the uplands. The lowest-lying areas may contain dry lakebeds called **playas**, which are commonly surfaced with silts and clays; when surfaced with evaporites the lakebed is called a **saltpan**, or **salina** (Figure 12-5). Playas are free of vegetation and are photographed in light tones when dry.

Alluvial fans occur along mountain fronts where sporadic flowing streams issue from steep canyons into a valley (Figures 3-5, 12-5, and 12-6). It is at these openings where velocity is suddenly diminished and the stream load is dropped and spread out in fan-shaped deposits. These fans generally slope about 1° to 10° toward the apex and are convex in cross section. Where mountain steams discharge close to each other, their fans may coalesce into a continuous sheet of aggraded sediments; this feature is called a **bajada**, or **alluvial apron** (Figure 12-5). Alluvial fans are easily identified on airphotos by their fan-shaped outlines and dichotomic drainage pattern.

Deltas are formed where rivers enter calm bodies of water such as lakes or seas. Reduced stream velocity results in buildups of sediments at the mouth of the river. One outstanding characteristic of deltas is a level surface. Differences in elevation caused by stream channels, natural levees, lakes, and backswamps are minor when the areal extent of the entire delta is considered. Slight slopes may occur in very small deltas or in deltas that are composed of coarse sediments.

The **arcuate delta** is most commonly observed. It is composed of coarse sediments, is triangular in shape, and always has a large number of distributaries (dichotomic drainage pattern). The delta of the Nile River is a prime example of the arcuate delta (Figure 5-10). The **birdfoot delta** is composed of very fine sediments, and the main channel divides into only a few distributaries. The delta of the Mississippi River is one of the best examples of a birdfoot delta. The **estuarine delta** develops at the mouth of a submerged river and assumes the general shape of the estuary (Figure 12-54).

Shoreline Features

The activity of moving water along the shorelines of oceans and large lakes, like the work of running water on land, can be erosional and depositional. These two processes

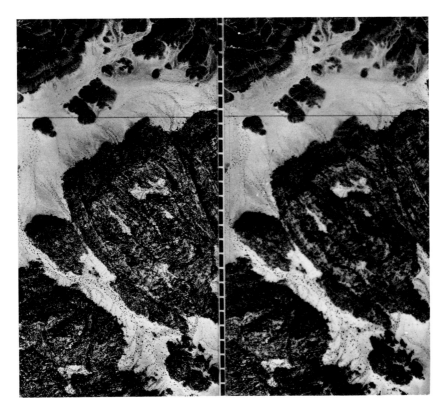

Figure 12-52 Stereogram of laminated schist of Precambrian age in Saudi Arabia. The rugged appearance seen in stereo results from alternating layers of hard and soft materials that are tilted upward, fractured, and weathered. Scale is about 1:40,000. (Courtesy U.S. Air Force.)

work hand in hand to produce a wide variety of shoreline features. Most features along a shoreline are very dynamic and are slowly and continually being changed by the normal actions of moving water. However, violent storms, such as hurricanes, with their accompanying strong wave and wind action, can cause profound changes to a shoreline in a few hours; an earthquake can bring about changes in a matter of seconds. Multidate airphotos can be used to identify and map these changes (Figures 12-55 and 12-56).

Depositional features along a shore are constructed of material eroded by the waves and transported (1) by **longshore currents** or **shore drifts**, (2) by material brought down by the streams from the landmasses, and (3) by material deposited by the wind. Currents and drifts may build up several types of sand ridges at the shoreline (Figure 12-54). When a linear sand ridge terminates in open water, it is called a **spit**; if it is curved, it is called a **hook**. When a spit extends from one headland to another, a **bar** results. Behind the bar, which now becomes a new shoreline, a shallow lake, or **lagoon**, forms. These features are illustrated in Figure 12-54.

On gently sloping shores, storm waves build up a **barrier beach**, also called a **beach ridge** or **storm beach**, which is a relatively low, narrow wall of sand that parallels the coast and is above the reach of normal waves (Figures 12-57 and 12-58). When formed in a series, each beach ridge represents an equilibrium line of a former shore that is indicative of an emerging coast or receding sea (Figure 12-58). The older ridges are often vegetated; the depressions

between the ridges (**swales**) can be occupied by small water bodies or swamps (Figure 12-58).

Tidal flats are formed in low-lying areas that are protected from direct wave action by bars, spits, and barrier beaches (Figures 12-57 and 12-58). Tidal flats have an imperceptible amount of relief, becoming totally or partially submerged at high tide. Tidal flats are of three types: (1) **Tidal marshes** are identified by their dense vegetation cover and the unique drainage pattern of wide, wandering dendritic channels; (2) **mud flats** are devoid of vegetation and have a similar drainage pattern, except that there are many small hairlike appendages; and (3) **sand flats** have neither vegetation nor a well-developed drainage system.

Coasts that have been uplifted are a type of **highland coast** characterized by narrow beaches, steep bluffs or cliffs, deep water close to the shore, and elevated wave-cut platforms called **marine terraces**. Along portions of the California coast are a series of wave-cut terraces, some as much as 400 m above sea level, that record a series of tectonic uplifts that occurred during the Pleistocene Epoch (Figure 12-59). Highland coasts are fully exposed to the surf, making erosion the dominant force; the only depositional feature is the narrow beach.

During the Pleistocene, many of the intermontane basins in the western United States held deep lakes. Remnant shoreline features of these large ancient lakes can be seen today around the margins of some of these basins. The expression of shorelines, or **strandlines**, of Pleistocene Searles Lake is shown in Figure 12-60.

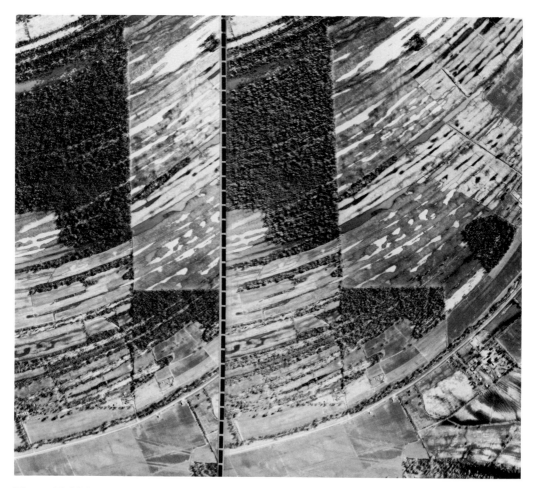

Figure 12-53 Stereogram showing a series of meander scars in Concordia Parish, Louisiana. This is a common pattern on floodplains such as that of the Mississippi River. Scale is about 1:22,000.

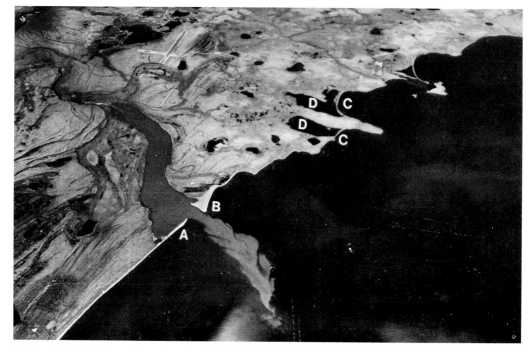

Figure 12-54 Oblique view of an estuarine delta forming in Llianna Lake, Alaska. Annotated features are as follows: (*A*) spit, (*B*) hook, (*C*) bar, and (*D*) lagoon. (Courtesy U.S. Air Force.)

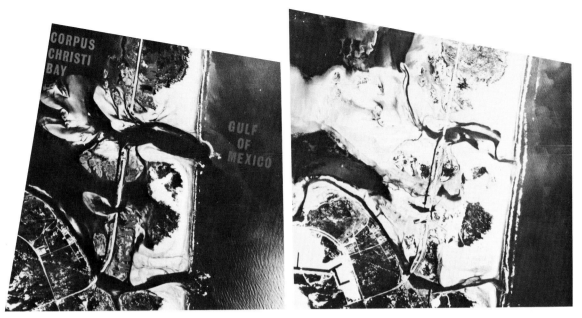

Figure 12-55 Vertical airphotos showing coastal changes after Hurricane Beulah damaged the Texas coast in September 1967. Major changes included the narrowing of stream channels and extensive infilling behind the barrier beach. (Courtesy U.S. Geological Survey.)

Figure 12-56 Vertical airphotos of the Hanning Bay fault at Fault Cove, Montague Island, Alaska, before (*top*) and after (*bottom*) the March 27, 1964, Alaskan earthquake (Richter magnitude = 8.4). The earthquake reactivated the reverse fault, uplifting the northwestern block some 5 m relative to the southeastern block. The bottom photo shows Fault Cove at low tide. Scale is 1:20,000. (Courtesy U.S. Geological Survey.)

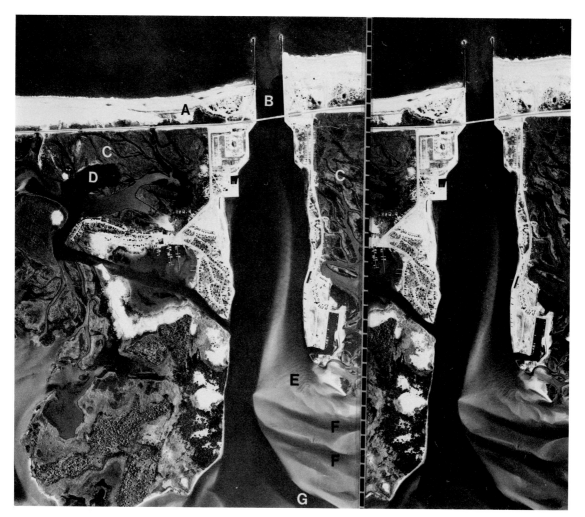

Figure 12-57 Stereogram of shoreline features at Rehoboth Bay, Sussex County, Delaware. Annotations are as follows: (*A*) barrier beach, (*B*) tidal inlet stabilized by rock jetties, (*C*) tidal flat, (*D*) lagoon, (*E*) oscillation ripples, (*F*) tide channel, and (*G*) main river channel. Note tonal signatures associated with different water depths. Scale is about 1:20,000.

Figure 12-58 (*Top, page 325*) Stereogram of shoreline features near Beaufort, South Carolina. Annotations are as follows: (*A*) present-day barrier beach or beach ridge, (*B*) young beach ridge becoming stabilized by vegetation, (*C*) old beach ridges stabilized by vegetation, (*D*) tidal flat, (*E*) offshore bar, and (*F*) shallow-water shoals. Scale is about 1:20,000.

Figure 12-59 (*Bottom, page 325*) Stereogram of the Pacific coast at Palos Verdes Hills, California, showing marine terraces of Pleistocene age, sea cliff, and narrow beach. These features are associated with uplifted coasts. Scale is 1:30,000. (For best viewing results, rotate stereogram slightly clockwise.)

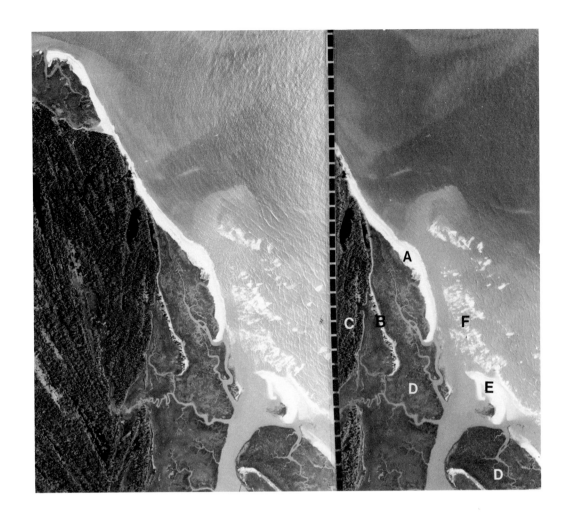

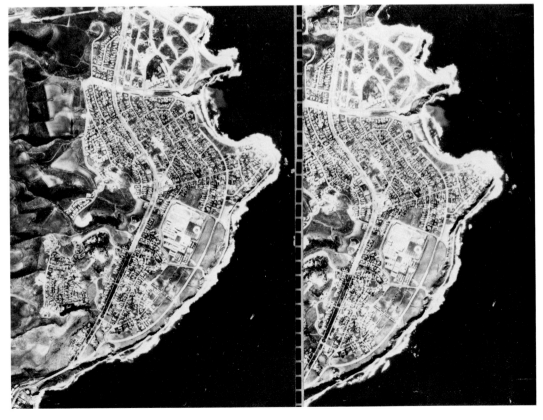

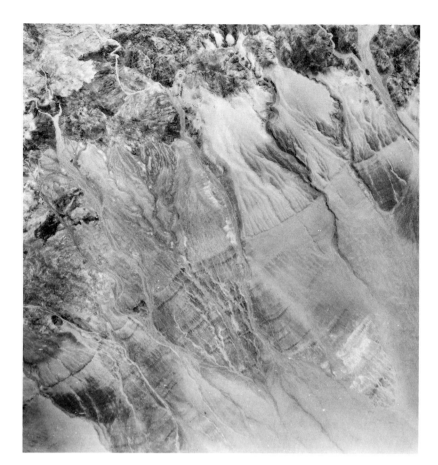

Figure 12-60 (Left) Vertical airphoto showing shorelines or strandlines of Pleistocene Searles Lake, California, on bedrock and alluvial fan deposits of Cenozoic age. Scale is 1:37,400.

Figure 12-61 (Below) Vertical airphoto of active alpine glaciation in the vicinity of Mount Cook, South Island, New Zealand. Annotations are as follows: (*A*) cirque in which alpine glaciers form, (*B*) arête, (*C*) horn, (*D*) hanging valley, (*E*) ice fall, (*F*) valley glacier with lateral moraines, (*G*) valley glacier with lateral and medial moraines, and (*H*) valley glacier completely covered with rock debris. (Courtesy New Zealand Department of Lands and Survey.)

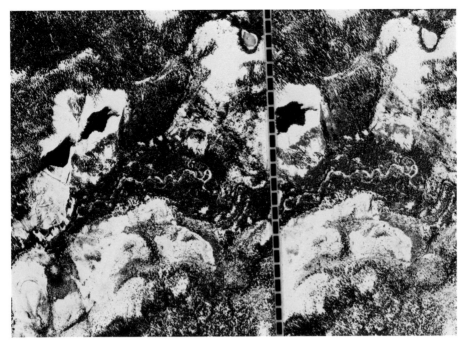

Figure 12-62 Stereogram of a glacial trough in the Sierra Nevada, California. Several hanging valleys are discernible when viewed in stereo. Scale is 1:47,000.

Glacial Features

Glaciation forms its own distinctive erosional and depositional features through the actions of moving ice and meltwater. There are two main types of glaciation: (1) **alpine glaciation**, which occurs today in most of the high mountain ranges of the world, and (2) **continental glaciation**, which occurs today at a reduced scale in only Greenland and Antarctica. Continental glaciation was widespread in the higher latitudes of the Northern Hemisphere during four glacial advances that occurred during the Pleistocene Epoch (**Great Ice Ages**). The last big **ice sheet**, or **continental glacier**, disappeared about 10,000 years ago from the northern states.

Several features of active alpine glaciation that are readily discernible in airphotos are shown in Figure 12-61. A **cirque** is a bowl-shaped valley head in a mountainside; glaciers create these amphitheaters by pulling and scraping rocks from their heads and sides. An **arête** is a narrow mountain ridge formed by the intersecting walls of two opposing cirques, whereas a **horn** is a pyramidal peak formed by the intersecting walls of several cirques. A **hanging valley** has a floor that is noticeably higher than the floor of its trunk valley, which is overdeepened by the scouring action of its larger glacier.

Airphotos can also depict the details of a glacier's surface (Figure 12-61). For example, an **icefall** is a maze of intersecting crevasses where a glacier encounters a steep valley slope, known as a **rock step**. Dark **morainal material**, or rock debris, is often observable along the edges (**lateral moraines**) or near the center (**medial moraines**) of a valley glacier. Where oversteepened walls occur, avalanches can drop rock debris over the entire top of a glacier.

After a change in climate has caused alpine glaciers to disappear, a striking erosional landform is revealed; this is the **U-shaped valley**, or **glacial trough** (Figure 12-62). A glacial trough is formed in a preglacial, V-shaped river valley that is widened, deepened, and straightened by glacial erosion. This leaves tributary valleys "hanging," with streams joining the major river via waterfalls.

During the Pleistocene, ice-sheet action and the enormous quantities of meltwater greatly changed the appearance of many parts of the northern continents. In general, highlands were subjected to erosion and lowlands to deposition; the Precambrian shields of Canada and Scandanavia were primarily scoured and polished by ice erosion (Figure 12-51). Three common depositional landforms are briefly described here.

Till plains are dominant landforms of glaciated regions. Till denotes unsorted mixtures of clay, sand, gravel, and boulders that were deposited by the ice sheets as an unconsolidated mantle on the countryside. Extensive areas of the Midwest are covered by till deposits of varying thickness, which conceal the preglacial hills and valleys in many places. The result is a rather level, softly undulating plain (Figure 12-63). The soil pattern is strongly mottled, with the light spots being dry and sandy and the darker areas being moist and clayey; integrated drainage networks are absent (Figure 12-63).

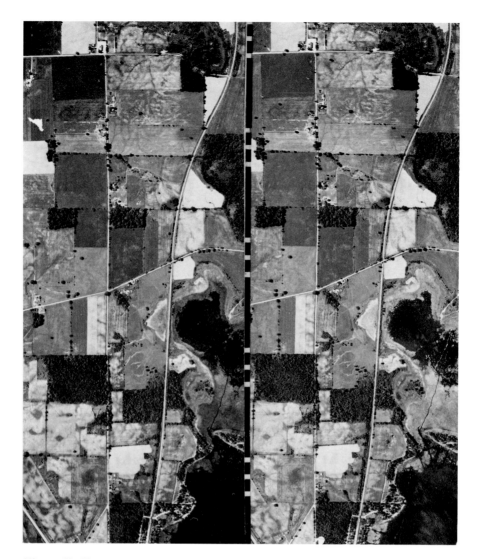

Figure 12-63 Stereogram of a young glacial till plain in La Porte County, Indiana. Light tones denote dry sandy deposits, and darker tones denote moist clayey deposits. Note the absence of an established drainage pattern. Scale is about 1:20,000.

A glacial landform of the water-sorted type is the **esker**, which is a narrow, snakelike ridge (Figure 12-64). Typically, eskers are 20 to 30 m high, 50 to 60 m wide, and up to several tens of kilometers in length; sideslopes are approximately 30°. These features are the sand and gravel fillings of channels and tunnels carved out by meltwater streams running under and within the ice sheet. Even when mantled by dense vegetation, eskers can be easily recognized on airphotos because of their unique topographic expression (Figure 12-64).

Drumlins are smooth, asymmetrical hills 15 to 45 m high, 150 to 300 m wide, and up to 2 km long (Figure 12-65). They are composed of unsorted sand and gravel mixed with clay and are oriented with their long axes parallel to the direction of the former glacial flow. The end that faced the glacier (forward, or stoss, end) is steeper, wider, and slightly higher than the more tapered lee end. Drumlins usually occur in groups called **swarms**. Generally, there is no drainage development on drumlins, but there may be a few gullies on the steeper slopes. Cultivated fields on drumlins are long and narrow, accentuating the linear pattern of the topography; where the slopes are overly steep, a drumlin may be densely timbered or in pasture (Figure 12-65).

Figure 12-64 Stereogram of an esker in Aroostosk County, Maine. Scale is about 1:20,000.

Eolian Features

Wind-deposited (**eolian**) materials are commonly classified as **sand dunes** or **loess deposits**. Dunes most often form along sandy coasts under the influence of onshore winds or in inland deserts, where dunes can be so numerous and extensive as to form **dune fields** and even-larger **sand seas**. The most common dune material is a medium-grained quartz sand, but dunes of gypsum and calcite particles or volcanic ash may also form. Dunes attain different shapes and sizes that reflect several environmental factors, including (1) the strength and direction of the wind, (2) the sand supply, (3) distance from the source, and (4) physical barriers, such as topographic obstacles and water bodies. Several of the basic types of sand dunes are described here.

Barchan dunes, the most common type of dune, are crescent-shaped, with their **horns**, or **arms**, pointed downwind (Figure 12-66). The windward slope of a barchan dune

is gentle, whereas the lee slope, or **slipface**, on the inside of the crescent is usually at the **angle of repose** of dry sand (commonly 30° to 34°). These dunes form with a limited supply of sand and when the dominant wind direction is from a single direction.

With an increase in the sand supply, individual barchan dunes will coalesce to form **barchanoid ridges** (Figure 12-66). They are oriented transverse to the prevailing wind direction and incorporate gently dipping windward slopes and steeply dipping lee slopes.

Linear dunes, also called **longitudinal dunes** or **seifs**, are straight to slightly sinuous sand ridges with slipfaces on both sides; the length of a linear dune is always many times greater than its width (Figure 12-67). These dunes can occur as isolated ridges, but they are normally found in parallel sets, separated by **interdune corridors** (Figure 12-67). The origin of linear dunes is controversial, but their axial alignment is commonly parallel to the prevailing direction of the regional wind.

Star dunes are pyramidal sand mounds, roughly star-shaped or resembling a pinwheel, with three or more arms radiating in various directions from the high central part of

Figure 12-65 Stereogram of a drumlin swarm in Jefferson County, Wisconsin. Arrow points to area being excavated for sand and gravel. Scale is about 1:20,000.

the dune (Figure 12-68). They accumulate in areas having multiple wind directions. Star dunes tend to grow vertically rather than migrating laterally and are, consequently, the highest of all dune types. For example, their height surpasses 400 m in the Taklimakan Desert, People's Republic of China.

Dome dunes are low circular or oval mounds of sand that develop when winds are sufficiently strong to bevel the normal upward growth of dune crests (Figure 12-69). When elongated, there is a prevalence of a particular wind direction over other wind directions. These dunes normally lack external slipfaces and occur on the upwind margins of some dune fields or sand seas. Dome dunes up to 150 m in height

dominate large tracts of the sand seas in Saudi Arabia (Figure 12-69).

Loess is a soft accumulation of wind-laid particles of silt size or smaller that are deposited in blanket form. Loess regions are found today to the leeward side of certain deserts (e.g., People's Republic of China, South America) and in retracting ice-sheet regions where out-blowing winds deposited fine "glacial dust" during the Pleistocene (e.g., central Europe, central United States). Loess may attain thicknesses exceeding 150 m.

Loess is a highly erodable material and is readily exposed to denudation. This is most conspicuous along streams, where the loess blanket is cut into a maze of valleys,

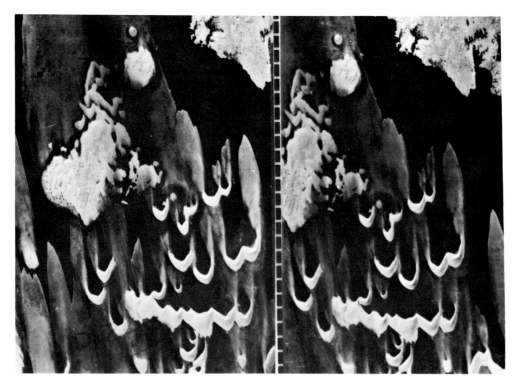

Figure 12-66 Black-and-white infrared stereogram of barchan dunes and barchanoid ridges in the Jafurah sand sea, Saudi Arabia. The dark areas denote a moist playa surface. (Courtesy Khattab Al-Hinai, King Fahd University of Petroleum and Minerals.)

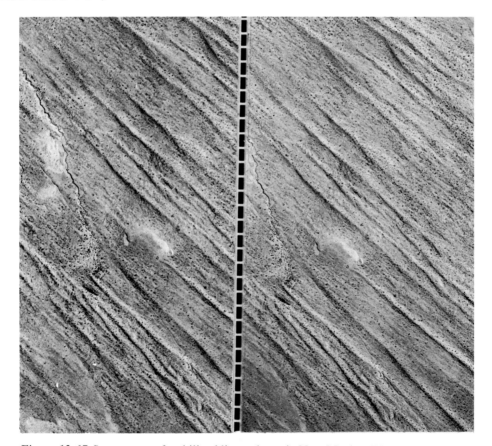

Figure 12-67 Stereogram of stabilized linear dunes in New Mexico. These dunes are of Pleistocene age. Scale is 1:20,000.

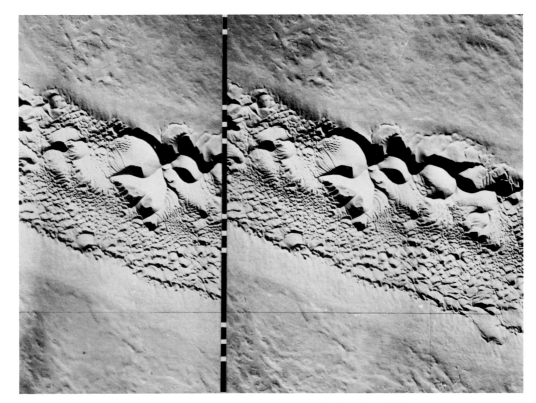

Figure 12-68 Stereogram of star dunes surrounded by complex barchanoid ridges in the Erg Er Raoui, Algeria. Scale is 1:40,000 (Courtesy U.S. Air Force).

Figure 12-69 Vertical airphoto of giant dome dunes in the Ad Dahna sand sea, Saudi Arabia. Crescentic ridges appear as fine texture on the dune summits. (Courtesy Khattab Al-Hinai, King Fahd University of Petroleum and Minerals.)

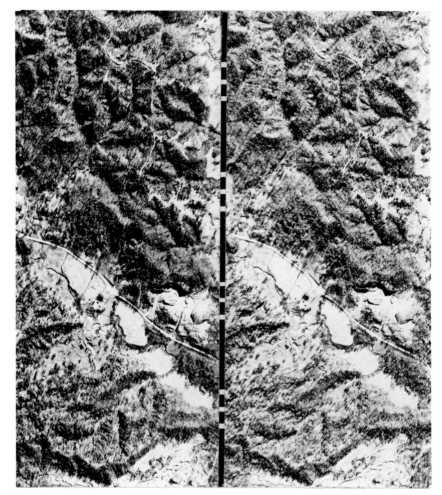

Figure 12-70 Stereogram of eroded loess bluffs near Natchez, Mississippi. Scale is about 1:22,000.

ravines, and countless gullies. Dissected loess forms vertical or near-vertical valley walls along major drainages; for this reason loess is called a **bluff formation** in the Mississippi valley (Figure 12-70). Highly dissected loess deposits have a fine-textured pinnate drainage pattern with pear-shaped headwater basins (Figure 12-33).

Nonphotographic Data

The newer remote sensors, such as electro-optical, radar, and sonar systems, are providing new sources of data for many types of geologic investigations. These sensors enable one to "see" the geologic environment well beyond the limits of the narrow visible spectrum (Figure 1-5). For example, spectral signatures in the mid- and thermal-infrared regions can help to identify specific lithologies or surface alteration zones associated with certain mineral deposits (Chapter 15). In addition, sonar images are providing totally new views of the world's seafloors, and L-band radars are providing views of landscapes buried beneath dry surficial deposits (Chapter 7).

Orbital images, such as those from the Landsat and SPOT sensors (Chapter 6), plus airborne and spaceborne radar images, enable diverse and complex terrain to be studied from a regional perspective because of their **synoptic views**. Landforms and structural features can often be followed for their entire lengths on a single image (Figures 12-71 and 12-72). To take advantage of shadow enhancement, Landsat and SPOT images incorporating low sun-angles are normally used to study geomorphic and structural features (Figures 12-73 and 12-74). The value of space-acquired images for many types of geologic investigations are found in the excellent publication *Geomorphology from Space* by Short and Blair (1986).

Figure 12-71 *Landsat 4* Multispectral Scanner (MSS) band 2 image (0.6–
0.7-μm) of a portion (185 × 185 km) of the Basin and Range Province in Cali-
fornia and Nevada, acquired November 17, 1982. Annotations are as follows:
(*A*) Panamint Ranges, (*B*) Searles Lake (playa), (*C*) Avawatz Mountains,
(*D*) Owshead Mountains, (*E*) Death Valley, (*F*) Amargosa Valley, (*G*) Pahrump
Valley, and (*H*) Spring Mountains. (Courtesy EROS Data Center, U.S. Geological
Survey.)

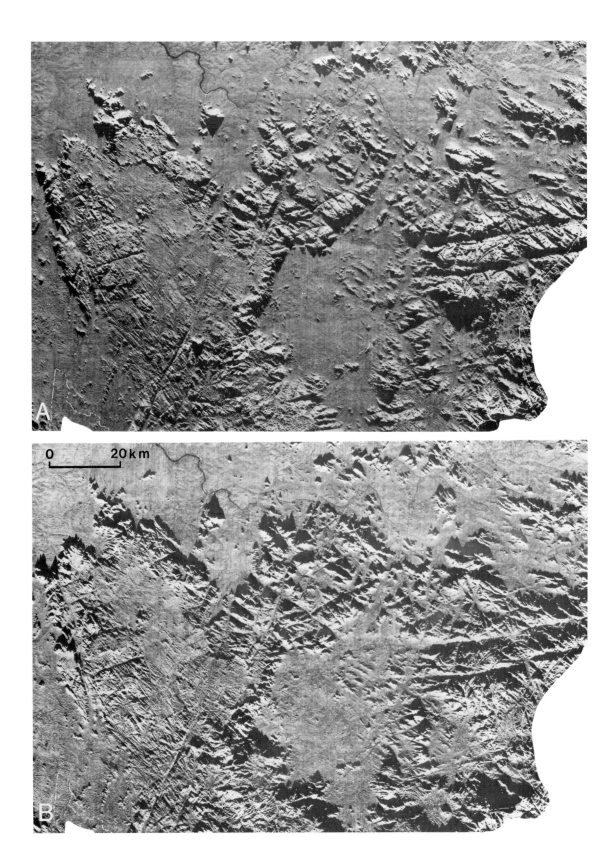

Figure 12-72 Motorola X-band (3.2-cm) real-aperture radar mosaics (opposite looks) reduced from an original scale of 1:250,000. Shown is an 18,500-km² area in southeastern Nigeria; (*A*) north look and (*B*) south look. Note that some geologic features are more readily visible in one look than the other. (Courtesy Ron Gelnett, MARS Associates, Inc.)

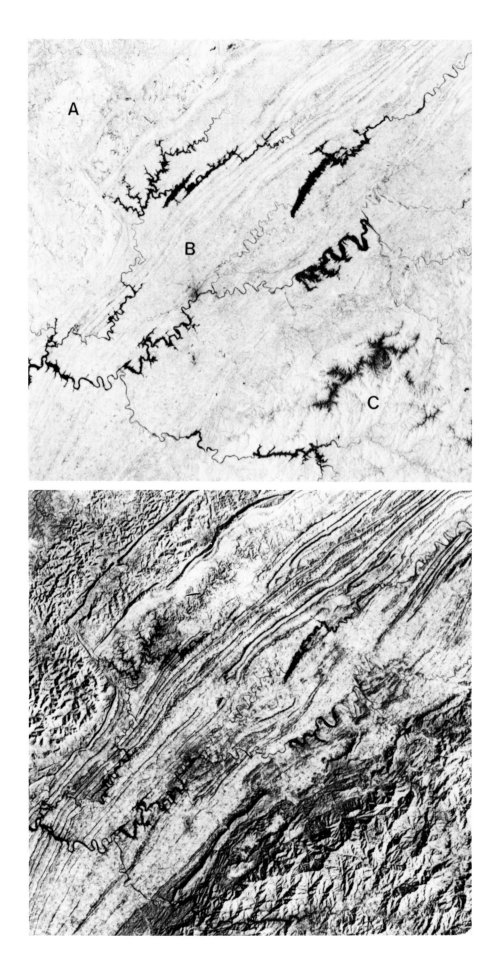

Figure 12-73 (*Left, page 336*) *Landsat 2* Multispectral Scanner (MSS) band 7 subscene images (0.8–1.1-μm) showing portions of the Cumberland Plateau (*A*), the Ridge and Valley Province (*B*), and the Great Smoky Mountains (*C*) in Kentucky, Tennessee, and North Carolina. *Top* image incorporates a sun elevation angle of 53° (May 21, 1976), whereas the *bottom* image incorporates a sun elevation angle of 27° (February 6, 1977). (Courtesy Kevin C. Horstman, University of Arizona.)

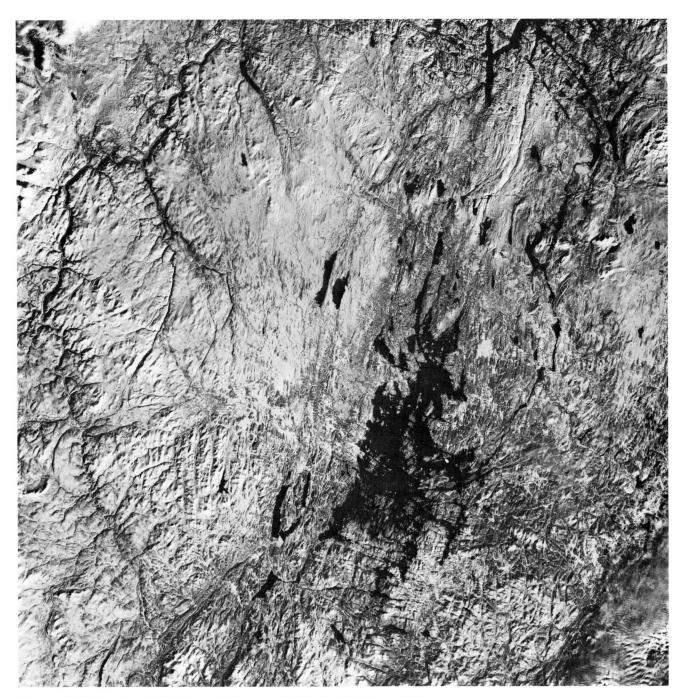

Figure 12-74 *Landsat 2* Multispectral Scanner (MSS) band 7 image (0.8–1.1-μm) of the Inari and Utsjoki regions (185 × 185 km) in Finland, acquired October 12, 1979; the sun elevation angle is 11°. Note how the fractured terrain is accentuated due to strong shadows and lighted slopes. (Courtesy Jussi Aarnisalo, Outokumpu Oy Exploration.)

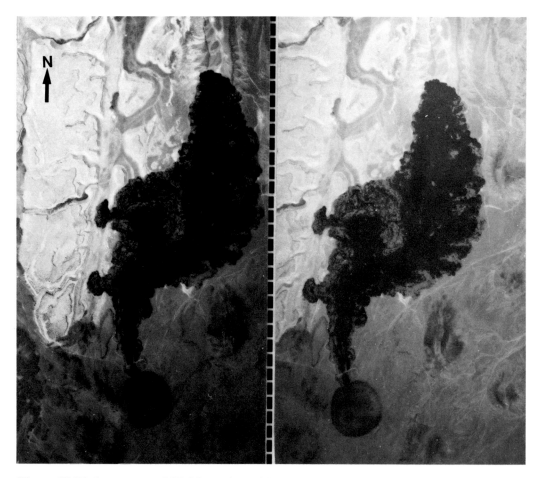

Figure 12-75 Stereogram of SP Mountain and lava flow (basaltic andesite) of late Pleistocene age (70,000 ± 4,000 years). Scale is 1:80,000.

Questions

1. Determine the bearing and azimuth of the joints shown in Figure 12-14.

	Bearing	Azimuth
Dominant trend:	_____	_____
Subordinate trend:	_____	_____

2. Examine Figure 12-57 and determine if the sand movement along the coast is from left to right or right to left. Explain your reasoning.
3. Examine Figure 12-65 and determine if the ice movement was from left to right or right to left. Explain your reasoning.
4. Examine Figure 12-66 and determine if the prevailing wind direction is from top to bottom or bottom to top. Explain your reasoning.

5. Examine the stereogram of SP Mountain and lava flow in Figure 12-75 and answer the following questions:

   a. What is the drainage pattern on the side of SP Mountain?

   b. Explain what caused the two lobes to form along the west side of the lava flow.

   c. Explain the two tonal signatures associated with the lava flow.

   d. Which formed first, the lava flow or the cinder cone? Explain. (Compare with Figure 12-29.)

   e. Is there any evidence suggesting the presence of an older lava flow in the local area? Explain.

   f. An elongated cinder cone is seen to the northeast of SP Mountain. What does the linearity of this feature suggest? Is there any additional evidence to the north of SP flow? Explain.

Bibliography and Suggested Readings

Berlin, G. L. 1980. *Earthquakes and the Urban Environment* (3 volumes). Boca Raton, Fla.: CRC Press.

Bunnett, R. B. 1968. *Physical Geography in Diagrams.* New York: Frederick A. Praeger.

Curran, H. A., P. A. Justus, D. M. Young, and J. B. Garver, Jr. 1984. *Atlas of Landforms,* 3d ed. New York: John Wiley & Sons.

Hackman, R. J. 1966. *Time, Shadows, Terrain and Photointerpretation.* NASA Technical Letter 22. Washington, D.C.: U.S. Geological Survey.

Kiefer, R. W. 1967. Landform Features in the United States. *Photogrammetric Engineering* 33:174–82.

Lattman, L. H., and R. G. Ray. 1965. *Aerial Photographs in Field Geology.* New York: Holt, Rinehart and Winston.

O'Leary, D. W., J. D. Friedman, and H. A. Pohn. 1976. Lineament, Linear, and Lineation: Some Proposed New Standards for Old Terms. *Geological Society of America Bulletin* 87:1463–69.

Ray, R. G. 1960. *Aerial Photographs in Geologic Interpretation and Mapping.* Geological Survey Professional Paper 373. Washington, D.C.: U.S. Government Printing Office.

Short, N. M., and R. W. Blair, Jr., eds. 1986. *Geomorphology from Space.* NASA Special Paper 486. Washington, D.C.: U.S. Government Printing Office.

Siegal, B. S., and Gillespie, A. R., eds. 1980. *Remote Sensing in Geology.* New York: John Wiley & Sons.

Strandberg, C. H. 1967. *Aerial Discovery Manual.* New York: John Wiley & Sons.

Tator, B. A., and others. 1960. Photo Interpretation in Geology. In *Manual of Photo Interpretation,* edited by R. N. Colwell, 169–342. Falls Church, Va.: American Society of Photogrammetry.

von Bandat, H. F. 1962. *Aerogeology.* Houston: Gulf Publishing Co.

Way, D. S. 1978. *Terrain Analysis: A Guide to Site Selection Using Aerial Photographic Interpretation,* 2d ed. Stroudsburg, Pa.: Dowden, Hutchinson & Ross.

Williams, R. S., Jr., and others. 1983. Geological Applications. In *Manual of Remote Sensing,* edited by R. N. Colwell, 2d ed., 1667–951. Falls Church, Va.: American Society for Photogrammetry and Remote Sensing.

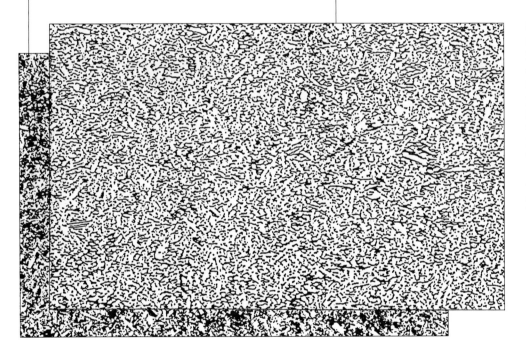

Chapter 13

Engineering Applications

required (e.g., vegetation patterns, soil moisture, or soil, rock, and water color). Although there has been a rapid increase in the use of nonphotographic images for engineering studies since the 1970s (see Chapters 6 and 7), airphotos continue to be the most widely used medium.

Some of the areas in which aerial photographs are used in the engineering field are as follows: (1) surveys of construction materials, (2) location of routes for transportation systems, (3) selection of potential site locations for critical engineered structures such as dams, nuclear power plants, and tunnels, (4) investigations of landslides, (5) surveys of disaster damage, (6) investigation of water pollution, (7) monitoring of mine-disturbed land, and (8) inventories of stockpiles. These areas are discussed in this chapter; comprehensive discussions on the use of airphotos in the engineering field are given by Mintzer et al. (1983) and Kennie and Matthews (1985).

Construction Material Surveys

The availability of suitable *construction materials* in sufficient quantities and within economically feasible hauling distances is crucial to the planning and execution of most construction projects. One common technique for locating construction materials through the stereoscopic study of airphotos is the **terrain analysis approach** (Mintzer et al. 1983). Landforms are first identified and delineated in their regional setting; a knowledge of landform origins leads to a general prediction of the type of materials to be expected. The most promising landforms are then analyzed in detail in terms of topography, drainage pattern, erosion, soil tone or color, vegetation cover, and land use for selecting the most favorable sites for field investigation. Photographic scales required for **construction material surveys** vary from one region to the next, but panchromatic exposures at a scale of 1:15,000 to 1:25,000 are often suitable.

Granular materials of various particle sizes are the most common type of construction material sought by the engineer. These materials are used as aggregate for asphalt and concrete, ballast for railroad beds, surfacing material for secondary roads, base or subbase for primary roads and railroads, and fill for earth dams and embankments. The most important particles and their respective diameters are

Overview of Engineering Applications

For more than four decades, both qualitative and quantitative data derived from aerial photographs have been used in a wide variety of **engineering projects** by engineering geologists and civil engineers. Panchromatic photographs in a stereoscopic format are used for most applications, but normal color, color infrared (IR) and black-and-white IR photographs may be preferred when specific information is

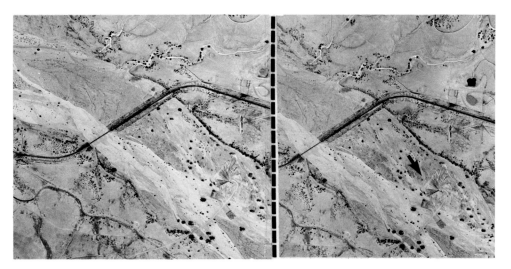

Figure 13-1 Stereogram of a sand and gravel pit in the bed of Granite Creek near Prescott, Arizona. Note how the borrow pit is partially protected from flooding by the upstream railroad levee. Scale is about 1:16,000.

Figure 13-2 Stereogram of sand and gravel pits in glacial till deposits in Michigan. Scale is about 1:20,000.

sand (0.06 to 2 mm), **granules** (2 to 4 mm), **pebbles** (4 to 64 mm), and **cobbles** (54 to 256 mm). These materials can be found as naturally occurring **unconsolidated deposits** or as **consolidated deposits**, which are crushed to provide the required particle sizes. The terrain analysis approach can be used to locate both types of deposits.

Unconsolidated granular materials are associated with deposition by *water, wind, ice,* and *gravity.* Running water has the ability to **sort** and **stratify** materials as a function of velocity changes. Important sources of water-sorted materials are often found along and within river beds. Along mature floodplains, for example, sand and fine-textured gravel deposits are associated with abandoned channels, point bars, and alluvial terraces (Figure 12-32). The beds of braided

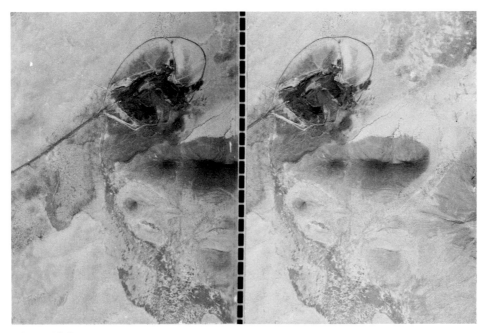

Figure 13-3 Stereogram of a volcanic cinder cone being excavated for its granular material.

streams in arid environments are usually good sources of coarse sand and gravel (Figures 12-31 and 13-1). Fluvial-glacial landforms, such as eskers and outwash plains, are often excellent sources of sand and gravel (Figure 12-64).

Wind sorting is highly selective, producing individual grains of a fairly uniform size. Thus, when sand dunes are recognized on airphotos, one would expect to find a fairly uniform gradation of sand-sized particles (Figures 12-66 to 12-69). Dunes are most often composed of well-rounded quartz grains, which can easily be excavated with light equipment.

Sorting by ice and gravity (mass wasting) is poor, supplying a large range of sizes of unstratified granular material. Consequently, the material may have to be mechanically separated into appropriate-size classes following excavation. Of the glacial deposits, young till and drumlins can be good sources of granular material (Figures 12-63, 12-64, and 13-2). The constituents of till generally reflect the bedrock composition in the immediate vicinity. For example, New England tills tend to have coarse textures because they were derived from crystalline rocks, whereas tills in the lower Midwest are finer in texture because they originated from softer sedimentary rocks. It is also possible to infer the textural composition of drumlins by observing their slopes and shapes. Generally, the forward, or stoss, end (steepest slope) of a drumlin contains coarser granules than the tail, or lee, end. In addition, smooth, broad shapes are indicative of fine-textured material, whereas long, narrow shapes can indicate coarser-textured material.

Talus or scree is an important gravity-sorted deposit that accumulates along the base of cliffs, scarps, and steep mountain sides (Figure 12-49). Talus deposits are composed mainly of coarse, angular rock fragments, which, when hard and dense, make good aggregate and fill material. Crushing is required when the fragments are overly large.

Volcanic cinder cones are a special type of gravity landform that are a source of granular material for aggregate, as a surfacing material for secondary roads, and as ballast for railroad beds (Figures 12-29 and 13-3). Cinder cones are easily recognized in airphotos; when seen in stereo, steep slopes (e.g., > 30°) are indicative of large fragments, whereas the gentler slopes usually indicate the presence of smaller fragments.

Airphoto interpretation techniques can be used to identify broad classes of consolidated rock formations (i.e., "consolidated landforms") that offer the best potential for crushed construction materials. The most important rock types are tough and durable (harder than concrete) and include limestone, dense sandstone, basalt, and quartzite. Typical airphoto recognition elements for these rock types are illustrated in Chapter 12.

Highway Route Locations

The proper location of new highways, railroads, canals, pipelines, or electrical transmission lines requires several analysis phases and the careful consideration of many physical and cultural factors. Airphotos are routinely used at appropriate times during the *planning, location,* and *construction phases* of a transportation project. A highway project, for example, may require as many as seven different photographic flights (Meyer and Gibson 1980). Airphotos fulfill dual needs by readily providing (1) qualitative information by photographic interpretation and (2) quantitative information by photogrammetric techniques (Figures 13-4 and 13-5). Although **route-selection techniques** employing

Figure 13-4 Vertical airphotos (different scales) showing a portion of Flagstaff, Arizona, before and during the construction of Interstate 40. Note how the divided highway skirts built-up areas and the large canyon and crosses the principal drainages nearly at right angles.

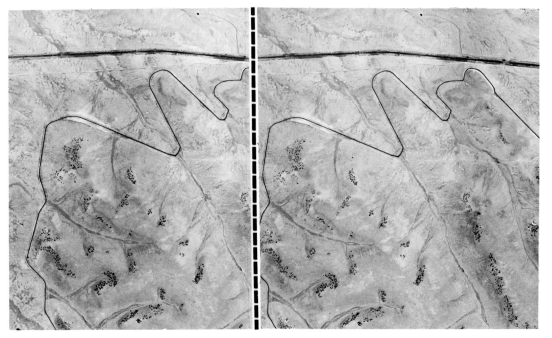

Figure 13-5 Stereogram showing a gravity-flow irrigation canal that was laid out to fit topographic conditions and minimize cut-and-fill requirements. Shown is the Chino Valley Irrigation Ditch in Yavapai County, Arizona. Scale is about 1:16,000.

Figure 13-6 Vertical airphoto of Carolina Bays in Cumberland County, North Carolina, showing primary and secondary roads skirting the swampy depressions.

airphotos are applicable to all transportation systems, locating a new highway is used as the illustrative example in this section.

The basic objective in route selection is fitting the new highway to the natural and cultural features of an area in a manner that is environmentally sound, is cost effective in terms of construction and operation, and causes minimal sociological impact (Figure 13-4). To accomplish this, the selection of a new highway route may be divided into three distinct phases: (1) the **reconnaissance survey**, (2) **feasible route comparisons**, and (3) **detailed survey of the best route**. Each phase helps to refine the route location more precisely than the preceding phase, and airphotos play an important role in each phase.

The Reconnaissance Survey

The purpose of the reconnaissance survey is to obtain an overall look of the entire area of land through which the route must pass to connect two terminal points and to select several general route corridors that appear to be *technically feasible;* each corridor may be from 0.5 to 2 km in width. This requires consideration of the following landscape elements: (1) landforms, (2) drainage patterns and textures, (3) soil and rock types, and (4) land use–land cover patterns. Much of the information for these elements can be gathered from the stereoscopic examination of airphotos. When available, published topographic maps, public utility maps, geology and soil maps, and land use–land cover maps are also analyzed.

The scale of the photography used for the reconnaissance survey may range from about 1:20,000 to 1:80,000. Existing government photography is usually adequate at this early stage of the project if it is not outdated. It is desirable to use the smallest practical scale of photography because fewer stereopairs are needed to encompass a given area, with commensurate savings in time and money.

The layout of a proposed highway route that will connect two terminal points is rarely a straight line. Rather, each layout will be influenced by a multitude of **location controls** in the intervening area that *attract* or *repel* the paths of the proposed routes. The following are important types of location controls that can be identified on the stereopairs: (1) terrain conditions (e.g., level, rolling, or mountainous) as they affect the economical attainment of desired grades and curvatures; (2) unstable ground, such as clay areas, which shrink and swell with changes in moisture content, or areas susceptible to landslides; (3) physical barriers, such as a dune field, swamp, or lake (Figure 13-6); (4) areas susceptible to flooding and avalanches; (5) unique features, such as mountain passes, tunnel sites, and narrow river crossings (Figure 13-7); (6) availability of construction materials; (7) crossings of existing highways and railroads; (8) intervening towns and cities; (9) areas requiring clearing (e.g., timber stands) or cut-and-fill operations; (10) ecologically sensitive areas; and, of course, (11) distance.

Because the "lay of the land" is so important in route selection studies, **form-line sketches** are often compiled where topographic maps are not available (e.g., in developing countries). **Form lines** are approximate contours that can be drawn with the aid of standard stereoscopes to show the general configuration of the terrain. Form lines on individual prints are usually transferred to photo index sheets or uncontrolled mosaics (Figure 4-29) to show a composite view of the tentative corridors in their respective topographic settings.

More accurate form lines can be drawn with **portable stereoplotters**, such as the **Zeiss Stereotope**, which are built around floating-mark lenses, stereometer, and tracing pantograph (Figure 13-8). This type of instrument is designed to be used with paper prints and for making topographic maps at scales ranging from about 1:25,000 to 1:100,000. If high-quality, vertical airphotos are available on low-shrink paper, accuracy of results will be comparable to that shown in Figure 13-9.

Figure 13-7 Stereogram of steel truss-frame railroad bridge (*A*) and dual-lane concrete highway bridges (*B*) spanning the Maumee River near Toledo, Ohio. Whenever possible, rivers are crossed where their widths are narrow and at approximate right angles. Note that the supporting piers for all three spans are aligned to allow for easy passage of small boats. Power transmission towers are circled on opposite river banks (*C, D*). Scale is about 1:8,000.

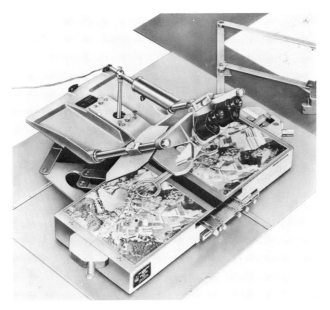

Figure 13-8 Zeiss Stereotope showing movable photo carriages, floating mark lenses, and tracing pantograph. (Courtesy Carl Zeiss, Oberkochen.)

Feasible Route Comparisons

For the second phase, feasible route comparisons, large-scale strip photographs are obtained for all practical routes joining the terminal points. The scale of the photography is generally between 1:2,500 and 1:5,000. The photographs are used to (1) closely examine the physical and cultural conditions along each corridor in three-dimensional views, (2) calculate heights, elevations, and grades by measurement of image parallax on stereopairs, and (3) prepare large-scale **topographic strip maps** with a contour interval of 2 to 3 m for each corridor (Figure 13-10). Glass diapositives are normally used with precision stereoplotters to produce the maps (see Chapter 4).

The location controls evaluated in the reconnaissance survey are reexamined in greater detail for each tentative route. This procedure usually narrows the choice of location to perhaps two or three. These remaining routes are then assessed in terms of land ownership, prevailing property values, and construction-cost estimates to select the most promising route for the new highway.

Figure 13-9 Vertical airphoto of a scene in West Germany showing superimposed contours plotted with the Zeiss Stereotope (Figure 13-8). Contour interval is 10 ft (about 3 m). (Courtesy Carl Zeiss, Oberkochen.)

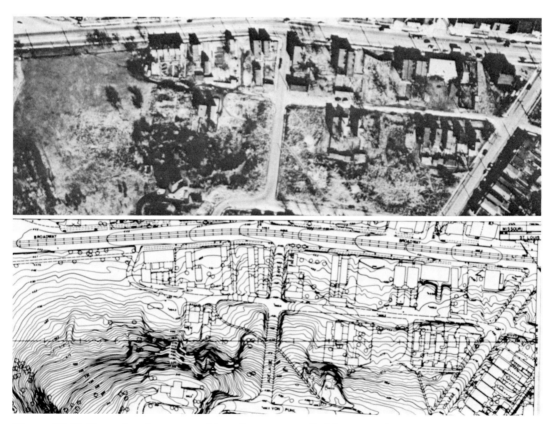

Figure 13-10 Example of a topographic strip map compiled for an engineering survey. The map (*bottom*) covers the same area as the aerial photograph (*top*). (Courtesy Jack Ammann Photogrammetric Engineers, Inc.)

Figure 13-11 Stereogram showing construction of a divided, limited-access highway in Michigan. Several loaded dump trucks and grading equipment are discernible. Note that bridges or overpasses are usually completed before road surfacing. Scale is about 1:6,000.

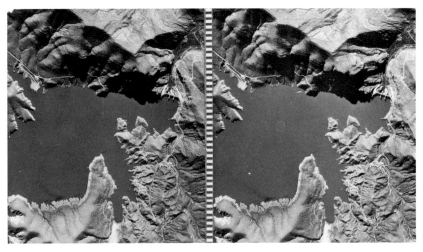

Figure 13-12 Stereogram of Coolidge Dam and San Carlos Lake, Arizona. The masonry dam is of the arch-gravity type. Scale is about 1:42,000.

Detailed Survey of the Best Route

Following public hearings and right-of-way acquisitions, new photography at a very large scale (e.g., 1:500) is obtained to assist in the third phase—detailed survey of the best route. In addition to interpretation tasks, the photographs are used to prepare precise topographic strip maps with a very small contour interval (e.g., 0.5 m) along the path of the proposed highway. The primary purpose of the photos and maps is to assist the engineering team in precisely positioning the highway alignment within the lateral limits of the right-of-way. They also serve as a base for plotting traffic lanes and intersections or interchanges. These are then

surveyed and staked out on the ground, which sets the stage for the actual construction to begin (Figure 13-11).

During construction of the new highway, the airphotos and the topographic strip maps may be used for a number of tasks. These include (1) making minor realignments to avoid obstacles not seen in the preconstruction phase, (2) estimating cut-and-fill volumes, (3) finding additional sources of nearby construction materials, and (4) determining the precise locations for underdrains and culverts.

Through the combination of aerial photographs and topographic strip maps, the construction engineer can be confident that all feasible routes have been given due consideration. Such assurances were acquired only via tedious

Figure 13-13 Stereogram of an earth dam and reservoir in a semiarid region. A road on top of the dam and a concrete spillway are clearly discernible.

and time-consuming ground survey methods before the adoption of photo interpretation and photogrammetric surveys by highway designers and engineers.

Dam Site Investigations

Airphoto interpretation is an established reconnaissance method for selecting potential site locations for dams; the most attractive sites are then evaluated by extensive fieldwork and laboratory testing for determining the best technical site. Dams may be classified as **masonry** or **earth**, and each type has an important bearing on the site location (Figures 13-12 and 13-13). For example, narrow and deep valleys, strong bedrock foundations, and large stream flows favor the selection of masonry dams, whereas wide valleys, weak bedrock foundations, and small stream flows favor earth dams (Mintzer et al. 1983).

Many physical factors determine the feasibility of *dam sites;* the major ones that can be identified and evaluated by airphoto interpretation include the following: (1) characteristics of the valley, including landforms, soil types, and dimensions; (2) characteristics of the watershed, including drainage network and vegetative cover; (3) characteristics of the area to be flooded, including transportation systems that must be replaced and land types (e.g., cropland, forestland, and built-up areas) for which acquisition costs must be paid; (4) identification and consideration of both regional and local geologic features; (5) access to the dam site by road or

rail; (6) location and extent of landslides or slopes susceptible to landslides near the proposed dam site; and (7) availability and suitability of construction materials to be used for building the dam, providing an impervious bottom seal for preventing leakage, or for constructing highways and railroads that must be relocated (Way 1978 and Belcher et al. 1960).

The lithologic and structural nature of sites proposed for the placement of large masonry dams warrant special investigation by airphoto interpretation followed by extensive field study. Being able to differentiate rock types and their physical properties is of paramount importance because the **bedrock base** must be able to support the weight of the dam, and the **abutment rocks** must be able to withstand the stresses caused by the water in the reservoir. In addition, faults and joints must be carefully mapped and field-verified because when they are present in the abutment rocks, they can be conduits for a flow of water that can affect the safety of the dam. It is also virtually impossible to construct dams over an active fault zone and not have them damaged or destroyed if substantial horizontal or vertical displacement occurs.

In general, most igneous rocks (e.g., granite) are as strong as—or stronger than—concrete and are among the most satisfactory materials for foundations and abutments if they are not highly fractured. Breccias and tuffs, however, do not offer suitable properties for the construction of dams because they are weak and permeable. Metamorphic rocks (e.g., quartzite) generally resemble igneous rocks in terms of strength, but most schists are excluded because they are too soft and weak. The most favorable sedimentary rocks are thick-bedded, flat-lying sandstones and limestones, both

Figure 13-14 Panchromatic (*top*) and infrared (*bottom*) photographs of the Briones (*A*) and San Pablo (*B*) Reservoirs in Costa County, California. The Briones Reservoir has a new earth dam and a low degree of sedimentation. Also discernible are: (*C*) exposed soil, (*D*) annual grasslands, (*E*) Monterey pine, (*F*) concrete spillway, and (*G*) mixed hardwoods. Normal color and color infrared photographs of this same area are shown in Plate 4. (Courtesy U.S. Forest Service Remote Sensing Project, Berkeley, California.)

Figure 13-15 Stereogram of an earth dam in Yavapai County, Arizona, showing a severely damaged downstream face. Scale is about 1:16,000.

Figure 13-16 Vertical airphotos of the Daly City, California, area in 1956 and 1966. The dashed lines approximate the San Andreas fault zone, and the solid line approximates the surface trace formed by the 1906 San Francisco earthquake. The Mussel Rock landslide is active and large, covering about 45 ha; the landslide deposits are friable marine sandstones of the Pliocene Merced Formation (Smelser 1987). (Courtesy National Science Foundation.)

of which are normally stronger than concrete. A chief problem of some sandstones is that they allow excess seepage to occur through intergranular pores and along joints, which are difficult to seal with cement grout. A problem sometimes presented by limestone is its solubility, with associated solution cavities. Shales or tilted sandstones and limestones interbedded with thin layers of clay or shale are avoided because they can fail when overloaded.

Following construction, **time-sequential aerial photographs** offer an efficient and cost-effective means of detecting or monitoring erosion below spillways, structure damage, seepage zones, siltation patterns in the reservoir, water levels, shoreline erosion, and land use–land cover changes along the shoreline (Figures 13-14 and 13-15). Panchromatic and normal color photographs usually provide good water information, whereas shorelines, small tributaries, and moisture anomalies are more distinctive on color IR and black-and-white IR photographs.

Landslide Investigations

The actual or potential downslope movement of earth and rock debris via landslides is a significant concern in any area being considered for engineering works, including highways, railroads, and dams. **Landslides** are usually recogniz-

able in aerial photographs by their characteristic shapes and internal features (Figure 13-16). Panchromatic photographs are generally suitable for recognizing most landslide deposits, but normal color and infrared photographs may be useful locally in observing variations in ground moisture, age and state of vegetation, or cultural modifications (Nilsen et al. 1979). Photographs at scales of 1:15,000 to 1:30,000 are usually adequate for recognizing and classifying landslides, but if an area is complicated or the landslides are small, airphotos at scales of 1:5,000 to 1:10,000 are preferable for analysis (Belcher et al. 1960).

Landslides are **slope failures** that are generated whenever the forces promoting movement exceed those resisting it (i.e., when the **shear strength** of the material is exceeded by the induced **shear stress**). As the mass becomes free, it moves downslope by **gravity transport** with as much velocity as is possible against the resistance it meets. Conditions favorable to the causation of landslides include the following: (1) steep slopes (e.g., cliffs and overhangs, canyons, glaciated valleys, fault scarps, stream banks, and the sides of artificial cuts and fills); (2) slippery, impermeable rocks (e.g., shale, mudstone, siltstone, slate, and highly foliated schist); (3) fractured bedrock, especially when the fractures parallel or intercept slopes; (4) interbedded strata tilted toward the toe of the slope; (5) unconsolidated surficial deposits having low shear strength; (6) clayey soils, which are movement-prone when moist; (7) the uppermost layer of permafrost if it thaws in summer; and (8) hillsides denuded of vegetation or where vegetation is shallow-rooted.

Coupled with the gravitational force, a number of human activities and natural processes contribute to the initiation of landslides. These **triggering mechanisms** include vehicle vibrations, blasting, sonic booms, excavation and infilling operations, volcanic eruptions, earthquakes (ground displacements or ground shaking), repeated freeze-thaw activity, and water saturation from heavy rains or rapid snowmelt. Water plays an especially critical role because it adds weight, reduces shearing resistance (e.g., lubricates bedding planes), and eliminates surface tension.

There are four basic types of landslides, and each type is illustrated in Figure 13-17. Certain landslides may exhibit characteristics associated with more than one of these four types; these are known as **complex landslides**.

A **debris slide** is composed of discrete blocks of earth and rock that move rapidly downslope as a rigid mass (Figures 13-17A and 13-18). This mass slides on a relatively flat and inclined, basal slip-plane, which can be a bedding plane or rock surface. Debris slides generally occur on steep slopes

that have a thick mantle of loose material. Morphologically, they can be recognized by the track they leave on a hillside, by a curved scarp or open area at the head, and by the irregular or hummocky mass of debris on their surfaces.

A **slump** is composed of a series of intact masses of earth and rock (**slump blocks**) that move downslope along a curved slip-plane (Figures 13-17B and 13-19). The downward rotary movement results in the backward tilting of the slump blocks. Slumping is most likely to occur where the slope material is homogeneous and where slopes have been oversteepened through natural processes (e.g., stream or wave erosion) or through construction activities. A slump is characterized by a steep headward scarp, steep flanking walls, and internal tilt blocks, which often appear as steplike terraces; water may be ponded behind the slump blocks.

An **earthflow** is a viscous mass of material that moves downslope mechanically as a liquid (Figures 13-17C and 13-20). Earthflows generally involve water-saturated surficial materials and not much, if any, of the underlying bed-

Figure 13-17 Sketches illustrating the four basic types of landslides: (*A*) debris slide, (*B*) slump, (*C*) earthflow, and (*D*) rockfall. (Adapted from Nilsen et al. 1979.)

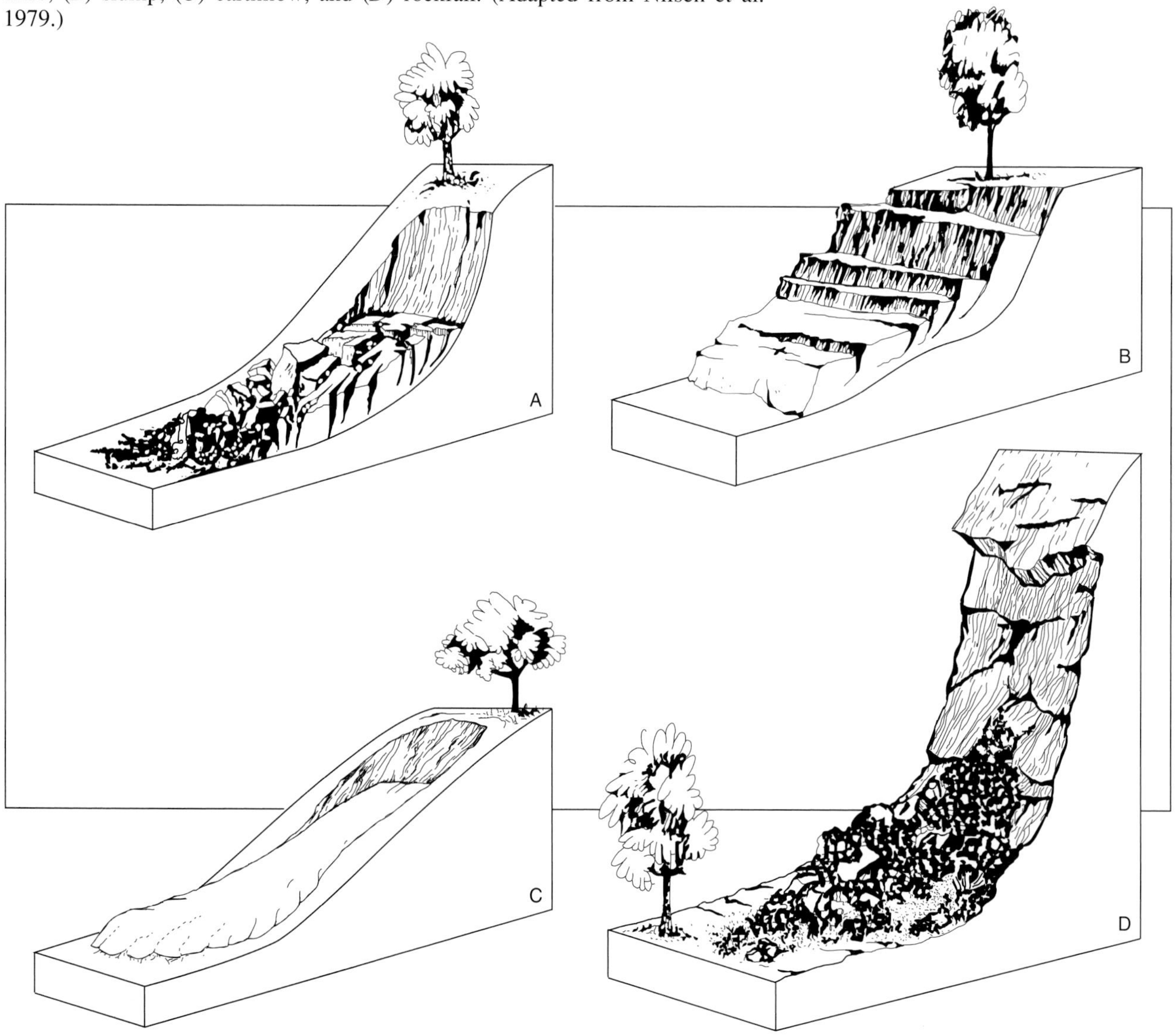

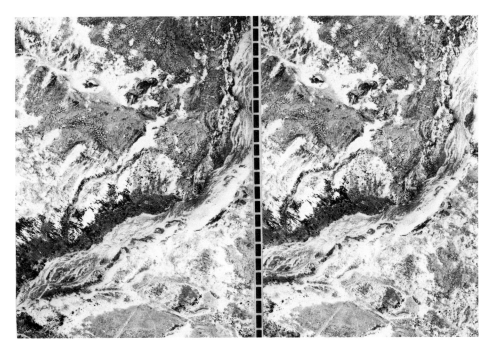

Figure 13-18 Stereogram of a debris slide in Alaska. Scale is 1:12,000.

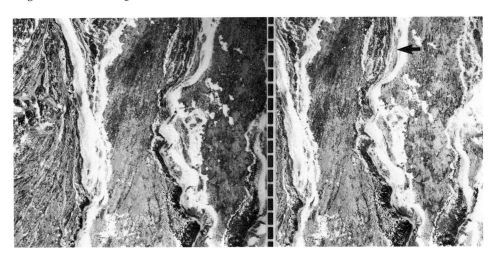

Figure 13-19 Stereogram of a slump in Alaska. Scale is 1:12,000.

rock. Most earthflows occur during or after periods of heavy rainfall or rapid snowmelt, when the amount of water present is more than that required to overcome the internal resistance of the mass. This permits downhill flowage even on fairly gentle slopes. Earthflows move in a wide range of velocities from a few centimeters per month to several kilometers per hour, depending upon the slope angle, properties of the material, and the water content. The surface of an earthflow may display large cracks and is usually very hummocky with low mounds and hollows; its toe is commonly characterized by a raised rim, whereas its head is bounded by a scarp.

A **rockfall** consists of comparatively large blocks of rock moving as individual parts of a much larger mass (Figures 13-17D and 13-21). This mass of rock debris moves suddenly and very rapidly by free-falling or bouncing; enormous amounts of rock may be transported great distances in

a matter of seconds. Rockfalls are most common along vertical or overhanging cliffs and very steep rock slopes, including bedrock cuts for transportation systems. Movement commonly occurs along preexisting fractures or bedding planes. The fallen rocks build up piles of debris called **talus**.

Landslides are recognizable in aerial photographs by the following **slide-formed features** or **conditions:** (1) hillside scars, (2) disturbed or disrupted soil and vegetation patterns, (3) distinctive changes in slope or drainage pattern, (4) irregular, hummocky surfaces, (5) small, undrained depressions, (6) steplike terraces, and (7) steep hillside scarps. In general, fewer of these characteristics are observable for small landslides, and the characteristics become less distinct as the landslide becomes older.

Aerial photographs are also valuable aids for recognizing **landslide-susceptible terrain**. Typical locations that

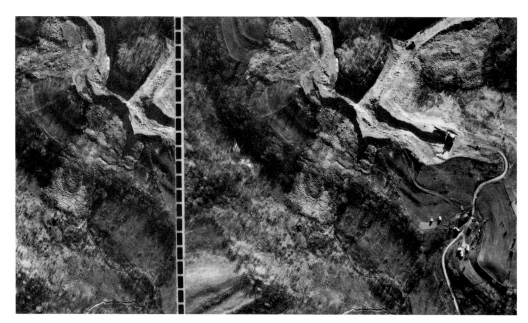

Figure 13-20 Stereogram of an earthflow caused by a coal-stripping operation near Middletown, Ohio. Scale is about 1:8,000.

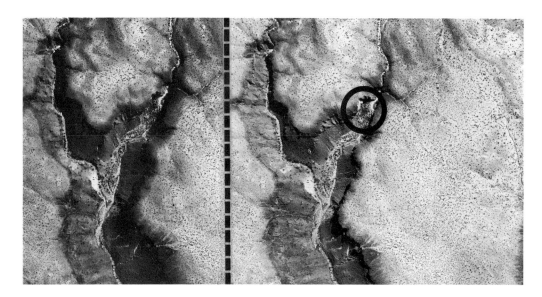

Figure 13-21 Stereogram of a rockfall in New Mexico. Here a basalt lava flow overlies gently dipping beds of sedimentary rock. Scale is about 1:6,000.

have a *potential* for landslides include (1) steep hillsides characterized by clayey soils, fractured bedrock, tilted strata, seeps or springs, gully erosion, or deforested slopes; (2) cliffs and banks undercut by streams and waves; (3) crescent-shaped or linear cracks in the surface soil; (4) tilted utility poles and trees; and (5) old landslides, which indicate previous unstable conditions that may still exist (Figures 13-22 and 13-23). Because the key features may be rather small, large-scale photographs (e.g., greater than 1:10,000) have been found to be the most useful (Way 1978).

Postdisaster Damage Surveys

Aerial photography combined with photo interpretation is an established method of gathering and analyzing data following large **natural disasters** that result in physical damage (e.g., hurricanes, tornadoes, major floods, volcanic

Figure 13-22 Stereogram of Sobrante Ridge (*A*), San Pablo Ridge (*B*), San Pablo Reservoir (*C*), the active, right-lateral Hayward Fault (*D*), and Richmond, California (*E*). The hillsides have great potential for developing future landslides because (1) they have steep slopes, (2) the underlying bedrock is poorly consolidated (Tertiary sandstones, shales, and conglomerates), and (3) landslides have occurred here in the past. Under the effects of high rainfall and/or seismic shaking, the hillsides are likely to be extremely unstable. Scale is about 1:60,000.

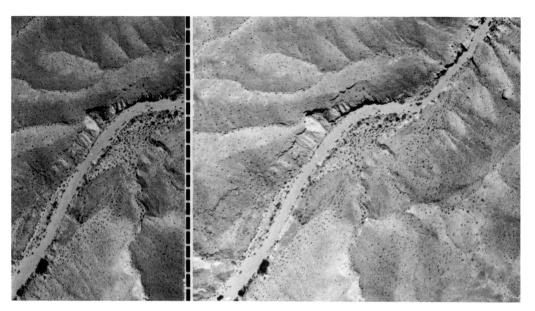

Figure 13-23 Stereogram of a debris slide along a river bank. The adjacent undercut wall (note shadow) is a likely site for a future slope failure.

Figure 13-24 Vertical airphoto of the Xenia, Ohio, region after the passage of a tornado in April 1974. Note how the 1:73,000-scale photo shows an overview of the tornado path. This photograph can be enlarged about 10 times without loss of image detail. (Rush et al. 1977; U.S. Air Force photograph.)

eruptions, and earthquakes). This method enables several types of **damage surveys** to be made *quickly* following a natural disaster. For example, initial estimates of damage to property and facilities for a disaster with a large areal extent can often be made in one or two days. By comparison, it could take several weeks to complete a similar survey by field observation, assuming that all areas were accessible.

Panchromatic film, used in conjunction with mapping cameras (see Chapter 2), is preferred for damage assessments because of its (1) wide exposure latitude, which makes it well suited for a variety of lighting conditions; (2) ability to penetrate different levels of atmospheric haze when properly filtered; (3) excellent resolution (some films can be mag-

nified more than 15 times without loss of image detail); and (4) processing and printing convenience. Because stereoscopic pairs disclose much more detail than single photographs of the same scale and quality, it is desirable to have complete stereoscopic coverage of the entire area to be interpreted.

Aerial photographs at different scales are used for various types of damage assessments. Small-scale photographs (e.g., 1:50,000 or smaller) permit the interpreter to study a large area in overview, isolating the damaged zones (Figure 13-24). When the photographs are optically magnified or the originals photographically enlarged, the experienced interpreter can usually determine the geographical distribution of

Figure 13-25 Vertical airphotos showing damaged and collapsed freeway structures on the day following the February 9, 1971, San Fernando, California, earthquake (Richter magnitude = 6.4): *left*—Route 5/210 interchange; *right*—Route 5/14 interchange. The photo acquisition scale was 1:60,000; the prints are shown at a scale of 1:20,000. Compare with Figure 13-26.

damage in broad categories, such as minor or severe. In addition, severely damaged transportation facilities and dams can normally be identified, but damage to individual buildings may be impossible to identify (Figures 13-25 to 13-27). Small-scale photographs can also reveal areas where larger-scale coverage is needed for more detailed investigations.

Medium-scale aerial photographs (e.g., 1:20,000 to 1:50,000) render a fairly detailed overview of the stricken area. Damaged areas can usually be identified quickly, and it is often possible to delineate a damaged area into subareas with different degrees of damage; this capability enables priorities to be set in terms of damage severity (Rush et al. 1977).

In addition to *area* analysis, it is possible to determine the general degree of damage to *individual* buildings and structures. For example, 1:24,000-scale photographs were used by Rush et al. (1977) to categorize the degree of damage sustained to individual structures in Corpus Christi, Texas, as a result of Hurricane Celia (August 3, 1970). Four categories of damage were assigned as follows:

Light: Up to 20 percent damage
Moderate: 20 to 40 percent damage
Heavy: 40 to 60 percent damage
Destroyed: Over 60 percent damage

Two different enlargements from one of the 1:24,000-scale frames are shown in Figures 13-28 and 13-29.

Medium-scale photographs are also useful as an aid for the following types of assessments: (1) determining which transportation routes are open for evacuations and the delivery of rescue equipment and personnel; (2) identifying crucial areas or facilities that may be affected by secondary effects such as fires, landslides, tsunamis (seismic sea waves), and aftershocks; (3) identifying accessible, structurally safe shelters for short-term housing (e.g., schools and warehouses); (4) identifying open sites, such as parks and athletic fields, for temporary housing units (e.g., tents or mobile homes); and (5) locating areas of potentially stagnant water that might constitute a health hazard (e.g., breeding habitat for mosquitoes).

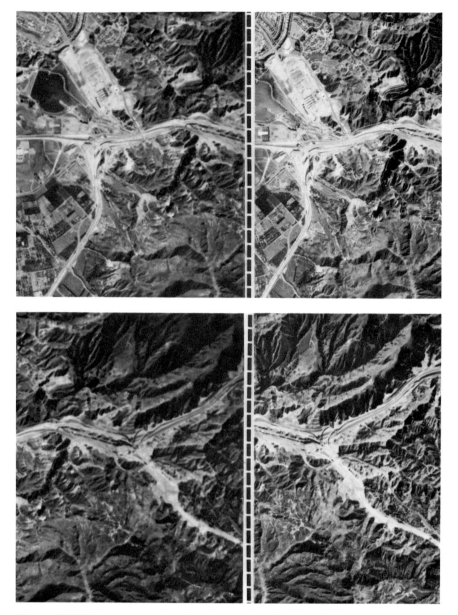

Figure 13-26 Stereograms showing damaged and collapsed freeway structures on the day following the February 9, 1971, San Fernando, California, earthquake (Richter magnitude = 6.4): *top*—Route 5/210 interchange; *bottom*—Route 5/14 interchange. The photo acquisition scale was 1:100,000; the stereograms are shown at a scale of 1:55,000. Compare with Figure 13-25.

Large-scale photographs (e.g., 1:10,000 or larger) are most often used for detailed damage assessments of (1) critical structures, such as hospitals, schools, residential housing, water-purification plants, petrochemical facilities, sewage-treatment facilities, and dams (Figure 13-30); (2) transportation systems and facilities, such as airport buildings, taxiways, and runways; and (3) utilities, such as electricity, gas, and telephone service systems. Also, damage to certain types of structures may be detected indirectly; examples include fires associated with ruptured gas lines, surface-expressed leaks from buried sewer and water lines, and spillage from ruptured storage tanks.

Some of the additional items that may be evaluated through the analysis of large-scale photographs are (1) soil and terrain damage, including the location and distribution of fault ruptures and landslides; (2) search-and-rescue op-

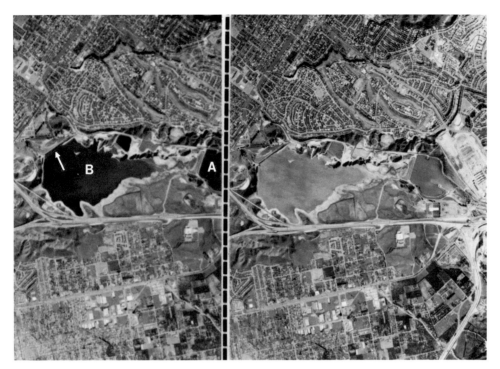

Figure 13-27 Stereogram of Upper (*A*) and Lower (*B*) Van Norman Reservoirs on the day following the February 9, 1971, San Fernando, California, earthquake (Richter magnitude = 6.4). The stereogram shows that a portion of the upstream, concrete face of Lower San Fernando Dam is missing (*arrow*). The photo acquisition scale was 1:100,000; the stereogram is shown at a scale of 1:55,000.

erations; (3) potential health hazards, such as ponded water and animal carcasses; (4) general support and logistical planning for relief teams moving into affected areas; and (5) reconstruction planning.

Water-Pollution Investigations

Water-pollution investigations are concerned with *detecting* and *assessing* the changing characteristics of water that render it undesirable or unfit for human consumption, aquatic life, and industrial or agricultural use. Among the principal types of water pollutants are (1) partially treated and untreated sewage, (2) industrial wastes and certain by-products, (3) mine and mineral processing wastes, (4) agricultural wastes (e.g., fertilizers and pesticides washed from fields and orchards), (5) decaying vegetation, (6) crude oil and petrochemicals, (7) artificially heated water (**thermal pollution**), and (8) excessive suspended sediments (**sediment pollution**).

Pollutants can enter bodies of water from highly localized **point sources**, such as sewage or industrial outfalls (i.e., piped outlets), or from relatively large **nonpoint sources**, such as strip-mine spoil banks and agricultural

fields. When pollutants enter natural water bodies from a point source, there is typically a **dispersal plume**.

Many pollutants are poisonous or toxic to animal and plant life when present in excessive amounts; certain substances may also contain infectious agents that can transmit disease. Thermal pollution can cause serious problems to aquatic life, including thermal shock (i.e., temperature changes that exceed an organism's tolerance limit, which leads to death) and oxygen deprivation (as the temperature of water increases, its ability to hold dissolved oxygen decreases). The addition of excessive quantities of suspended sediments to a water body reduces light penetration for photosynthesis. When the particles sink to the bottom, they smother and kill bottom-dwelling organisms and eliminate fish spawning areas.

Some forms of water pollution can be detected *directly* on aerial photographs and others, *indirectly*. Direct detection is possible when the pollutant has a spectral signature within the photographic spectrum (Figure 1-5) that is measurably different from the receiving water. Because **water color** can be an important indicator of the amounts and kinds of materials in suspension or solution, normal color photographs at scales of 1:5,000 to 1:10,000 are widely used for direct-detection studies (Plates 2 and 4). Panchromatic photographs often show a distinct tonal contrast between an injected effluent and the receiving water and between clear water and water containing suspended solids (Figures 13-31 through

Figure 13-28 Vertical airphoto (part of a frame) of an area in Corpus Christi, Texas, following Hurricane Celia (August 3, 1970). Annotations correspond to four categories of structure damage: *L* = light: up to 20 percent damage; *M* = moderate: 20 to 40 percent damage; *H* = heavy: 40 to 60 percent damage. Black line delineates damaged area; the subarea (white boundary) is shown in Figure 13-29. The photo acquisition scale was 1:24,000; original in color. (Rush et al. 1977; NASA photograph.)

Figure 13-29 Enlargement of a small section of Corpus Christi, Texas, following Hurricane Celia (August 3, 1970), showing different degrees of damage to housing units (compare with Figure 13-28). The photo acquisition scale was 1:24,000; original in color. (Rush et al. 1977; NASA photograph.)

Figure 13-30 Vertical airphoto showing the severely damaged Lower San Fernando Dam on the day following the February 9, 1971, San Fernando, California, earthquake (Richter magnitude = 6.4). Most of the upstream concrete face and crest of the dam were shaken loose, slipping into Lower Van Norman Reservoir. Because of the imminent danger, about 80,000 people were evacuated from the downstream residential area; the reservoir was permanently drained following the earthquake. Scale is about 1:10,000. Compare with Figure 13-27. (Rush et al. 1977; NASA photograph.)

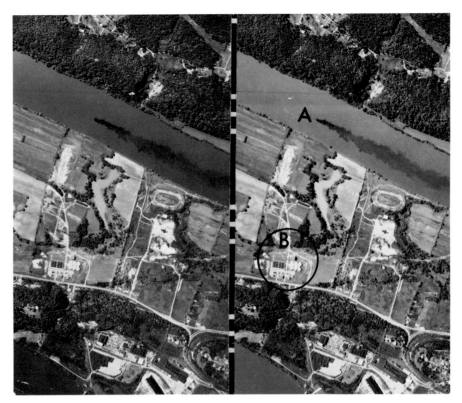

Figure 13-31 Stereogram of a sewage plume (*A*) extending downstream from a subsurface outfall in the Tennessee River near Chattanooga; the sewage treatment plant is circled (*B*). Because of dilution, note how the plume becomes less distinct as the distance increases from the outfall. Scale is about 1:20,000. (Courtesy U.S. Department of Agriculture.)

13-33). Because of significant reflectance differences between clear water and oil films in the 0.3- to 0.4-μm spectral region, ultraviolet (UV) photographs are especially well suited for detecting floating crude oil and refined petroleum products (Figures 2-45 and 2-46). Although the source and areal extent of many water pollutants are clearly depicted on aerial photographs, it is important to realize that measurements of such parameters as chemical compositions or concentrations must be obtained by field sampling and laboratory analysis.

Indirect detection is concerned with observing changes in the aquatic environment that are caused by "invisible" polluting agents. Examples include stressed or dead shoreline vegetation, algae **blooms,*** and dense mats of aquatic weeds, including water hyacinth, water chestnut, water fern, and water lettuce. The population explosion of algae and aquatic weeds is often caused by an overabundance of phosphorus in natural waters. The largest contributors of phosphorus to the aqueous environment are domestic wastes (especially laundry detergents) in sewage effluent and fertilizer runoff from agricultural land.

**An algal population is called a* bloom *when there are 500 or more individuals per milliliter of water.*

With relatively small populations of algae, nonturbid water is normally clear and blue, but with large populations, its color turns green. These conditions are often detectable on natural color photographs (Plate 6). However, when the algae become abundant enough to form a bloom, the concentration is best defined on color IR photographs (Plate 6).

Aquatic weeds can be easily identified in color IR photographs for control or removal operations (Plate 5). This is extremely important because in many tropical and subtropical regions of the world, aquatic weeds have multiplied explosively, interfering with navigation, fishing, irrigation, and the generation of hydroelectric power.

Sulfuric acid mine drainage is one of the most serious forms of surface-water pollution in the United States. It is most common in Appalachia, where abandoned coal mines are concentrated. The acid is formed by seepage water reacting with iron sulfides, especially pyrite, in coal seams and in waste piles containing coal particles; the reaction also produces iron oxide. When these compounds enter surface streams in sufficient concentrations, the water turns rusty red, which is easily detected in normal color photographs. This condition appears as a light tone in panchromatic photographs and as green or yellow in color IR photographs (see Chapter 2). Acid runoff can be lethal to aquatic life, contaminate water supplies, and cause corrosion to piers and boats.

Figure 13-32 Stereogram showing the discharge of petrochemical wastes into a large river near Charleston, South Carolina. Scale is about 1:20,000.

Figure 13-33 Vertical airphoto showing sediment-transport patterns in San Pablo Bay, California. Note how the jettys disrupt the natural flow patterns.

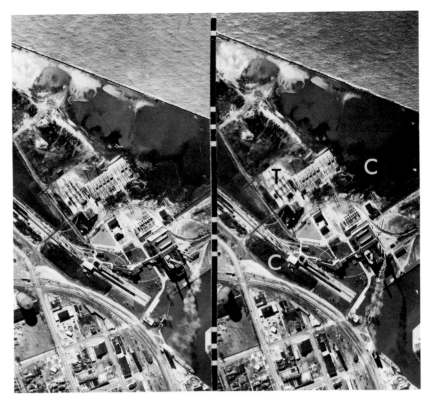

Figure 13-34 Stereogram of a fossil fuel power plant along the shore of Lake Michigan. Piles of coal (*C*) and the transformer unit (*T*) are easily picked out. This plant uses a large amount of water for cooling; the water is returned to the lake at a temperature higher than it was when it was removed. Scale is about 1:16,000.

For investigation of thermal pollution, aerial photographs provide a means for identifying the potential **producer** of artificially heated water but *not* the **discharged plume**. Facilities discharging large quantities of heated water include steel mills and fossil-fuel and nuclear power plants (Figures 3-15 and 13-34). Because injected heated water has a higher temperature than the receiving water, thermal IR images represent the best medium for detecting and monitoring thermal plumes (Figures 6-24 and 6-53 and Plate 8).

Images from airborne electro-optical sensors (Chapter 6) are being used increasingly in water quality investigations. This general class of sensor offers the following advantages over photographic film: (1) Some systems produce real-time images for immediate analysis (Figures 6-1, 6-2, and 6-34 and Plates 7 and 8); (2) these sensors are capable of operating in multiple channels, both within and beyond the relatively narrow confines of the photographic spectrum (Figures 1-5 and 6-6 and Table 6-1); (3) systems operating in the thermal IR region have a day/night capability (Figures 6-24 and 6-25); (4) calibrated thermal IR systems produce accurate radiant temperature images (Plate 8); and (5) digital systems enable subtle patterns in the original images to be computer enhanced (Chapter 15).

As resolutions have improved and spectral-band sensitivities have expanded, images from spaceborne electro-optical sensors are finding application in water-quality studies (e.g., Lathrop and Lillesand (1989) and Gibbons et al. (1989)). The most widely used images are from the Landsat Thematic Mapper (TM) and the SPOT High Resolution Visible (HRV) sensors (Plates 10 and 11). These systems are described in Chapter 6.

Monitoring Mine-Disturbed Land

Airphoto interpretation, especially with stereopairs, is an effective technique for monitoring **mine-disturbed land**, both contemporary and historical. Most *surface* mining operations exhibit characteristic patterns or signatures that permit their identification on aerial photographs. In addition, aerial photographs can be used to monitor mine expansion, to inventory abandoned mine lands, to assess the environmental impact of mine-disturbed land on surrounding land and water, and to plan and assess postmining reclamation efforts that may be required by state or federal statutes.* Satellite photographs and images can be used to monitor surface mining if the disturbed area is sufficiently extensive.

An excellent review of federal mining laws applicable to private and public lands in the United States is given by Tank (1983).

Figure 13-35 Stereogram of an open-pit bauxite mine in Saline County, Arkansas. Note the light tones produced by the aluminum ore. Scale is about 1:20,000. (Courtesy U.S. Department of Agriculture.)

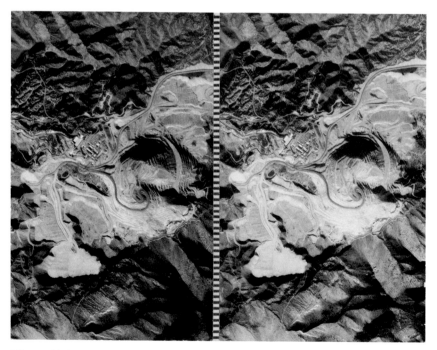

Figure 13-36 Stereogram of an open-pit copper mine near Miami, Arizona. Scale is about 1:42,000.

Figure 13-37 Stereogram of an area strip mine (coal) in Walker County, Alabama. A power shovel working a seam of coal is circled. Scale is about 1:20,000. (Courtesy U.S. Department of Agriculture.)

The principal surface methods employed in modern mining operations are classified as (1) **open-pit mining** (also called **open-cast mining**), (2) **strip mining**, and (3) **dredging**. The key factors that influence the choice of a mining method include (1) size and shape of the deposit, (2) value of the material, (3) type and thickness of the overburden, (4) topographic setting, and (5) legal or environmental constraints.

Open-pit mining is exemplified by (1) rock quarries supplying building stone or crushed stone for aggregate (e.g., limestone, sandstone, quartzite, and marble); (2) borrow pits for the extraction of granular materials such as sand, gravel, and volcanic cinders (Figures 13-1 to 13-3); and (3) large excavations opened to extract metallic ores such as bauxite, iron, copper, and disseminated gold (Figures 13-35 and 13-36). Deep, open-pit mines have a distinctive pattern of benches, arranged in spirals or as levels with connecting ramps; haulage is generally by truck or rail (Figures 13-35 and 13-36). Abandoned or inactive pits often contain standing water, which may be contaminated.

Strip mining is generally done in coal seams or other flat-lying bedded deposits, such as phosphate, that have a relatively thin overburden. There are two types of strip mining, **area strip mining** and **contour strip mining**. Area strip mining is practiced on relatively level terrain (Figures 13-37 and 13-38). For large operations, a trench, called a **cut**, is made through the overburden by a dragline or bucketwheel excavator to expose the deposit, which is then removed by power shovels. The overburden from the first cut is placed on unmined land immediately adjacent to the cut.

A second cut is then made parallel to the first, and the overburden is deposited in the cut previously excavated. This procedure is repeated until the final cut leaves an open trench bounded on one side by the last overburden deposit (called **spoil**) and on the other side by the undisturbed **highwall**. The final cut may be several kilometers or more from the first cut, and without reclamation the resulting landscape is a series of elongated piles of spoil (Figures 13-37 and 13-38).

Contour strip mining is carried out in hilly terrain (Figure 13-39). It consists of removing overburden from the seam by starting on the slope and following the contour of the topography. After the uncovered segment of the seam is removed, successive cuts are made into the hillside until the depth of the overburden becomes too thick for the economic retrieval of the deposit. This type of mining creates a **shelf**, or **bench**, along the hillside that is bordered by a highwall and a precipitous slope created by the overburden being cast down the hillside (Figure 13-39). These spoil banks are subject to severe erosion and landslides (Figure 13-20).

Dredging operations utilize suction equipment or various mechanical devices, such as ladder or chain buckets or draglines, mounted on floating barges. Dredge mining is done in excavated ponds or lagoons, in lakes and rivers, and offshore. This method of mining is used to recover sand and gravel, shell deposits, and **placer minerals**, such as gold and tin. **Dredge tailings** from operations to recover placer minerals have a configuration similar to spoil piles at area strip mines (Figure 13-40). Dredging can affect the environment in numerous ways, including the following: (1) tailing piles

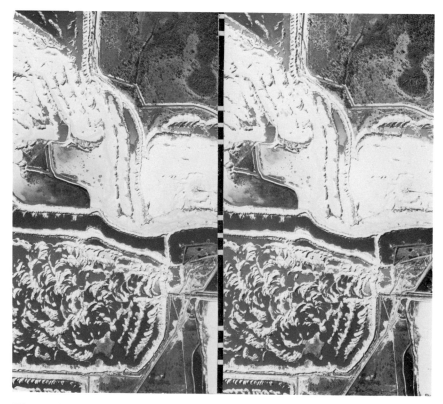

Figure 13-38 Stereogram of an area strip mine (phosphate) in Polk County, Florida. Unless drastic reclamation measures are taken here, these scars will blot the landscape for years. Scale is about 1:20,000. (Courtesy U.S. Department of Agriculture.)

are subject to severe erosion affecting adjoining lands and streams; (2) flora, fauna, and spawning sites for fish in the areas being dredged are destroyed; and (3) large quantities of sediment brought into suspension can affect aquatic environments and municipal water supplies at considerable distances from the actual mine site.

Numerous studies have established that aerial photographs can be used for monitoring surface-mining operations. For example, Mamula (1978) used panchromatic and color IR aerial photographs of various scales to identify and delineate several categories of surface-mine operations and concurrent stages of reclamation for a large coal-mining area near Colstrip, Montana. The categories included the following: (1) active and abandoned highwall and bench areas, (2) ungraded spoil piles, (3) graded and recontoured areas, (4) first- and second-year revegetated areas, (5) third- and fourth-year revegetated areas, and (6) natural and impounded surface-water features. Mroczynski and Weismiller (1982) used 1:30,000-scale color IR photographs to identify and inventory different types of abandoned or derelict land associated with coal strip-mining operations in a 20-county area in southwestern Indiana. The categories of derelict land were (1) barren spoil, (2) gob piles (coarse refuse removed during the first coal-cleaning process), (3) slurry deposits (fine-grained material, mostly coal fines, removed during the final coal-cleaning process), and (4) surface-water bodies contaminated by acid runoff from mined land.

Some *underground* mining operations exhibit associative surface features that permit their identifications on aerial photographs. Some of these surface features include processing facilities (e.g., smelters), headframes with hoist or winder house, haul roads, waste piles, tailing ponds or lagoons, and stockpiles of raw materials (Figure 13-41).

Petroleum extraction may produce a variety of surface patterns. For example, in forested regions, a patchwork mosaic of small clearings connected by a grid system of access roads is a common indicator of such activity (Figure 8-16). Drilling equipment, oil derricks, pump jacks, and storage tanks also aid in identification (Figure 13-42).

Stockpile Inventories

Huge stockpiles of raw materials such as coal, limestone, sand and gravel, mineral ores, and fertilizer must be periodically measured for inventory and cost accounting (Figure 13-41). In earlier days, such inventories were accomplished by laborious plane-table surveys or ground-cross sectioning. Today, cubic volumes of raw-material stockpiles may be determined accurately and efficiently by photogrammetric methods.

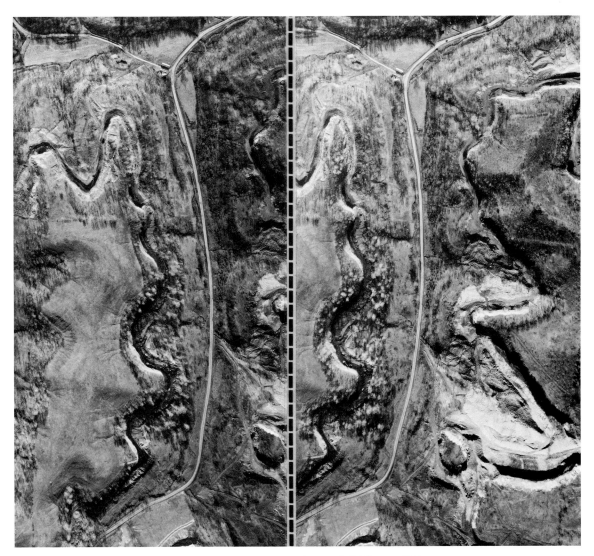

Figure 13-39 Stereogram of contour strip mines (coal) near Middleton, Ohio.
Scale is 1:8,400.

	TABLE 13-1 Approximate Weights of Selective Stockpile Materials	
Material	Kilograms per Cubic Meter	Pounds per Cubic Yard
Coal, anthracite	753–930	1,270–1,567
Coal, bituminous	641–866	1,080–1,460
Clay, dry	1,009	1,700
Clay, damp	1,691	2,850
Limestone, crushed	1,543–1,955	2,600–3,295
Sandstone, crushed	1,442	2,430
Sand and gravel, dry, loose	1,442–1,679	2,430–2,830
Sand and gravel, dry, packed	1,602–1,922	2,700–3,240
Sand and gravel, wet	2,017	3,400

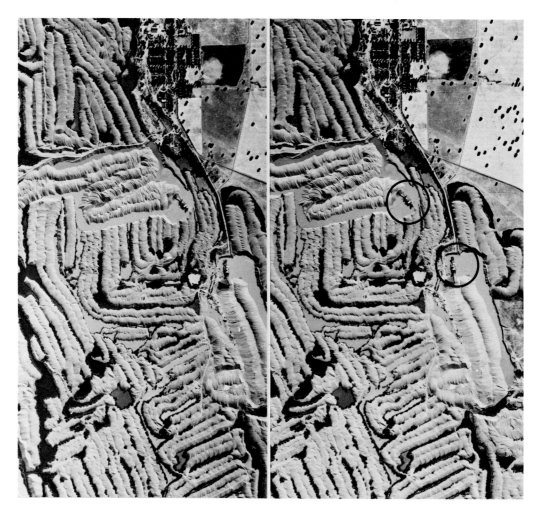

Figure 13-40 Stereogram showing the placer mining of gold as it was once practiced in Yuba County, California. Dredges may be seen in the circled areas. Scale is about 1:20,000. (Courtesy U.S. Department of Agriculture.)

With this technique, piles are contoured at 0.5-m intervals on the slopes with a 25-cm auxiliary interval on the tops. This is accomplished by stereoscopic plotting of the contours of each pile from large-scale stereopairs (e.g., about 1:400). After contouring, the area of each contoured layer or slice is determined by planimetry and the volume is computed.

When weight conversions per cubic meter or per cubic yard are known, volumes of piles may be converted to weight values (Table 13-1). Corrections are made for variations in density for different piles of the same materials, because settling or compaction will result in significant changes in volume-weight ratios. In summary, the photogrammetric method of stockpile inventory has these advantages:

1. It is accurate, economic, and convenient.

2. Inventories can be set for one date, because all photographs can be obtained in a single day.

3. Ground control needs to be established only once.

4. The method provides a permanent record of the size of the pile at the time the photographs were taken; a

volume can be checked at any future time if any question arises as to the accuracy of the record.

5. No bulldozing or pile dressing is required, whereas these operations are usually necessary in the cross-section method.

Figure 13-41 (*Top, page 371*) Stereogram of a lead-zinc mining operation in Ottawa County, Oklahoma. Underground deposits are being brought to the smelter shown at the top of the stereogram. Scale is about 1:20,000

Figure 13-42 (*Bottom, page 371*) Stereogram of oil extraction in Lafayette Parish, Louisiana. Pictured here are a floating drilling rig (*A*) and oil storage tanks (*B*). In areas where canals are common, drilling equipment is easily hauled by barge from one location to another. Scale is about 1:20,000. (Courtesy U.S. Department of Agriculture.)

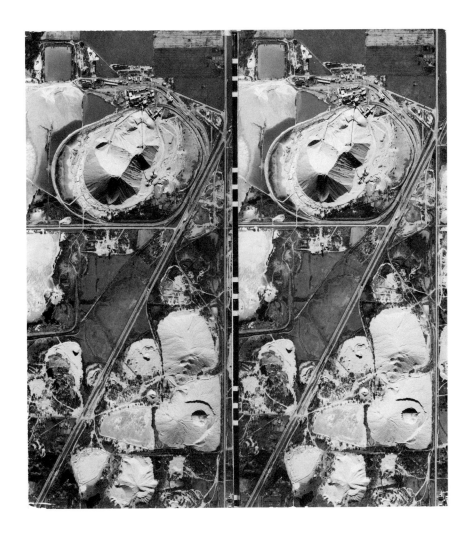

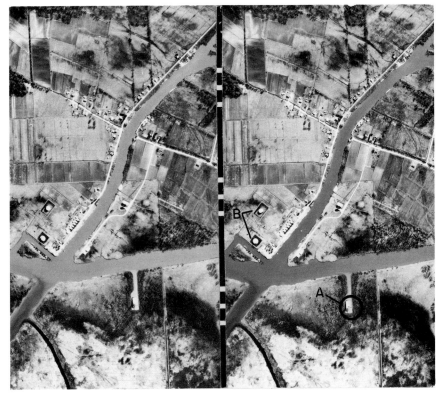

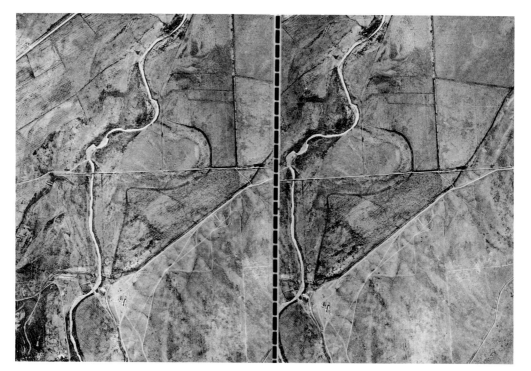

Figure 13-43 Stereogram of a meandering intermittent stream in Yavapai County, Arizona. Scale is about 1:16,000.

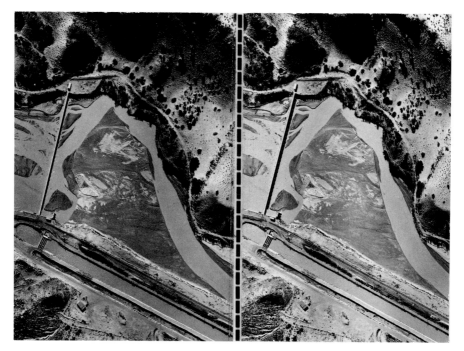

Figure 13-44 Stereogram of a masonry dam on the Rio Grande in New Mexico. In this dry region, stream gradients are low and occasional flash floods result in severe erosion. Scale is about 1:6,000. (Courtesy Abrams Aerial Survey Corp.)

Questions

1. Examine the panchromatic airphoto in Figure 13-33 and explain why the suspended sediment appears lighter in tone than the clear water.
2. Which one of the following types of pollution is least amenable to detection by a photographic film: (a) sewage effluent, (b) suspended sediment, (c) waste discharge from a nuclear power plant, or (d) an oil slick? Explain your reasoning.

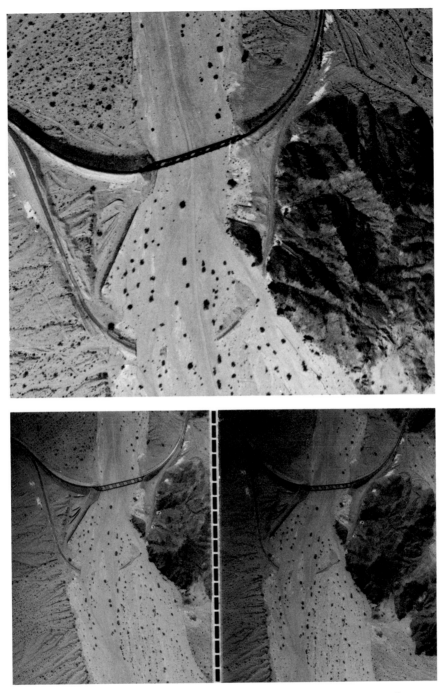

Figure 13-45 Vertical airphoto and stereogram of a railroad bridge spanning an intermittent stream in Riverside County, California.

3. Examine Figure 13-43 and select a site that would likely be a good source for sand and gravel. Explain your reasoning.
4. Examine Figure 13-44 and describe a problem that is hampering the dam's usefulness.
5. Examine Figure 13-45 and answer the following questions.

a. Is there any evidence supporting the existence of an earlier stream crossing? Explain.
b. Explain how the bridge has been effectively heightened to protect it against flash floods.
c. Which direction does the water flow? Explain your reasoning.

Figure 13-46 Vertical airphoto of mine-disturbed land near Middleton, Ohio.

6. Examine Figure 13-46 and match the appropriate numbered feature with the following:

    _____ Vegetated spoil piles

    _____ Potentially clear water

    _____ Active contour strip mine

    _____ Graded spoil

    _____ Potentially polluted water

    _____ Barren spoil piles

    _____ Site of potential acid mine drainage

    _____ Abandoned contour strip mine

    _____ Highwall

    _____ Bench

Bibliography and Suggested Readings

Aniya, M. 1985. Landslide-Susceptibility Mapping in the Amahata River Basin, Japan. *Annals of the Association of American Geographers* 75:102–14.

Belcher, D. J., T. D. Lewis, J. D. Mollard, and W. T. Pryor. 1960. Photo Interpretation in Engineering. In *Manual of Photo Interpretation,* edited by R. N. Colwell, 403–56. Falls Church, Va.: American Society of Photogrammetry.

Brabb, E. E., and B. L. Harrod, eds. 1989. *Landslides: Extent and Economic Significance.* Rotterdam: A. A. Balkema.

Branch, M. C. 1972. Los Angeles Earthquake Damage. *Photogrammetric Engineering* 38: front cover and 36.

Garofalo, D., and F. J. Wobber. 1974. Solid Waste and Remote Sensing. *Photogrammetric Engineering* 40:45–59.

Geraci, A. L. 1981. Remote Sensing Techniques Aid in Water Pollution Evaluation. *Sea Technology* 10:20–21.

Gibbons, D. E., G. E. Wukelic, J. P. Leighton, and M. J. Doyle. 1989. Application of Landsat Thematic Mapper Data for Coastal Thermal Plume Analysis at Diablo Canyon. *Photogrammetric Engineering and Remote Sensing* 55:903–909.

Johnson, R. W., and R. C. Harriss. 1980. Remote Sensing for Water Quality and Biological Measurements in Coastal Water. *Photogrammetric Engineering and Remote Sensing* 46:77–85.

Kennie, T.J.M., and M. C. Matthews, eds. 1985. *Remote Sensing in Civil Engineering.* New York: Halsted Press.

Klooster, S. A., and J. P. Scherz. 1974. Water Quality by Photographic Analysis. *Photogrammetric Engineering* 40:927–35.

Lathrop, R. G., and T. M. Lillesand. 1989. Monitoring Water Quality and River Plume Transport in Green Bay, Lake Michigan with SPOT-1 Imagery. *Photogrammetric Engineering and Remote Sensing* 55:349–54.

Mamula, N. 1978. Remote Sensing Methods for Monitoring Surface Coal Mining in the Northern Great Plains. *Journal of Research, U.S. Geological Survey* 6:149–60.

Massa, W. S. 1958. Inventory of Large Coal Piles. *Photogrammetric Engineering* 24:77–81.

Meyer, C. F., and D. W. Gibson. 1980. *Route Surveying and Design,* 5th ed. New York: Harper & Row.

Mintzer, O., and others. 1983. Engineering Applications. In *Manual of Remote Sensing,* edited by R. N. Colwell, 2d ed., 1955–2109. Falls Church, Va.: American Society for Photogrammetry and Remote Sensing.

Mroczynski, R. P., and R. A. Weismiller. 1982. Aerial Photography: A Tool for Strip Mine Reclamation. In *Remote Sensing for Resource Management,* edited by C. J. Johannsen and J. L. Sanders, 331–37. Ankeny, Ia.: Soil Conservation Society of America.

Nilsen, T. H., and others. 1979. *Relative Slope Stability and Land-Use Planning in the San Francisco Bay Region, California.* Geological Survey Professional Paper 944. Washington, D.C.: U.S. Government Printing Office.

Pallamary, M. 1990. Surveying the San Francisco Earthquake. *Professional Surveyor* 10:4–12 and 52–53.

Piech, K. R., and J. F. Walker, 1972. Outfall Inventory Using Airphoto Interpretation. *Photogrammetric Engineering* 38:907–14.

Robinson, G. D., and A. M. Spieker, eds. 1978. *Nature to be Commanded.* Geological Survey Professional Paper 950. Washington, D.C.: U.S. Government Printing Office.

Rush, M., A. Holguin, and S. Vernon. 1977. ''Potential Role of Remote Sensing in Disaster Relief Management'' (Unpublished report). Houston: University of Texas Health Science Center.

Smelser, M. G. 1987. Geology of Mussel Rock Landslide. *California Geology* 40:59–66.

Strandberg, C. H. 1967. *Aerial Discovery Manual.* New York: John Wiley & Sons.

Tank. R. W. 1983. *Legal Aspects of Geology.* New York: Plenum Press.

Watkins, J. S., M. L. Bottino, and M. Morisawa. 1975. *Our Geological Environment.* Philadelphia: W. B. Saunders Co.

Way, D. S. 1978. *Terrain Analysis: A Guide to Site Selection Using Aerial Photographic Interpretation.* 2d ed. Stroudsburg, Pa.: Dowder, Hutchinson & Ross.

Wobber, F. J. 1971. Imaging Techniques for Oil Pollution Survey Purposes. *Photographic Applications in Science, Technology and Medicine* 6:16–23.

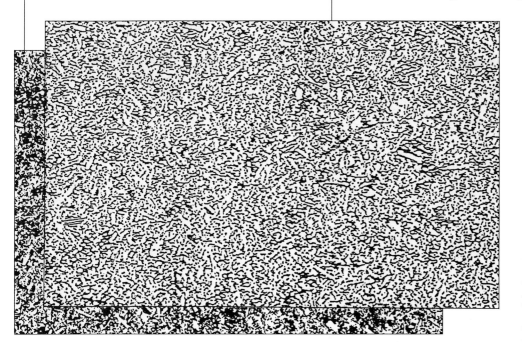

Chapter 14

Urban-Industrial Applications

building sites (e.g., shopping centers and schools), and planning for urban development. **Sequential aerial photographs** are used to reconstruct the historic sequence of changes in urban growth and to predict and plan for future changes (Figure 14-1). **Photogrammetric techniques** are used for making accurate distance, size, and height measurements as well as preparing planimetric and topographic maps.

Vertical stereopairs from mapping cameras are used almost exclusively for urban investigations (Figure 14-1). However, **oblique airphotos** are useful as educational devices for presenting problems and proposed solutions to administrators, city councils, planning and zoning commissions, and the general public. Oblique airphotos tend to provide more spectacular or illustrative views and show heights and relative heights better than vertical photographs (Figure 14-2).

Minus-blue panchromatic and color infrared (IR) aerial photographs are preferred for most types of urban studies. Both have the ability to penetrate atmospheric haze, which effectively sharpens the images of all depicted objects (i.e., increases the amount of detail shown in the photographs). Plate 1 illustrates the loss in detail for a hazy-day, normal color photograph. Color IR photographs have the additional advantage of enabling the interpreter to differentiate objects by color variation. In areas of *rapid change*, it is imperative that *recent photography* be used, regardless of type.

The optimal choice of photographic scale—and hence the level of detail required—is a function of the study objective. For example, small-scale photographs (e.g., 1:120,000) are best suited for the general examination of large urban areas, while large-scale photographs (e.g., 1:10,000 and larger) are necessary for studies in which detail is needed. At the latter scale, for example, small buildings as well as different types of vehicles are clearly evident. It is important to remember that using aerial photographs at a scale *smaller* than is needed results in less accurate work, while a scale that is *larger* than necessary causes an additional expense and additional work in analyzing the extra photographs (Davis 1966). Consequently, new contract photography may have to be obtained when existing photography (e.g., from government agencies) is not available at the necessary scale.

Information regarding photo scales and the purposes for which the scales are suitable is presented in Table 14-1. Panchromatic airphotos at various scales are presented in Figures 14-3 through 14-8 to show the level of urban detail discernible at each scale.

The Airphoto Approach to Urban Studies

Aerial photographs are routinely used by planners, resource managers, geographers, and engineers for a number of diverse urban applications. Typical **photo interpretation applications** include land use and cover mapping, parking and transportation studies, environmental monitoring, real estate assessment, outdoor recreational surveys, evaluation of housing quality, damage surveys, choosing possible

Figure 14-1 Comparative stereograms illustrating land use changes in the vicinity of Calhoun, Tennessee. The top view was made in 1937; the lower view was made in 1967. Scale is about 1:24,000. (Courtesy Tennessee Valley Authority.)

A Systematic Guide to Urban Analysis

A series of 11 topical guides designed to aid geographers in the systematic study of aerial photographs has been prepared by Stone (1964). These guides are postulated on the theory that photo interpretation is largely a *deductive* rather than an *inductive* process and, therefore, analyses should proceed from the known parts of a topic to the unknown. If interpretation activities are organized for working from general patterns toward specific identifications or inferences, stereoscopic study will begin with the smallest scale of photography and end with prints of the largest scale available. A topical outline for the interpretation of urban

features is presented here. In using prepared guides of this type, the interpreter must realize that positive identifications are rarely possible for all urban patterns; thus, listings of uncertain areas should be accompanied by several possible identifications based on the concept of associated features.

1. Outline built-up areas having urban characteristics.

2. Mark the major land and water transportation routes passing through the city (Figure 14-9).

3. Mark the principal commercial airports (Figure 14-10).

4. For the built-up area, outline subareas to show types of water bodies, drainage systems, terrain configuration, and natural vegetation.

5. Divide the built-up area into subareas based on differences in street patterns.

Figure 14-2 Oblique airphotos of downtown Ottawa, Canada, in 1928 (*top*) and in 1964 (*bottom*). Parliament buildings are near top center in each photograph. (Courtesy Surveys and Mapping Branch, Canada Department of Energy, Mines, and Resources.)

Figure 14-3 Vertical airphoto of the New York City region. Annotations are as follows: (A) Hudson River, (B) Central Park on Manhattan Island, (C) Bronx, (D) East River, (E) Brooklyn, (F) Queens, (G) Flushing Bay, and (H) La Guardia Airport. Scale is 1:80,000. A 4× enlargement of La Guardia Airport is shown in Figure 14-4.

6. Outline the older and newer parts of the city.

7. Identify the principal transportation routes within the city.

8. Mark the minor land and water transportation routes passing through the city.

9. Circle the places where there is a change in the type of transportation (Figure 14-11).

10. Outline the primary commercial subareas in the central business district and in the suburbs.

11. Outline principal industrial subareas, including municipal utilities.

12. Outline subareas of warehouses and open storage.

13. Mark the recreational areas.

14. Mark the cemeteries.

15. Outline sections of the residential subareas by differing characteristics of the residences and lots and their relative locations to other functional subareas.

16. Mark the principal administrative and government buildings.

17. Mark the secondary commercial centers.

18. Mark the isolated industrial plants.

19. Mark the probable locations of light industrial establishments.

Figure 14-4 Vertical airphoto of La Guardia Airport, New York City. This 1:20,000-scale view is a 4× enlargement of Figure 14-3.

Information accumulated for a given urban area becomes the basis for more detailed studies. For example, in planning for future expansion of public facilities, correlations might be established between information on population density, number of automobiles per dwelling unit, or water use per capita and such planning multiplier factors as roadway capacity or area of recreational land per thousand persons.

Parking and Transportation Studies

Special photographic flights during peak traffic times are ideal for finding bottlenecks in automobile flow patterns.

Figure 14-5 Stereogram of Oakland, California, and environs. Scale is 1:60,000.

Figure 14-6 Stereogram of Niagara Falls, New York. Scale is about 1:20,000.

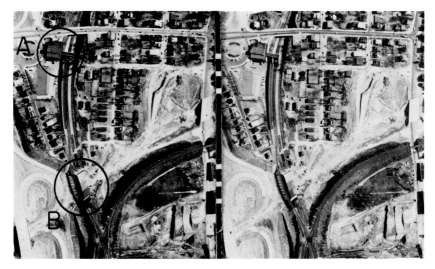

Figure 14-7 Stereogram of a railroad passenger station (*A*) and construction of a highway overpass for trains (*B*). Note planned extension of the new highway through the residential area. Scale is about 1:7,900.

TABLE 14-1   Airphoto Scale Ranges and Examples of Use in the Urban Environment

Scale	Use
1:70,000–1:130,000	General examination of large urban areas; generalized inventories of land use and land cover (e.g., Level II categories in Table 8-1); overview surveys for new highways (Chapter 13); overview surveys of postdisaster damage (Chapter 13).
1:30,000–1:70,000	Detail discernible at preceding scales; inventories of Level III land use and land cover (e.g., Table 8-4); postdisaster damage assessments to major facilities (Chapter 13); water pollution studies (Chapter 13).
1:10,000–1:30,000	Detail discernible at preceding scales; inventories of Level IV land use and land cover (Chapter 8); post-disaster damage assessments to individual buildings and structures (Chapter 13); housing market analyses; site selections for shopping centers and schools; identification of industrial groups.
1:10,000 and larger	Detail discernible at preceding scales; highway route selections (Chapter 13); detailed postdisaster damage assessments (Chapter 13); determination of housing quality; parking and transportation studies; classification and inventory of buildings; identification of specific industries.

Figure 14-8 Vertical airphoto of a new residential area new Las Vegas, Nevada.
Scale is about 1:5,400.

Figure 14-9 Vertical airphoto of an interstate highway bypassing the central business district of Chattanooga, Tennessee. Although some added parking areas are evident between the highway interchanges and the downtown section, shopping centers, such as the one circled, pose an economic threat to downtown merchants. This type of problem confronts many cities. Scale is about 1:20,000.

Similarly, coverage of congested business districts can quickly reveal diurnal parking patterns and the locations of districts having shortages or surpluses of parking spaces during each hour of the day. Law enforcement officers have also found sequential, low-altitude photographs to be of assistance in pinpointing areas where cars are habitually parked in restricted zones.

Individually painted parking spaces can be easily discerned and counted at a photographic scale of 1:6,000; at larger scales, the size and type of vehicle can also be assessed (Figure 14-12). In a few cities, aerial surveys of parking facilities have revealed that there is sufficient unused space in vacant backyards and alleys within the business district to make available more than double the existing parking capacity. Photographs also revealed that traffic would be able to reach parking lots behind shops if only a few new access streets were opened to handle the traffic flow. In other urban areas, aerial photographs have indicated that more parking facilities than were originally allotted will have to be provided; additional spaces are often needed to supply legal parking for vehicles that previously used loading zones and other nonallocated spaces.

When the time interval between successive, overlapping aerial exposures is known, speeds of vehicles imaged on adjacent prints can be computed (Figure 14-13). Such information can be of considerable utility in the analysis of traffic flows during rush hour.

Patterns of Residential Development

Many urban planners believe that the only answer to the control of haphazard urban sprawl is a rigidly administered property zoning system. Certainly there are valid arguments favoring the orderly regulation of community development; otherwise, smoky industrial plants may force down property values in exclusive suburbs, and taverns might be constructed adjacent to school buildings.

The suburbanite who wishes to reside in an area free from polluted air, speeding automobiles, and supersonic aircraft may be hardpressed to find solace in today's metropolis. Nevertheless, interpreters of urban features have found that residential property is one of the key indicators of a family's socioeconomic status and that a person's address often reveals much more about an individual than just where

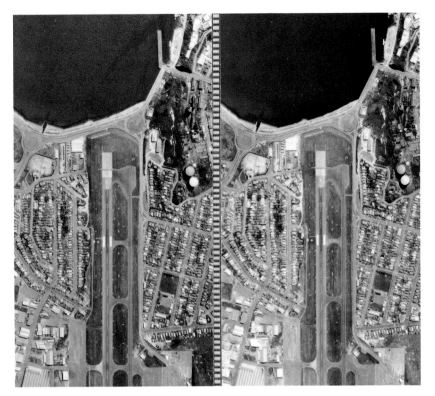

Figure 14-10 Stereogram showing a portion of the municipal airport at Welling-
ton, New Zealand. Scale is about 1:20,000. (Courtesy New Zealand Department
of Lands and Survey.)

Figure 14-11 Vertical airphoto of Lyttleton Harbor, near Christchurch, New Zealand. Such areas constitute a focal point for water, rail, and auto transportation.
(Courtesy New Zealand Department of Lands and Survey.)

Figure 14-12 Large-scale stereoview of parking lots in Youngstown, Ohio, at about 11 A.M. Note concentrations of vehicles in some lots and surplus spaces in others. Buses, trucks, and compact cars can be distinguished from standard-sized automobiles. Scale is 1:3,960.

he or she lives. One might, for example, reflect upon the social or economic status associated with a residence on San Francisco's Nob Hill around 1900. Residence location has meaning not only in terms of real estate cost or rental, but also in terms of occupation, educational level, income class, nationality group, cultural attributes, and even religious preferences.

Even though the typical single-family dwelling in America has grown larger and more luxurious, the high cost and scarcity of building sites result in more and more houses being built on smaller parcels of real estate. The confining atmosphere and unimaginative landscaping that result are painfully illustrated by Figures 14-14 and 14-15.

Indicators of Housing Quality

Information is required periodically on residential housing quality for evaluation and mapping of neighborhoods that need remedial action and for allocation of housing improvement funds. Among the factors obtainable from aerial photographs, the following are considered to be impor-

tant for discriminating between various housing quality classes (Joyce 1974; Jensen et al. 1983):

Open Space

1. Type (e.g., park, golf course, vacant, right-of-way)
2. Amount of land
3. Condition of vegetation
4. Degree of landscaping

Housing Inventory

1. Trailer parks
2. Type of dwelling unit (multiple or single-family)
3. Density of dwelling units per unit area
4. Relative age of housing
5. Dwelling size

Parcel Inventory

1. Parcel area
2. Percentage coverage by structures
3. Percentage of lot landscaped
4. Presence and condition of front, back, and side yards
5. Presence of convenience structures (e.g., swimming pools, patios, courtyards, etc.)

Figure 14-13 The speed of the circled vehicle may be computed from a knowledge of the time interval between photo exposures. Scale is 1:6,000. (Refer to Question 1.)

6. Presence of sidewalks, curbs, and walkways
7. Presence of litter and rubbish

Community Facilities

1. Proximity of schools, churches, hospitals, public buildings, and recreational waterways
2. Presence of overhead utilities
3. Pavement type, width, and condition

Hazards and Nuisances

1. Street traffic
2. Railroads and switchyards
3. Airports
4. Areas subject to surface flooding
5. Swamps and marshes
6. Sewage treatment plants and power stations
7. Industrial areas

Most evaluations of such diagnostic characteristics have used large-scale (1:5,000 to 1:10,000) panchromatic or color IR aerial photographs.

Urban Recreational Planning

The failure of many large cities to make early provision for parks, golf courses, and other outdoor recreational facilities has resulted in tremendous pressures on existing lands as populations have increased. Notable examples of foresight by city planners would include such urban oases as Central Park in Manhattan, Rock Creek Park in Washington, D.C., and City Park in New Orleans (Figure 14-3). In many other heavily populated regions, however, carefully

Figure 14-14 Stereogram of large ranch-style residences being constructed close together on small lots in Milwaukee, Wisconsin. As is commonly seen in new subdivisions, sod, topsoil, and trees have been scraped away to make for more efficient materials handlings. The monotomy of this scoured landscape will remain for years to come. Scale is about 1:4,800.

maintained havens of grass and trees may be sorely inadequate or wholly lacking.

When open areas are not reserved for public use at an early stage in a city's growth, rising real estate values may effectively block the establishment of large municipal parks and athletic facilities at a later date. As a result, private country clubs, concentrated tourist attractions, or spectator sports may offer the only alternatives to local residents (Figures 14-16 through 14-19).

Surveys of population pressures on existing recreational areas and inventories of potential recreational sites are often aided by intensive study of large-scale aerial photographs. Diagnostic factors that are commonly evaluated to indicate areas of high recreational potential include the following:

1. Population factors, such as building density, existing recreational opportunities, kind and direction of urban expansion.

2. Current land use, including factors that might limit or prohibit recreational development (e.g., undesirable industries nearby).

3. Characteristics of potential water-based recreational sites (e.g., size, shape, shoreline configuration, depth, water quality).

4. Existing and potential roads for access to new sites and for use by hikers, skiers, or horseback riders.

5. Character and appeal of vegetation on potential sites.

Legal Applications of Remote Sensing

Following is a partial listing of the types of legal problems for which aerial photography might play a vital part:

1. Discovery and assessment of taxable property.

2. Establishment of boundary lines in ownership disputes.

3. Appraisal of lands to be condemned under states' rights of eminent domain.

4. Discovery and evaluation of the illegal deposition of fill dirt or waste materials on private property.

5. Auto, railway, and airline accidents.

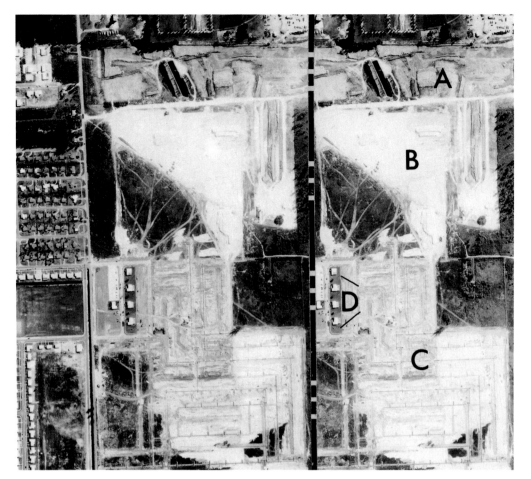

Figure 14-15 In many seacoast areas, one may dig a canal, use excavated materials to fill in lowlands, and create a subdivision with waterfront residences. Here in Dade County, Florida, excavated sand (*A*) is transported to adjacent development areas (*B, C*). In anticipation of advance residential sales, four model homes are already open to customers (*D*). The final result will seemingly appear somewhat similar to the mill-town arrangement at the extreme left edge of the photograph. Scale is about 1:10,000.

6. Inventory of damages from fires, hurricanes, floods, and other disasters.

7. Evaluation of vegetation killed by noxious fumes from industrial point sources.

The discovery and assessment of taxable property (e.g., improvements to single-family residences) may be cited as an application of remote sensing techniques. Recent improvements shown on property records can be checked against ''apparent'' improvements as determined from the interpretation of large-scale aerial photographs; discrepancies are then cross-checked against building permit files. Irregularities that cannot be reconciled are subsequently scheduled for ground check or appraisal. The kinds of improvements most easily found on aerial photographs are additions to existing residences, new garages, swimming pools, apartment buildings, and commercial blacktop areas.

Recognition of Industrial Features

The general classification or specific identification of certain **industrial features** is of vital concern to photo interpreters engaged in urban planning, control of water and air pollution, or military target analysis. In a few instances, unique structures or rooftop signs can make the task exceedingly simple (Figures 14-20 and 14-21). In other cases, however, the correct categorization may require sound knowledge of industrial components, a high degree of deductive reasoning, and one or more photo interpretation keys. The more one knows about industrial processing methods, the more success one will have in recognizing those same activities on vertical aerial photographs.

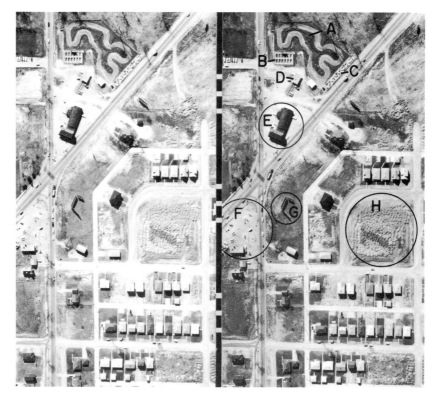

Figure 14-16 Stereogram of recreational and related service facilities near a residential development in Milwaukee, Wisconsin. Items designated are (A) a go-cart track, (B) trampoline pits, (C) miniature golf, (D) a drive-in restaurant, (E) a roller rink or dance hall, (F) a gas station, and (G) billboards. The dumped fill material (H) possibly came from basements exacavated for new houses. Note that lots in the lower part of the picture are so narrow that many homes are of the "shotgun" design, with no space for driveways. Scale is about 1:5,000.

Figure 14-17 Stereogram of a stadium in Jacksonville, Florida, where the annual Gator Bowl football game is held (near the center of photograph). Scale is about 1:16,000.

Figure 14-18 Hotels along Miami Beach, Florida, represent intensively developed recreational facilities for tourists. Scale is about 1:10,000.

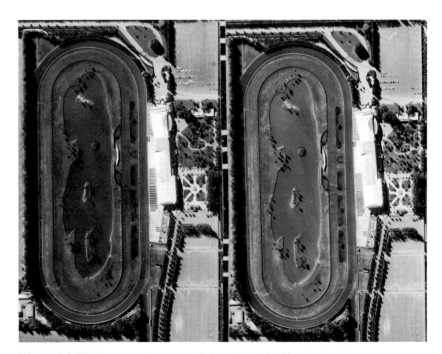

Figure 14-19 Horse racing, one of America's leading spectator sports, attracts many thousands to Hialeah Park in Miami, Florida, each year. Palm trees and a central lake add to the aesthetic value of the location. Scale is about 1:10,000.

Figure 14-20 Stereogram of a cigarette-manufacturing plant (*top of photo*) and tobacco warehouses (*bottom of photo*) in Winston-Salem, North Carolina. Scale is 1:7,920.

Figure 14-21 If one can recognize a railroad turntable (*A*) and locomotives (*B*), then it can be deduced that the engines are being shunted into a repair or maintenance shop (*C*). This heavy fabrication industry is located at New Haven, Indiana. Scale is 1:7,920.

As pointed out by Chisnell and Cole (1958), each type of industrial complex has a unique sequence of raw materials, buildings, equipment, end products, and waste that typifies the industry. Many of these components can be seen *directly* on aerial photographs; others (those that are obscured or are inside structures) must be detected by *inference* from the images of minor associated components. By studying the distinctive shapes, patterns, or tones of raw materials, for example, one may frequently deduce the kinds of processes or equipment that are hidden from view. Arrangements of chimneys, stacks, boilers, tanks, conveyors, and overhead cranes may also provide essential identification clues. And, finally, the finished product can occasionally be seen as it emerges from an assembly line or is stored in open yards awaiting shipment.

A number of photo interpretation guides for use in identification of general classes of industries have been compiled by or for various military agencies. One of these selective keys (Figure 14-22) is based on general industrial categorizations of **extraction, processing**, and **fabrication**. If industries are imaged on photographs at a scale of about 1:20,000 or larger, it has been shown that relatively unskilled interpreters can use such a key to categorize various industries, even though a specific identification may not be possible. Because industrial components tend to exhibit common images irrespective of geographic location, this key is applicable in many parts of the world.

To use the various recognition features to categorize an industry from its image components, follow the procedure recommended here:

1. Decide whether it is an extraction, processing, or fabrication industry.

2. If it is a processing industry, decide whether it is chemical, heat, or mechanical processing—in that order.

3. If it is a fabrication industry, decide whether it is light or heavy fabrication.

Examples of Industrial Categories

Extraction industries, typified by oil drilling, rock quarries, gravel pits, and mining operations, are among the easiest types of industries to classify. They may be recognized by the presence of excavations, ponds, mine shafts, and earth-moving equipment; buildings are usually small and often of temporary construction. Frequently, such operations appear to be rather disorganized as viewed on aerial photographs, even though extracted materials are mechanically handled by conveyors or stored in ponds, tanks, or bins. In some cases, the interpreter must exercise special care in distinguishing waste piles from usable materials. The surface mining patterns that were illustrated in Chapter 13 provide appropriate examples of the extraction industries.

Processing industries are divided into three subclasses: **mechanical, chemical**, and **heat processing**. Mechanical processing industries are those that size, sort, sepa-

---

INDUSTRIAL CLASSIFICATION KEY

Extraction industries are characterized by these features: excavations, mine headframes, ponds, and derricks; piles of waste; bulk materials stored in piles, ponds, or tanks; handling equipment (e.g., conveyors, pipelines, bulldozers, cranes, power shovels, or mine cars); buildings that are few and small.

Processing industries are characterized by these features: facilities for storage of large quantities of bulk materials in piles, ponds, silos, tanks, hoppers, and bunkers; facilities for handling of bulk materials (e.g., conveyors, pipelines, cranes, and mobile equipment); large outdoor processing equipment (e.g., blast furnaces, cooling towers, kilns, and chemical-processing towers; provision for large quantities of heat or power as evidenced by boiler houses); oil tanks, coal piles, large chimneys or many smokestacks, or transformer yards; large or complex buildings; piles or ponds of waste. Three types of processing industries may be recognized:

1. *Mechanical processing* is typified by few pipelines or closed tanks, little fuel in evidence, few stacks, and no kilns.

2. *Chemical processing* is typified by many closed or tall tanks, gas-holders, pipelines, and much large, outdoor processing equipment.

3. *Heat processing* is typified by few pipelines or tanks, large chimneys or many stacks, large quantities of fuel, and kilns.

Fabrication industries are characterized by these features: few facilities for storing or handling bulk materials; a minimum of outdoor equipment except for cranes; little or no waste; buildings may be large or small and of almost any structural design.

1. *Heavy fabrication plants* are typified by heavy steel-frame, one-story buildings, storage yards with heavy lifting equipment, and rail lines entering buildings.

2. *Light fabrication plants* are typified by light steel-frame or wood-frame buildings and wall-bearing, multistory structures, lack of heavy lifting equipment, and little open storage of raw materials.

Figure 14-22 Sample classification key for use in identifying general classes of industries. (From Thomas C. Chisnell and Gordon E. Cole, 1958, Industrial Components—A Photo Interpretation Key on Industry, *Photogrammetric Engineering* 24:590–602. Copyright 1958 by the American Society of Photogrammetry. Reprinted with permission.)

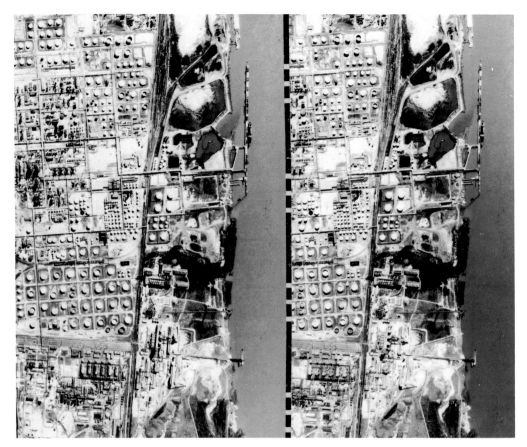

Figure 14-23 Stereogram of a petroleum refinery along the Mississipi River at Baton Rouge, Louisiana. Although this chemical processing industry is pictured at a scale of about 1:25,000, it is a sufficiently distinctive complex to be categorized by use of the industrial classification key. (See Figure 14-22.)

rate, or change the physical form or appearance of raw materials. Industries that are typical of this category are sawmills, grain mills, and ore concentration plants (Figure 13-41); utilities in the same grouping would include hydroelectric plants and water purification and sewage disposal installations (Figure 13-31).

Chemical processing industries are those that employ chemicals to separate, treat, or rearrange the constituents of raw materials. Among representative chemical processing industries are those for sulfuric acid production, aluminum production, petroleum refining, wood pressure treatment, and by-products coke production (Figures 14-23 through 14-25).

Heat processing industries use heat to refine, divide, or reshape raw materials or to produce energy from raw materials. Large quantities of fuel are required, waste piles are common, and blast furnaces or kilns are often in evidence. Thermal electric power production is included in this category, along with cement production, clay products manufacturing, iron production, and copper smelting (Figures 14-26 through 14-28).

Fabrication industries are those that use the output of processing plants to form or assemble finished products. Although a majority of all industries are of the fabrication type, they are the most difficult to identify specifically because most of the activities are hidden from view by well-constructed buildings and enclosures. There is little outdoor equipment in evidence except for large cranes; bulk materials, waste piles, and storage ponds are usually absent. Heavy fabrication industries include structural steel production, shipbuilding, and the manufacture or repair of railroad cars and locomotives (Figure 14-29). Typical of the light fabrication industries are aircraft assembly, canning and meatpacking, small-boat construction, and the manufacture of plastics products (Figures 14-30 and 14-31).

Most interpreters find that the best way to identify an industrial complex is by looking for components that are *key recognition features* of the classification in which the industry falls. Although one or two components may not provide a specific identification, associations with other features will usually provide the missing link in the recognition chain. Knowledge of the photographic scale and the exact geographic locale of photography are additional factors that are of primary importance in recognition of industries. When available, current sets of county maps, topographic quadrangle sheets, stereogram files of representative industries, and a recent commercial atlas are valuable reference aids to photo interpreters.

Figure 14-24 This chemical processing plant near Milwaukee, Wisconsin, is engaged in the preservative treatment of wood materials with creosote. Both untreated stacks of wood (light tones) and treated materials (dark tones) are visible in the storage yard. Tram cars of untreated materials (*A*) may be seen lined up for movement into the pressure cylinder (*B*). Liquid preservatives are stored in cylindrical tanks (*C*). Scale is about 1:4,800.

Figure 14-25 Coal by-products production in Youngstown, Ohio. Components are (*A*) pushing rams to move coke from ovens into quenching cars, (*B*) long coke ovens, (*C*) tank storage of tars and other coal by-products, (*D*) buildings where coal is washed, and (*E*) towers where coal is fed into coke ovens. Scale is 1:3,960. Although this would usually be classed as a chemical processing industry, heat processes are also in evidence at installations such as this.

Figure 14-26 Both extraction and heat processing industries are pictured in this stereotriplet of a Michigan cement plant.

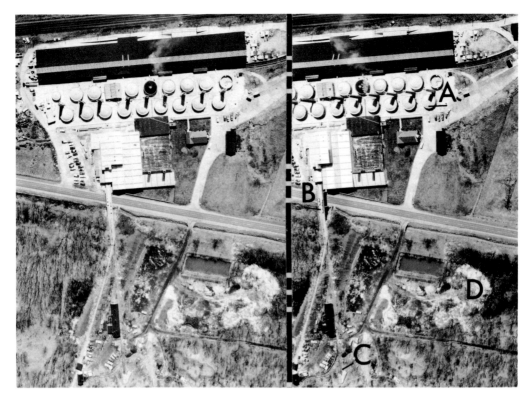

Figure 14-27 Clay products constitute the output of this heat processing industry at Newcomerstown, Ohio. Key components include: (*A*) clay kilns, (*B*) tram cars of raw materials, (*C*) a hillside tunnel to a clay mine, and (*D*) waste materials. Scale is about 1:3,000. (Courtesy Abrams Aerial Survey Corp.)

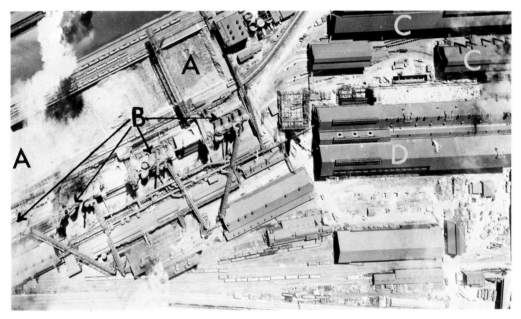

Figure 14-28 Heat processing industry at Youngstown, Ohio. Designated here are: (*A*) limestone and iron-ore storage bins, (*B*) blast furnaces that use coke, iron ore, and limestone to produce pig iron, (*C*) buildings housing open-hearth furnaces, and (*D*) iron-rolling mills. Scale is 1:3,960.

Figure 14-29 Pullman railroad cars are manufactured at this heavy fabrication plant in Michigan City, Indiana. Visible are storage yards with overhead cranes (*A*) and rail lines entering the building (*B*). Painted on the roof is the name of the city and its longitude and latitude. Scale is 1:15,840.

Capacities of Industrial Storage Tanks

The capacity of any **cylindrical storage tank** seen on vertical photography of known scale can be determined from measurements of its diameter (*d*) and height (*h*). The diameter is determined by a direct photo measurement and conversion to a linear distance in meters or feet based upon scale. The height can be computed using stereoscopic parallax, object displacement, or shadow length; these methods for computing heights are described in Chapter 4. Once *d* and *h* are determined, their values are merely substituted into the formula for the volume of a cylinder (*V*):

$$V = \frac{\pi d^2}{4}(h) = 0.7854d^2(h). \qquad (14\text{-}1)$$

When cylinder diameters and heights are in meters, volumes will be in cubic meters. When the diameters are in feet, volumes will be in cubic feet.

Satellite Images and Photographs for Urban Studies

Of the image data available from the various earth resources satellites (see Chapter 6), the spatial resolutions afforded by the Landsat Thematic Mapper (TM) and the SPOT High Resolution Visible (HRV) sensors provide the most detailed views of the urban environment (e.g., Bernstein et al. 1984; Colwell and Poulton 1985; Khorram et al. 1987; Chavez and Bowell 1988; Howarth et al. 1988). The images from these sensors clearly show urban areas from nonurban areas and land that has been cleared for construction or urban expansion. There is normally a clear delineation of central business districts, major buildings and bridges, industrial and harbor facilities, transportation-system infrastructures, and parks and golf courses. The ability to identify individual urban features generally improves in the following order: TM images at 30-m resolution (six nonthermal bands), HRV multispectral images at 20-m resolution, and HRV panchromatic images at 10-m resolution.

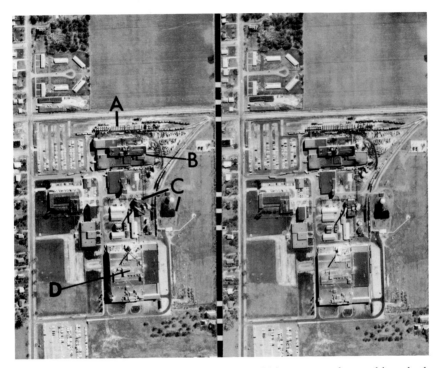

Figure 14-31 This plastics plant in Toledo, Ohio, apparently combines both chemical processing and light fabrication activities. Identified are (*A*) liquid chemical storage tanks, (*B*) a chemical-processing building, (*C*) a power plant and coal piles, and (*D*) a multistory fabrication building. Scale is 1:7,920.

Landsat TM subscene images (black-and-white and color composites) of the San Francisco Bay region are shown in Figures 6-41 and Plate 10. SPOT HRV panchromatic and multispectral subscene images of a portion of Long Island, New York, are shown in Figure 6-44 and Plate 11. A SPOT HRV panchromatic stereogram of a coastal region near Marseille, France, is presented in Figure 6-45.

An excellent example showing the utility of TM images for portraying urban classes of land use and land cover is given by Trolier and Philipson (1986). They report on the results of an interpretation test for recognizing land use and

cover classes on 1:70,000-scale TM single-band and color composite images of Rochester, New York. The following classes were identified on at least one of the images: central business district, commercial, heavy industry, high-density residential, medium-density residential, transportation, tended grass (parks, sports fields, large lawns, cemeteries, golf courses), developing urban (areas under construction), and mixed built-up (mixed urban land uses or covers).

The high spatial resolution and stereoscopic photographs produced by the Large Format Camera (LFC) during Space Shuttle Mission 41-G in 1984 have shown their utility

Figure 14-32 LFC subscene photograph of Ha'il, Saudi Arabia, and environs shown at a scale of about 1:115,000; the acquisition scale was about 1:900,000. Ha'il is one of the largest cities in north-central Saudi Arabia, with an estimated population of 100,000.

as an urban land use and land cover mapping tool. For example, Lo and Noble (1990), using Boston, Massachusetts, as a test site, determined that LFC photographs could be employed to produce urban land use and cover maps at a scale of 1:50,000 for the U.S. Geological Survey's Level II categories (Table 8-1).

Although LFC acquisition scales ranged from about 1:750,000 to 1:1,200,000, enlargements can exceed 10× with little loss of image quality (Figure 14-32). It is anticipated that future Space Shuttle missions will carry this camera. The LFC is described in Chapter 5, and LFC stereograms are shown in Figures 5-11 and 5-12.

Questions

1. Referring to Figure 14-13 and given a photo scale of 1:6,000, what would be the vehicle's rate of speed if the exposure interval was 12 seconds?
2. Describe the advantages of using color IR photographs over panchromatic exposures for discriminating between various housing quality classes.
3. Identify the labeled features in Figure 14-33.
   a. _____
   b. _____
   c. _____
   d. _____
   e. _____

   What two annoyances might be present in this area?
4. Examine Figure 14-34 and determine the type of industry. Which recognition features did you use in making the identification?
5. Calculate the storage capacity of the circled tank in Figure 14-34, given a photo scale of 1:7,900 and a tank height of 20 m.

Figure 14-33 Stereogram of a residential area in Monroe County, New York. Scale is about 1:6,000.

Figure 14-34 Stereogram of an industrial complex in the midwestern United States. Scale is 1:7,920.

Bibliography and Suggested Readings

Adeniyi, P. O. 1980. Land-Use Change Analysis Using Sequential Aerial Photography and Computer Techniques. *Photogrammetric Engineering and Remote Sensing* 46:1447–64.

Baker, R. D., J. E. deSteiguer, D. E. Grant, and M. J. Newton. 1979. Land-Use/Land-Cover Mapping from Aerial Photographs. *Photogrammetric Engineering and Remote Sensing* 45:661–68.

Bernstein, R., J. B. Lotspiech, H. J. Myers, H. G. Kolsky, and R. D. Lees. 1984. Analysis and Processing Landsat-4 Sensor Data Using Advanced Image Processing Techniques and Technologies. *IEEE Transactions on Geoscience and Remote Sensing* GE-22:192–221.

Chavez, P. S., Jr., and J. A. Bowell. 1988. Comparison of the Spectral Information Content of Landsat Thematic Mapper and SPOT for Three Different Sites in the Phoenix, Arizona, Region. *Photogrammetric Engineering and Remote Sensing* 54:1699–708.

Chisnell, T. C., and G. E. Cole. 1958. Industrial Components—A Photo Interpretation Key on Industry. *Photogrammetric Engineering* 24:590–602.

Colwell, R. N., and C. E. Poulton. 1985. SPOT Simulation Imagery for Urban Monitoring: A Comparison with Landsat TM and MSS Imagery and with High Altitude Color Infrared Photography. *Photogrammetric Engineering and Remote Sensing* 51:1093–101.

Davis, J. M. 1966. *Uses of Airphotos for Rural and Urban Planning.* Agriculture Handbook No. 315. Washington, D.C.: U.S. Government Printing Office.

Doyle, F. J. 1985. The Large Format Camera on Shuttle Mission 41-G. *Photogrammetric Engineering and Remote Sensing* 51:200 and cover.

Ehlers, M., M. A. Jadkowski, R. R. Howard, and D. E. Brostuen. 1990. Application of SPOT Data for Regional Growth Analysis and Local Planning. *Photogrammetric Engineering and Remote Sensing* 56:175–80.

Gautam, N. C. 1976. Aerial Photo-Interpretation Techniques for Classifying Urban Land Use. *Photogrammetric Engineering and Remote Sensing* 42:815–22.

Horton, F. E. 1971. The Application of Remote Sensing Techniques to Urban Data Acquisition. In *Proceedings of the International Workshop on Earth Resources Survey Systems,* 213–23. Washington, D.C.: U.S. Government Printing Office.

Howarth, P. J., L.R.G. Martin, G. H. Holder, D. D. Johnson, and J. Wang. 1988. SPOT Imagery for Detecting Residential Expansion in the Rural-Urban Fringe of Toronto, Canada. In *SPOT-1: Image Utilization, Assessment, Results,* 491–98. Toulouse, France: Cedadues-Editions.

Jensen, J. R., and others. 1983. Urban/Suburban Land Use Analysis. In *Manual of Remote Sensing,* edited by R. N. Colwell, 2d ed., 1571–666. Falls Church, Va.: American Society for Photogrammetry and Remote Sensing.

Joyce, R. 1974. A Practical Method for the Collection and Analysis of Housing and Urban Environment Data: An Application of Color Infrared Photography. In *Aerial Photography as a Planning Tool,* 15–20. Publication M-128. Rochester, N.Y.: Eastman Kodak Co.

Khorram, S., J. A. Brockhaus, and H. M. Cheshire. 1987. Comparison of Landsat MSS and TM Data for Urban Land-Use Classification. *IEEE Transactions on Geoscience and Remote Sensing* GE-25:238–43.

Lindgren, D. T. 1971. Dwelling Unit Estimation with Color-IR Photos. *Photogrammetric Engineering* 37:373–77.

Lindsay, J. J. 1969. Locating Potential Outdoor Recreational Areas from Aerial Photographs. *Journal of Forestry* 67:33–35.

Lins, H. F., Jr. 1979. Some Legal Considerations in Remote Sensing. *Photogrammetric Engineering and Remote Sensing* 45:741–48.

Lo, C. P., and W. E. Noble, Jr. 1990. Detailed Urban Land-Use and Land-Cover Mapping Using Large Format Camera Photographs: An Evaluation. *Photogrammetric Engineering and Remote Sensing* 56:197–206.

Rex, R. L. 1963. *Evaluation and Conclusions of Assessing and Improvement Control by Aerial Assessment and Interpretation Methods: A Case History.* Chicago: Sidewell Studio.

Richter, D. M. 1969. Sequential Urban Change. *Photogrammetric Engineering* 35:764–70.

Stone, K. H. 1964. A Guide to the Interpretation and Analysis of Aerial Photos. *Annals of the Association of American Geographers.* 54:318–28.

Trolier, L. J., and W. R. Philipson. 1986. Visual Analysis of Landsat Thematic Mapper Images for Hydrologic Land Use and Cover. *Photogrammetric Engineering and Remote Sensing* 52:1531–538.

Wray, J. R. 1971. Census Cities Project and Atlas of Urban and Regional Change. In *Proceedings of the International Workshop on Earth Resources Survey Systems,* 243–59. Washington, D.C.: U.S. Government Printing Office.

Wray, J. R., and others. 1960. Photo Interpretation in Urban Area Analysis. In *Manual of Photo Interpretation,* edited by R. N. Colwell, 667–716. Falls Church, Va.: American Society of Photogrammetry.

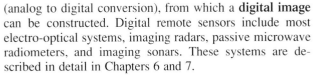

Chapter 15

Digital Image Processing

(analog to digital conversion), from which a **digital image** can be constructed. Digital remote sensors include most electro-optical systems, imaging radars, passive microwave radiometers, and imaging sonars. These systems are described in detail in Chapters 6 and 7.

Because the information content of a digital image is expressed in *numerical form,* analyses and manipulations can be accomplished by *mathematical means.* Consequently, there has been a steadily increasing reliance on computers and auxiliary equipment to process, analyze, display, and store the voluminous amounts of data now being routinely acquired by digitally based remote sensor systems.

This newest and fastest growing branch of remote sensing is called **digital image processing.** It encompasses four major areas of computer operation:

1. **Image Restoration** or **Preprocessing:** Computer routines correct a degraded digital image to its intended form, usually a precursor to the steps that follow.

2. **Image Enhancement:** Computer routines improve the detectability of objects or patterns in a digital image for visual interpretation.

3. **Image Classification:** Quantitative decision rules classify or identify objects or patterns on the basis of their multispectral radiance values (as such, the normal output is analogous to an image map requiring little or no visual interpretation).

4. **Data-Set Merging:** Computer routines integrate multiple sets of data from the same location such that congruent measurements can be made (representative types of information include geographical, geological, geophysical, geochemical, and multispectral radiance data).

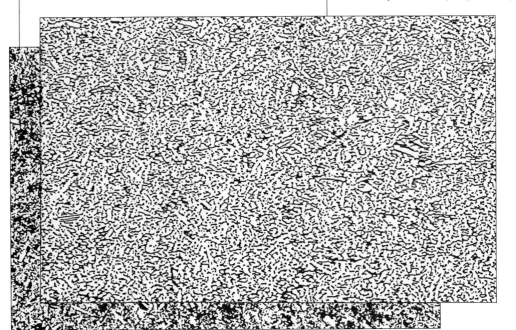

This chapter describes and illustrates the basic principles of digital image processing and the techniques commonly used for earth science applications. Most of the techniques are illustrated with Landsat examples, although other types of digital image data are equally applicable. The material is largely developed from research and demonstration projects of the U.S. Geological Survey's (USGS's) Image Processing Facility in Flagstaff, Arizona, and the Earth Resources Observations Systems (EROS) Data Center's Applications Branch in Sioux Falls, South Dakota. For detailed discussions of digital image processing and its mathematical concepts, the reader is referred to books by Moik (1980), Rosenfeld and Kak (1982), Schowengerdt (1983), Jensen (1986), and Muller (1988).

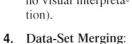

The Scope of Digital Image Processing

Today, many types of remote sensing images are recorded in **digital form.** This is accomplished by sophisticated imaging systems that collect and measure radiance intensities at *n* points in a rectilinear pattern. The sampled intensities are first translated into proportional electronic signals and then encoded into numerical equivalents as integers

TABLE 15-1 Digital Parameters for Landsat and SPOT Images

Sensor	Resolution	Number of Samples	Number of Lines	Pixels/Band (10⁶)	Number of Bands	Total Pixels (10⁶)
Landsat[a]						
MSS	~80 m	3,548	2,953	10.58	4	42.33
TM	30 and 120 m[b]	6,967	5,965	41.56	7	290.91
SPOT[c]						
HRV multispectral	20 m	3,000	3,000	9.0	3	27.0
HRV panchromatic	10 m	6,000	6,000	36.0	1	36.0

[a]*Full scene measures 185 km across-track by 170 km along-track.*
[b]*TM bands 1–5 and 7 = 30 m; band 6 (thermal IR) = 120 m; band 6 is digitally enlarged to the same number of pixels as the other six nonthermal bands.*
[c]*Full nadir scene measures 60 km across-track by 60 km along-track.*

Digital Image Characteristics

In physical form, a **digital image** is a two-dimensional array of small areas called **pixels**, or **pels** (contractions of *picture elements*), that correspond spatially to relatively small terrain areas called **ground resolution cells** (e.g., Figure 6-5). The *horizontal rows* of pixels are called **lines**, and the *vertical columns* of pixels are termed **samples**. Hence, the array consists of *j* lines running from top to bottom and *s* samples running from left to right. Because of this ordering, the *origin* of a grid referencing system for a digital image is always the *upper left pixel;* its coordinates are line 1, sample 1.

Because a single digital image can contain several thousand lines and samples, pixel counts run into the millions. Table 15-1 summarizes the digital parameters for the Landsat MSS and TM systems plus the SPOT HRV multispectral and panchromatic sensors.

Digital Numbers

In quantitative terms, a digital image that represents some portion of the electromagnetic spectrum is a numerical representation of radiance amplitudes emanating from ground resolution cells (assuming that atmospheric scattering and sensor noise components have been removed). Numerical representation is in the form of positive integers that are referred to as **digital numbers**, or simply **DNs**. Suitably formatted DN arrays are placed on magnetic tapes and magnetic disks as a series of magnetized-nonmagnetized spots. Tapes holding correctly formatted digital data are called **computer-compatible tapes** or **CCTs**.*

Suppliers of image CCTs and their addresses are found in Chapters 6 and 7.

Digital numbers cannot be entered into a computer unless they are first expressed as a coded series of **bits** (an abbreviation for *binary digits*). A bit can be only one of two absolute values, 0 or 1. Binary digits become readable in a digital computer by a bistable (two-state) switching device: switch "on" signifies 1, switch "off" equals 0.

Bit Scales

Digital images from most remote sensing systems are usually *quantized* into a 6-, 7-, or 8-bit scale for magnetic tape and disk storage. The 8-bit scale has become the standard for the newer electro-optical sensors, imaging radars, passive microwave radiometers, and imaging sonars. Table 15-2 equates three different bit scales used in digital image processing to DN ranges and gray-scale divisions. CCT bit scales for the Landsat and SPOT sensors are presented in Table 15-3.

For many computer systems, 6- and 7-bit data are often expanded to the 0–255 scale. From Table 15-2 it can be seen that 6- and 7-bit data can be expanded to an 8-bit scale by *quadrupling* or *doubling* DN values, respectively. This expansion is called **rescaling**. Most computer processing is done in an 8-bit mode because image contrast is related to the range of gray-tone densities (quantization levels) available for image construction; the greater the range, the greater the contrast (Schowengerdt 1983).

TABLE 15-2 Bit Scales, DN Ranges, and Gray-Scale Divisions

Bit Scale	DN Range	Gray-Scale Divisions (Quantization Levels)
6 (2⁶)	0–63	64
7 (2⁷)	0–127	128
8 (2⁸)	0–255	256

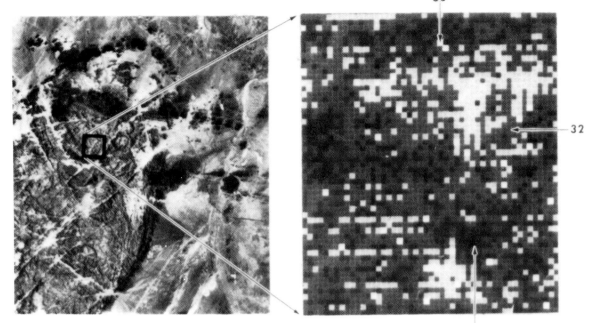

63

32

0

Figure 15-1 Digital image, enlarged subscene showing pixel matrix, and three digital numbers representing high, intermediate, and low radiance amplitudes.

In black-and-white form, a digital image is analogous to a checkerboard pattern of gray tones, with each tone representing the DN value of the pixel at the same location in the digital array (Figure 15-1). Regarding tonal extremes, black equals 0 DN (low scene radiance) and white (high scene radiance) represents DNs 63, 127, and 255 in 6-, 7-, and 8-bit scales, respectively.

DN Histograms and Statistics

It is usually necessary to know the defining statistical characteristics of DN distributions before digital image data are subjected to computer enhancement and classification routines. **Histograms** or **column diagrams** are used to graphically portray DN frequency distributions, whereas the major variabilities of the data are defined by several measures: (1) the **arithmetic mean** and **median** (measures of central tendency), (2) the **standard deviation** (measure of dispersion), and (3) the **skew coefficient** (measure of the departure from symmetry).

A part of a *Landsat 1* MSS band 7 image for an area in northern Arizona is shown in Figure 15-2, and for comparison purposes a histogram of the reflectance data for this image is shown in Figure 15-3. Referring to the histogram, the vertical height of the bars (* symbols) represents the relative number of pixel counts per DN class. The maximum count of pixels is displayed as 100 percent of the vertical axis; bars representing all other DN pixel counts are adjusted in height as percentages to the maximum vertical dimension.

TABLE 15-3 Bit Scales for Landsat and SPOT Sensors

Satellite	Sensor/Band	Bits/Pixel (CCTs)
Landsats 1, 2, 3	MSS 4, 5, 6	7
	MSS 7	6
Landsat 3	RBV panchromatic	7
Landsats 4, 5	MSS 1–4	8
	TM 1–7	8
SPOT	HRV 1–3	8
	HRV panchromatic	8

The horizontal axis plots DN levels. For MSS band 7, the unexpanded data are in a 6-bit scale, and the range is thus 0–63. If values are not contained in the distribution, they are omitted from the horizontal axis (0 DN in Figure 15-3).

The USGS's Image Processing Facility adds the following information to the histogram: (1) number of pixels per DN class, (2) percentage of each DN pixel count to the total population, and (3) DN pixel cumulative percentages. The median of the DN distribution is identified from the cumulative percentage column (i.e., the DN closest to 50 percent). The percent symbols within the histogram represent a cumulative percentage curve (Figure 15-3). The DN distribution in Figure 15-3 appears to be **bell-shaped** (trending toward a **Gaussian** or **normal distribution**).

Quantitative measures (computer calculated) for the distribution in Figure 15-3 are mean, 30.0; median, 30.0;

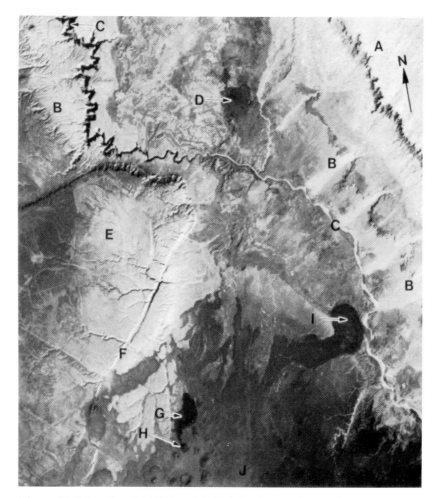

Figure 15-2 *Landsat 1* MSS band 7 digital subscene image for a volcanic and sedimentary region in northern Arizona obtained on October 29, 1973. The image contains approximately 610,000 picture elements. Scale is 1:500,000. Annotated features include: (*A*) Moenkopi Plateau, (*B*) Painted Desert, (*C*) Little Colorado River, (*D*) Shadow Mountain, (*E*) Coconino Plateau, (*F*) Mesa Butte fault zone, (*G*) SP Flow, (*H*) SP Mountain, (*I*) Black Point Flow, and (*J*) San Francisco Plateau. (Courtesy U.S. Geological Survey.)

standard deviation, 10.9; and skew, 0.00. Agreement of the mean and median indicates a normality, and zero skewness categorizes the distribution as symmetrical.

Interactive and Batch Processing

There are two computer-based methods for processing digital image data—interactive and batch. With **interactive processing**, an operator-analyst has direct control over the computer while a job is running; the system must be responsive to the analyst's input messages. This enables an analyst to review intermediate or final processing results as black-and-white or color images on a TV monitor as they are created. The major disadvantage of this method of operation is that the processing is usually performed on relatively small digital arrays to assure completion in a reasonable time.

With **batch processing**, a set of instructions for an image-processing task must be prepared in total before the job is submitted to the computer. The job is then run with the user having no further access to it. Processing results are normally available shortly after a job has been submitted, but the work can take hours, or even days, when there are many jobs to run. Moreover, because there is no human-computer interaction during a computing job, some procedures can take many iterations before a satisfactory result is achieved.

Even with these shortcomings, batch processing is extremely useful if large volumes of data must be handled efficiently. For example, once an analyst has developed an optimum processing routine on a representative data sample, the routine can be developed into a batch-processing operation on large digital arrays and on additional images without further analyst participation. Therefore, the most versatile image-processing system is one of a **hybrid** or **complementary configuration**, incorporating *both* interactive and batch-processing capabilities.

DN	%	Cumulative %	Frequency
1	2.095	2.09	12901.*%*******************************
2	0.109	2.20	672.**
3	0.095	2.30	583.**
4	0.077	2.38	476.*%
5	0.025	2.40	156. %
6	0.226	2.63	1392.**%*
7	0.118	2.75	725.**%
8	0.349	3.09	2148.**%***
9	0.578	3.67	3558.***%*****
10	0.750	4.42	4618.***%********
11	0.408	4.83	2514.****%**
12	0.648	5.48	3992.****%*****
13	0.616	6.09	3796.*****%****
14	0.846	6.94	5208.******%*******
15	0.478	7.42	2945.******%*
16	2.333	9.75	14369.*********%*****************
17	2.353	12.10	14487.**********%****************
18	1.285	13.39	7913.************%********
19	3.089	16.48	19020.**************%*****************************
20	1.614	18.09	9939.***************%********
21	4.214	22.31	25950.****************%********************************
22	0.856	23.16	5269.************* %
23	5.166	28.33	31810.******************%************************************
24	3.377	31.70	20798.********************%*********************
25	1.763	33.47	10854.************************* %
26	4.869	38.34	29986.***********************%**************************
27	1.032	39.37	6355.**************** %
28	4.791	44.16	29501.**************************%***********************
29	1.011	45.17	6226.**************** %
30	6.213	51.38	38260.****************************%******************************
31	4.210	55.59	25926.*********************************%************
32	2.113	57.71	13012.*************************** %
33	4.047	61.75	24923.******************************%***
34	2.073	63.83	12768.*************************** %
35	4.002	67.83	24642.*********************************%*
36	2.893	70.72	17817.*********************************** %
37	3.953	74.68	24341.*********************************** %
38	3.628	78.30	22339.********************************** %
39	2.418	80.72	14893.************************** %
40	2.505	83.23	15426.*************************** %
41	3.032	86.26	18673.********************************** %
42	1.156	87.41	7117.****************** %
43	2.027	89.44	12481.*********************** %
44	0.921	90.36	5670.************** %
45	2.568	92.93	15813.*************************** %
46	1.549	94.48	9539.********************** %
47	0.722	95.20	4449.************ %
48	1.239	96.44	7630.******************* %
49	0.520	96.96	3201.******** %
50	0.385	97.35	2373.****** %
51	0.738	98.08	4544.************ %
52	0.740	98.82	4556.************ %
53	0.162	98.99	1000.*** %
54	0.285	99.27	1756.***** %
55	0.110	99.38	677** %
56	0.183	99.56	1125.*** %
57	0.061	99.62	376.* %
58	0.132	99.76	814.** %
59	0.084	99.84	517.* %
60	0.013	99.85	78. %
61	0.049	99.90	301.* %
62	0.012	99.91	72. %
63	0.086	100.00	530.* %

Figure 15-3 Histogram of DN values for the *Landsat 1* MSS band 7 image shown in Figure 15-2. (Courtesy U.S. Geological Survey.)

Image Processing Systems and Software

Many different **image processing systems** (*hardware* and *software*) are available on the commercial market. Prices range from about $5,000 to $50,000 for **microcomputer-based systems** having a limited level of capability to more than $500,000 for state-of-the-art **minicomputer-based systems**. A directory of suppliers of image processing systems is available from the Earth Observation Satellite Co. (EOSAT), Customer Service Department, 4300 Forbes Boulevard, Lanham, MD 20706.

Most interactive systems have a built-in expansion and modification capability based upon **modular construction**, or the **building-block concept**. This type of architecture enables a system to be expanded from a single-user to a multiuser configuration and state-of-the-art hardware and software to be incorporated into the existing system to increase capacity or keep pace with expanding requirements.

The **software**, or **program package**, is an essential element of an image processing system, and a system with comprehensive capabilities may contain several hundred routines. Software packages are usually *expandable*, en-abling existing programs to be modified or special routines to be added by the user. If there are no in-house programmers to accomplish this, most commercial suppliers of image processing systems provide a custom programming service. Programs can also be purchased from **commercial software suppliers** or from the **Computer Software Management and Information Center (COSMIC)**, a federally funded facility that sells computer programs developed under government sponsorship. The address is as follows: COSMIC, The University of Georgia, 382 East Broad Street, Athens, GA 30602. A list of commercial software suppliers (batch and interactive) is available from EOSAT.

Many systems make use of a software **interactive menu** or **functions directory**. This mode of operation allows an analyst to select appropriate image processing and display options from lists presented on a display monitor. Beyond menus, many systems use **interactive prompts**, which are of particular value to users who have no programming knowledge because they lead the person step by step through each required routine; this feature is sometimes termed *user friendly*. Some companies supply menus in several languages, including English, French, and Spanish. Also, some interactive systems are **hardwired** (i.e., programs wired permanently into the computer's circuitry), enabling a number of common operations to be executed on input digital data by simply depressing function keys.

Figure 15-4 Single-user hardware configuration of a minicomputer-based image processing system capable of processing digital data in both batch and interactive modes.

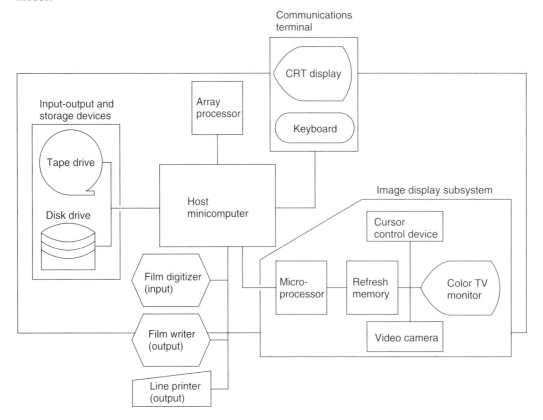

Minicomputer-Based Image Processing Systems

A generalized hardware configuration of a minicomputer-based image processing system is shown in Figure 15-4. Specific characteristics and functions of its principal components are described in the following paragraphs.

Minicomputer and Array Processor

The heart of an interactive image processing system is the **host minicomputer**, which performs software analyses, controls all input-output, and supervises the operation of the entire system. Minicomputers used for image processing are typically configured with from 10 to 30 megabytes (Mb) of **main** or **internal memory**.

For certain operations, significant savings in processing throughout can be realized if an **array processor** is interfaced to the host computer. The array processor is capable of performing several time-consuming operations simultaneously or in parallel (e.g., arithmetic computations, address indexing, memory fetches, storing) at very high speeds rather than serially or sequentially as is the case with a standard computer's **central processing unit (CPU)**. For a given task, an array processor may be 100 times faster than the host CPU. Once tasks for the array processor are assigned by the host computer, it can proceed to perform other processing tasks.

Communications Terminal

The **communications terminal** or **operator console** is composed of a cathode-ray tube (CRT) monitor and a typewriterlike keyboard. It is through the communications terminal that an analyst interacts directly with the computer. For the hardware configuration shown in Figure 15-4, the communications terminal is used both as the *system-operation station* (including batch-processing job setups) and the *analyst station* for the image-display subsystem. The CRT monitor enables various types of information to be presented to the analyst during a processing session; these include input-output messages, statistical information and graphical presentations for a given image, program menus, and queries regarding erroneous or incomplete user commands. An interactive system can be configured with several terminals if there are to be multiple users.

Data Storage

The **magnetic disk** and associated **disk drive** are direct-access devices for the storage of image data and software. For interactive systems, memory capacities generally range from 20 to 1,200 Mb per disk. Large interactive units usually use separate disks to store two types of data: (1) the **system disk** contains the operating system and applications software, and (2) **multiple-image storage disks** are used as an extension of the host computer's main memory to store on-line digital images as they are manipulated by the applications programs.

The most widely used medium for the external, or off-line, storage of digital image information is **nine-track magnetic tape**. A **magnetic tape drive** is used (1) to load images stored on tape to the direct-access storage disk unit for processing and (2) for writing image data onto tape after processing. Today, tape drives are designed to read CCTs having densities of 1,600 and/or 6,250 bits/in. (BPI).* All images kept on-line for processing are stored on disk rather than tape because data access and readback are much faster from disk than from tape.

Image Display Subsystem

The **image display subsystem** consists of a **microprocessor, refresh memory, color TV monitor**, and a **cursor-control device**. In large-scale systems, the microprocessor is capable of internally performing certain image processing operations on digital data loaded by the host CPU. The microprocessor in less sophisticated systems may perform only control operations for the image display subsystem. The microprocessor can be controlled by the analyst from the communications terminal and architecturally is a peripheral device to the host minicomputer.

Refresh-memory channels are used to store multiple images and graphics or graphic overlays (e.g., alphanumeric image annotations, map information, polygonal masks, histograms of image-related statistical information) that are to be displayed on the TV screen. The memory media are usually **random-access memory (RAM) semiconductor silicon chips** that each hold millions of electronic circuits. Data stored in refresh memory can be extracted many times per second and displayed on the TV monitor. The amount of refresh memory available on existing interactive systems varies between 3 and 16 image planes or channels and 1 to 16 graphics planes. Normal pixel depths for each image plane range from 4 to 8 bits. The depth is usually a single bit for a graphics plane.

Two common video output configurations are these: (1) Any three refresh-image channels can be used simultaneously to produce blue, green, and red refresh-image signals for a color composite display on the screen, and (2) each refresh-memory plane can be used to produce individual video signals for either black-and-white or pseudocolor displays. **Pseudocolor** is a processing technique that arbitrarily assigns color to image brightness; it can provide dramatic improvements in image contrast because small changes in intensity can be transformed to abrupt changes in color. Because the video signals are refreshed up to 60 times per second, a stable, flicker-free image can be produced on the TV screen.

The color TV monitor is the principal device enabling the analyst to view images in near real time rather than waiting for the results to be "written" on film and developed. Common display resolutions are 512 × 512 pixels and 1,024 × 1,024 pixels. These pixel resolutions dictate that the display of an entire scene (e.g., MSS or TM) requires **sampling** or **subsetting** (i.e., using every *n*th line and sample). Sampling is sometimes termed **overview**, and although

*The dimensional parameters of a magnetic tape are universally expressed in customary, not metric, units. In this context, the standard nine-track tape is 0.5 in. wide and 2,400 ft long.

there is a considerable loss in resolution, it does permit the analyst to visually select image subareas for detailed analyses at full or improved pixel resolution.

A cursor is an electronic targeting device that defines pixel coordinates. It can be positioned on any pixel in the displayed image by some type of control (e.g., joystick, trackball, mouse, or a special stylus called a light pen). The cursor can be displayed on the TV screen in various shapes, but common forms are crosshairs, arrows, and rectangles. To enhance the detectability of the cursor against the image background, many vendors offer features such as cursor blinking, color cursors, and variable-intensity cursors. This enables an analyst, for example, to position a cursor over individual pixels to access intensity data and select control points for geometric corrections (**fixed cursor**) or to outline image subareas for special processing (**window cursor**).

Hardcopy Output Devices

There are various types of **hardcopy output devices** for producing permanent records of image-processing results. The most common nonphotographic unit is a **line printer**, which records letters, numbers, and a range of symbols on computer printout paper. All information displayed on the operator's CRT (e.g., programs, statistical information, and DN histograms for a given image) can be printed. The unit may also be used to (1) print spatial arrays of DNs, enabling an analyst to see exact DN values at given pixel locations, and (2) generate gray-scale images by overprinting. With the latter output, gray-level approximations are formed by repetitive printing of different characters per pixel location. Because a single printout page can accommodate only 1,250 to 7,200 pixels, data for small image subareas are commonly printed at full or slightly compressed resolution. More recently, matrix printers have been introduced that produce dot graphics that are superior to character overprinting and allow for more pixels to be placed on each page.

The most desirable form of hardcopy is some type of film product. The simplest procedure is to photograph the TV screen with a detached 35- or 70-mm camera. However, because these *off-the-screen photographs* incorporate **barrel**, or **curvature**, **distortion**, they are best suited for quick-look documentation purposes. Distortion-free color and black-and-white hardcopy can, however, be obtained by a photographic system that intercepts video signals traveling from the refresh-image channels to the color TV monitor. With the Polaroid Videoprint system, a videoprocessor converts video signals into images on a flat-faced CRT (eliminates barrel distortion); an internal camera system is then used to photograph the raster display. The system is capable of producing negatives, Polaroid prints, or 35-mm color transparencies.

High-resolution hardcopy images suitable for interpretation and mapping can be produced by a **film writer** or **plotter**. With the Optronics' rotating-drum film writer, film copy is provided by attaching unexposed film (commonly 25.4 × 25.4 cm) to the outside of a rotating drum and allowing a very small beam of modulated light to sequentially expose the film pixel by pixel in proportion to the DN values that are being read from refresh memory, disk, or magnetic tape. There are 256 gray-scale modulation levels for 8-bit data. Common pixel resolutions are 12.5 μm, 25 μm,

50 μm, and 100 μm. A film writer is especially well suited to generating images with a dense matrix of values because more than 350 million pixels can be recorded on a single piece of 25.4 × 25.4-cm film.

Optronics, an Intergraph Division, offers models that can use both black-and-white and color films or panchromatic film only. For the panchromatic medium, images can be produced as negatives or positive transparencies. Three film positives that have been punch-registered to a piece of unexposed color-transparent film can be used to produce a color composite image by contact printing each transparency onto the color film using blue, green, and red filtered light (see Chapter 6).

Because many remotely sensed images are recorded directly on film (aerial photographs, thermal infrared, and radar images), state-of-the-art interactive systems are normally configured with some type of **film digitizer** (e.g., drum scanner, flatbed scanner, TV-video digitizer). A film digitizer scans color or monochrome films and converts film density variations into digital numbers for computer processing, a process known as **digitization**. An Optronics drum scanner can digitize an image to 256 DN levels at selectable sampling rasters or spot sizes as small as 12.5 × 12.5 μm. For a color product, the film is scanned sequentially through blue, green, and red filtered light to extract density data from each of the film's three dye layers (Plate 1).

EOSAT and SPOT CCT Products

EOSAT offers unenhanced Landsat TM and MSS image data on standard nine-track CCTs that incorporate **radiometric** and **geometric corrections**. Radiometric corrections take into account such things as the calibration factors for the detectors (e.g., gain and offset), whereas geometric corrections account for such things as earth rotation and variations in viewing angle. The image data are presented in one of three cartographic projections: (1) Space Oblique Mercator, used when no special projection is specified by the user; (2) Universal Transverse Mercator, available when scene nadir lies between 65° N and 65° S latitude; and (3) Polar Stereographic, available when scene nadir lies beyond 65° N or 65° S latitude.* The standard data format for the CCTs is **Band Sequential** (**BSQ**), wherein each band's data are stored sequentially in a separate image file for all scan lines in the image array.

Because of the huge digital arrays for a **TM full scene** (Table 15-1), EOSAT also offers **TM quadrant scenes**, which are each about one fourth of a full scene. In addition, **movable digital scenes** are available (e.g., 50 × 100 km, N-S by E-W; 100 × 50 km, N-S by E-W). These scenes can be positioned anywhere within a given Landsat path, allowing users to frame closely their area of interest.

SPOT Image Corporation offers unenhanced HRV panchromatic and multispectral image data on CCTs that incorporate different levels of radiometric and geometric

---

The Hotine Oblique Mercator was the standard Landsat 1, 2, *and 3 MSS and RBV projection.*

| | TABLE 15-4   Landsat and SPOT Standard CCT Products

	Number of Tapes	
Sensor—Coverage	1,600 bits/in.	6,250 bits/in.
Landsat		
MSS, 4 bands at ~80-m resolution		
Full scene, 185 × 170 km	1	1
TM, 7 bands at 30-m and 120-m resolution		
Full scene, 185 × 170 km	12[a]	3–4
Quadrant scene, ~46 × 42 km	3[a]	1
SPOT		
HRV multispectral, 3 bands at 20-m resolution		
Full nadir scene, 60 × 60 km[b]	2–3	1
HRV panchromatic, 1 band at 10-m resolution		
Full nadir scene, 60 × 60 km[b]	2–3	1

[a]*Available with a 10 percent surcharge.*
[b]*The across-track width can reach 80 km for off-nadir viewing.*

corrections. Rectified image data can be presented in Lambert Conformal, Universal Transverse Mercator, Oblique Equatorial, Polar Stereographic, or Polyconic projections. In addition to **SPOT full scenes**, several special products are available in CCT formats. These include **SPOT Quad-Maps,**™ which correspond to any 1:24,000-scale USGS topographic map; **SPOT MetroViews,**™ which can be centered on any major city; and **SPOT BasinViews,**™ which can cover any geologic basin in the world.

Factors such as sensor type, product format, tape density, and study-area dimensions determine the number of CCTs needed for a particular study. Table 15-4 summarizes the number of CCTs needed for standard MSS, TM, and HRV scenes.

Microcomputer-Based Image Processing Systems

The relatively recent revolution in microprocessor technology has led to the introduction of inexpensive microcomputer-based image processing systems (Figure 15-5). Primarily because of limitations in CPU speeds and data storage, a basic microsystem can process only small digital data sets. Even with this limitation, however, they are adequate for many research endeavors, and they are rapidly becoming important teaching tools in many academic institutions and government agencies.

For systems without a tape drive, DNs are stored on **floppy disks** (also called "**floppies**") rather than larger-capacity CCTs. One common method of placing image data on floppies is by interfacing a microcomputer to a minicomputer having tape drives; in this configuration, subscenes from a CCT are downloaded to the microcomputer's floppy-disk drives for reformatting and storage.

Figure 15-5 Interactive image processing system showing (*A*) tape drive, (*B*) line printer, (*C*) control monitor, (*D*) cabinet containing the microcomputer, hard disks, and disk drives, (*E*) control keyboard, (*F*) cursor control device (mouse), and (*G*) color TV monitor. (Courtesy Brett A. Borup and Margaret Moore, Northern Arizona University.)

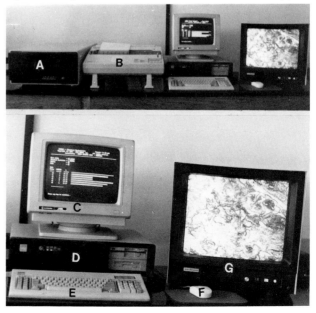

EOSAT offers Landsat TM digital images to the public on 5¼-in. floppy disks. Each scene covers an area 512 pixels square (15 × 15-km scenes); seven floppy disks are provided, one for each TM band. The image data are formatted in a simple, easy-to-use PC/MS-DOS format. At this time, Landsat MSS and SPOT HRV image data are not offered by EOSAT or SPOT Image Corporation.

Image Restoration

Image restoration, or **preprocessing**, is concerned with correcting a degraded (i.e., distorted, noisy) digital image to its intended form. Image correction (**clean-up**) is one of the most important stages of digital processing, because many enhancement and classification operations will emphasize image imperfections to such an extent that useful information can be obscured. For this reason, preprocessing usually *precedes* other types of image processing.

A series of computer algorithms have been developed to recognize and remove several types of errors and distracting effects from digital images. They include:

1. Geometric distortions.

2. Noise patterns.

3. Variations in solar illumination angle.

4. Atmospheric haze.

All the algorithms are applicable to multispectral images (e.g., Landsat MSS), but a radar image might need to be subjected to only the first two correction routines. An image that has undergone the appropriate preprocessing is called a **data base**.

Geometric Correction

Numerous **systematic** and **nonsystematic geometric distortions** are inherent in raw digital images. Because systematic distortions are constant over time, they are predictable, and geometric transformations are relatively simple to

design and inexpensive to run. The purpose of these transformations is to correct pixel locational errors, thereby placing ground features in their correct positions throughout the image.

Skew. **Skew** is one type of systematic distortion associated with the Landsat sensors. Skew distortion is introduced as a result of the earth's rotation and the satellite's orbital movements as an image is being acquired by the MSS or TM. Each scene is **deskewed** by an algorithm that shifts scan lines by a calculated number of pixels—a number dependent on the estimated latitude for the scan line being processed with respect to the starting line of the image (Rohde et al. 1978). For example, at 40° N latitude the right-hand shift between the top and bottom lines of an MSS frame is 122 pixels (Moik 1980). Skew correction produces the familiar parallelogram shape of a full MSS or TM scene (Figure 6-39 and Plate 9).

Nonsystematic Distortions. Altitude and attitude variations (roll, pitch, yaw) and topographic elevation differences are responsible for **nonsystematic (random) distortions** in a digital image. The correction process depends on the scene and makes use of well-distributed **ground control points** that are identifiable in both the distorted image and a reference map or control base (e.g., highway intersections, airport runways, and other landmarks). Each ground control point is identified by line-sample (image) and latitude-longitude (map) coordinates. These values are then used to establish the geometric transformations required to match the image to the map. The reference coordinate system for correcting Landsat MSS and TM images can be based on several map projections (e.g., Space Oblique Mercator, Universal Transverse Mercator, and Polar Stereographic).

Resampling. Because the new locations of transformed pixels will rarely coincide with the locations of input or source pixels, DNs for the transformed pixels must be interpolated from the neighborhood surrounding the source pixel. This process is called **resampling**, and three algorithms can be used (U.S. Geological Survey 1983; Moik 1980) (Figure 15-6). They are as follows:

1. **Nearest-Neighbor Resampling**: DN equal to that of the nearest input pixel is assigned to the output pixel. Because the true location may be offset by as much as

Figure 15-6 Three resampling options for determining digital numbers for transformed pixels (Adapted from Short 1982).

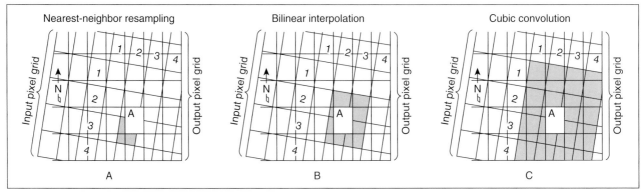

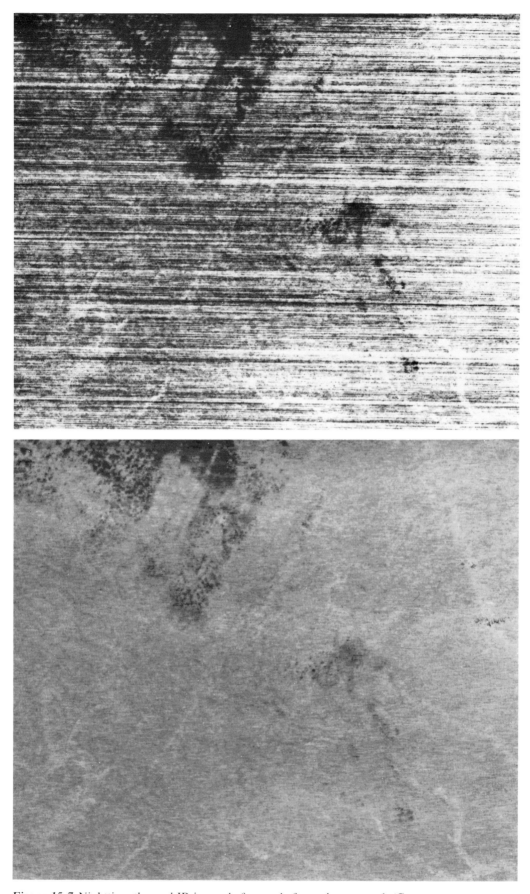

Figure 15-7 Nighttime thermal IR image before and after noise removal. (Courtesy Pat S. Chavez, Jr., U.S. Geological Survey.)

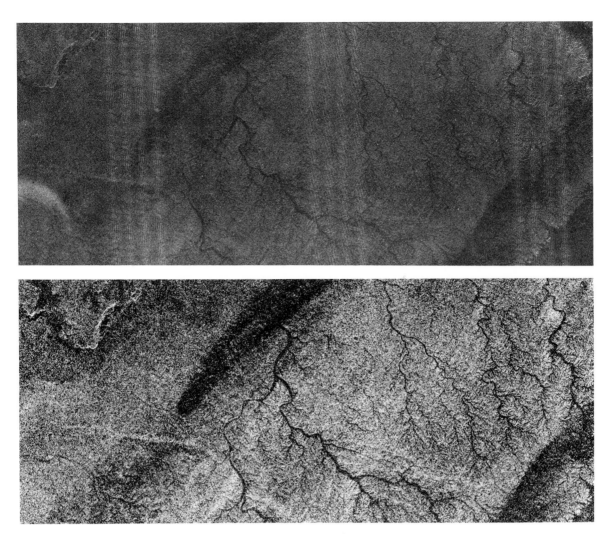

Figure 15-8 SIR-B radar image before and after noise removal.

one-half pixel in the output matrix, linear features in particular will have a blocky, steplike appearance. Computational requirements are relatively low because only one input value is required to determine a resampled value.

2. **Bilinear Interpolation**: Output DNs are determined by taking a proximity-weighted average of input DNs from the four nearest pixels. There will be a loss in image resolution because of the smoothing or blurring caused by DN averaging. Three to four times more computation time is required than with the nearest-neighbor technique.

3. **Cubic Convolution**: Output DNs are assigned on the basis of a weighted average of input DNs from 16 surrounding pixels. This resampling approach has virtually replaced bilinear interpolation because its blurring effect is much less noticeable on a processed image. Cubic convolution requires the most computation (it takes 10 to 12 times longer to run than the nearest-neighbor technique).

Noise Correction

The sensing and recording process can introduce **electronic noise** to an image—noise that is independent of the information transmitted from the scene. Both **random** and **periodic noise** may be found in a digital image. Periodic noise masks radiance data, in essence producing "two-component" DNs (i.e., some combination of valid and nonvalid data). Random noise represents only nonvalid image data. An example of periodic noise is six-line banding (six-line scanner noise), commonly associated with uncorrected Landsat MSS images. Two types of random noise are bit errors (speckle or spike noise) and line drops. The dramatic effects of noise removal for a thermal infrared (IR) image and a SIR-B radar image are shown in Figures 15-7 and 15-8.

MSS Six-Line Banding. The MSS system uses a six-detector array for each of the four spectral bands, enabling six ground lines to be imaged simultaneously during each sweep of the mirror (Figure 6-38). However, when the de-

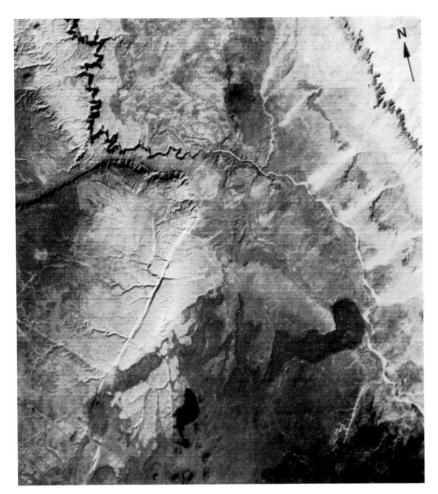

Figure 15-9 Unprocessed *Landsat 1* MSS band 7 subscene image for a volcanic and sedimentary region in northern Arizona obtained October 29, 1973. Scale is 1:500,000. This image incorporates six-line banding and several lines of dropped data. Compare with computer-corrected image shown in Figure 15-2. (Courtesy U. S. Geological Survey.)

tectors have slightly different output responses, the resulting raw images often exhibit a *striping effect,* known as **six-line banding**, that can (1) reduce the accuracy of image classification statistics because some DNs are contaminated with a noise component or (2) reduce the aesthetic appearance of an image (Figure 15-9, Plate 14A). Several **destriping** methods have been devised to remove or suppress this noise pattern. For example, the EROS Data Center uses a two-pass, "through-the-image" operation; the following description of the EROS technique is from Rohde et al. (1978).

In the first pass, DN data for each detector are normalized with the relationship:

$$\text{DN}_{O(j,s)} = \text{DN}_{I(j,s)} \left(\frac{S_A}{S_I} \right) + M_A - M_I \left(\frac{S_A}{S_I} \right), \quad (15\text{-}1)$$

where: $\text{DN}_{O(j,s)}$ = digital number of output pixel at line j, sample s,

$\text{DN}_{I(j,s)}$ = digital number of input pixel at line j, sample s,

S_A = standard deviation of entire scene,

S_I = standard deviation of individual detector ($I = 1$ to 6),

M_A = mean DN value of entire scene, and

M_I = mean DN value of individual detector.

If local statistics differ markedly from those derived from the total scene, the one-pass algorithm may not be totally effective. A second-pass algorithm is then used to perform a local averaging (smoothing) adjustment to remove the residual striping from the affected subregions. With the second pass, the digital data are processed in six-line groups. The first line of each group is selected as a "good data," or reference line (DNs will remain unchanged), and each succeeding line is processed to be similar to the line above. A local average along line I (LOCAV$_I$) of 75 pixels in each direction from the pixel being processed in a line is compared to the corresponding local average from the preceding line, $I - 1$ (LOCAV$_{I-1}$). If this difference is less than a predetermined threshold value, the pixel's digital number is modified by the difference:

$$\text{DN}_{O(j,s)} = \text{DN}_{I(j,s)} - D, \quad (15\text{-}2)$$

where: $\text{DN}_{O(j,s)}$ = digital number of output pixel at line j, sample s,

$\text{DN}_{I(j,s)}$ = digital number of input pixel at line j, sample s, and

$D = \text{LOCAV}_I - \text{LOCAV}_{I-1}$.

The 75-pixel window is moved at 1-pixel increments through the image from left to right and from top to bottom. The visual effect of destriping is shown in Figure 15-2.

Bit Errors and Line Drops. **Bit errors** are represented by isolated spikes (each of 1 pixel) of high or low DNs that

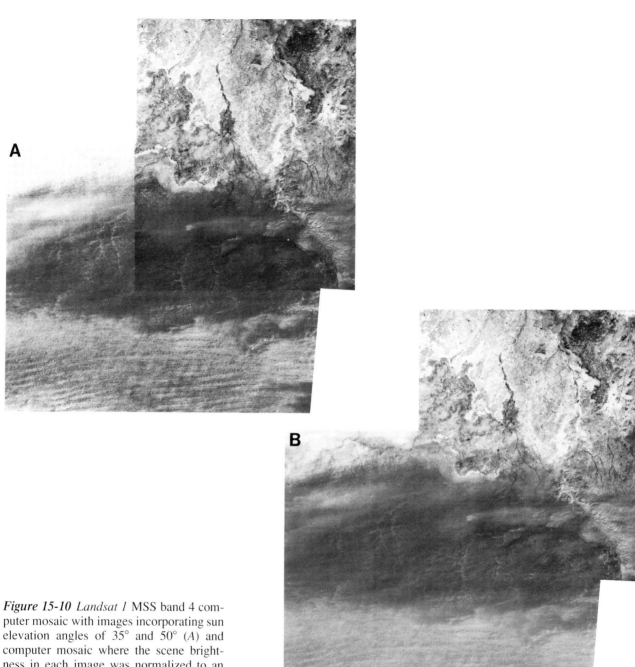

Figure 15-10 *Landsat 1* MSS band 4 computer mosaic with images incorporating sun elevation angles of 35° and 50° (*A*) and computer mosaic where the scene brightness in each image was normalized to an elevation angle of 90° (*B*).

deviate significantly from values of surrounding pixels, and horizontal lines of anomalous DN values represent **line drops** (Figure 15-9). Rather than adjusting DN values as with six-line banding, restorations for bit and line noise use **artificial data**, and the improvement is, therefore, *cosmetic*.

Bit errors can be removed from a digital image by comparing each pixel value with its neighbors; if the difference exceeds a certain threshold value, the pixel is considered a noise point and the aberrant value is replaced by the average of neighboring pixel values. Commonly used neighborhoods or windows have dimensions of 3 × 3 pixels or 5 × 5 pixels.

Lost lines of data can be replaced (1) by average values calculated from the line immediately above and the line immediately below the lost line or (2) by DNs from the preceding line. Line replacement by the averaging method for an MSS image is shown in Figure 15-2 (compare with Figure 15-9).

Sun-Angle Correction

If images (e.g., Landsat MSS) generated during different times of the year are to be used for mosaics or compared for change detection, the seasonal effects of variations in solar illumination angle should be normalized. Corrections can be made by *dividing DNs for each pixel by the cosine of the illumination angle.* Scene brightness for each image, regardless of season, is normalized to a solar-illumination angle of 90°. Because the function assumes a smooth, flat surface, topographic shadows are retained. Figure 15-10 shows the value of normalizing the solar illumination angles for two Landsat MSS images acquired during different seasons.

Correction for Atmospheric Scattering

Whenever multispectral images are generated from the visible and near infrared spectral regions, they are affected unequally by atmospheric scattering. Because Rayleigh and mie scattering are inversely proportional to wavelength, the images representing the shorter wavelength regions are affected the most by haze (Figure 1-11). Regarding the MSS system on *Landsats 1, 2,* and *3,* band 4 (0.5–0.6 μm) will have the highest component of scattered light and band 7 (0.8–1.1 μm) the lowest component (MSS bands 1 and 4 on *Landsats 4* and *5* (Figure 6-39).

The scattering process contributes radiance from the atmosphere (haze), which reduces scene contrast; water vapor is the major controlling parameter (Figure 1-14). A black-and-white image so affected will have a washed-out or fogged appearance (band 4 in Figure 6-39). For many uncorrected MSS color composite images (e.g., bands 4, 5, 7), the effect of scattering is manifested by an overall bluish cast caused by the dominance of haze in band 4.

A common **haze correction technique** uses histograms of MSS images that contain deep water bodies, topographic shadows, or cloud shadows. The method assumes band 7 is essentially free of haze, and thus deep water or shadow pixels are black. A histogram of band 7 will therefore contain 0 DNs or, at most, DN 1 values, but the histograms for the other three bands will not. Rather, the bars for these histograms are displaced to the right (the shorter the wavelength, the larger the displacement), and at some DN level there is an abrupt increase in DN pixel frequencies. The general lack of DNs below this level is attributed to the scattering component. Therefore, the DN where the abrupt increase occurs is assumed to be the haze component. This value is *subtracted* from all pixel values in that band, thus producing a haze removal correction at a first-order approximation.

Histograms showing the frequency distribution of MSS DNs before adjustment for atmospheric scattering for an area in northern Arizona (Figure 15-2) are presented in Figure 15-11. Haze DN values for this scene are as follows: band 4 = 12, band 5 = 7, band 6 = 4, and band 7 = 0.

This method was also used to remove the atmospheric haze from the digitized natural color photograph shown in Plate 1. The haze DNs were as follows: blue plane = 24, green plane = 11, and red plane = 6. Note how the bluish cast was successfully removed, rendering natural colors without haze effects.

Chavez (1988) has introduced another histogram technique for atmospheric scattering correction that is applicable to both MSS and TM data. Unlike the technique previously described, the haze values for each band are not selected independently. Rather, the user selects one of five atmos-

pheric scattering models (very clear = λ^{-4}, clear = λ^{-2}, moderate = λ^{-1}, hazy = $\lambda^{-0.7}$, and very hazy = $\lambda^{-0.5}$) to predict the haze DNs for *all* the spectral bands from a haze value identified on only *one* of the band's histograms. The method then normalizes the predicted haze values for the different gain and offset parameters of the MSS and TM. This method is especially useful for producing "haze-free" TM natural color images (e.g., Plate 10).

Because atmospheric scattering is wavelength-dependent and varies with time, obtaining haze-free values is extremely important for spectral ratio analysis and normalizing multitemporal images. If data base images, excluding TM natural color images, are to be used only for visual interpretations, atmospheric correction is not needed because bias subtraction is accomplished when the DN data are enhanced by contrast stretching; this enhancement technique is described in a later section.

Image Enhancement

The goal of **image enhancement** is to improve the detectability of objects or patterns in a digital image for visual interpretation. However, because certain enhanced images do not resemble their original forms, the user must understand changes caused by the processing if interpretation is to be correct. Enhancement can be divided into the following categories:

1. Contrast stretching.

2. Spatial filtering.

3. Edge enhancement.

4. Directional first differencing.

5. Multispectral band ratioing.

6. Simulated natural color.

7. Linear data transformations.

These operations are applied to digital image data after the appropriate preprocessing steps have been completed.

Contrast Stretching

Contrast stretching is designed to accentuate the contrast between features of interest in a digital image. It is accomplished by redistributing a range of input digital numbers to fill a larger output scale. The redistribution can be linear (**uniform expansion**) or nonlinear (**nonuniform expansion**). Several types of contrast stretches can be applied to data bases; which will be the most effective is determined by the range and variation of the original DNs and the nature of the investigation.

Linear Stretches. Increasing the contrast in a single digital image, while preserving original radiance relationships, can be accomplished by a **linear stretch.** Linear contrast enhancement is done by assigning new DNs to each pixel with the linear relationship:

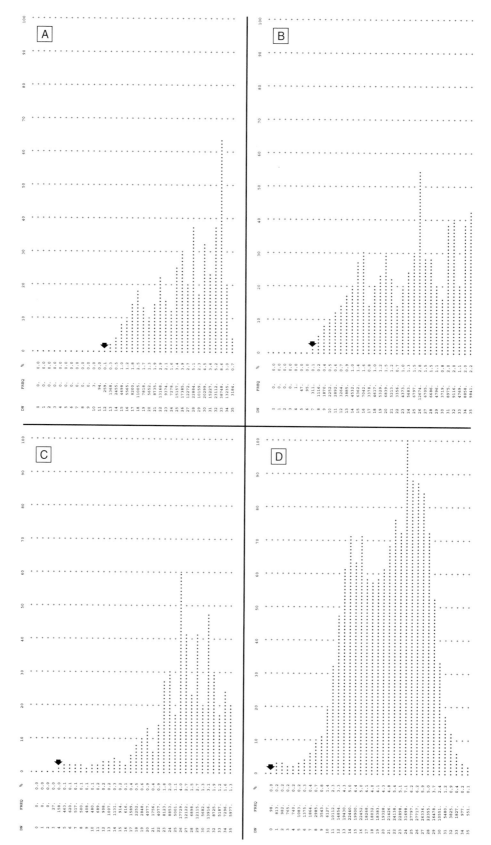

Figure 15-11 Histograms showing distribution of *Landsat 1* MSS DNs (bands 4–7) before adjustment for atmospheric scattering. DNs represent radiance values for a volcanic and sedimentary landscape in northern Arizona (see Figure 15-9). Haze DN values (arrows) are as follows: band 4 = 12, band 5 = 7, band 6 = 4, and band 7 = 0. (Courtesy U.S. Geological Survey.)

$$DN_{O(j,s)} = \left(\frac{DN_{I(j,s)} - MIN}{MAX - MIN}\right)255, \qquad (15\text{-}3)$$

where: $DN_{O(j,s)}$ = output digital number at line j, sample s,

$DN_{I(j,s)}$ = original digital number of input image at line j, sample s,

MIN = minimum DN parameter in input image (user selected), and

MAX = maximum DN parameter in input image (user selected).

All pixels with DN values equal to or less than MIN are reassigned the value 0, and pixels with DNs equal to or greater than MAX are reassigned the value 255. All DNs between MIN and MAX have a multiplier applied to them that linearly expands the range of DNs between 0 and 255. This increases the difference between DNs of image features, and as this increases, the contrast between these features also increases.

The simplest linear stretch is one that expands 6- and 7-bit data to an 8-bit scale (Figure 15-12A). DN input-output parameters for computer entry can be written as shown in Table 15-5. Both transformations would increase image contrast because gray-level or DN separations have been expanded by factors of two (7-bit) and four (6-bit) as illustrated in Table 15-6.

The contrast between features can be further increased if the lower and upper tails of a frequency distribution are *truncated,* or *trimmed,* before enhancement, a process known as **histogram trimming**. Truncation allows a narrower range of input DNs to undergo expansion to the full

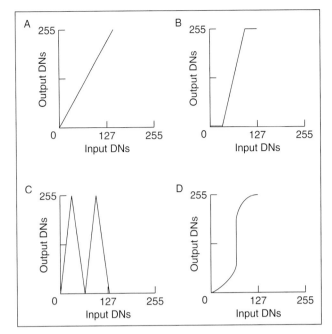

Figure 15-12 Basic types of contrast sketches: (*A*) linear without saturation, (*B*) linear with saturation, (*C*) sinusoidal, or sine, and (*D*) nonlinear. Note that input DNs are in a 7-bit format and output DNs are in an 8-bit format.

0–255 scale (Figure 15-12B). This type of linear stretch is especially applicable to a Gaussian distribution because relatively few DN pixel counts occupy the tails. However, it is advantageous at times to select the truncation parameters

TABLE 15-5 Expanding 6- and 7-Bit Data to an 8-Bit Scale by Linear Stretches

7- to 8-Bit Expansion			6- to 8-Bit Expansion		
Input DN		Output DN	Input DN		Output DN
0	(goes to)	0	0	(goes to)	0
127	(goes to)	255	63	(goes to)	255
255[a]	(goes to)	255[a]	255[a]	(goes to)	255[a]

[a]"255 (goes to) 255" statement is used to terminate the stretch algorithm.

TABLE 15-6 Expansion of 6- and 7-Bit Data to an 8-Bit Scale with Resultant DN Separations

	DN Sequence						
7-Bit DNs	0	1	2	3	4	5	6
to							
8-Bit DNs	0	2	4	6	8	10	12
6-Bit DNs	0	1	2	3	4	5	6
to							
8-Bit DNs	0	4	8	12	16	20	24

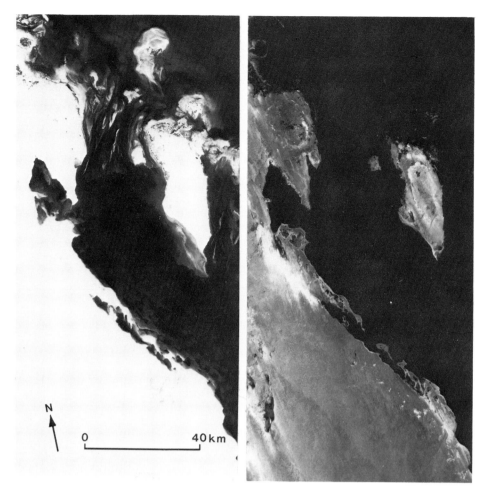

Figure 15-13 Two versions of a *Landsat 1* MSS band 4 subscene image of the western Arabian Gulf region obtained September 4, 1972. Each image incorporates a different area-specific stretch to enhance bottom reflectance patterns of the Arabian Gulf (*left*) and reflectance patterns of the land surface in Saudi Arabia and Bahrain (*right*). Many of the light-toned features in the gulf are coral reefs. Band 4 was selected to highlight bottom features because its wavelengths (0.5–0.6-μm) penetrate water to a maximum depth of about 20 m. Band 5 (0.6–0.7-μm) penetrates only about 2 m; bands 6 and 7 (0.7–0.8 and 0.8–1.1-μm, respectively) do not penetrate water. (Courtesy Pat S. Chavez, Jr., U.S. Geological Survey.)

by trial and error on an interactive image analysis system to ensure that there is not a significant loss of meaningful information.

Plate 14B shows a portion of a *Landsat 1* MSS 4, 5, 7 data-base color composite image that incorporates this type of linear contrast enhancement. The MIN/MAX values for each band were determined by interactive analysis. The stretch parameters are listed in Table 15-7.

Although the image data for Plate 14B were preprocessed by the histogram correction algorithm to remove atmospheric haze effects, the lower DN truncation parameter for this type of linear stretch produces, in essence, a subtraction bias that will correct for atmospheric haze. For this reason, the histogram correction algorithm for haze removal can often be omitted in the preprocessing stage if the images are to be used for color composites (e.g., MSS bands 4, 5, 7).

The linear stretch is the most commonly used technique to enhance the contrast in digital images, regardless of type. For example, the Landsat TM and SPOT HRV images shown in Chapter 6 and the digital radar and sonar images presented in Chapter 7 all incorporate linear contrast stretches. This is also the case for the TM and HRV color composite images shown in Plates 10 and 11.

Area-Specific Stretch. The **area-specific stretch** (linear form) is used to enhance a portion of an image that is of interest to the analyst. For example, if a data-base image contained both land and water surfaces, an area-specific stretch could enhance subtle patterns in the water. Most of the land information would likely be lost because its DNs would lie outside the DN range of water. Likewise, a second area-specific stretch could be implemented on the same data to enhance only the land patterns.

TABLE 15-7 Linear Stretches on MSS Bands 4, 5, and 7 Using Histogram Trimming[a]

Band 4		Band 5		Band 7	
Input DN	Output DN	Input DN	Output DN	Input DN	Output DN
0	0	0	0	0	0
3	0	7	0	5	0
67	255	111	255	57	255
255	255	255	255	255	255

[a]Color composite image incorporating these stretches is presented in Plate 14B.

TABLE 15-8 Area-Specific Stretches on a *Landsat 1* MSS Band 4 Image for Bottom Reflectance and Surface Reflectance

Bottom Reflectance (Water)[a]		Surface Reflectance (Land)[a]	
Input DN	Output DN	Input DN	Output DN
0	0	0	0
45	255	45	0
255	255	90	255
		255	255

[a]Images incorporating these stretches are presented in Figure 15-13.

TABLE 15-9 Sinusoidal Stretches on MSS Bands 4, 5, and 7[a]

Band 4		Band 5		Band 7	
Input DN	Output DN	Input DN	Output DN	Input DN	Output DN
0	0	0	0	0	0
18	255	19	255	15	255
35	0	40	0	29	0
56	255	63	255	44	255
72	0	79	0	63	0
255	255	102	255	255	255
		255	255		

[a]Color composite image incorporating these stretches is presented in Plate 14C.

Figure 15-13 shows two *Landsat 1* MSS band 4 images incorporating area-specific stretches for bottom reflectance enhancement for water and surface reflectance enhancement for land. The stretch parameters are given in Table 15-8.

Sinusoidal Stretch. A **sinusoidal**, or **sine**, **stretch** is designed to enhance subtle differences within "homogeneous" units such as a forest stand, volcanic field, or dune field. The stretch parameters are usually determined from the form of the DN distribution (histogram interpretation). The distribution is divided into several intervals or ranges and each of these is expanded over the output range. Through trial and error it has been found that an image will not become overly "busy" if there are six or fewer intervals and each interval is 15 to 20 DNs wide. Range boundaries are established where there are breaks or diminutions in the DN distribution.

Because several different input DNs can be mapped to one output DN, sinusoidal stretches are usually applied to three multispectral images that are to be color combined. This reduces the possibility that two different features will have the identical color output. The reason this stretch is called sinusoidal is that when input and output DN stretch parameters are plotted against each other, a sinusoidal curve is formed (Chavez et al. 1977b) (Figure 15-12C).

A *Landsat 1* MSS 4, 5, 7 color composite image incorporating sinusoidal stretches is shown in Plate 14C. Note that when compared to a conventional MSS 4, 5, 7 color composite image (Plate 14B), there is little or no correlation in color for the same ground features. Note also that the sinusoidal-stretched image shows details within volcanic areas and surficial deposits that are not discernible in the conventional color infrared image. The sinusoidal stretch parameters for the image shown in Plate 14C are listed in Table 15-9.

Nonlinear Stretches. **Nonlinear stretches** have flexible parameters that are controlled by DN pixel frequencies and the shape of the original distribution (Figure 15-12D). Two types are the **uniform distribution stretch** and the **Gaussian stretch**. With the uniform distribution stretch, original DNs are redistributed on the basis of their frequency of occurrence; the greatest contrast enhancement occurs within the range with the most original DNs. The Gaussian stretch forces a skewed frequency distribution of input data to a normal or nonskewed distribution. This stretch is useful if distributions are skewed in such a way that features could become abnormally light or dark when stretched linearly. The Gaussian stretch prevents saturation while enhancing overall scene contrast.

Comparison of Linear- and Gaussian-Stretched RBV Images. The effects of linear and Gaussian stretches on a *Landsat 3* Return Beam Vidicon (RBV) image of the eastern Grand Canyon are illustrated in Figure 15-14. Because the original 7-bit digital data represent a positive skewness frequency distribution (steep slope for low DN tail, shallow slope for high DN tail), the linear stretch expands the distribution but does not alter its shape. Hence, the enhanced image lacks contrast in the dark-toned areas. However, the Gaussian stretch expands the low DN tail preferentially so the output distribution resembles a normal curve. This has the effect of preferentially increasing contrast in the darker areas. This nonlinear adjustment is made possible by forcing the input median DN to the output median DN or 127 for an 8-bit scale (see Table 15-10).

From the Gaussian stretch parameters (Table 15-10), it can be seen that the first 50 percent of the input data lie between DNs 0 and 50, or 51 levels, and these are being redistributed over 128 levels (0 to 127). Therefore, the separation or spacing between DNs in the output will vary between 2 and 3 (average separation of 2.5 DN). However, 77 DNs are included in the second 50 percent, and, consequently, the DN separation for this part of the distribution will vary between 2 and 1 (average separation of 1.7 DN).

Spatial Filtering

Filtering a digital image in the spatial domain is designed to enhance different scales of tonal or DN "rough-

ness" (i.e., different **spatial frequencies**). Unlike enhancement by contrast stretching, **spatial filtering** depends not only upon the value of the pixel being processed, but also on the pixel values surrounding it (i.e., its **neighborhood**). In this regard, spatial filtering is an **area operation**, whereas contrast stretching is a **point operation**. Spatial filters are used to either emphasize or de-emphasize abrupt changes in pixel DNs, thereby altering an image's textural appearance.

A digital image contains both low- and high-frequency spatial information; their sum constitutes the original image. The **low-frequency component** (LFC) represents gradual DN changes over a relatively large number of pixels (i.e., *low tonal variance*). The LFC, therefore, defines the "smooth" areas of an image (e.g., forest cover, lava field, or water). The **high-frequency component** (HFC) represents rapid DN changes over a short space (i.e., *large tonal variance*). The HFC defines the "rough" areas or the details of an image (e.g., slope attitude contrasts, lithologic contacts, drainage networks, lineaments, and cultural linear features).

Algorithms that perform spatial-frequency enhancement are called filters because they pass or emphasize certain spatial frequencies and suppress others. Spatial filters that pass high frequencies, emphasizing the details of an image, are called **high-pass filters** (HPF). Conversely, **low-pass filters** (LPF) produce image smoothing by suppressing the high spatial frequencies.

The spatial filter is simply a subarray (**box** or **window**) of N by M pixels that is moved through the larger image array. The filter is usually of an odd-integer dimension along each side so that a central pixel exists for DN reassignment based upon surrounding pixel values. Filter shapes can be square or rectangular. **Square filters** (e.g., 31 lines by 31 samples) have uniform weights, ensuring that enhancement is equal in all image directions (**uniform-weight filters**). **Rectangular filters** provide maximum enhancement to features trending perpendicular to the long axis of the filter (**directional filters**). Two common forms of directional filters are the **horizontal-line filter** (e.g., 1 line by 31 samples) and the **vertical-line filter** (e.g., 31 lines by 1 sample). Care must be exercised in interpreting directionally filtered images because the filtering operation may produce artificial linear features that must be distinguished from actual features.

Low-Pass Spatial Filtering. The **low-pass spatial filter** is implemented by calculating a local DN average of an N-by-M digital array or window centered around the pixel being

Figure 15-14 *Landsat 3* RBV subscene image of the eastern Grand Canyon, Arizona, incorporating a standard linear contrast stretch (*top*) and a Gaussian contrast stretch (*bottom*). Acquisition date: March 22, 1981. (Courtesy Pat S. Chavez, Jr., U.S. Geological Survey.)

processed. The DN average of the filter box (A) is considered to be the *low-frequency component*:

$$\text{LFC}_{(j,s)} = \frac{S_{(j,s)}}{N_{(j,s)}} = A, \qquad (15\text{-}4)$$

where: $\text{LFC}_{(j,s)}$ = low-frequency component at line j, sample s,

$S_{(j,s)}$ = sum of valid DNs in the filter box centered at line j, sample s, and

$N_{(j,s)}$ = number of pixels with a valid DN value in the filter box centered at line j, sample s.

(*Note:* Invalid data could include cloud and shadow information.)

Thus, the output digital number (DN_O) centered at line j, sample s, is equal to the local DN average computed from N-by-M pixels (A) and centered at line j, sample s:

$$\text{DN}_{O(j,s)} = A_{(j,s)}. \qquad (15\text{-}5)$$

For a 3×3-pixel box, for example, the central pixel's original value would be replaced by the DN average of 9 pixels if there was a deviation (Figure 15-15).

The LPF moves in 1-pixel increments from left to right and from top to bottom until all pixel values in the original image have been replaced with window averages. So that the edge pixels of an image can be center points in the filter box, internal pixels are unfolded vertically, horizontally, and diagonally along the sides of the image (Figure 15-16). For image output, the LPF DNs are usually subjected to a contrast stretch operation to expand the data distribution.

The low-pass algorithm reduces deviations from the local average and thus produces image *smoothing* or *blurring*; as filter size increases, smoothing increases. Because of the blurring effect, low-pass filtering is useful for reducing certain noise patterns (e.g., bit errors or salt-and-pepper noise) and for smoothing blocky image data before visual interpretation or numerical analysis (Figure 15-17). Filter sizes to accomplish this are usually 3×3 or 5×5 pixels.

High-Pass Spatial Filtering. From an earlier discussion, it will be remembered that the sum of the LFC and HFC equals the original image. Therefore, the *high-frequency component* of a digital image can be determined by the following relationship:

$$\text{DN}_{O(j,s)} = \text{DN}_{I(j,s)} - A_{(j,s)} + K, \qquad (15\text{-}6)$$

where: $\text{DN}_{O(j,s)}$ = output digital number at line j, sample s,

$\text{DN}_{I(j,s)}$ = original digital number of input image at line j, sample s,

A = local DN average computed from N by M pixels and centered at line j, sample s, and

K = constant to keep all values positive; default = 127.

Because the subtraction process in Equation 15-6 can produce either positive or negative integers, a constant (K) is added to keep all values positive and centered between 0 and 255; this output is then assigned to the central pixel. The constant used is normally the median of the output data range (e.g., 127 for 8-bit data). A linear contrast stretch is then applied to the output DNs to increase scene contrast. The visual effect of Equation 15-6 is illustrated in Figure 15-18.

High-pass filtered images are normally used for identifying and mapping structural geologic features, including faults, fractures, and monoclines, that are characterized by different spatial frequency ranges. In general, a **high-pass spatial filter** enhances features that are less than half the size of the window being used while de-emphasizing features that are more than half the window size. For this reason, filters with different sized windows are used to highlight small-, intermediate-, and large-scale structures; filter sizes to accomplish this could be 11×11 pixels, 51×51 pixels, and 101×101 pixels, respectively (Figure 15-19).

Three-band, color composite images can be produced when black-and-white filtered images are available for multiple bands (e.g., Landsat MSS bands 4, 5, and 7). Compositing is done so that potential enhancement differences from three separate HPF images can be incorporated into a single, false color image (Plate 14D).

Edge Enhancement

The **edge enhancement** algorithm is designed to enhance rapid changes in DN levels from one pixel to the next. These very abrupt changes represent high spatial frequencies called **edges**. An edge can be a *boundary* separating two different features or a *line* that differs from the features on both its sides. Edge enhancement has the effect of producing a "sharper" image because the high-frequency information is strengthened by high-pass filtering.

Edge enhancement corrects the **Modulation Transfer Function (MTF)** response of a digital imaging system (e.g., Landsat MSS and TM sensors). Because an imaging system samples a continuous function at discrete intervals, high-frequency information cannot be recorded with the same precision as the lower-frequency data. Thus, spatial frequencies representing fine detail or edge information are suppressed. Edge enhancement, to a first-order approximation, corrects the MTF response by emphasizing the higher spatial frequencies (i.e., high frequencies are recorded with the same precision as low frequencies).

Basically, the enhancement of edges in a digital image is accomplished as follows:

1. Generate an HPF image using a 3×3-pixel to 9×9-pixel window; the output is the edge component.

2. Add the original data base image back into the HPF image; the output is edge enhanced.

3. Increase scene contrast in the edge-enhanced output by applying a linear stretch.

An image produced in this manner contains *both* radiometric or low-frequency information and exaggerated local contrast or high-frequency information (Figure 15-20).

Figure 15-15 Concept of low-pass spatial filtering for a 3- $\times$ 3-pixel window. Note that the low-frequency component (A') is equal to the window's DN average.

Pixel	DN	Input				Output		
A	42							
B	38	B	C	D				
C	41				$A' = \dfrac{A + B + C + \cdots + I}{9}$			
D	40	E	A	F			A'	
E	37				$A' = \dfrac{360}{9}$			
F	44	G	H	I				
G	38				$A' = 40$			
H	42							
I	38							

Figure 15-16 Concept of unfolding image pixels to position B as the central point for a 3- $\times$ 3-pixel low-pass filter operation.

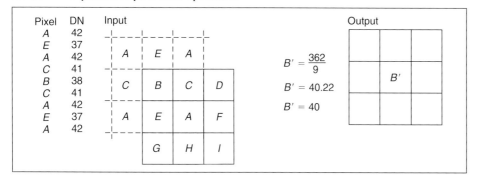

Pixel	DN	Input				Output		
A	42							
E	37	A	E	A				
A	42							
C	41				$B' = \dfrac{362}{9}$			
B	38	C	B	C	D		B'	
C	41				$B' = 40.22$			
A	42	A	E	A	F			
E	37				$B' = 40$			
A	42	G	H	I				

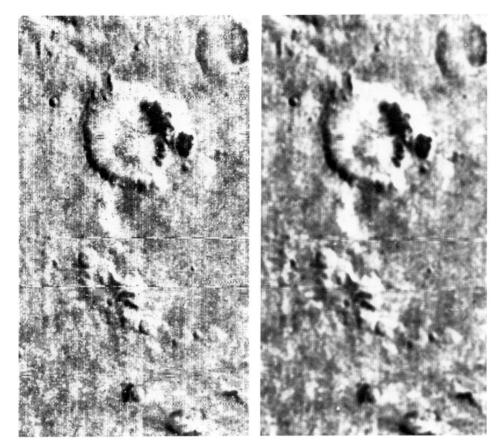

Figure 15-17 Mars-Mariner 9 subscene images: contrast-stretched image (*left*) and an image produced by a 3- × 3-pixel low-pass filter (*right*). (Condit and Chavez 1979.)

The edge enhancement algorithm can be expressed by the relationship

$$DN_{O(j,s)} = K(1 - X)$$
$$+ DN_{I(j,s)}(1 + X) - A_{(j,s)}, \quad (15\text{-}7)$$

where: $DN_{O(j,s)}$ = output digital number at line j, sample s,

K = constant to keep all values positive; default = 127 (median of output range for 8-bit data),

X = fraction of input digital number (DN_I) to be added back to the high-frequency component, and

$A_{(j,s)}$ = local DN average computed from N by M pixels and centered at line j, sample s.

Note that if 100 percent of the original image data are added back to the HPF output (i.e., $X = 1$, or 100 percent add-back), Equation 15-7 reduces to:

$$DN_{O(j,s)} = 127(1 - 1)$$
$$+ DN_{I(j,s)}(1 + 1) - A_{(j,s)}$$
$$= 0 + DN_{I(j,s)}(2) - A_{(j,s)}$$
$$= 2\, DN_{I(j,s)} - A_{(j,s)}. \quad (15\text{-}8)$$

From Equation 15-8 it can be seen that pixel DNs that are larger than the local average become still larger after edge enhancement, and DNs less than the local average become smaller. This relationship can be illustrated as follows:

Line number	480	481
Sample number	612	613
DN_I	44	42
A	42	44
DN_O	46	40

If an edge occurred between samples 612 and 613, note that the DN_I difference is 2, but the DN_O difference is 6. This type of DN adjustment would *exaggerate* the local contrast.

Varying the X parameter in Equation 15-7 enables the analyst to control the amount of original image data that is *added back* to the high-frequency component. Such an option makes it possible to reduce the dynamic range between light and dark image areas in direct proportion to diminutions in X. This can permit greater recognizability of high-frequency objects because low-frequency albedo or brightness masking effects are reduced. For example, Berlin et al. (1982) evaluated MSS 4, 5, 7 color infrared edge-enhanced images with different "add-back" percentages and found that the one with 30 percent add-back was superior for highlighting small stands of phreatophytic vegetation in northern Saudi Arabia.

As was previously mentioned, filter dimensions for the edge enhancement algorithm, applicable to Landsat MSS data, can vary from about 3 × 3 to 9 × 9 pixels. For TM data at 30-m resolution, filter dimensions range from about 3 × 3 to 23 × 23 pixels. Filter size is dependent upon the "busyness" (i.e., number of edges) of the image being processed. Generally, the smallest filters are used for "rough" images, whereas the larger filters are best suited for use with "smooth" images (Figures 15-21 and 15-22). This is because an HPF enhances features that are less than half the size of the filter window and de-emphasizes features that are more than half the window size. Chavez and Bauer (1982) have developed a quantitative technique that automatically selects the optimum filter size for enhancing the edges in a particular digital image.

Because of the aesthetic improvements created by edge enhancement, multiple-band images are commonly subjected to this type of processing for color compositing.

An edge-enhanced *Landsat 1* MSS 4, 5, 7 color infrared image (3 × 3-pixel window) is presented in Plate 14E; a companion image without edge enhancement is shown in Plate 14B for comparison.

The edge enhancement algorithm can also be used to produce "sharper" image details in photographs that have been digitized. This is illustrated in Figure 15-23, which shows a Large Format Camera (LFC) photograph before and after edge enhancement. Note how the urban details are visually more vivid in the enhanced version. The LFC is described in Chapter 5.

Spatial Filtering to Remove TM Scan-Line Noise

Special spatial-filtering routines can be used to remove sensor striping, scan striping, or a combination of both, which is sometimes present in Landsat TM images. The noise anomaly can be expressed as a **16-line striping pattern** that is due to slight errors in the internal calibration

Figure 15-18 Result of high-pass filtering a *Landsat 1* MSS band 5 subscene image of northern Arizona: (*A*) original data-base image, (*B*) 31- × 31-pixel LPF image, and (*C*) 31- × 31-pixel HPF image. Essentially, the HPF image is the original image minus the LPF image (image *C* = image *A* − image *B*). The images have been contrast stretched to emphasize their differences. See Figure 15-2 for feature identifications. (Courtesy U.S. Geological Survey.)

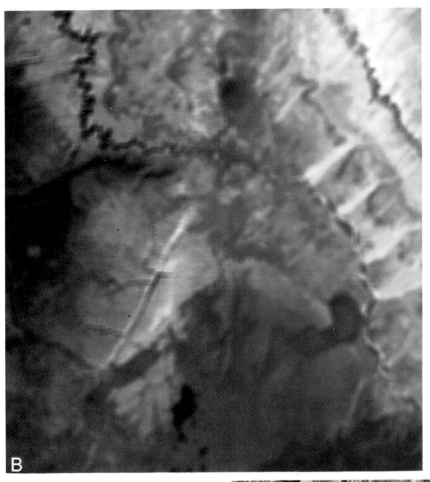

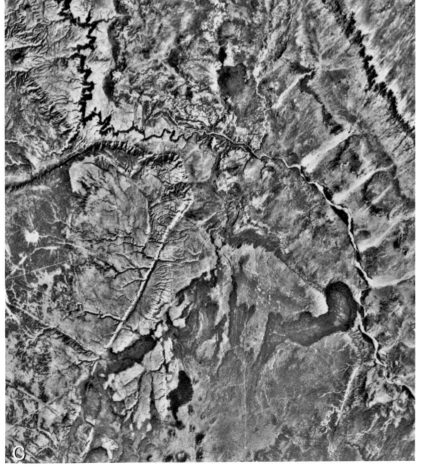

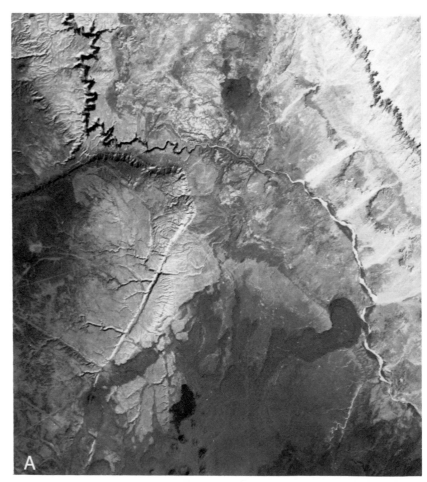

Figure 15-19 High-pass filtering a *Landsat 1* MSS band 5 subscene image of northern Arizona: (*A*) original data base image, (*B*) 11- × 11-pixel HPF image, (*C*) 51- × 51-pixel HPF image, and (*D*) 101- × 101-pixel HPF image. Note that as the window size is increased, fewer low-frequency components are removed from the original image and more high-frequency components are retained. See Figure 15-2 for feature identifications. (Courtesy U.S. Geological Survey.)

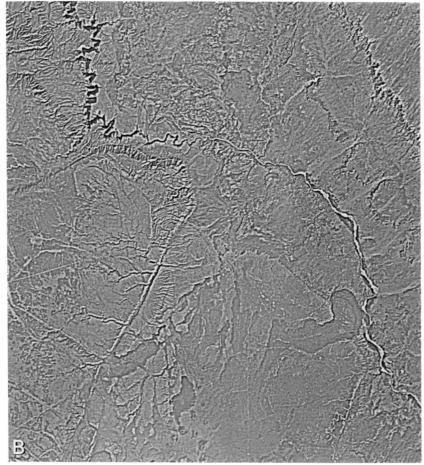

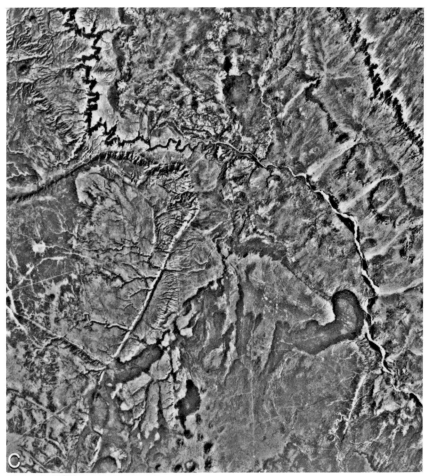

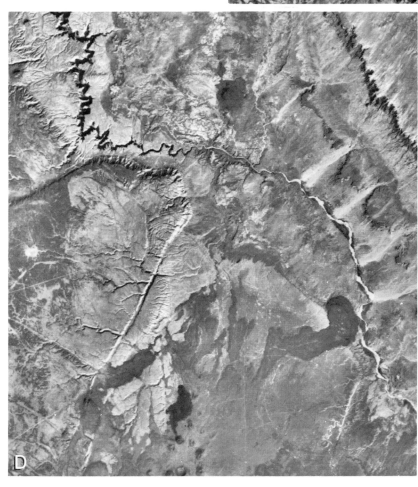

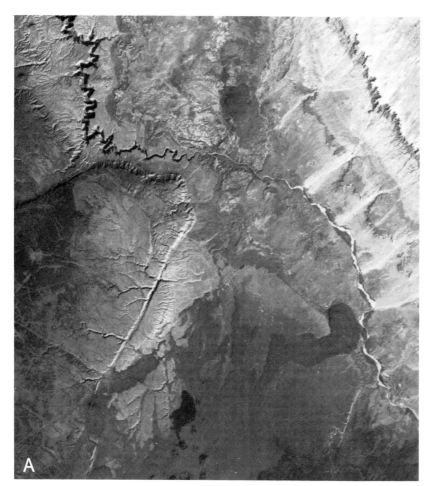

A

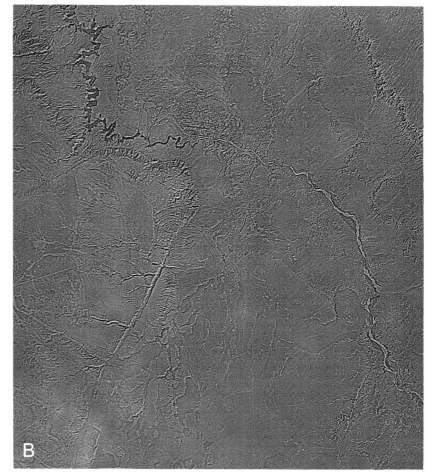

B

Figure 15-20 Edge-enhanced processing for a *Landsat 1* MSS band 5 subscene image of northern Arizona: (A) original data base image, (B) 5- × 5-pixel HPF image, and (C) edge-enhanced image. The edge-enhanced image is the sum of the HPF and original images (image C = image A + image B). The images have been contrast stretched to emphasize their differences. See Figure 15-2 for feature identifications. (Courtesy U.S. Geological Survey.)

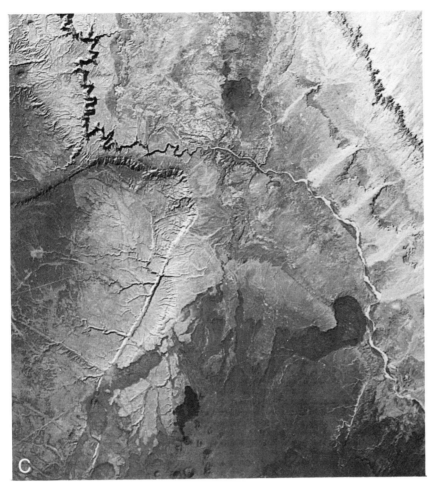

system or variation in the response of the 16-element detector arrays used for each of the six reflective bands (see Chapter 6). The procedure developed by Crippen (1989) uses standard spatial filters and arithmetic routines that are already present in most image processing systems. It comprises four steps:

1. A 101-sample by 1-line low-pass filter is applied to the original image. This operation largely isolates the low-frequency signal plus the scan-line noise from the high-frequency signal.

2. A 33-line by 1-sample high-pass filter is applied to the output from Step 1. This operation separates the high-frequency and cyclic scan-line noise from the low-frequency signal.

3. A 31-sample by 1-line low-pass filter is applied to the output from Step 2. This operation suppresses artifacts introduced by the high-pass filter in Step 2 and thus isolates the noise component.

4. The "noise image" produced in Steps 1, 2, and 3 is subtracted from the original image, leaving the "cleaned image."

A constant (e.g., 127) is normally added in Steps 2 and 4 to keep DN values within the 8-bit quantization range.

Directional First Differencing

Analogous to edge enhancement by high-pass spatial filtering, the **directional first-differencing algorithm**, which approximates the *first derivative*, is designed to highlight the edge information in a digital image. First differencing enhances edges on a pixel-to-pixel scale by simple DN subtraction. The algorithm developed at the USGS's Image Processing Facility (IPF) produces the first difference of the image input in the *horizontal, vertical,* and *diagonal directions* (Chavez et al. 1977a). DN differences are computed at the pixel $X_{j,s}$ by the following equations:

j s	j $s + 1$
$j + 1$ s	$j + 1$ $s + 1$

Horizontal: $DN_O = DN_I(X_{j,s})$

$$- DN_I(X_{j,s+1}) + K, \qquad (15\text{-}9)$$

Vertical: $DN_O = DN_I(X_{j,s})$

$$- DN_I(X_{j+1,s}) + K, \qquad (15\text{-}10)$$

Diagonal: $DN_O = DN_I(X_{j,s})$

$$- DN_I(X_{j+1,s+1})$$

$$+ K, \qquad (15\text{-}11)$$

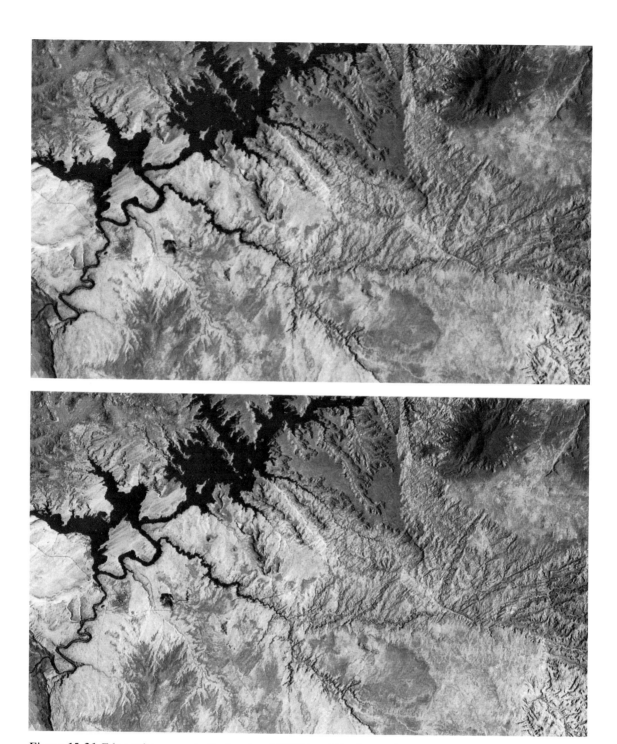

Figure 15-21 Edge enhancement of a *Landsat 2* MSS band 7 subscene image of Lake Powell in Arizona and Utah: original data base image (*top*) and the edge-enhanced image with a 3- × 3-pixel window (*bottom*). Both images incorporate a linear contrast stretch. (Courtesy Pat S. Chavez, Jr., U.S. Geological Survey.)

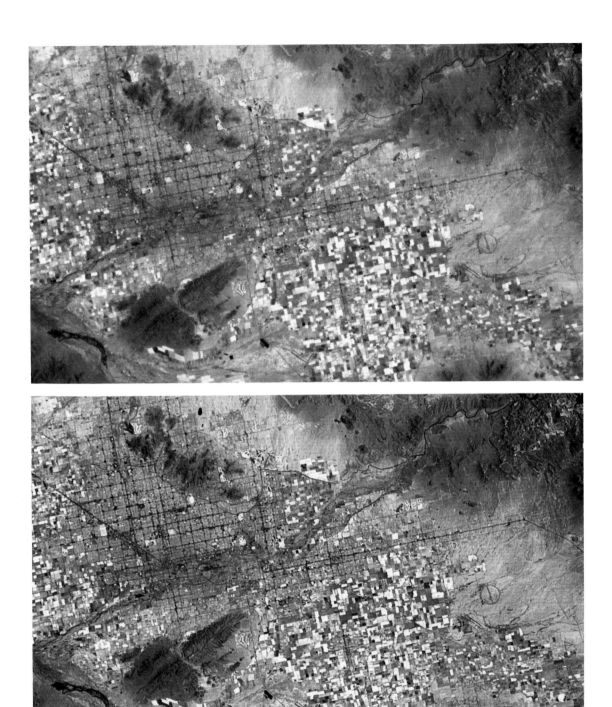

Figure 15-22 Edge enhancement of a *Landsat 2* MSS band 7 subscene image of Phoenix, Arizona, and environs: original data-base image (*top*) and the edge-enhanced image with a 7- × 7-pixel window (*bottom*). Both images incorporate a linear contrast stretch. (Courtesy Pat S. Chavez, Jr., U.S. Geological Survey.)

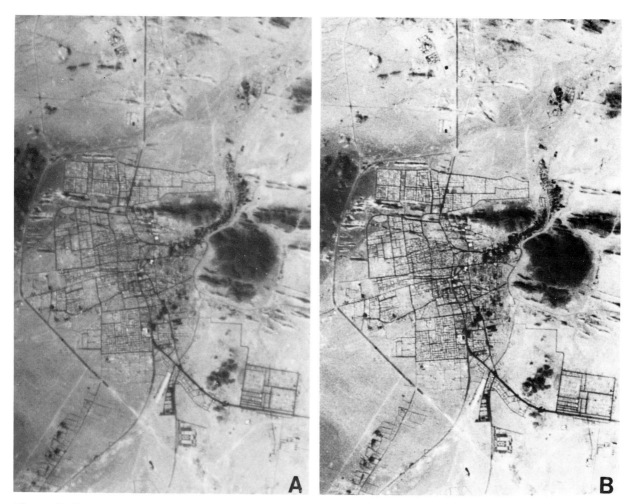

Figure 15-23 Edge enhancement of a portion of a LFC photograph of Ha'il, Saudi Arabia, and environs: (A) original photograph and (B) edge-enhanced photograph with an 11- × 11-pixel window and incorporating a linear contrast stretch. The photograph was taken during the Space Shuttle Mission 41-G in 1984. Scale is about 1:115,000.

where: DN_O = output digital number,
DN_I = input digital number,
K = constant to keep all values positive; default = 127,
j = line number, and
s = sample number.

The subtraction output can be either negative or positive; consequently, a constant (K) is added to make all values positive and centered between 0 and 255. The IPF software uses the median of the output data range as K, or DN 127 for 8-bit data: 127 = 0 difference, 128 = +1 difference, 126 = −1 difference, and so on. Output DNs cluster around midrange because most pixel-to-pixel changes are relatively small.

For an image display, the first-difference DNs are contrast stretched to accentuate edge amplitudes. The MIN-MAX stretch limits are usually selected so as to be equidistant from midrange to ensure that an intermediate gray signature in the image indicates no DN difference between adjacent pixels. Three Landsat MSS band 5 first-difference images, along with a standard reference image, are shown in Figure 15-24.

Three directional algorithms are used because any edge trending in the same direction as the first-difference vector will be suppressed. Edge de-emphasis occurs because the differencing shows no pixel-to-pixel DN change as the subtraction process moves over the long axis of the edge; an edge would be detected only at its ends. Edges trending normal to the direction of the first difference are enhanced the most because of different DNs on adjacent sides of the edge. The preferential highlighting of edges is illustrated in Figure 15-24.

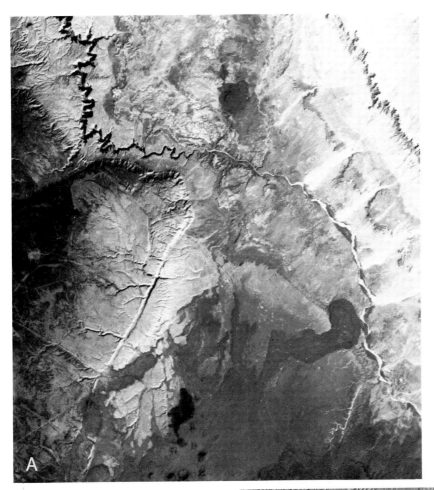

Figure 15-24 Result of first differencing a *Landsat 1* MSS band 5 subscene image of an area in northern Arizona: (*A*) original data-base image, (*B*) horizontal first-difference image, (*C*) vertical first-difference image, and (*D*) diagonal first-difference image. Note that edges in the first-difference images are dark toned when the subtraction process moves from light to dark areas in the input image and light toned when the subtraction process moves from dark to light areas in the input image. All images incorporate linear contrast stretches. See Figure 15-2 for feature identifications. (Courtesy U.S. Geological Survey.)

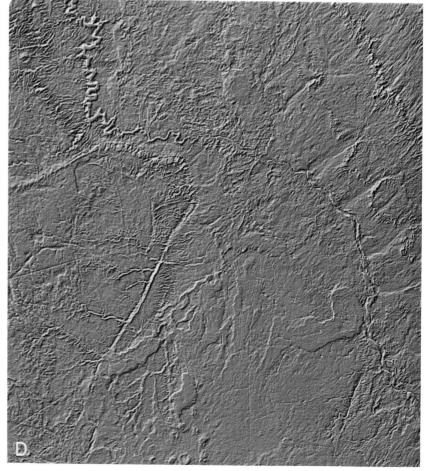

Figure 15-24 (*Continued*).

Multispectral Band Ratioing

The **interband ratioing of multispectral images** enhances subtle *spectral-reflectance* or *color differences* between surface materials that are often difficult to detect in standard images (i.e., single-band images and color composites). Ratioing accentuates color differences while removing first-order brightness or albedo variations caused by topography (i.e., sunlit or shadowed slopes). Thus, interband ratioing effectively normalizes spectral data by removing brightness contrasts and emphasizing the color content of the data.

Figure 15-25 Concept of spectral band ratioing, in this case the four bands of the Landsat Multispectral Scanner (MSS). (Adapted from Taranik 1978.)

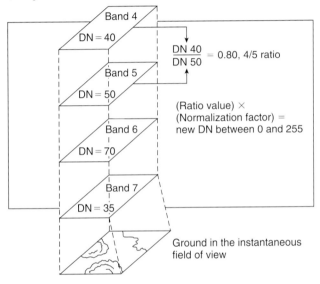

Figure 15-26 Concept of removing illumination differences by interband ratioing. (Adapted from Taranik 1978.)

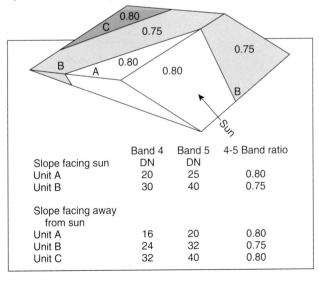

	Band 4 DN	Band 5 DN	4-5 Band ratio
Slope facing sun			
Unit A	20	25	0.80
Unit B	30	40	0.75
Slope facing away from sun			
Unit A	16	20	0.80
Unit B	24	32	0.75
Unit C	32	40	0.80

Ratioing is accomplished by dividing the data base DNs in one spectral band by the data base DNs in a second spectral band for each spatially registered pixel pair. The quotients are then converted to 8-bit integers using a multiplication factor (Figure 15-25). The ratio algorithm can be expressed in the form:

$$DN_{O(j,s)} = \left(\frac{DNX_{(j,s)}}{DNY_{(j,s)}}\right)K, \qquad (15\text{-}12)$$

where: $DN_{O(j,s)}$ = output digital number at line j, sample s,

$DNX_{(j,s)}$ = input digital number of band X at line j, sample s,

$DNY_{(j,s)}$ = input digital number of band Y at line j, sample s, and

K = normalization factor for converting quotients to 8-bit integers; default = 100.

The output DN distribution is expanded by a contrast stretch before image construction. The DN redistribution is usually linear (i.e., uniform distribution) to preserve the original ratio relationships. The extreme tones in a ratio image represent the largest spectral-reflectivity differences.

Preprocessing for noise removal is most important because ratioing will enhance noise patterns. In addition, the atmospheric haze component must be removed before the ratioing of interband DNs. If the atmospheric scattering component is present, ratioing will produce incorrect values because haze is an *additive component*, rather than a *multiplicative component*, and thus is dependent on wavelength (i.e., the unequal effects of haze are not canceled by interband division, Equation 15-12). Topographic suppression does not occur when the haze component remains in the preratioed data.

The removal of illumination differences in a ratio image is illustrated in Figure 15-26. Note that a uniform material (unit A or unit B) has the same ratio value on both the sunlit and shadowed slopes. Thus, the material would be depicted in a common tone in the ratio image. The sensation of relief would therefore be removed because the image would show no brightness or albedo differences caused by topography.

Figure 15-26 also illustrates a potential disadvantage of spectral-band ratioing. Although two different materials may have different absolute values (units A and C), they can have identical ratio values (e.g., 16/20 = 0.8 and 32/40 = 0.8). Under these circumstances, there would be a potential loss of information because the units could not be distinguished on the basis of image tone (perhaps by shape differences). The problem would be compounded if the units were contiguous because they would appear as a homogeneous material in the ratio image. These same units would probably be separable in standard band images.

MSS and TM Ratio Combinations. The number (n) of possible ratio combinations for a multispectral sensor with P bands is $n = P (P - 1)$. Thus, for the four bands of the MSS, there are 12 different ratio combinations—5/4, 6/4, 6/5, 7/4, 7/5, 7/6, and their six reciprocals (Table 6-6). For the TM's six nonthermal bands (Table 6-7), there are 30 different ratio combinations—15 original and 15 reciprocal.

Figure 15-27 Contrast-stretched *Landsat 1* MSS subscene images of an area in northern Arizona acquired October 29, 1973: (*A*) band 4, 0.5–0.6-μm; (*B*) band 5, 0.6–0.7-μm; (*C*) band 6, 0.7–0.8-μm; and (*D*) band 7, 0.8–1.1-μm. Compare with Figure 15-28. Note the redundancy of data in these images. See Figure 15-2 for feature identifications. (Courtesy U.S. Geological Survey.)

441

Most Landsat ratio applications have been with MSS data because of their availability since 1972. Four *Landsat 1* MSS bands for an area in northern Arizona (Figure 15-27) were digitally divided in pairs to produce the six ratio images shown in Figure 15-28.

The general utility of MSS ratios is described next; several of their uses are illustrated in Figure 15-28.

1. The 4/5, 4/7, 5/7, and 6/7 ratios are important for characterizing soil and rock units. Such an ordering would have vegetation depicted in dark tones.

2. The 5/4 ratio is especially sensitive to the presence of iron oxide or ferric iron. In Figure 15-28A, the light image tones at sites 1 and 2 are associated with the Triassic Moenkopi Formation (reddish-brown mudstone, siltstone, and sandstone) and the Triassic Moenave Formation (reddish-orange sandstone and sandy siltstone), respectively. In a 4/5 image, these same units would be depicted in dark tones.

3. The 6/4, 6/5, 7/4, and 7/5 ratios are useful for highlighting vegetation patterns because of the large differences in reflectance between the infrared bands (6 and 7) and visible bands (4 and 5). In Figure 15-28B the primary vegetation types are pinon-juniper (area 1) and tamarisk or salt cedar (area 2).

4. The 7/5 ratio is the most useful of the MSS ratios for assessing the relative greenness of vegetation (e.g., stressed plants versus unstressed) and for estimating biomass.

5. The 5/6 or 6/5 ratio is most often used for distinguishing general material types of soil and rock, vegetation, and water. The ratio, however, is not very useful for discrimination within any one material class. Observe in Figure 15-28C that the 6/5 ratio separates vegetation from soil and rock, but that there is very little tonal variation within either material class. Chavez et al. (1977b) demonstrated how the 5/6 ratio, with the appropriate contrast stretch, would be used for producing thematic maps of soil and rock, vegetation, and water (Figure 15-29).

As was previously mentioned, there are 30 different ratio combinations for the TM's six nonthermal bands. Several TM ratios and their general utility are described here.

1. The 3/1 (red/blue) and 3/2 (red/green) ratios are important for delineating ferric iron-rich rocks (light tones) and ferric iron-poor rocks (dark tones).

2. The 5/7 ratio is useful for identifying clay-rich rocks (light tones) because clay minerals exhibit strong absorption in the 2.2-μm region (band 7) and high reflectance in the 1.6-μm region (band 5).

3. The 4/3 ratio (near IR/red) uniquely defines the distribution of vegetation. Generally, the lighter the tone, the greater the amount of vegetation present.

4. The 5/2 ratio (mid IR/green) is useful for distinguishing different types of vegetation. Its reciprocal is useful for identifying water bodies and wetlands.

5. The 3/7 ratio (red/mid IR) is useful for observing differences in water turbidity.

Color Composite Ratio Images. The ability to distinguish between different surface materials can be significantly increased by selectively combining three black-and-white ratio image sets into color composites using the three primary colors—blue, green, and red (Plate 14F–G). The increase in discrimination occurs for two primary reasons: (1) spectral-reflectivity information from three different ratio images is combined and (2) the colors created by compositing permit a wider range of visual discriminations because of the eye's sensitivity to subtle color changes.

Selecting the three-ratio combination for color compositing that will allow for the optimum discrimination of material classes in a scene can be both difficult and time-consuming because of the large number of possible combinations. For example, excluding reciprocals, 20 combinations can be made from the six MSS ratios, taken three at a time, and 455 different combinations are possible with the 15 nonthermal TM ratio images (see Equation 6-12). To help overcome the selection problem, Chavez et al. (1982) devised a quantitative technique, called the **Optimum Index Factor (OIF)**, that ranks all possible ratio combinations according to the amount of correlation and total variance present between the various ratios under consideration. The three-ratio combination with the highest ranking should contain the most information with the least amount of duplication.

False color MSS ratio images have been used successfully in many geological investigations for lithology mapping and identifying alteration halos and mineralized areas. For example, the ratio combination 4/5, 5/6, 6/7 was used by Rowan et al. (1974) for discriminating rock types and detecting hydrothermally altered areas in south-central Nevada, by Blodget et al. (1978) for the discrimination of rock types in southwestern Saudi Arabia, and by Chavez et al. (1982) for identifying different rock types and surficial deposits in north-central Arizona. The ratio combination 4/5, 4/6, 6/7 was used by Raines et al. (1978) to identify facies related to uranium deposits in the Powder River basin in Wyoming.

Regarding TM ratio composites, Sabins (1987) describes the utility of the ratio combination 3/1, 3/5, 5/7 for differentiating sedimentary rocks in the Thermopolis area, Wyoming, and identifying hydrothermally altered rocks at Goldfield, Nevada. Davis and Berlin (1989) found that the ratio combination 4/1, 5/4, 7/5, along with other TM color composite images (e.g., bands 1, 4, 7), enabled discrimination among unaltered and altered basalt and rhyolite, granite/syenite, monzogranite, sandstone, and complex metamorphic units in the Jabal Salma region of Saudi Arabia.

In some cases, albedo information lost in the ratioing process (e.g., units A and C in Figure 15-26) can be restored by color-combining a single spectral band image with two ratioed images. The resulting **hybrid-ratio** false color image will enhance both albedo and spectral color differences (Chavez et al. 1977a). However, topographical detail is restored with the introduction of the albedo information (Plate 14H).

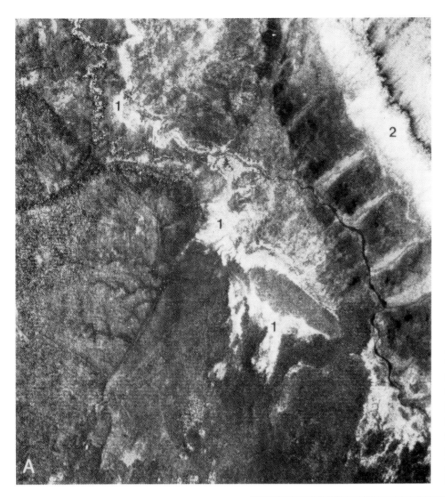

Figure 15-28 Contrast-stretched MSS ratio images of an area in northern Arizona: (*A*) 5/4 ratio, (*B*) 6/4 ratio, (*C*) 6/5 ratio, (*D*) 7/4 ratio, (*E*) 7/5 ratio, and (*F*) 7/6 ratio. Refer to text for explanation of numerical annotations. Note that the images are almost free of brightness contrasts caused by topography. Compare with Figure 15-27. (Courtesy U.S. Geological Survey.)

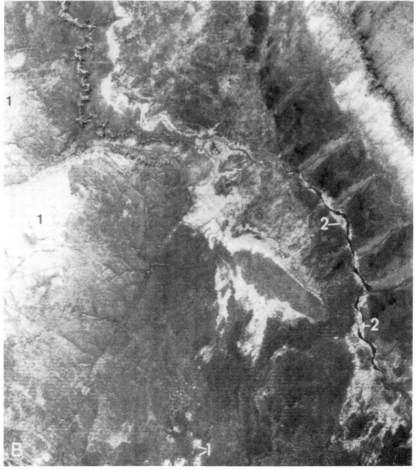

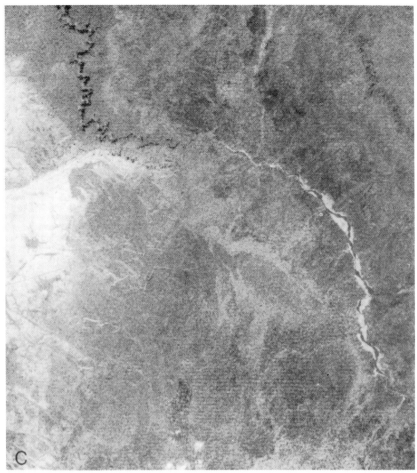

C

Figure 15-28 (*Continued*).

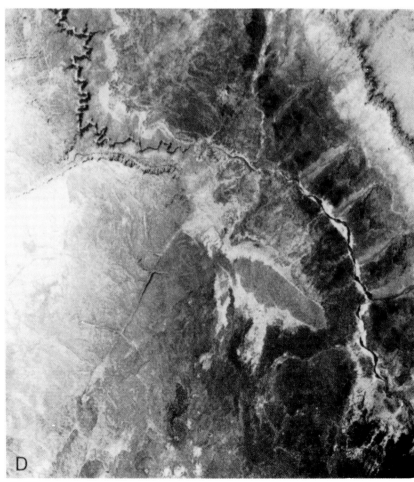

D

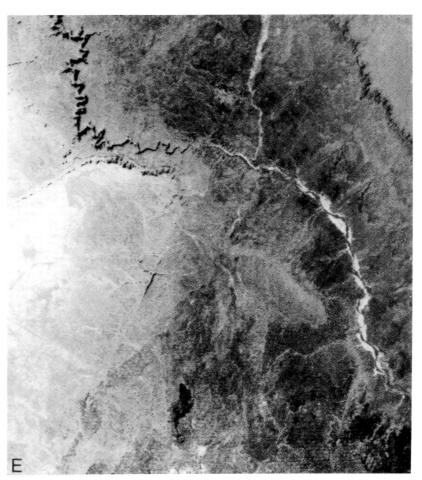

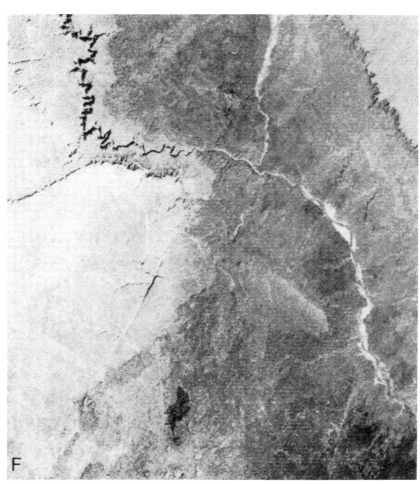

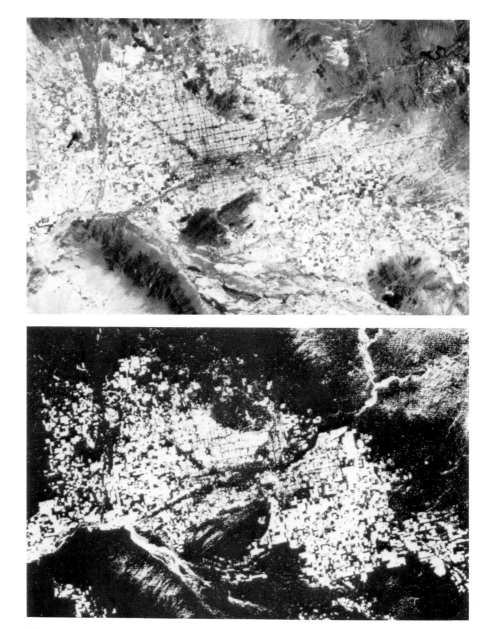

Figure 15-29 *Landsat 2* MSS band 7 subscene image of Phoenix, Arizona, generated May 15, 1974 (*top*) and a thematic map of vegetation produced from the MSS band 5/6 ratio (*bottom*). For the thematic map image, a contrast stretch was used to map the 5/6 "vegetation DNs" to 255 or white and the remainder to 0 DN or black. (Courtesy Pat S. Chavez, Jr., U.S. Geological Survey.)

Simulated MSS Natural Color Images

The USGS's Image Processing Facility has developed a technique for producing a modified Landsat MSS image in which colors appear as the human eye would perceive them from orbital altitudes, but in the absence of an intervening atmosphere (Plate 14I). The **simulated natural color image** is devoid of atmospheric haze and accentuates subtle color differences. Essentially the procedure is to predict the blue part of the spectrum, which is not sensed by the MSS. The blue component is derived from the shape of the spectral response in the green, red, and near IR MSS data (Eliason et al. 1974).

A schematic diagram illustrating how simulated natural color is achieved is shown in Figure 15-30. The procedure can be summarized as follows:

1. The MSS 5/6 ratio is used to identify pixels as belonging to the major surface categories of vegetation (ratio value equals 0.45 ± 0.2), soil and rock (ratio value equals 0.95 ± 0.25), and water (ratio exceeds 1.45).

2. Material-dependent algorithms are applied to the real MSS data (i.e., bands 4 through 7) to determine the digital numbers for the pixels in the new "blue band" or "band 3."

3. To produce a color composite image, band 3 is projected through blue light, band 4 through green light, and band 5 through red light (Plate 14I).

Spectral Ratios for Vegetation Monitoring

Various forms of ratio combinations in the wavelength range 0.7 to 1.1-μm (near IR) to those in the 0.6 to 0.7-μm range (red) have been developed for vegetation monitoring (e.g., assessing biomass or **leaf area index** and discriminating between stressed and nonstressed vegetation) with different remote sensors and vegetation conditions (e.g., Tucker 1979; Jackson 1983; Philipson and Teng 1988; Teng 1990). These ratio combinations are known as **vegetation indices**.

One commonly used index is the **Normalized Difference Vegetation Index** (**NDVI**); it is defined by the following general equation:

$$\text{NDVI} = \frac{\text{near IR band} - \text{red band}}{\text{near IR band} + \text{red band}}. \qquad (15\text{-}13)$$

The summed denominator largely compensates for changing illumination conditions, surface slope, and viewing aspects. In image form, the lighter tones are associated with *dense* coverages of *healthy* vegetation.

Red and near IR data from the following satellite sensors can be used for the NDVI:

1. Landsat MSS—bands 5 (0.6–0.7-μm) and 6 (0.7–0.8-μm) or 7 (0.8–1.1-μm); bands 2, 3, and 4, respectively, for *Landsat 4* and *Landsat 5*.

2. Landsat TM—bands 3 (0.63–0.69-μm) and 4 (0.76–0.90-μm).

3. SPOT HRV multispectral—bands 2 (0.61–0.68-μm) and 3 (0.79–0.89-μm).

4. NOAA AVHRR—bands 1 (0.58–0.68-μm) and 2 (0.72–1.0-μm).

An NDVI image using MSS bands 5 and 7 is shown in Figure 15-31. The operational use of NDVI data from the AVHRR sensor for monitoring and assessing crop vigor and locating areas of drought, desertification, and deforestation is discussed in Chapter 6. Plate 11 shows AVHRR NDVI color-coded image mosaics of the northern Great Plains for early June 1987 (normal year) and early June 1988 (drought year); the 1988 growing season was one of the driest and hottest in U.S. history.

Linear Data Transformations

The individual bands of a multispectral data set are often observed to be highly correlated or redundant in their informational content, that is, they are visually and numerically similar (Figure 15-27). Two mathematical transformation techniques, **principal components analysis** (PCA) and **canonical analysis** (CA), are often used to minimize this spectral redundancy, while reducing or compressing the dimensionality of the data (i.e., fewer channels are needed to accurately describe the original data set).

The techniques are similar in that they both compute a set of new, transformed variables called **components**, with each component largely independent of the others (uncorrelated). Geometrically, the components represent a set of mutually orthogonal and independent axes that are fitted to the original data such that the first new axis contains the highest percentage of the total variance or scatter in the data set, with each succeeding (lower-order) axis containing less variance (Figure 15-32).

Figure 15-33 shows the results of principal components processing of highly correlated *Landsat 1* MSS band pairs. Figure 15-34 shows two *Landsat 4* TM subscene images representing band 5 (1.55–1.75-μm) and band 7 (2.08–2.35-μm) and their corresponding principal component images (PC1 and PC2). Because the two input bands were so highly correlated, 98.4 percent of the total variance of the two input images was contained in PC1. PC2 contained the remaining variance (1.6 percent), which included

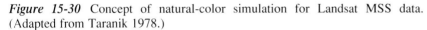

Figure 15-30 Concept of natural-color simulation for Landsat MSS data. (Adapted from Taranik 1978.)

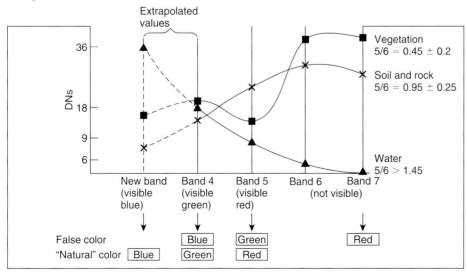

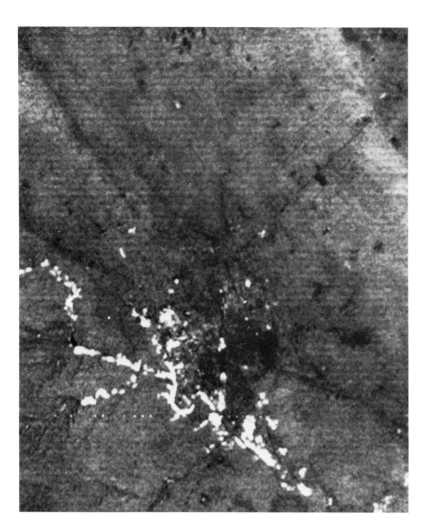

Figure 15-31 Normalized Difference Vegetation Index (NDVI) image of Riyadh, Saudi Arabia, and environs using *Landsat 3* MSS bands 5 and 7. Light tones denote stands of irrigated vegetation, principally date palms. (Courtesy U.S. Geological Survey.)

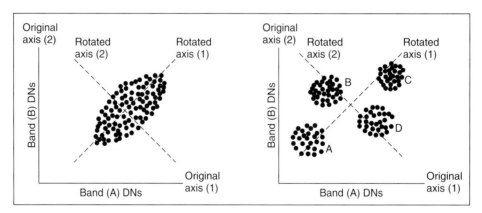

Figure 15-32 Rotation of axes in two-dimensional space for a hypothetical two-band data set by principal components analysis (*left*) and canonical analysis (*right*). The first rotated axis (first principal component, or PC1; first canonical component, or CC1) is positioned to account for the largest percentage of the total variance, with the second rotated axis (second principal component, or PC2; second canonical component, or CC2) containing the remaining variance. Principal components analysis uses DN information from the total scene, whereas canonical analysis uses the spectral characteristics of categories defined within the data to increase their separability (in this case, four material classes, *A–D*). (Adapted from Jensen and Waltz 1979.)

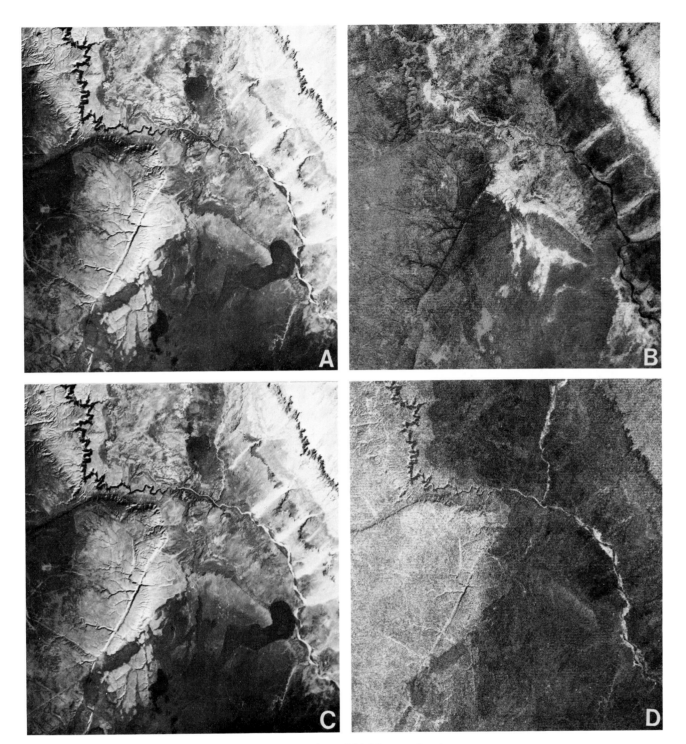

Figure 15-33 Results of principal components processing of highly correlated *Landsat 1* MSS band pairs: (*A*) PC1 of bands 4 and 5, (*B*) PC2 of bands 4 and 5, (*C*) PC1 of bands 6 and 7, and (*D*) PC2 of bands 6 and 7. Note how the PC1 images closely resemble their respective band images shown in Figure 15-27. Note also how the PC2 image of bands 4 and 5 highlights different rock types (compare with the 5/4 ratio image in Figure 15-28A) while the PC2 image of bands 6 and 7 highlights vegetation (compare with the 7/6 ratio image in Figure 15-28F). All images incorporate linear contrast stretches. (Courtesy U.S. Geological Survey.)

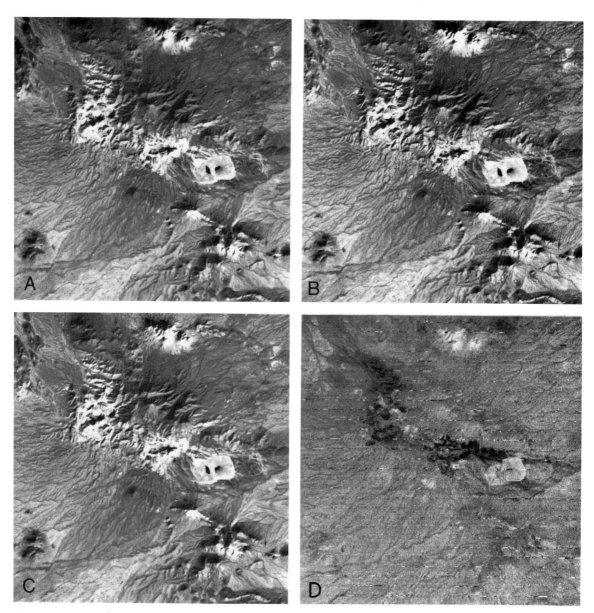

Figure 15-34 Results of principal components processing a highly correlated *Landsat 4* TM band pair: (*A*) TM band 4 subscene image of the Silver Bell district, Arizona; (*B*) TM band 7 subscene image; (*C*) first principal component (PC1) image of TM bands 5 and 7; and (*D*) second principal component (PC2) image of TM bands 5 and 7. In image *D*, the dark signatures closely correspond to zones of hydrothermal alteration associated with a porphyry copper deposit. All images incorporate linear contrast stretches. (Courtesy Pat S. Chavez, Jr., U.S. Geological Survey.)

most of the noise but also zones of hydrothermal alteration associated with a porphyry copper deposit. This "selective" technique of using the most correlated image pairs for principal components analysis was developed by Chavez et al. (1982; 1984) to maximize the amount of information that would be contained in a single, three-component, color composite image when using either six MSS ratios or seven TM bands as input variables.

Canonical analysis, also referred to as **multiple-discriminant analysis**, differs from principal components

analysis in that it produces a series of data transformations based upon the unique spectral characteristics of a set of user-defined land cover or feature categories established for a given data base and application. The main objective of canonical analysis is to increase the separability of categories defined with the data while minimizing the differences occurring within each category. The linear transformation is such that the maximum category separability is placed on the first rotated axis (first canonical component, or CC1) with the succeeding or lower-order axes containing successively less category separability (Figure 15-32). Canonical analysis creates one less transformed component than the number of defined categories (Holm 1982).

Canonical analysis is most often used as a preprocessing step to image classification because of its potential (1) for improving classification accuracy by increasing the separability of categories defined within the data and (2) for improving computer processing efficiency by reducing data dimensionality. For example, Holm (1982) used canonical analysis to reduce a 12-band MSS data set (Landsat MSS bands 4 through 7 from three dates) to six transformed canonical channels or components without the loss of any cover-type information (50 percent reduction). Cover types included wheat, alfalfa, potatoes, corn, soybeans, rangeland, and water. The percentage of cover-type separability for each of the six transformed canonical components is as follows: CC1, 47.21 percent; CC2, 27.43 percent; CC3, 14.61 percent; CC4, 4.90 percent; CC5, 4.67 percent; and CC6, 1.18 percent. Images of the six canonical components are shown in Figure 15-35, and a graphic representation of the cover-type separability for each channel is shown in Figure 15-36. Color infrared images and a "ground truth" map for the study area are presented in Plate 15.

Image Classification

Image enhancement improves the detectability of objects or patterns in a digital image for **visual interpretation**. Automated digital analysis or **image classification** is an information extraction process (**machine** or **automated interpretation**) that involves the application of pattern recognition theory to multispectral images. Simply stated, image classification analyzes the spectral properties of various surface features (e.g., crops) in a multiband image and sorts the spectral data into spectrally similar categories by the use of predefined, numerical decision rules.

The general procedure for image classification follows five basic steps:

1. **Training Class Selection.** The initial step involves defining image properties (i.e., pixel DN values) that represent a group of information or training classes.

2. **Generation of Statistical Parameters.** Statistical algorithms are used to define the unique spectral characteristics (**spectral signatures**) of the training classes. Typical statistical parameters include class means, standard deviations, covariance matrices, and correlation matrices. The statistical descriptions are used to "train" the classification algorithm.

3. **Data Classification.** The "trained" classification algorithm assigns each pixel composing the data set of interest to one of the training class categories.

4. **Evaluation and Refinement.** An assessment is made of the classification accuracy. Returning to Step 1 and repeating the process is required if the classification results are judged to be unacceptable. A common rea-

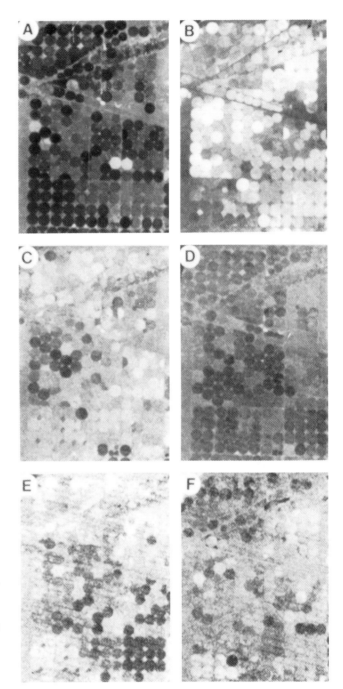

Figure 15-35 Six transformed canonical component images representing the Clarke, Oregon, region: (*A*) CC1, (*B*) CC2, (*C*) CC3, (*D*) CC4, (*E*) CC5, and (*F*) CC6. CC1 has the maximum separability between cover types, with succeeding components having progressively less cover-type separability. Compare with Plate 15. Average field size for the center-pivot irrigation system is 53 ha. (Courtesy Thomas M. Holm, Technicolor Government Services, Inc., EROS Data Center, Sioux Falls, S. Dak.)

son for classification refinement is the excessive spectral overlap between categories.

5. **Documentation.** Once a final classification is acceptable, the results are documented in the form of maps and tabular data summaries.

Classification Approaches

Two primary approaches can be used for defining training classes: *unsupervised* and *supervised* (Figure 15-37). **Unsupervised classification** uses an automatic clustering algorithm that analyzes the "unknown" pixels in the data base and divides them into a number of **spectrally distinct classes** based upon their natural groupings (i.e., **clusters**) in *n*-spectral dimensions. The analyst specifies the number of spectral classes into which the data are to be grouped. The spectral classes defined by the clustering algorithm are then used to classify the entire data set. After the classification is completed, the analyst must identify the cover type represented by each spectral class, using various types of reference information (e.g., color infrared images and photographs, published maps, field reconnaissance data, etc.). Only then are cover-type labels assigned to the spectral classes (Figure 15-38). Generally, as the number of spectral classes increases, the ability to identify each class becomes more difficult. A flow diagram of unsupervised classification is shown in Figure 15-37.

With **supervised classification**, the analyst selects areas of known cover type in the image and specifies these to the computer as **training areas**. Statistical measures are generated for the training areas and input to the classifier, which then determines other areas in the image that have similar spectral characteristics (Figure 15-39). Supervised classification usually requires less computer time to execute than unsupervised classification because it relies on explicit analyst guidelines, and it does not require a clustering step. A flow diagram of supervised classification is shown in Figure 15-37.

Classification Accuracies

Image classification using Landsat MSS and TM data has been the most successful for identifying various types of agricultural crops and natural vegetation. In general, geologic applications have met with less success, largely be-

Figure 15-36 Graphic representation of cover-type separability for six separate canonical transformed components, or channels, and a summary of the separability of all six channels combined. Graph illustrates where there was some degree of confusion in determining cover-type separability. (Courtesy Thomas M. Holm, Technicolor Government Services, Inc., EROS Data Center, Sioux Falls, S. Dak.)

Channel 1

Crop types	1	2	3	4	5	6	7
1	*	–	–	–	–	–	–
2	–	*	*	–	–	–	–
3	–	*	*	–	–	–	–
4	–	–	–	*	–	–	–
5	–	–	–	–	*	–	–
6	–	–	–	–	–	*	*
7	–	–	–	–	–	*	*

Channel 2

Crop types	1	2	3	4	5	6	7
1	*	–	–	–	–	–	–
2	–	*	–	–	–	–	–
3	–	–	*	–	*	–	–
4	–	–	–	*	–	–	–
5	–	–	*	–	*	–	–
6	–	–	–	–	–	*	–
7	–	–	–	–	–	–	*

Channel 3

Crop types	1	2	3	4	5	6	7
1	*	–	–	–	–	–	–
2	–	*	–	–	–	–	–
3	–	–	*	–	–	–	–
4	–	–	–	*	–	–	*
5	–	–	–	–	*	–	–
6	–	–	–	–	–	*	–
7	–	–	–	*	–	–	*

Channel 4

Crop types	1	2	3	4	5	6	7
1	*	–	–	–	–	–	–
2	–	*	–	–	–	*	*
3	–	–	*	–	–	–	–
4	–	–	–	*	*	–	–
5	–	–	–	*	*	–	–
6	–	*	–	–	–	*	*
7	–	*	–	–	–	*	*

Channel 5

Crop types	1	2	3	4	5	6	7
1	*	*	–	–	*	*	–
2	*	*	–	–	*	*	–
3	–	–	*	–	–	–	–
4	–	–	–	*	–	–	–
5	*	*	–	–	*	–	–
6	*	*	–	–	*	*	–
7	–	–	–	–	–	–	*

Channel 6

Crop types	1	2	3	4	5	6	7
1	*	*	–	–	–	–	–
2	*	*	–	–	–	–	–
3	–	–	*	*	–	*	–
4	–	–	*	*	–	*	–
5	–	*	–	–	*	–	–
6	–	–	*	*	–	*	–
7	–	–	–	–	–	–	*

All channels

Crop types	1	2	3	4	5	6	7
1	*	–	–	–	–	–	–
2	–	*	–	–	–	–	–
3	–	–	*	–	–	–	–
4	–	–	–	*	–	–	–
5	–	–	–	–	*	–	–
6	–	–	–	–	–	*	–
7	–	–	–	–	–	–	*

Legend
1 Wheat
2 Alfalfa
3 Potatoes
4 Corn
5 Soybeans
6 Rangeland
7 Water

–indicates cover types are separable
*indicates cover types are not separable

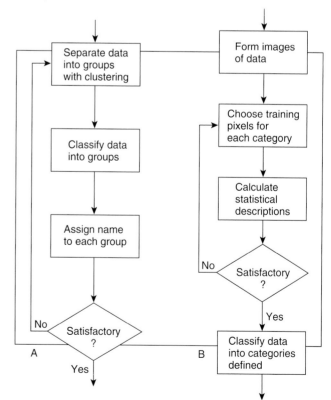

Figure 15-37 Flow diagrams representing unsupervised classification (*A*) and supervised classification (*B*). (Short 1982.)

cause lithologic units are rarely homogeneous, and they are commonly masked by various amounts of soil, vegetation, colluvium, and organic debris. Such factors make selection of representative training areas extremely difficult, especially in areas of rugged relief where spectral signatures are influenced by variations in illumination.

Improvements in classification accuracy can usually be realized by incorporating additional data sets into the classification algorithm. For example, when attempting to classify diverse crop types within a given area, a single Landsat MSS data set of four spectral dimensions may not have the necessary spectral contrasts for all crop types represented. However, by using several co-registered MSS digital images acquired on different dates in the growing season, it would be possible to track the phenologic changes of crop types and to incorporate these changes into the decision criteria for classification (Holm 1982). This process is called **multitemporal image classification**.

Multitemporal Image Classification and Canonical Analysis

An example of classifying Landsat MSS digital images for crop-type discrimination using multitemporal data and canonical analysis has been presented by Holm (1982; 1983) for the Clarke, Oregon, region (Plate 15). Canonical analysis was used to transform a 12-band, three-date Landsat MSS data set into 6 transformed channels. A comparison of the classification based on the 6 canonical channels (**minimum distance to mean classifier**) and the 12 MSS channels (**maximum likelihood classifier**) showed no significant difference in terms of classification accuracy—75.8 and 75.9 percent, respectively. While accuracies were nearly identical, computer processing unit (CPU) time was 14,871 CPU sec for the 12-band classification and 4,240 sec for the canonical-based classification. The notable difference in time was due to the reduction in data dimensionality and the differences in the complexity of the classifiers used.

Data-Set Merging

The **digital merging** or **co-registration** of different spectral and nonspectral data sets is becoming an increasingly important component of digital image processing because it allows for the *simultaneous analysis* (visual or machine aided) of many types of information for the same ground area. Merging is the spatial superposition of digital images taken at *different wavelengths,* at *different times,* or by *different sensors,* plus digital images representing *ancillary reference information* such that *congruent measurements* can be obtained for each corresponding pixel (Moik 1980). Data-set merging can be of benefit to both computer-based analysis, including multidimensional classification and change detection, plus visual interpretation, including stereoscopic and multisensor image analysis. Examples of these applications are discussed in a later section of this chapter.

Data-Set Preparation

To accomplish digital merging, every data set must be in the proper numerical and spatially ordered form. For analog images such as aerial photographs, a digitizer is used to convert density variations into a two-dimensional array of digital numbers. Nonimage spatial data, such as point and polygon observations, must be transformed from a vector to a raster or grid-cell structure (Figure 8-6). **Surface generation** (also referred to as **surface interpolation**) is a mathematical process for estimating values for every cell of a grid from a set of irregularly spaced point measurements. Essentially, surface interpolation fills in the gaps between data

Figure 15-38 Concept of clustering a hypothetical two-band data set. *Left:* Two-dimensional plot of pixel DNs that the analyst wants grouped into three spectral classes. *Right:* Results of cluster analysis; the clustering algorithm automatically groups the data into three clusters or spectral classes. An image or map showing the spatial attributes of spectral classes would have to be compared to ground reference data to determine their true identity and value. (Adapted from Landgrebe 1974.)

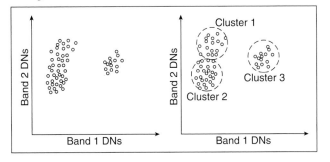

Figure 15-39 Concept of supervised classification for a hypothetical two-band data set. *Left:* Two-dimensional plot of pixel DNs representing three crop types and an unknown point *U*. *Right:* Result of minimum distance to means classification to determine which crop class point *U* is associated with. This particular supervised algorithm first determines the conditional centroid or center point of each class. Then the locus of points equidistant from the three centroids is plotted, resulting in three straight-line segments, or decision boundaries. In this example, point *U* would be associated with the soybean class as a result of its location with respect to the decision boundaries. (Adapted from Landgrebe 1974.)

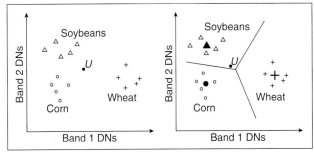

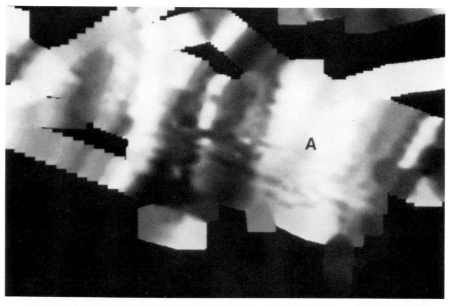

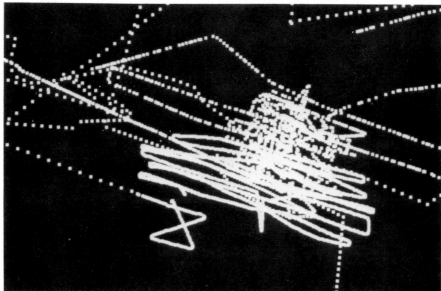

Figure 15-40 Interpolated image showing magnetic variations for the ocean floor in an area of the Mid-Atlantic Ridge (*A*). Image tone is related to the intensity of the magnetic field: light tones equal "high magnetics"; dark tones equal "low magnetics." The ship paths along which the original point data were collected are shown in the *bottom* image. Digital interpolation was used to fill in the gaps between the original points by a distance-weighing function. (Courtesy Pat S. Chavez, Jr., U.S. Geological Survey.)

points based upon some distance weighing function. For computer manipulation and image representation, the data are rescaled from their original range of values to the dynamic range of the image-processing system, for example, 0–255 DN for an 8-bit scale (Figure 15-40). Data depicted as polygons (e.g., soil and lithology maps) do not require surface interpolation. Rather, each bounded area is assigned a constant value corresponding to a particular unit (Eliason et al. 1983).

Once the data sets are in digital form, it is then necessary to reformat and scale them to match the geometry of a common reference system. Registration can be accomplished by selecting one image as a reference and spatially manipulating the others to coincide with it (the "**master-slave**" technique) or by registering all of the images to a reference map.

Change-Detection Images

Temporal information for a given area can be extracted from co-registered images that were acquired at different times by **multitemporal** or **multidate processing**. The

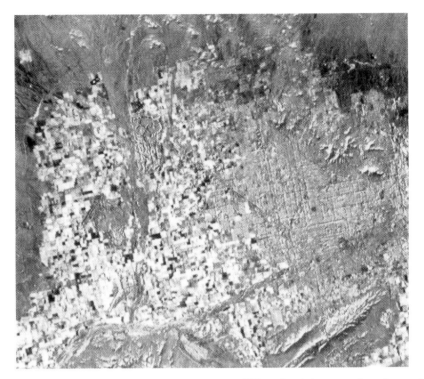

Figure 15-41 Temporal-difference image of Phoenix, Arizona, and environs: the MSS band 7 image obtained November 29, 1974, subtracted from the May 15, 1974, image. Most of the changes are associated with seasonal vegetation because band 7 was used. Light signatures represent maximum responses in May and dark signatures represent maximum responses in November. (Courtesy Pat S. Chavez, Jr., U.S. Geological Survey.)

change detection can be seasonal (e.g., snow cover, flooding, agricultural practices, natural vegetation growth cycles) or permanent (e.g., urbanization, forest clear-cutting, surface mining).

A common form of multitemporal processing is band differencing. For example, after two Landsat MSS images have undergone clean-up and digital registration, the DNs of one image can be subtracted from those of the second image. The resulting DNs can be positive, negative, or zero (no change). For image output, a readjustment value (127 DN for 8-bit data) is added to all "differences" so that medium gray represents no change and dark and light tones represent negative and positive differences, respectively. Contrast stretching is commonly employed to further emphasize the differences (Figure 15-41). This type of image is referred to as a **temporal-difference image** because of the subtraction of the image for one date from that of another date.

Similar results can be obtained when the same bands from two different dates are ratioed. Unlike differencing, however, the division process removes first-order brightness variations caused by topography (Chavez et al. 1977b). An image produced by this process is called a **temporal-ratio image**.

Multisensor Image Merging

The digital merging of images collected by different remote sensing instruments at different times is designed to *exploit* the strengths or advantages of each particular system in combined displays (black-and-white images or color composite images). Nondigital products, such as aerial photographs or optically correlated radar images, may be merged with digital images following the digitization process. **Multisensor image merging** may be done to exploit the following:

1. **Spatial and Spectral Resolutions** (e.g., SPOT HRV panchromatic and HRV multispectral images; Landsat TM and HRV panchromatic images; digitized airphotos and TM images).

2. **Chemical Properties and Physical Properties Such as Texture and Roughness** (e.g., MSS or TM images and radar images, including Seasat, SIR-A, and SIR-B).

3. **Surface Texture, Roughness, and Local Slope Properties** (e.g., radar images, including Seasat, SIR-A, and SIR-B).

In some cases, the merged images provide more information than the sum of the separate images; this is termed **synergism**.

Plate 16 shows the results of merging a digitized panchromatic photograph (spatial information) from the National High Altitude Program, or NHAP (see Chapter 5), and a 30-m Landsat TM color IR data set (spectral information) involving bands 2, 3, and 4 (Chavez 1986). The resolution of the airphoto was about 4 m after digitization. The resulting high-resolution color IR composite image was produced using a five-step process:

1. The TM data set was digitally expanded in both the X and Y directions by a factor of about 7 using pixel duplication. This essentially formatted the NHAP airphoto and TM data sets to the same "pixel size."

2. Because the digital enlargement by pixel duplication produces a "blocky image," the TM data were next smoothed by a 7×7-pixel low-pass filter (dimensions about equal to the digital enlargement).

3. Five corresponding control points were identified on both the NHAP airphoto and TM images and a second-order polynominal fit was used to register both data sets.

4. The two data sets were then digitally combined with pixel-by-pixel DN addition (band 2 + NHAP, band 3 + NHAP, and band 4 + NHAP) and contrast-stretched.

5. Three black-and-white positive transparencies were made of the three new bands, and a merged color IR composite was made using standard photographic laboratory color-compositing techniques (band 2 + NHAP printed blue, band 3 + NHAP printed green, and band 4 + NHAP printed red). As seen in Plate 16, the merged composite image incorporates the spectral resolution of the original TM component and the spatial resolution of the NHAP component.

Digital Elevation Data

Topographic information can be an important component in digital merging projects. Fortunately, **digital elevation data** stored on CCTs are available for the continental United States. The CCTs contain digital elevation data in a grid-cell format at two spatial resolutions. Digital terrain tapes produced by the Defense Mapping Agency (DMA) consist of elevation values spaced at 63.5-m intervals that were produced by digitizing the contours from 1:250,000-scale topographic maps. CCTs of **digital elevation models (DEMs)**, corresponding in coverage to standard 7.5-min topographic quadrangles, are produced for selected areas of the country by the USGS. The majority of the DEMs are created using digitizing-orthophoto equipment and consist of elevation values spaced at regular 30-m intervals.

Both DMA and DEM digital terrain tapes are distributed by the USGS. Specific information regarding digital elevation data and ordering information can be obtained from the User Services Section, National Cartographic Information Center (NCIC), U.S. Geological Survey, 507 National Center, Reston, VA 22092.

Special computer algorithms have been developed to calculate other topographic attributes from the digital elevation data. These include slope, aspect or slope direction, and solar illumination angle. These data, in turn, can be used to generate **elevation images, slope images,** and **shaded-relief images** on most computer systems (Figure 15-42).

Shaded-relief images convey an instant impression of the shape of the land (Figure 15-42C). In contrast to aerial photographs, for example, they contain no relief-induced distortions and the tonal variation is unambiguously identified with relief, rather than with snow, vegetation, or other albedo variations (Batson et al. 1975). Several shaded-relief images, incorporating different illumination angles and azimuths, can be generated for a common ground area (Figure 15-43). A stereoscopic pair can be made by using the original, undistorted image and a second image into which parallax has been introduced by a special computer program (Figure 15-44).

The USGS's Image Processing Facility has produced numerous shaded-relief image mosaics of individual states and regions covering several states. The shaded-relief image mosaic of Arizona is shown in Figure 15-45.

Multidimensional Image Classification

Improving the recognition accuracy in image classification can usually be achieved by combining multispectral image data with certain types of ancillary information, such as topographic information derived from digital tapes describing terrain. For example, Strahler et al. (1978) demonstrated that the accuracy in classifying forest cover from Landsat MSS images could be improved 27 percent by incorporating elevation and aspect information into the classification algorithm. Inclusion of the topographic information made it possible for the classifier to differentiate between species with similar spectral characteristics but different habitats. This technique of combining multispectral and ancillary information into the classification algorithm is called **multidimensional image classification**.

Synthetic Stereo Images

Stereoscopy is a powerful interpretation tool, but it has had only limited success with Landsat images. This is because the images exhibit a weak stereoscopic effect and stereo coverage is limited to small areas where there is overlap between orbital passes. However, by processing the images with digital elevation data, a computer algorithm can introduce **artificial parallax** as a linear function of terrain height at each picture element to produce a stereoscopic mate to an image that has no parallax (Batson et al. 1976) (Figure 15-46). A stereogram made from a single Landsat MSS band 7 image is shown in Figure 15-47.

Stereographic Landsat MSS images can also be generated by introducing relief displacement proportional to the magnitude of other data-set values, including **residual aeromagnetic data, gravity data,** and **geochemical data** (Figure 15-48). In fact, any continuous data set that is spatially related in X and Y directions and varies in the Z, or vertical, direction can be processed to create stereoscopic views showing the morphology of the data-set structure (Trautwein et al. 1982).

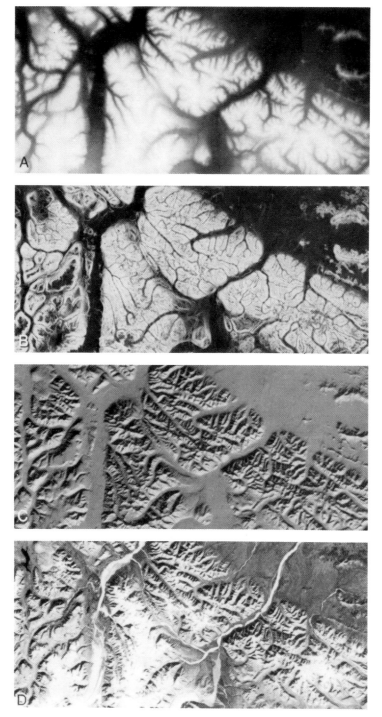

Figure 15-42 Computer-generated images derived from digital elevation data for a portion of the Nabesna quadrangle, Alaska, where tones are proportional to the ranges of the parameters. (*A*) Elevation image, where black represents the lowest elevation and white, the highest elevation. (*B*) Percent slope image, where black represents level terrain and white, the steepest terrain. (*C*) Shaded-relief image, where black represents surfaces facing away from the illumination source and white represents surfaces facing the illumination source. A Landsat MSS band 5 image, acquired September 18, 1973, is included as a reference (*D*). (Courtesy Charles M. Trautwein, U.S. Geological Survey.)

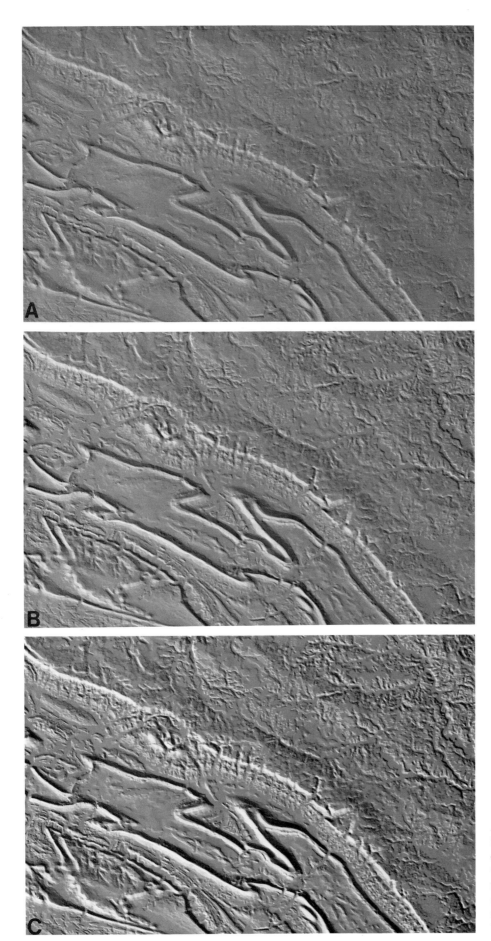

Figure 15-43 Shaded-relief images of the folded Appalachians near Altoona, Pennsylvania. Images incorporate three illumination angles: (*A*) 45°, (*B*) 30°, and (*C*) 15°. (Courtesy Pat S. Chavez, Jr., U.S. Geological Survey.)

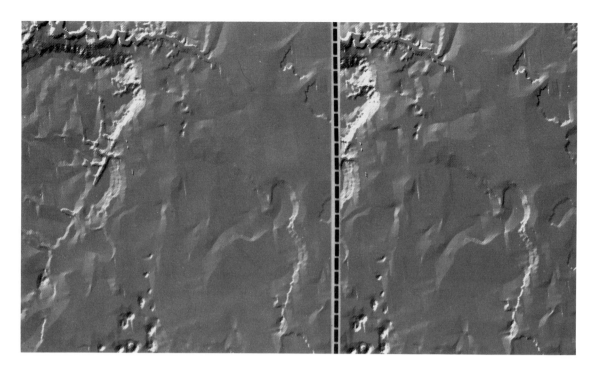

Figure 15-44 Computer-generated, shaded-relief stereogram of an area in northern Arizona. The sun is assumed to be 35° above the southeastern horizon; north is at top. Scale is 1:500,000. Compare with Figure 15-2. (Courtesy U.S. Geological Survey.)

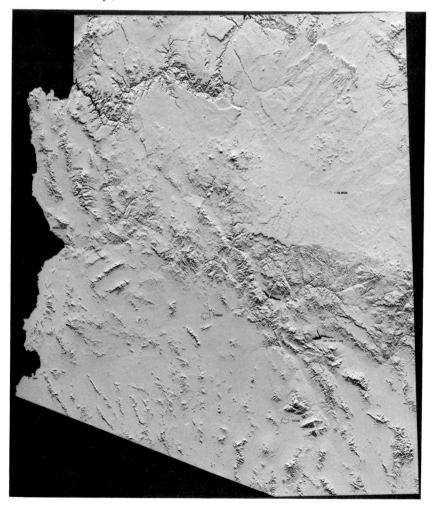

Figure 15-45 Shaded-relief image mosaic of Arizona. (Courtesy U.S. Geological Survey.)

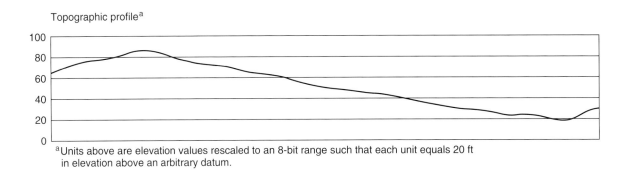

Topographic profile[a]

[a]Units above are elevation values rescaled to an 8-bit range such that each unit equals 20 ft in elevation above an arbitrary datum.

Elevations of Landsat pixels	71	74	78	81	84	81	78	74	71	67	64	57	55	52	47	44	41	38	36	32	28	25	25	21	20	25
Pixels in original image (right)	A	B	C	D	E	F	G	H	I	J	K	L	M	N	O	P	Q	R	S	T	U	V	W	X	Y	Z
Displacement	3 →	4 →	4 →	4 →	5 →	4 →	4 →	4 →	3 →	3 →	2 →	2 →	1 →	1 →	0	0	0	0	0	0	1 ←	1 ←	1 ←	2 ←	2 ←	1 ←
Pixels in parallax image (left)			A	∗	B	C	D	∗	E F	G	H I	J K	L M	N O	P	Q	R	S	T U	V	W X	Y	∗	Z		

∗ Pixel value determined by interpolation.
Double letters represent pixel overlap; right pixel replaces left pixel.

Figure 15-46 Generation of artificial parallax for stereographic viewing. The topographic profile shows elevations along a single line of data in the digital topography data set as illustrated in the matrix below the profile. Elevation values can be related to Landsat MSS pixels in the same spatial position. By application of a stereographic generation equation, MSS pixels are shifted to create relief displacement that is proportional to the relief in the topographic profile. This results in a distorted image that when viewed stereoscopically with the original image produces a three-dimensional effect. (Trautwein et al. 1982.)

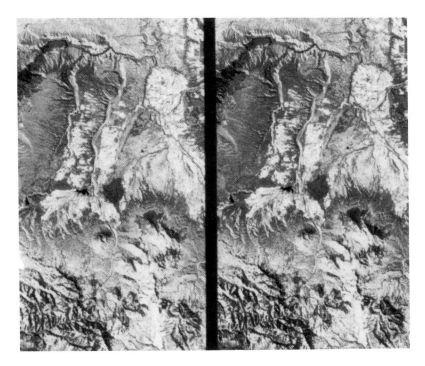

Figure 15-47 Landsat MSS band 7 synthetic stereogram of the Montrose, Colorado, region. Artificial stereoscopy was created from digital elevation data. The *left* (distorted) image contains pixels that are shifted (proportional to elevation) to create relief displacement. (Courtesy Pat S. Chavez, Jr., U.S. Geological Survey.)

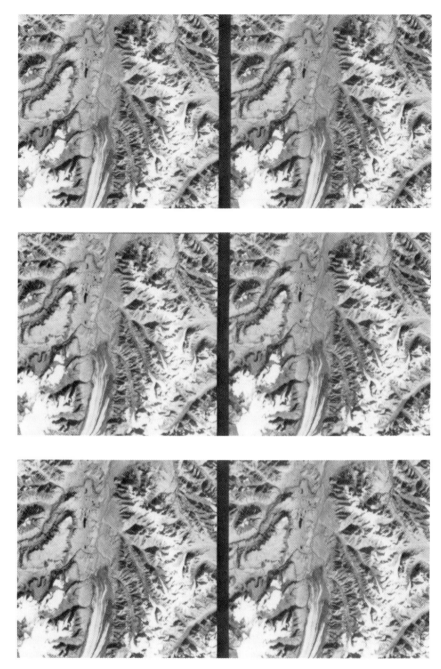

Figure 15-48 Stereoscopically combined data sets for the Nabesna quadrangle region, Alaska. Parallax was digitally introduced into a Landsat MSS band 7 image as a function of residual aeromagnetic data (*top stereogram*), gravity data (*middle stereogram*), and geochemical copper data (*bottom stereogram*). (Courtesy Charles M. Trautwein, U.S. Geological Survey.)

Figure 15-49 *Landsat 3* RBV subscene image of the San Francisco volcanic field in north central Arizona. Compare with Figures 15-50 and 15-51. (Courtesy U.S. Geological Survey.)

Perspective-View Images

Digital images may also be combined with digital topographic data and rotated to **oblique** or **perspective views** of the terrain. Parallax can then be introduced to the perspective-view image to produce a stereoscopic mate. These procedures are illustrated in Figures 15-49 through 15-51, which show a portion of the San Francisco volcanic field in north-central Arizona in **vertical**, **perspective**, and **stereo-perspective views** using a digital *Landsat 3* Return Beam Vidicon (RBV) image. In addition, digital sonar image data can be combined with bathymetric digital data to produce **perspective-view sonographs** (Figure 7-45).

Image Processing Services

There are currently more than 200 value-added companies worldwide that offer comprehensive **image processing services**. A directory of these organizations is available upon request from the Earth Observation Satellite Company (EOSAT), Customer Service Department, 4300 Forbes Boulevard, Lanham, MD 20706. The directory is updated periodically.

Figure 15-50 RBV perspective-view image of the San Francisco volcanic field in north-central Arizona. Compare with Figures 15-49 and 15-51. (Courtesy U.S. Geological Survey.)

Figure 15-51 RBV perspective-view stereogram of the San Francisco volcanic field in north-central Arizona. Compare with Figures 15-49 and 15-50. (Courtesy U.S. Geological Survey.)

Questions

1. Describe the advantages of an image-processing system that incorporates both interactive processing and batch processing.
2. Refer to Equation 15-3. If the minimum and maximum DN parameters are 16 and 106, what are the output (stretched) digital numbers when the original DNs are 13, 42, 94, and 111? What specific type of contrast stretch is indicated?
3. How does edge enhancement differ from high-pass spatial filtering?
4. Define the tone (light or dark) associated with the following MSS ratio images and surface materials.

Ratio	Surface Material	Tone
5/4	Ferric iron-rich rocks	———
7/6	Clear water	———
7/5	Healthy vegetation	———
5/7	Muddy water	———
6/4	Healthy vegetation	———
5/4	Ferric iron-poor rocks	———
5/6	Healthy vegetation	———

5. Describe the differences between image classification, multitemporal image classification, and multidimensional image classification.
6. Examine the DN frequency histogram in Figure 15-3 and explain why a contrast stretch incorporating truncation might be appropriate.

Bibliography and Suggested Readings

Batson, R. M., K. Edwards, and E. M. Eliason. 1975. Computer-Generated Shaded-Relief Images. *Journal of Research U.S. Geological Survey* 3:401–408.

———. 1976. Synthetic Stereo and Landsat Pictures. *Photogrammetric Engineering and Remote Sensing* 42:1279–84.

Berlin, G. L., M. A. Tarabzouni, and Z. M. Munshi. 1982. Vegetation Assessment of the Northern Arabian Shield for Ground-Water Exploration Using Edge-Enhanced MSS Images. In *Proceedings, 16th International Symposium on Remote Sensing of Environment, Second Thematic Conference*, 539–47. Ann Arbor: Environmental Research Institute of Michigan.

Blodget, H. W., F. J. Gunther, and M. H. Podwysocki. 1978. *Discrimination of Rock Classes and Alteration Products in Southwestern Saudi Arabia with Computer-Enhanced Landsat Data*. NASA Technical Paper 1327. Washington, D.C.: U.S. Government Printing Office.

Chavez, P. S., Jr. 1986. Digital Merging of Landsat TM and Digitized NHAP Data for 1:24,000-Scale Image Mapping. *Photogrammetric Engineering and Remote Sensing* 52:1637–46.

———. 1988. An Improved Dark-Object Subtraction Technique for Atmospheric Scattering Correction of Multispectral Data. *Remote Sensing of Environment* 24:459–79.

Chavez, P. S., Jr., and B. Bauer. 1982. An Automatic Optimum Kernal-Size Selection Technique for Edge Enhancement. *Remote Sensing of Environment* 12:23–38.

Chavez, P. S., Jr., G. L. Berlin, and A. V. Acosta. 1977a. Computer Processing of Landsat MSS Digital Data for Linear Enhancements. In *Proceedings, Second Annual William T. Pecora Memorial Symposium*, 235–50. Falls Church, Va.: American Society of Photogrammetry.

Chavez, P. S., Jr., G. L. Berlin, and W. B. Mitchell. 1977b. Computer Enhancement Techniques of Landsat MSS Digital Images for Land Use/Land Cover Assessments. In *Remote Sensing of Earth Resources*. Vol. 6, 259–75. Tullahoma: University of Tennessee Space Institute.

Chavez, P. S., Jr., G. L. Berlin, and L. B. Sowers. 1982. Statistical Method for Selecting Landsat MSS Ratios. *Journal of Applied Photographic Engineering* 8:23–30.

Chavez, P. S., Jr., S. C. Guptill, and J. A. Bowell. 1984. Image Processing Techniques for Thematic Mapper Data. In *Proceedings, 50th Annual Meeting*, 728–43. Falls Church, Va.: American Society of Photogrammetry.

Condit, C. D., and P. S. Chavez, Jr. 1979. *Basic Concepts of Computerized Digital Image Processing for Geologists*. Geological Survey Bulletin 1462. Washington, D.C.: U.S. Government Printing Office.

Crippen, R. E. 1989. A Simple Spatial Filtering Routine for the Cosmetic Removal of Scan-Line Noise from Landsat TM P-Tape Imagery. *Photogrammetric Engineering and Remote Sensing* 53:327–31.

Davis, P. A., and G. L. Berlin. 1989. Rock Discrimination in the Complex Geologic Environment of Jabal Salma, Saudi Arabia, Using Landsat Thematic Mapper Data. *Photogrammetric Engineering and Remote Sensing* 55:1147–60.

Eliason, E. M., P. S. Chavez, Jr., and L. A. Soderblom. 1974. Simulated True Color Images for ERTS Data. *Geology* 2:231–34.

Eliason, E. M., and A. S. McEwen. 1990. Adaptive Box Filters for Removal of Random Noise from Digital Images. *Photogrammetric Engineering and Remote Sensing* 56:453–58.

Eliason, P. T., T. J. Donovan, and P. S. Chavez, Jr. 1983. Integration of Geologic, Geochemical, and Geophysical Data of the Cement Oil Field, Oklahoma, Using Spatial Array Processing. *Geophysics* 48:1305–17.

Fleming, M. D., J. S. Berkebile, and R. M. Hoffer. 1975. *Computer-Aided Analysis of Landsat-1 MSS Data: A Comparison of Three Approaches*. LARS Information Note 072475. West Lafayette, Ind.: Purdue University.

Holm, T. M. 1982. *Canonical Analysis: The Use of Transformed Landsat Data for Crop Type Discrimination*. Sioux Falls, S. D.: U. S. Geological Survey.

———. 1983. *Canonical Analysis of Crop Type Discrimination*. In *Proceedings, Ninth International Symposium on Machine Processing of Remotely Sensed Data*, 216. West Lafayette, Ind.: Purdue University.

Hutchinson, C. F. 1982. Techniques for Combining Landsat and Ancillary Data for Digital Classification Improvement. *Photogrammetric Engineering and Remote Sensing* 48:123–30.

Jackson, R. D. 1983. Spectral Indices in N-Space. *Remote Sensing of Environment* 13:409–21.

Jensen, J. R. 1986. *Introductory Digital Image Processing*. Englewood Cliffs, N.J.: Prentice Hall.

Jensen, S. K., and F. A. Waltz. 1979. Principal Components Analysis and Canonical Analysis in Remote Sensing. In *Proceedings, 45th Annual Meeting*, 337–48. Falls Church, Va.: American Society of Photogrammetry.

Landgrebe, D. A. 1974. Machine Processing of Remotely Sensed Data. In *Syllabus, Workshop on Remote Sensing and Image Interpretation*, 1–36. Open File Report 75-196. Sioux Falls, S.D.: U.S. Geological Survey.

Lillesand, T. M., and R. W. Kiefer. 1987. *Remote Sensing and Image Interpretation.* 2d ed. New York: John Wiley & Sons.

Moik, J. G. 1980. *Digital Processing of Remotely Sensed Images.* NASA Special Paper 431. Washington, D.C.: U.S. Government Printing Office.

Muller, J. P. 1988. *Digital Image Processing in Remote Sensing.* Philadelphia: Taylor & Francis.

Philipson, W. R., and W. L. Teng. 1988. Operational Interpretation of AVHRR Vegetation Indices for World Crop Information. *Photogrammetric Engineering and Remote Sensing* 54:55–59.

Raines, G. L., T. W. Offield, and E. S. Santos. 1978. Remote Sensing and Subsurface Definition of Facies and Structure Related to Uranium Deposits, Powder River Basin, Wyoming. *Economic Geology* 73:1706–23.

Rohde, W. G., J. K. Lo, and R. A. Pohl. 1978. EROS Data Center Landsat Digital Enhancement Techniques and Imagery Availability. *Canadian Journal of Remote Sensing* 4:63–76.

Rosenfeld, A., and A. C. Kak. 1982. *Digital Picture Processing.* 2d ed. New York: Academic Press.

Rowan, L. C., P. H. Wetlaufer, A.F.H. Goetz, F. C. Billingsley, and J. H. Stewart. 1974. *Discrimination of Rock Types and Detection of Hydrothermally Altered Areas in South-Central Nevada by the Use of Computer-Enhanced ERTS Images.* Geological Survey Professional Paper 883. Washington, D.C.: U.S. Government Printing Office.

Sabins, F. F., Jr. 1987. *Remote Sensing Principles and Interpretation,* 2d ed. New York: W. H. Freeman & Co.

Santisteban, A., and L. Munoz. 1978. Principal Components of a Multispectral Image: Application to a Geological Problem. *IBM Journal of Research and Development* 22:444–54.

Schowengerdt, R. A. 1983. *Techniques for Image Processing and Classification in Remote Sensing.* New York: Academic Press.

Short, N. M. 1982. *The Landsat Tutorial Workbook.* NASA Reference Publication 1078. Washington, D.C.: U.S. Government Printing Office.

Skidmore, A. K. 1989. An Expert System Classifies Eucalypt Forest Types Using Thematic Mapper Data and a Digital Terrain Model. *Photogrammetric Engineering and Remote Sensing* 55:1449–64.

Strahler, A. H., T. L. Logan, and N. A. Bryant. 1978. Improving Forest Cover Classification Accuracy from Landsat by Introducing Topographic Information. In *Proceedings, 12th International Symposium on Remote Sensing of Environment,* 727–42. Ann Arbor: Environmental Research Institute of Michigan.

Taranik, J. V. 1978. *Principles of Computer Processing of Landsat Data for Geologic Applications.* Open File Report 78-117. Sioux Falls, S.D.: U.S. Geological Survey.

Teng, W. L. 1990. AVHRR Monitoring of U.S. Crops During the 1988 Drought. *Photogrammetric Engineering and Remote Sensing* 56:1143–46.

Trautwein, C. M., D. D. Greenlee, and D. G. Orr. 1982. *Digital Data Base Application to Porphyry Copper Mineralization in Alaska: Case Study Summary.* Open File Report 82-801. Sioux Falls, S.D.: U.S. Geological Survey.

Tucker, C. J. 1979. Red and Photographic Infrared Linear Combinations for Monitoring Vegetation. *Remote Sensing of Environment* 8:127–50.

U.S. Geological Survey. 1983. *20th International Remote Sensing Workshop, Quantitative Remote Sensing.* Sioux Falls, S.D.: EROS Data Center.

Williams, R. S., Jr., and others. 1983. Geological Applications. In *Manual of Remote Sensing,* edited by R. N. Colwell, 2d ed., 1667–953. Falls Church, Va.: American Society for Photogrammetry and Remote Sensing.

Index

GREEK ALPHABET

Name	Capital	Lowercase	Name	Capital	Lowercase
Alpha	A	α	Nu	N	ν
Beta	B	β	Xi	Ξ	ξ
Gamma	Γ	γ	Omicron	O	o
Delta	Δ	δ	Pi	Π	π
Epsilon	E	ϵ	Rho	P	ρ
Zeta	Z	ζ	Sigma	Σ	σ
Eta	H	η	Tau	T	τ
Theta	Θ	θ	Upsilon	Υ	υ
Iota	I	ι	Phi	Φ	ϕ
Kappa	K	κ	Chi	X	χ
Lambda	Λ	λ	Psi	Ψ	ψ
Mu	M	μ	Omega	Ω	ω

TRIGONOMETRIC FUNCTIONS IN A RIGHT TRIANGLE

If Known	Use These Formulas for Finding These Unknowns				
	x	y	z	a	b
x and y			$\sqrt{x^2 + y^2}$	arc tan $\frac{y}{x}$	arc tan $\frac{x}{y}$
x and z		$\sqrt{z^2 - x^2}$		arc cos $\frac{x}{z}$	arc sin $\frac{x}{z}$
y and z	$\sqrt{z^2 - y^2}$			arc sin $\frac{y}{z}$	arc cos $\frac{y}{z}$
x and a		$x \tan a$	$\dfrac{x}{\cos a}$		$90° - a$
x and b		$\dfrac{x}{\tan b}$	$\dfrac{x}{\sin b}$	$90° - b$	
y and a	$\dfrac{y}{\tan a}$		$\dfrac{y}{\sin a}$		$90° - a$
y and b	$y \tan b$		$\dfrac{y}{\cos b}$	$90° - b$	
z and a	$z \cos a$	$z \sin a$			$90° - a$
z and b	$z \sin b$	$z \cos b$		$90° - b$	

(From Naval Reconnaissance & Technical Support Center. Image Interpretation Handbook. *Washington, D.C.: U.S. Government Printing Office, 1967, p. B-9.)*